Atlas Florae Europaeae

II

ATLAS FLORAE EUROPAEAE

DISTRIBUTION OF VASCULAR PLANTS IN EUROPE

II

3
Salicaceae to Balanophoraceae

4
Polygonaceae

5
Chenopodiaceae to Basellaceae

Edited by
Jaakko Jalas & Juha Suominen
on the basis of team-work by European Botanists

The right of the University of Cambridge to print and sell all manner of books was granted by Henry VIII in 1534. The University has printed and published continuously since 1584.

CAMBRIDGE UNIVERSITY PRESS

Cambridge

New York New Rochelle Melbourne Sydney

Published by the Press Syndicate of the University of Cambridge
The Pitt Building, Trumpington Street, Cambridge CB2 1RP
32 East 57th Street, New York, NY 10022, USA
10 Stamford Road, Oakleigh, Melbourne 3166, Australia

First published in separate volumes by
The Committee for Mapping the Flora of Europe and Societas Biologica Fennica Vanamo, Helsinki
Volume 3 1976
Volume 4 1979
Volume 5 1980

First published as a compendium by Cambridge University Press 1988

Printed in Great Britain at the University Press, Cambridge

British Library cataloguing in publication data

Atlas florae Europaeae: distribution of Vascular plants in Europe
1. Plants – Europe – Maps
I. Jalas, Jaakko II. Suominen, Jaha
912′.158194 QK281

Library of Congress cataloguing in publication data

Atlas florae Europaeae: distribution of vascular plants in Europe / edited by Jaakko Jalas & Juha Suominen.
p. cm.
Contents: I. 1. Pteridophyta (Psilotaceae to Azollaceae). 2. Gymnospermae (Pinaceae to Ephedraceae)
– II. 3. Salicaceae to Balanophoraceae. 4. Polygonaceae. 5. Chenopodiaceae to Basellaceae.
– III. 6. Caryophyllaceae (Alsinoideae and Paronychicideae).
7. Caryophyllaceae (Silenoaideae)
ISBN 0 521 34270 8 (v. 1)
1. Phytogeography – Europe – Maps – Collected works. I. Jalas, Jaakko. II. Suominen, Juha.
QK281.A85 1988
581.94 – dc19 87 – 33834

ISBN 0 521 34271 6

GENERAL PREFACE

I am very pleased to be given the opportunity to help in introducing to a wider public these distribution maps, to which I have made such frequent and profitable reference over the past few years – all the more so since it enables me to justify to some small extent my nomination as a member of the advisory committee for the *Atlas*, of which I have hitherto been a rather inactive member.

Towards the end of 1961 I became conscious, in the course of my editorial work on the first volume of *Flora Europaea*, that in many contributions which were satisfactory taxonomically the data on geographical distribution were defective or inaccurate. Sometimes countries were omitted from mere inadvertence; sometimes they were given for a plant which had never been more than a casual and was long since extinct in the field but still flourishing in the Florae; very often the mistakes arose from a failure to take account of territorial changes in the period 1914–1945. It was astonishing how many experienced botanists seemed to imagine that the Carpathians were still in Hungary, or that Finland still had an arctic coast. To remedy this state of affairs I assumed, with the consent of the editorial committee, responsibility for revising geographical data in all accounts. But time was short, and when the first volume went to press I had been able to correct only some of the grosser errors. In the remaining volumes I believe the geographical data to be be more accurate.

The maps so far published in *Atlas Florae Europapeae* all relate to plants which are described in the first volume of *Florae Europaea*, and they are of especial value in serving to correct the rather numerous errors in the latter. They have been of enormous help to me in the preparation of the second edition of this volume. But their usefulness goes far beyond that. In *Flora Europaea* considerations of space limited us to a phrase of about 25 words at most to sum up the distribution. It is in most cases reasonably accurate, but the *Atlas*, besides correcting some minor inaccuracies, fills in a lot of detail that could not be included in the printed phrase. For *Stellaria holostea*, for example (I choose a species which is free from any taxonomic doubts or difficulties), *Flora Europaea* rightly says that it is 'rare in the Mediterranean region', but the *Atlas* shows how unevenly distributed is this rarity: the plant is completely absent from southern Spain, extremely rare in Greece and Albania, but relatively frequent in Italy. Or again, take *Spergularia salina* (= *S. marina*), for which *Flora Europaea* gives 'Coasts of Europe and inland saline areas'. Few readers can have realized how extensive these inland saline areas are until the *Atlas* revealed them.

The *Atlas*, of course, is not perfect; its editors would be the last to claim that it is. But most of its shortcomings arise from the fact that the editors are largely dependent on the reports they receive from the various countries of Europe; and although they have built up an impressive team of collaborators (far more complete than that achieved by the editors of *Flora Europaea*), it is inevitable that they are sometimes supplied with questionable or defective data, or, worse still, with none at all. So we find occasional species which are native throughout Spain but alien throughout Portugal; species which are present in every square in Bulgaria but curiously rare in all the neighbouring countries; and numerous species for which the sparse distribution in the more remote parts of the USSR probably tells us more about the distribution of Russian botanists than of plants.

More disconcerting is the occasional failure of some regions to supply any data at all; in recent issues there have been alarming blanks in Romania and Ukraine for several species. In these circumstances dots for relatively rare species can be provided, at least in part, from the herbaria and literature, but for very common species this is impossible. It is difficult to know what the editors can do in such a situation; my only suggestion is that they should indicate rather more clearly than at present (even at the sacrifice of a little tact) which of the blanks indicate absence of plants and which of them absence of data.

I mention these limitations of the maps not for the sake of carping, but to enable the reader by bearing in mind these difficulties to interpret the maps with more discrimination, and also because these limitations have been dealt with at greater length in a recent article by Dr Holub (Norrlinia 2: 107–115 (1984)). Although it contains some judicious praise of the *Atlas*, it does not hesitate to suggest a large number of ways in which it might be improved. The breaking down of all aggregates into their segregates (according to whose taxonomy?); the treatment of apomictic genera on a heroic scale (with maps of 'at least 200 or 300 species of *Rubus*'); the modification of the grid system in certain areas; the ironing out of differing views on native, alien or casual status; greater detail in data on recent increase or decline; fuller citations from literature; the immediate preparation (out of sequence) of maps for the most critical genera – these are some of the improvements which he suggests. They are all desirable, but they would need the services of about a hundred more floristic botanists, at least a dozen extra critical taxonomists, and at least one trained diplomat with a generous travelling allowance. Holub, one suspects, wishes, as did Omar Khayyam, that he could 'with Fate conspire, to take this sorry scheme of things entire . . . and then remould it nearer to his heart's desire'.

Pending this consummation, let us be thankful for what we have got, and realize that in this imperfect world we are not likely to get much more, and let us trust the editors to make the best decision when faced with the many awkward problems which beset them.

The thanks of all European botanists (and of many outside Europe) are due to the editors of the *Atlas* for their enterprise in initiating the project and for their tenacity in continuing it. Thanks are also due to the government of Finland for the generous and far-sighted support which it has provided. It gives me special pleasure as an Irishman to see a country not so very much more populous than my own, and nearly as isolated on the periphery of Europe, playing such an important role in forwarding the scholarship of Europe as a whole.

June 1987

D. A. Webb
Trinity College, Dublin

3

SALICACEAE TO BALANOPHORACEAE

CONTENTS

Index to Vol.3 is at the end of Vol. 5
Base map is at the end of Vol.5

THE COMMITTEE FOR MAPPING THE FLORA OF EUROPE

SECRETARIAT, HELSINKI

J. Jalas (Chairman of the Committee)
J. Suominen (Secretary General)

ADVISERS

T. W. Böcher, København
A. R. Clapham, Arkholme
E. Hultén, Stockholm
D. A. Webb, Dublin

COMMITTEE MEMBERS ACTING AS REGIONAL COLLABORATORS

Albania (Al)	I. Mitrushi, Tiranë	
Austria (Au)	H. Niklfeld, Wien Assisted by collaborators in Mapping the Flora of Central Europe	
Belgium (Be, excl. Luxembourg)	E. van Rompaey, Antwerpen	
British Isles (Br, Hb, Channel Islands)	F. H. Perring, Abbots Ripton	D. A. Webb, Dublin (Hb)
Bulgaria (Bu)	S. Kožuharov, Sofia Assisted by: A. V. Petrova, Sofia	
Czechoslovakia (Cz)	J. Futák, Bratislava Assisted by: D. Magic, Bratislava F. Mladý, Průhonice	J. Holub, Průhonice B. Slavík, Průhonice
Denmark (Da, Fa)	A. Hansen, København	
Finland (Fe)	J. Suominen, Helsinki Assisted by: A. Kurtto, Helsinki I. Kytövuori, Helsinki	 T. Ulvinen, Oulu
France (Co, Ga)	P. Dupont, Nantes Maps checked by several French botanists	

German Democratic Republic (Ge: DDR)	H. Meusel, Halle A. Buhl, Halle Assisted by: W. Fischer, Potsdam F. Fukarek, Greifswald	E. Weinert, Halle W. Hempel, Dresden
German Federal Republic (Ge: BRD)	H. Ellenberg, Göttingen H. Haeupler, Göttingen Assisted by collaborators in Mapping the Flora of Central Europe	P. Schönfelder, Regensburg
Greece (Cr, Gr)	P. Critopoulos, Athínai Assisted by: G. Lavrentiades, Thessaloniki	W. Greuter, Genève H. Runemark, Lund (Aegean Islands)
Hungary (Hu)	R. v. Soó, Budapest	A. Terpó, Budapest
Iceland (Is)	E. Einarsson, Reykjavík Assisted by: H. Björnsson, Öraefi I. Davídsson, Reykjavík H. Guttormsson, Neskaupstadur H. Hallgrímsson, Akureyri	 B. Jóhannsson, Reykjavík I. Óskarsson, Reykjavík S. Steindórsson, Akureyri
Italy (It, Sa, Si)	E. Nardi, Firenze	M. Raffaelli, Firenze
Jugoslavia (Ju)	P. Fukarek, Sarajevo Assisted by: D. Trpin, Ljubljana	E. Mayer, Ljubljana
Luxembourg (Be, excl. Belgium)	L. Reichling, Luxembourg	
Netherlands (Ho)	J. Mennema, Leiden	
Norway (No, Sb)	K. Faegri, Bergen Assisted by: K. Høiland, Oslo	
Poland (Po)	J. Kornaś, Kraków Assisted by: M. Ciaciura, Wrocław D. Fijałkowski, Lublin J. Jasnowska, Szczecin K. Kępczyński, Toruń M. Kopij, Warszawa T. Krzaczek, Lublin J. Mądalski, Wrocław R. Olaczek, Łódź	M. Gostyńska-Jakuszewska, Kraków L. Olesiński, Olsztyn H. Rutowicz, Łódź Z. Schwarz, Gdańsk A. Sokołowski, Białowieża R. Sowa, Łódź A. Zając, Kraków E. U. Zając, Kraków W. Żukowski, Poznań
Portugal (Az, Lu)	J. do Amaral Franco, Lisboa Assisted by: M. L. da Rocha-Afonso, Lisboa	

Romania (Rm)	N. Boșcaiu, Cluj	
Spain (Bl, Hs)	E. F.-Galiano, Sevilla	P. Montserrat, Jaca
	R. Folch i Guillèn, Barcelona (Bl and Cataluña)	
	Assisted by:	
	B. Cabezudo, Sevilla	B. Valdés, Sevilla
Sweden (Su)	B. Jonsell, Uppsala	
	Assisted by:	
	M. Thulin, Uppsala	B. Widén, Lund
Switzerland (He)	O. Hegg, Bern	M. Welten, Bern
	Assisted by:	
	R. Sutter, Bern	
Turkey (European part; Tu)	H. Demiriz, Diyarbakir (Oct. 1973 —)	D. A. Webb, Dublin (— Sept. 1973)
U.S.S.R. (Rs (N, B, C, W, K, E))	A. I. Tolmachev, Leningrad	
	V. I. Czopik, Kiev	V. N. Tikhomirov, Moskva
	Assisted by:	
	J. E. Alekseev, Moskva	N. B. Oktyabreva, Moskva
	I. J. Fatare, Riga	T. Plieva, Leningrad
	I. Grudzinskaya, Leningrad	A. Rasiņš, Riga
	G. B. Kļaviņa, Riga	N. K. Shvedchikova, Moskva
	L. Laasimer, Tartu	A. K. Skvortsov, Moskva
	A. N. Laschenkova, Syktyvkar	O. Svyazeva, Leningrad
	A. Lekavičius, Vilnius	

CONSULTANTS ON TAXONOMY AND NOMENCLATURE

J. Chmelař, Brno (Salix)
R. Hendrych, Praha (Thesium)

FINNISH CONSULTATION COMMITTEE, HELSINKI

T. Ahti	I. Kukkonen
P. Isoviita	H. Luther
J. Jalas (Chairman)	R. Ruuhijärvi
A. Kalela	J. Suominen (Secretary)

EXPLANATORY NOTES

MAPPING TECHNIQUES

The area mapped is that of Flora Europaea. The Azores and Spitsbergen appear as insets.

The mapping unit used is the 50-km square of the UTM (Universal Transverse Mercator) grid maps, covering Europe on a scale of 1 : 1 000 000. Europe comprises ca. 4 400 squares, which are used without respect to political boundaries. Along the compensation zones of every sixth meridian, the breadth (W-E) of the squares varies between 40 and 60 km.

A base map, 1 : 10 000 000, is used in collecting the primary data in the various countries, and in compiling and publishing the final maps, which have a scale of ca. 1 : 31 000 000. A sample specimen of this base map was distributed as an appendix to Vol. 1.

Deviations from the UTM grid system. The deviations are of three kinds. a) In coastal areas the squares with less than 10 % (250 km^2) of land are included in a neighbouring mainland square. However, islands far from the coast and long peninsulas have their own squares. b) Mapping symbols for isolated islands are put on the actual position of the island, irrespective of the grid. These cases are marked with an asterisk (*) in the list below. c) Individual mountain tops forming distinct biogeographical entities are not divided between two or more squares, despite of the grid.

As the situation is particularly complicated in Greece (Gr and Cr), a special map showing the deviations in this area is added.

British Isles

The entire S. part of the Isle of Lewis is in PE2.
All of North Uist (etc.) is in *ND3.
All of S. South Uist (Barra, etc.) is in *ND4.
Dunvegan Head is in PD3.
All of the Isle of Man is in *UF4.
All of the Scilly Isles are in PR4.
UA2 is included in *UA1.
UA4 is included in UA3.
Guernsey and Sark are in WV1.
Alderney and Jersey are in WV3.

Denmark

All of Laesö is in *PJ1.
All of Bornholm is in *VB4.
Faeröerne are in *FaeN and *FaeS, the limit between them follows Skopen Fjord.

Germany

Helgoland is in ME1.
VA2 contains only E. Rügen.

Greece

See the map.

Iceland

All of Snaefellsnes is in *AN4.

Italy

NP4 is included in PP2.
*PN2 contains: all of Elba, Capraia, Pianosa, Montecristo, and islands W. of it. (Mainland of PN2 is included in PN4.)
UF2 contains also islands of UF4.
All of Isole Eolie o Lìpari are in *VC3.
All of Pantelleria is in *QF3.
All of Isole Pelagie are in *TV1.
(All of Malta is in *VV1.)

Netherlands

All of Vlieland is in FU1.
All of Terschelling is in FV4.

Norway

Kjelmesöy and the peninsula E. of it are in VC2.
*Björnöya and *Jan Mayen form a square of their own.
Kvitöya is in *NK4.
Kongsöya is in *NH1.
Svensköya is in *MH4.
Hopen is in *MF4.

Portugal

Each island of the *Azores forms a square of its own.

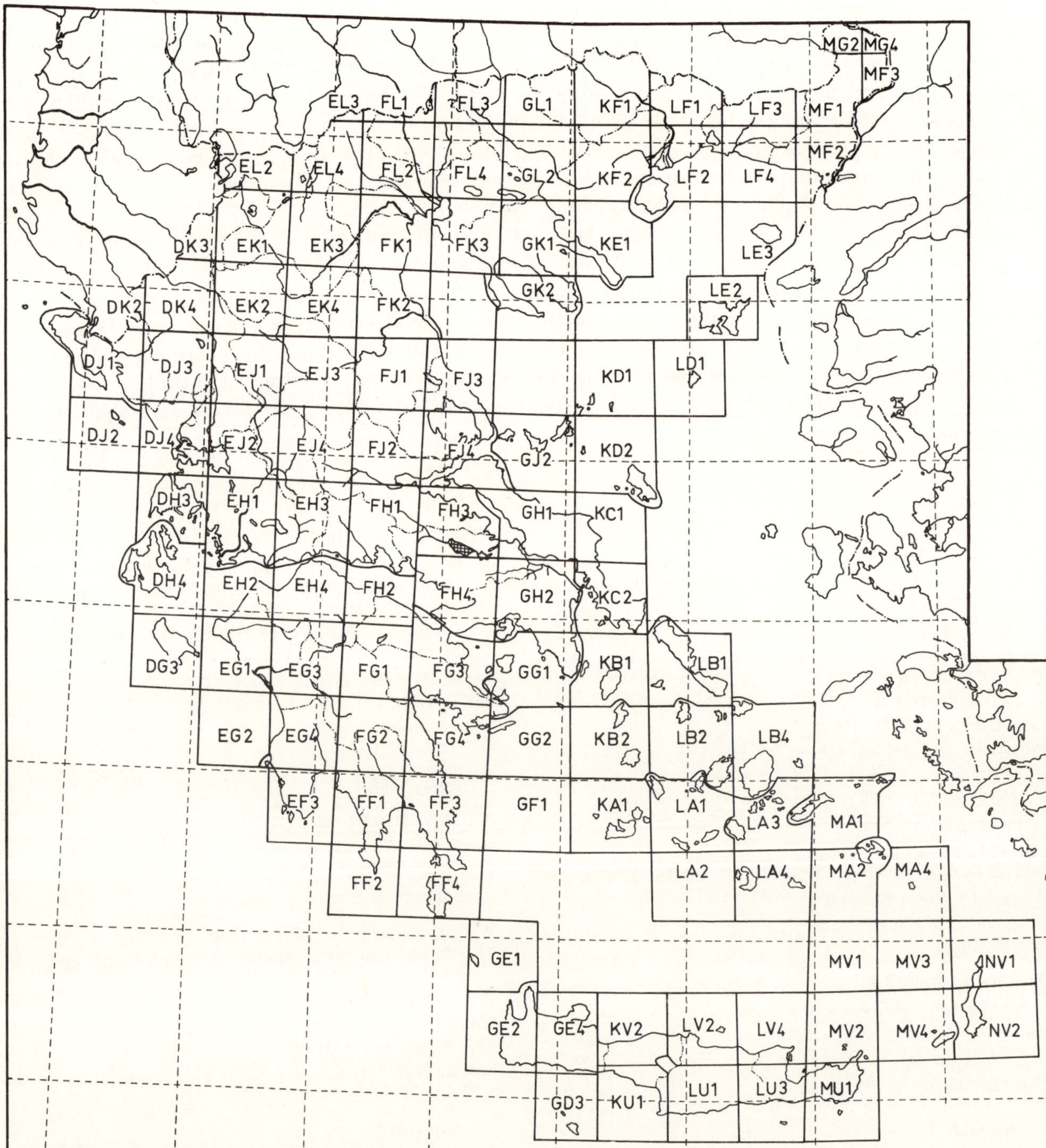

Deviations from the UTM grid in Greece (Gr and Cr). See the text.

Spain

BD2 is included in *BC1.
All of Isla de Ibiza is in *CD4.
The entire W. part of Isla de Mallorca is in *DD3, the E. part in *ED1.
All of Isla de Menorca is in *EE4.
I. Alboran is in *VE3.

Sweden

All of Gotska Sandön is in *CK3.
All of Central Gotland is in *CJ1.
W. edge of XD1 is in WD3, N. Öland is in XD2.

Turkey (European part)

All of Imroz Ada is in ME2.
PF4 is included in *PF3.

U.S.S.R.

All of Poluostrov Rybachiy is in VC4.
NW. Kanin is in *MB4.
Ostrov Matveyev is in EB3, the rest of EC4 is included in EC3.
All of Sur Sari is in NG1.
Ostrov Malyy Tyutyarsari is in NG2.
Kolka is in EJ3.
Ostrov Dzharylgach is in VS4, Rybnoye in VR4.

The mapping symbols are:

- ● native occurrence
- ○ introduction (established alien)
- ◒ status unknown or uncertain
- + extinct
- × probably extinct, or, at least, not recorded since 1930
- ? record uncertain as to identification or locality

For the second and third taxon in a map the symbols ▲ and ★ are used. The special symbols (○, ◒, +, ×, ?, etc.) belong to the taxon marked by ●, if not otherwise stated.

Any reliable record was used when entering a symbol in a square, not only those based on herbarium specimens but also reliable published or unpublished sight records.

DEVIATIONS FROM FLORA EUROPAEA

The species and subspecies included are according to Flora Europaea (Vol. 1, 1964), with some deviations, mainly caused by new data published after 1964.

Included, although not given in Fl. Eur., are a number of new taxa first described in or after 1964: *Salix apennina, S. vinogradovii, Thesium sommieri, T. kyrnosum, T. hispanicum, T. macedonicum, T. corsalpinum, Cytinus hypocistis* subsp. *pityusensis*. An additional group consists of species recorded as new for Europe after 1964; here belong *Thesium refractum* and *Aristolochia parvifolia*, as well as the established aliens *Myrica conifera, Humulus scandens, Urtica morifolia* and *Parietaria pensylvanica*. The remaining taxa that have been added are mostly mentioned in the textual comments of Fl. Eur., or otherwise known as cases subject to diverging taxonomical interpretations: *Salix neotricha, S. glauca* subsp. *callicarpaea, S. recurvigemmis, S. jenisseensis, Quercus crenata, Urtica dioica* subsp. *sondenii*.

Omitted, although given in Fl. Eur.: *Aristolochia pontica*, which is not considered to belong to the native European flora. Other taxa also possibly not belonging are *Salix alba* subsp. *micans* and *S. aegyptiaca*.

Purely *nomenclatural deviations* from Fl. Eur. include *Salix myrsinifolia* Salisb. (instead of *S. nigricans* Sm.), *Urtica membranacea* Poiret (instead of *U. dubia* Forskål), *Parietaria judaica* L. (instead of *P. diffusa* Mert. & Koch) and *Thesium pyrenaicum* subsp. *grandiflorum* (A. DC.) Hendrych (instead of subsp. *alpestre* Schwarz). Nomenclatural changes are more frequently caused by differences in the *taxonomical delimitation* of the taxa involved. Examples of major changes of this kind are *Salix nummularia, S. burjatica* and *Parietaria serbica*.

PROBLEMS REGARDING TAXONOMY AND STATUS

A *collective map* was often the only possibility for difficult species consisting of two or more subspecies, or groups consisting of two or more species, even though the taxa in question are recognized as separate entities in Fl. Eur., and even though a more or less satisfactory map could be produced to demonstrate the European area of some of the taxa included. Normally the use of a collective map is not meant to express any particular taxonomical opinion, but in some cases the difficulties encountered in preparing separate maps may indicate the need for taxonomical revaluation.

From the very beginning of the European mapping project it was agreed to omit rare and/or ephemeral casuals, short-distance and/or inconstant escapes from cultivation, and plantings, including large-scale plantings of forest trees. In Vol. 2 (Gymnospermae) this decision was made still more precise as concerns *cultivated trees and shrubs*, which were disregarded irrespective of the purpose of cultivation, whether ornamental or commercial, and whether the plantations had given rise to self-propagating stands or not. Besides the complete exclusion of certain non-European taxa, this decision also resulted in the omission of a large number of records representing established introductions of European taxa outside their native range: the symbol for established aliens, an open circle, was not used in Vol. 2 at all. However, the present volume includes a wide variety of woody plants that differ greatly in their history in Europe, their relationship to human exploitation and their degree of naturalization. This made a rigid application of the instructions concerning non-native occurrences extremely difficult, and a revaluation of the entire policy for plants of this category seemed unavoidable.

The difficulties involved are especially well illustrated in the maps of *Castanea sativa*, *Ficus carica* and *Juglans regia*. Some of the most striking discrepancies between the data of neighbouring countries on the status of different square records could admittedly be smoothed down in special correspondence with the respective Committee members during preparation of the final maps. Nevertheless, there is little possibility that an exact distinction can be made between truly native occurrences and introductions (possibly dating back to prehistorical times), and it would certainly be unwise in these cases to delete the open circles from the final maps. On the contrary, it has become clear that to do so would mean an unnecessary loss of data relevant to the proper evaluation of the distributional patterns concerned.

Considerations of this kind led to the decision to include introductions (open circles) for most of the plants in question, and to make special mention of their omission under »Notes» in the few cases in which they have been deleted.

A fairly liberal policy also seemed to be indicated for introduced non-European woody plants. Each case has been decided separately on its own merits, but we are fully aware that general agreement can hardly be achieved.

THE TEXTUAL COMMENTS

The *synonyms* are those given in the text and index of Fl. Eur., supplemented with a selection of combinations formed in the most recent literature or otherwise representing and elucidating different views on the status and affinities of the taxon involved.

Under the headings *Taxonomy* and *Nomenclature* are given new recently published views and facts, differing from or completing the treatment in Fl. Eur.

Notes give important new or omitted records (mainly indicated with the territorial abbreviations), especially when they complete or correct Fl. Eur. When the distributional information given by Fl. Eur. is not particularly detailed, as tends to be the case at the level of subspecies, all the pertinent territories are listed for the taxa mapped. When possible, reasons have been given for the major additions and omissions at country level by reference to the relevant literature. The Committee members could no doubt help to bring these references more up-to-date than they are in the present volume.

As in Fl. Eur., the following signs are used in connexion with the two-letter geographical abbreviations: [] = not native; * = status doubtful, possibly native; ? = occurrence doubtful.

Total range is treated in two different ways, although this has caused some misunderstanding in the past. First, in the case of European endemics, no references are given to distribution maps published earlier. Taxa endemic to areas within continental Europe are simply given as »endemic to Europe», and more detailed information is provided for those restricted to geographically well-separated island areas, e.g. »endemic to Sicilia». Second, when the total range exceeds the European boundaries, original total range maps or maps covering a major part of the total range (e.g. Eurasian part of a circumpolar range) have been cited, if any. Maps published previous to Hegi 1957, Hultén AA 1958 and MJW 1965 (for the abbreviations, see below) are generally not given, except for the more important forest trees, the modern mapping of which essentially dates back to Schmucker 1942. The aim is to provide a number of references to reliable total range maps readily accessible to users of this Atlas living in different parts of the continent and belonging to different language groups. The hopeless task of providing a complete list has not been attempted. It may be added that this time, Index holmensis (ed. by H. Tralau) has been of only limited help, since its Vol. IV (303 pp. Zürich 1974), the first one dealing with Dicotyledones, covers only the beginning of the alphabet, not extending beyond B.

The following abbreviations are used for frequently quoted literature:

Čermák 1955 = K. Čermák et al., Lesnický a myslivecký Atlas. — 15 pp. + text 91 pp. + maps 120 pp. Praha 1955.

Fenaroli 1967 = L. Fenaroli, Gli Alberi d'Italia. — 320 pp. Milano 1967.

Fl. Eur. = T. G. Tutin, V. H. Heywood et al. (ed.), Flora Europaea 1. — xxxiv + 464 pp. Cambridge 1964.

Hegi 1957, 1958 = G. Hegi, Illustrierte Flora von Mitteleuropa, 2. Aufl. 3 (1): 1—240 (1957), 241—452 (1958). München 1957, 1958 (maps by H. Meusel).

Hendrych 1968 = R. Hendrych, Abriss der chorologischen Entwicklung der Gattung Thesium (Santalaceae) in der Paläoholarktis. Thesium-Studien XXI. — Preslia 40: 28—51. 1968.

Hultén AA 1958 = E. Hultén, The Amphi-Atlantic Plants. — Kungl. Svenska Vet.-Akad. Handl., Ser. 4, 7 (1): 1—340. 1958.

Hultén Alaska 1968 = E. Hultén, Flora of Alaska and neighboring territories. — 1008 pp. Stanford 1968.

Hultén CP 1971 = E. Hultén, The Circumpolar Plants. II. Dicotyledons. — Kungl. Svenska Vet.-Akad. Handl., Ser. 4, 13 (1): 1—463. 1971.

MJW 1965 = H. Meusel, E. Jäger & E. Weinert, Vergleichende Chorologie der zentraleuropäischen Flora. — Text 583 pp., Karten 258 pp. Jena 1965.

Neumann & Polatschek 1972 = A. Neumann & A. Polatschek, Cytotaxonomischer Beitrag zur Gattung Salix. — Ann. Naturh. Mus. (Wien) 76: 619—633. 1972.

Schmucker 1942 = Th. Schmucker, La distribution des espèces arborescentes de la zone septentrionale tempérée. — Die Baumarten der nördlich-gemässigten Zone und ihre Verbreitung. — The tree species of the Northern Temperate zone and their distribution. — Las especies arbóreas de la zona templada septentrional y su difusión. — Le specie arboree della zona nordico-temperata e la loro diffusione. — Silvae Orbis 4: I—X + 1—156 + Cartes 1—250. Berlin—Wannsee 1942.

Skvortsov 1968 = A. K. Skvortsov, Willows of the USSR. A taxonomic and geographic revision. — Proc. Study Fauna Fl. USSR, N.S., Sect. Bot. 15 (XXIII): 1—262. 1968.

ACKNOWLEDGEMENTS

We are greatly obliged to the members of the Committee for Mapping the Flora of Europe, whose contributions have made possible the production of another volume of Atlas Florae Europeae. In addition to taking care of the collection of the material for their respective territories, a number of Committee Members also gave valuable taxonomic and nomenclatural advice, and provided the Secretariat with essential material from outside their own countries. Generous help was also received from numerous other botanists consulted on special problems.

The expenditure of the Secretariat was covered by annual allowances from the Ministry of Education of Finland and the Finnish National Research Council for Sciences. The publication of the present volume, like that of the preceding ones, was undertaken by the Finnish Biological Society Vanamo.

As earlier, working facilities for the Secretariat have been provided by the Department of Botany, University of Helsinki (Head: Professor HANS LUTHER).

Technical assistance in preparing this volume has been given by Miss MARJA ISOVIITA, Mrs. IRMA JÄRVINEN, M.A., Mrs. AUNE KOPONEN, M.A., Mrs. ULLA KURIMO, M.A., Mrs. LIISA MÄKELÄ, M.A., Mrs. PAULA OKSANEN, Miss MARKETTA PYHÄLAMMI and Mrs. MARJA UOTILA, M.A.

ANGIOSPERMAE

DICOTYLEDONES

SALICACEAE

Salix pentandra L. — Map 201.

Notes. ?Gr added (one single doubtful record; not given in Fl. Eur.).
Total range. Čermák 1955: 76; MJW 1965: map 113a; Fenaroli 1967: 88; Skvortsov 1968: 106.

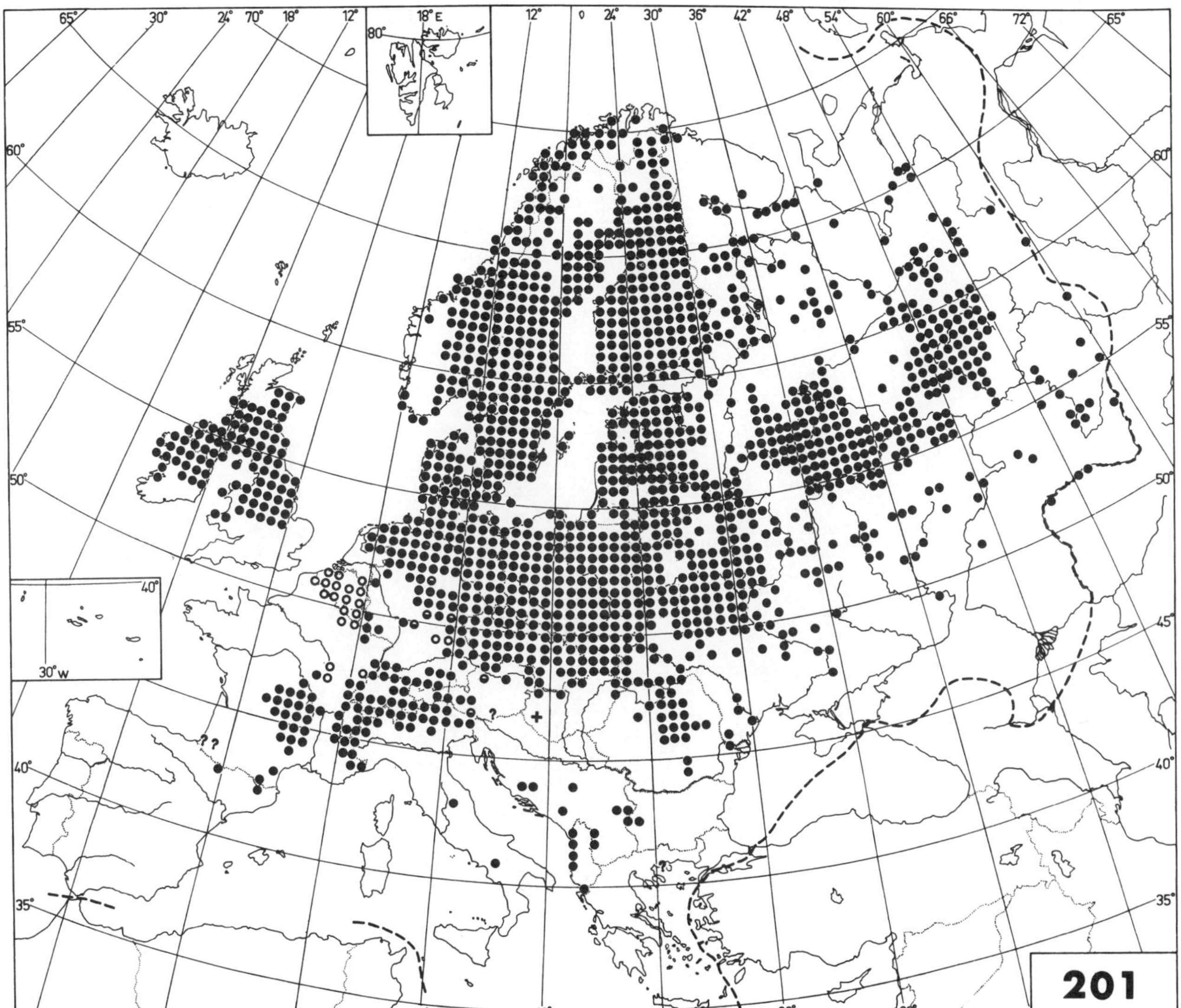

Salix pentandra

Salix fragilis L. — Map 202.

Taxonomy. Neumann & Polatschek 1972: 621.
Notes. Introductions not mapped. There is much additional confusion with the hybrid *Salix alba* × *fragilis* (*S.* × *rubens* Schrank); the actual native area of the true *S. fragilis* may be considerable smaller than given on the map. Br, Da, Hb, Lu, No, Rs(N) and Su omitted (given in Fl. Eur.), Tu confirmed (?Tu in Fl. Eur.).
Total range. Skvortsov 1968: 112 (incl. *Salix* × *rubens*).

Salix excelsa S. G. Gmelin

Salix australior N. J. Andersson

Taxonomy. Skvortsov 1968: 110—111.
Notes. Mentioned in Fl. Euı. without indication of range. Native European occurrences seemingly unknown.
Total range. The extra-European range mapped by Skvortsov 1968: 110.

Salix neotricha R. Görz

Taxonomy. Possibly an independent species (Ncumann & Polatschek 1972: 621).
Notes. Hs, Lu (the species not mentioned in Fl. Eur.). Not mapped.

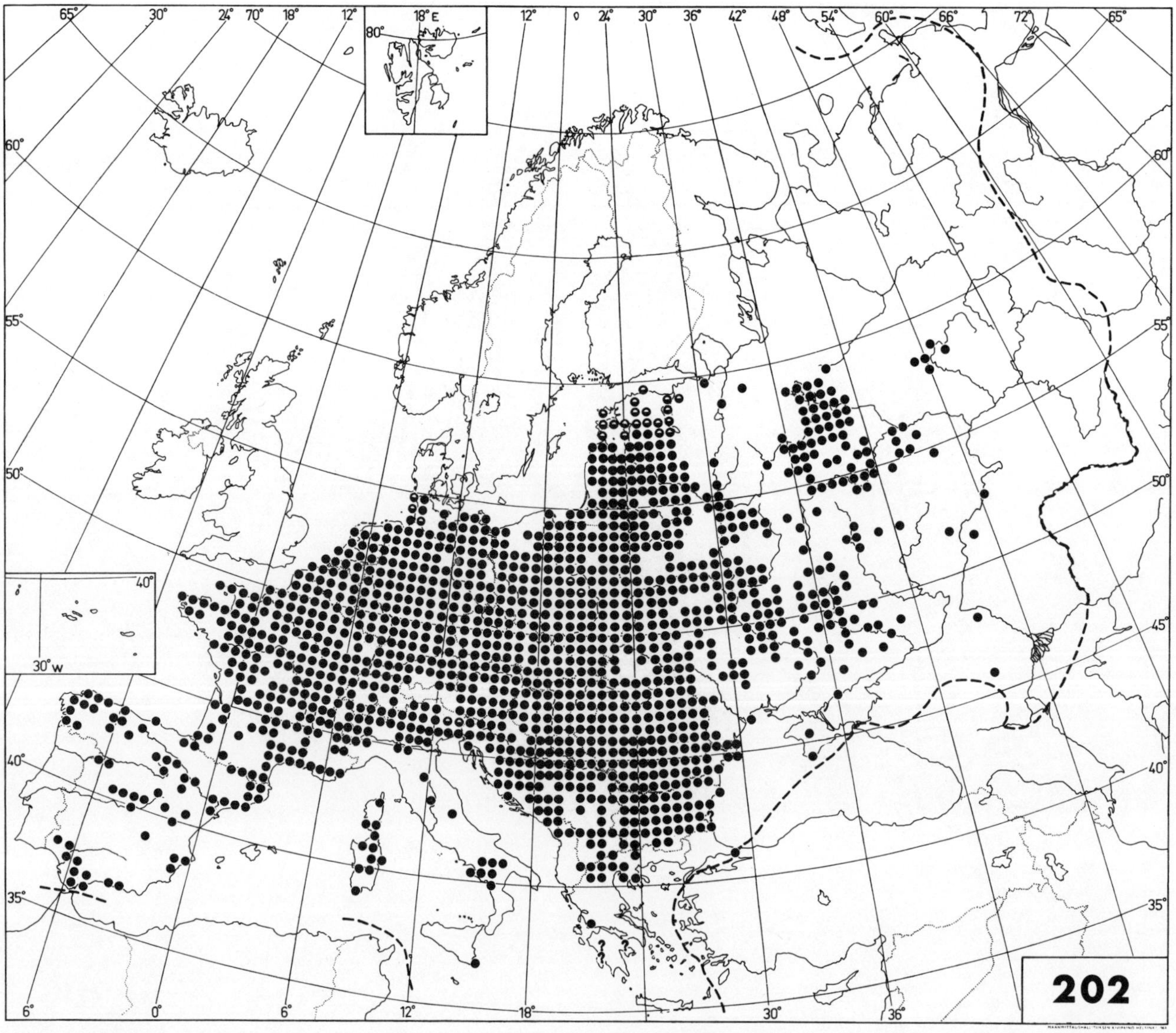

Salix fragilis

Salix alba L. — Map 203.

Notes. Introductions not mapped. Native status doubtful in Br and Cr (given as native in Fl. Eur.), Da and Hb omitted (given in Fl. Eur.). Mapped as one collective entity. A number of peripheral dots probably relate to occurrences of cultivated or naturalized plants. The exact boundaries of the true native range seem almost impossible to ascertain.

Total range. Čermák 1955: 76; MJW 1965: map 113b; Fenaroli 1967: 78; Skvortsov 1968: 110.

S. alba subsp. **alba**

Notes. Of the two subspecies considered native in Europe (Fl. Eur.), subsp. *alba* covers most of the records, if not all.

S. alba subsp. **coerulea** (Sm.) Rech. fil.

Notes. »Frequently planted, occasionally naturalized» (Fl. Eur.). Not mapped.

S. alba subsp. **vitellina** (L.) Arcangeli

Notes. »Frequently planted» (Fl. Eur.). Not mapped.

S. alba subsp. **micans** (N. J. Andersson) Rech. fil.

Taxonomy. K. H. Rechinger, Österr. Bot. Zeitschr. 110: 338 (1963); Skvortsov 1968: 108—109.

Notes. »Greece» (Fl. Eur.). However, at the present stage of knowledge it is not possible to map subsp. *micans* separately, especially as Skvortsov 1968: 108—110 does not mention it from Europe at all.

Total range. Given as endemic to Caucasia in Skvortsov 1968: 110.

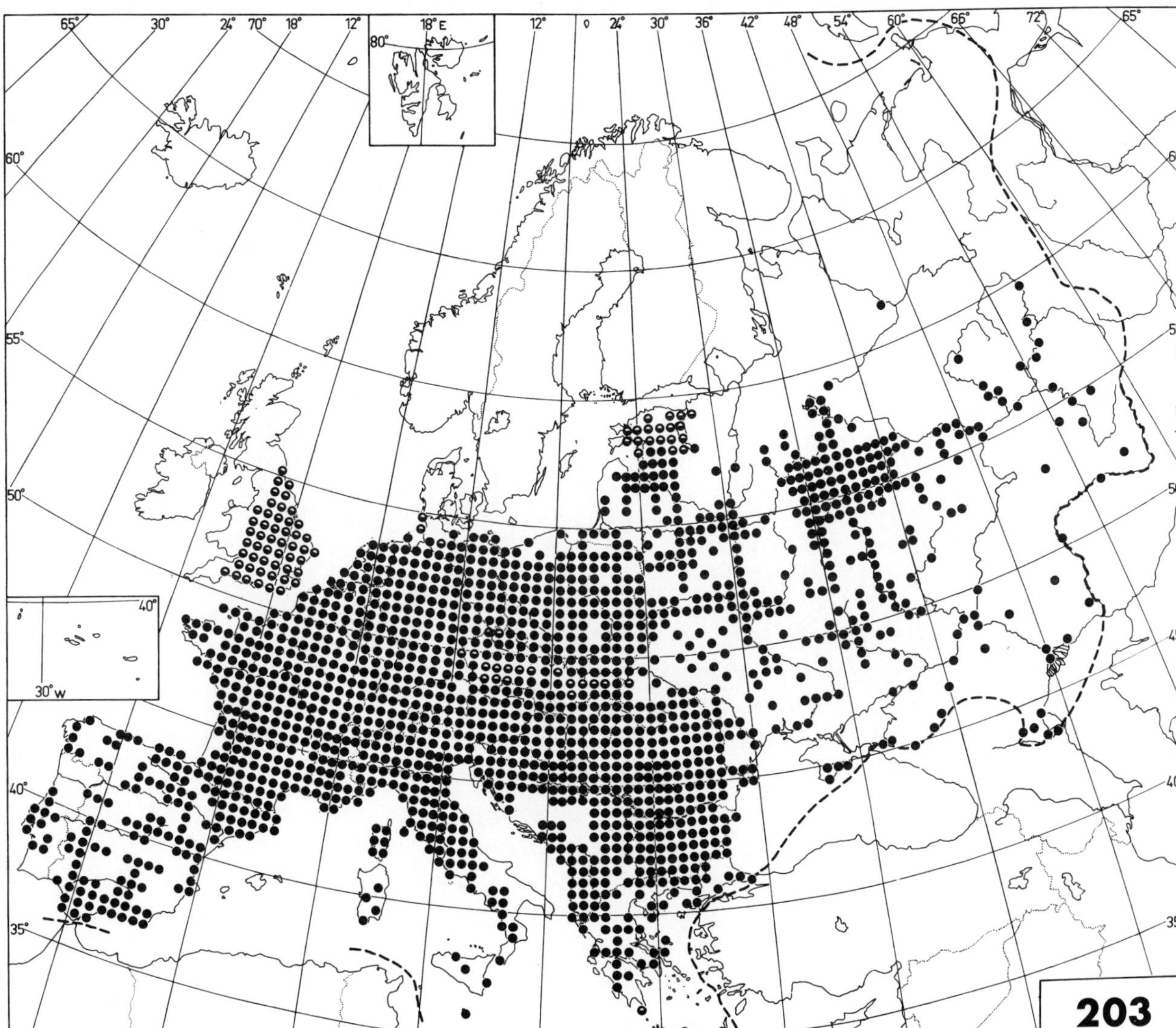

Salix alba

Salix triandra L. — Map 204.

Salix amygdalina L.

Taxonomy. The two subspecies mentioned in Fl. Eur. have not been recognized in Skvortsov 1968: 100—101, and could not be mapped separately on the basis of the material available. However, cf. the diverging opinion expressed by Neumann & Polatschek 1972: 622.

Notes. Ju added (not given in Fl. Eur.), Sa omitted (given in Fl. Eur.).

Total range. Schmucker 1942: maps 131—132; Čermák 1955: 76; MJW 1965: map 113c; Fenaroli 1967: 90; Skvortsov 1968: 102.

S. triandra subsp. **triandra**

Salix triandra subsp. concolor (Koch) A. Neumann ex Rech. fil.

Notes. »Common in the lowlands of N.W. Europe» (Fl. Eur.).

S. triandra subsp. **discolor** (Koch) Arcangeli

Taxonomy. Neumann & Polatschek 1972: 622—623.

Notes. »Common in the S. & E. of the area of the species and in the valleys of the Alps» (Fl. Eur.).

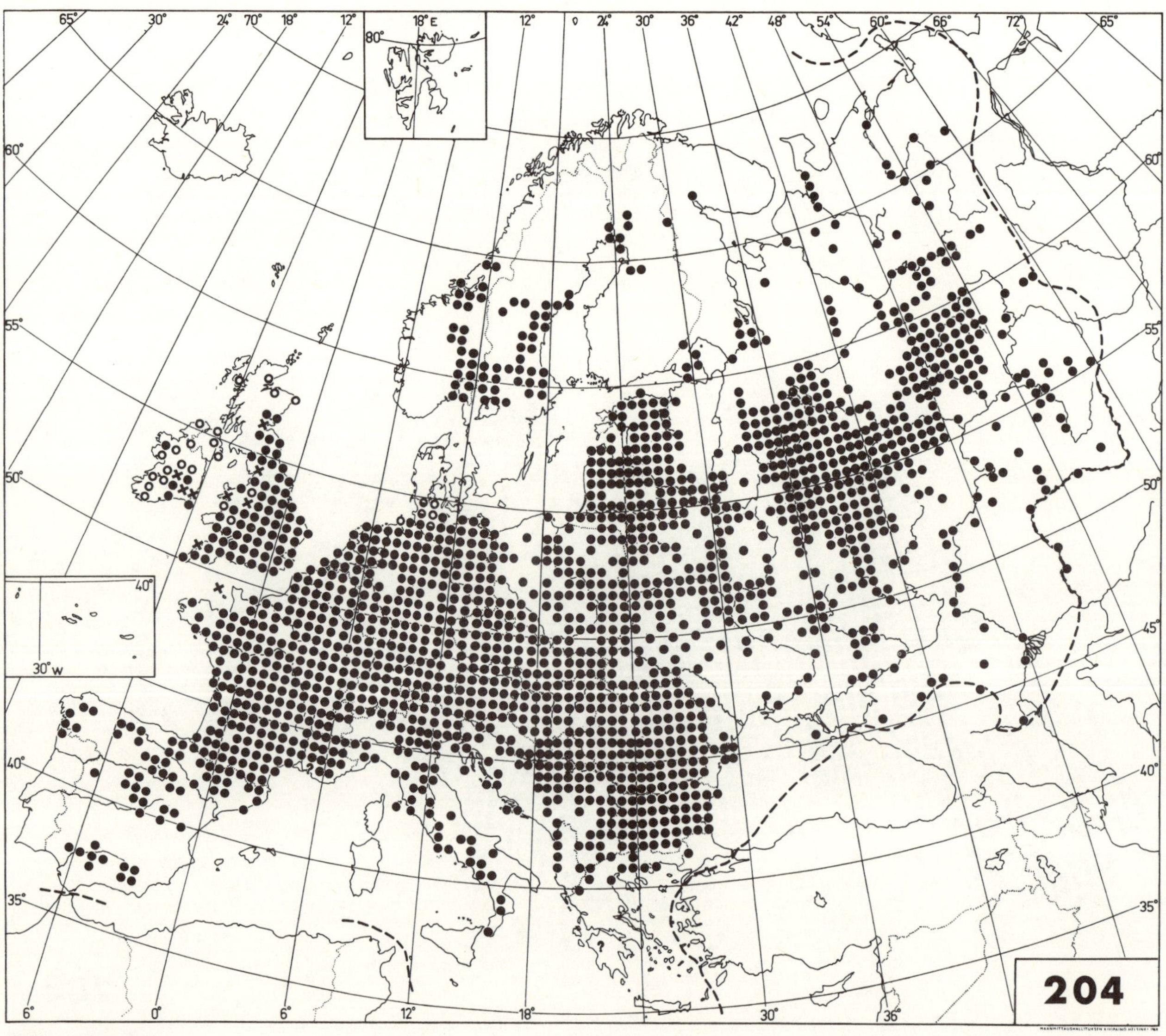

Salix triandra

Salix reticulata L. — Map 205.

Notes. Al added (not given in Fl. Eur.), Rs(W) omitted (given in Fl. Eur.).

Total range. MJW 1965: map 113d; Skvortsov 1968: 102 (Eurasian part); Hultén CP 1971: map 39; K. Browicz & Cz. Kaczmarek, Atlas Distr. Trees Shrubs Poland 11: map 36 (Warszawa & Poznań 1972).

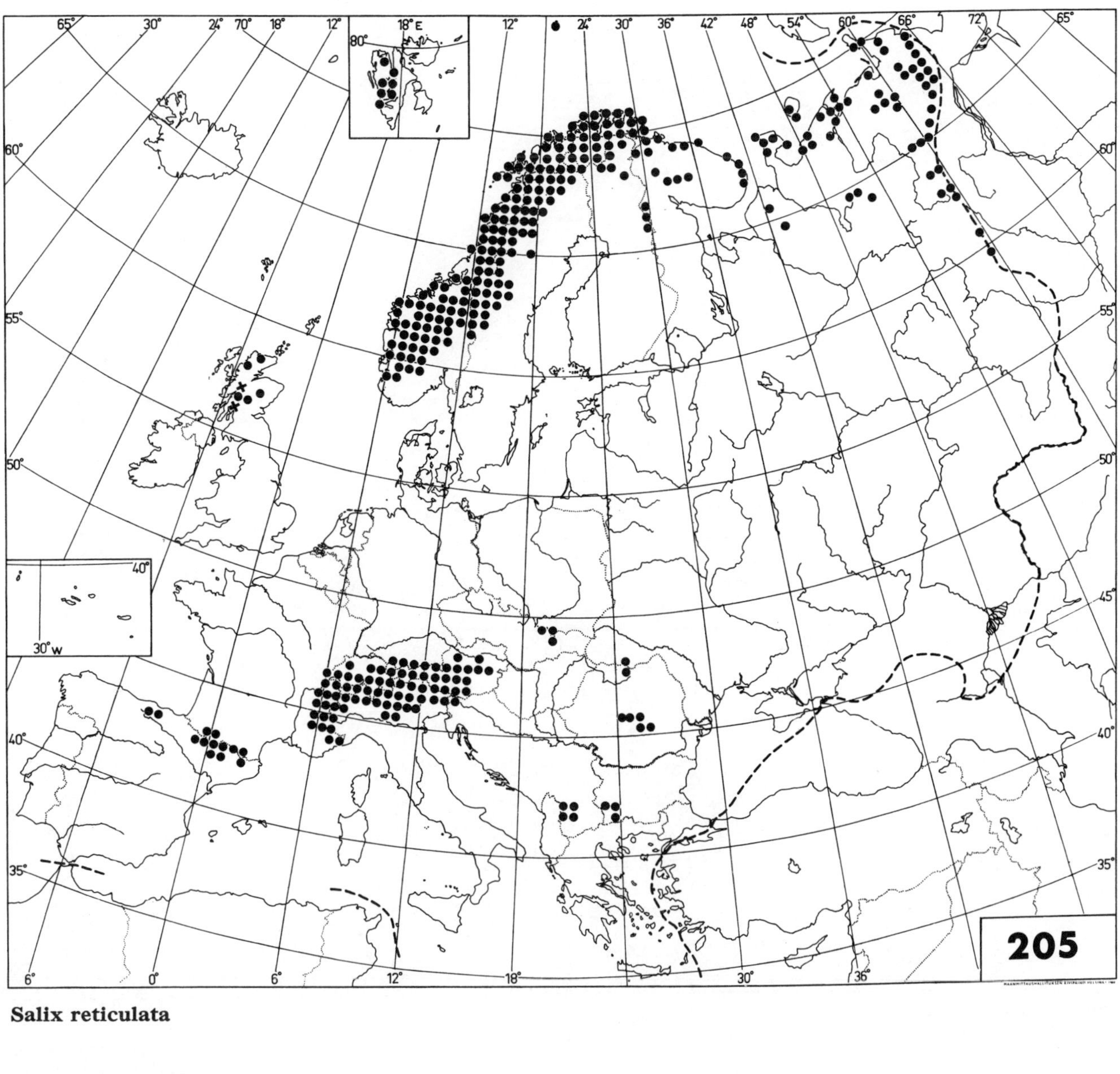

Salix reticulata

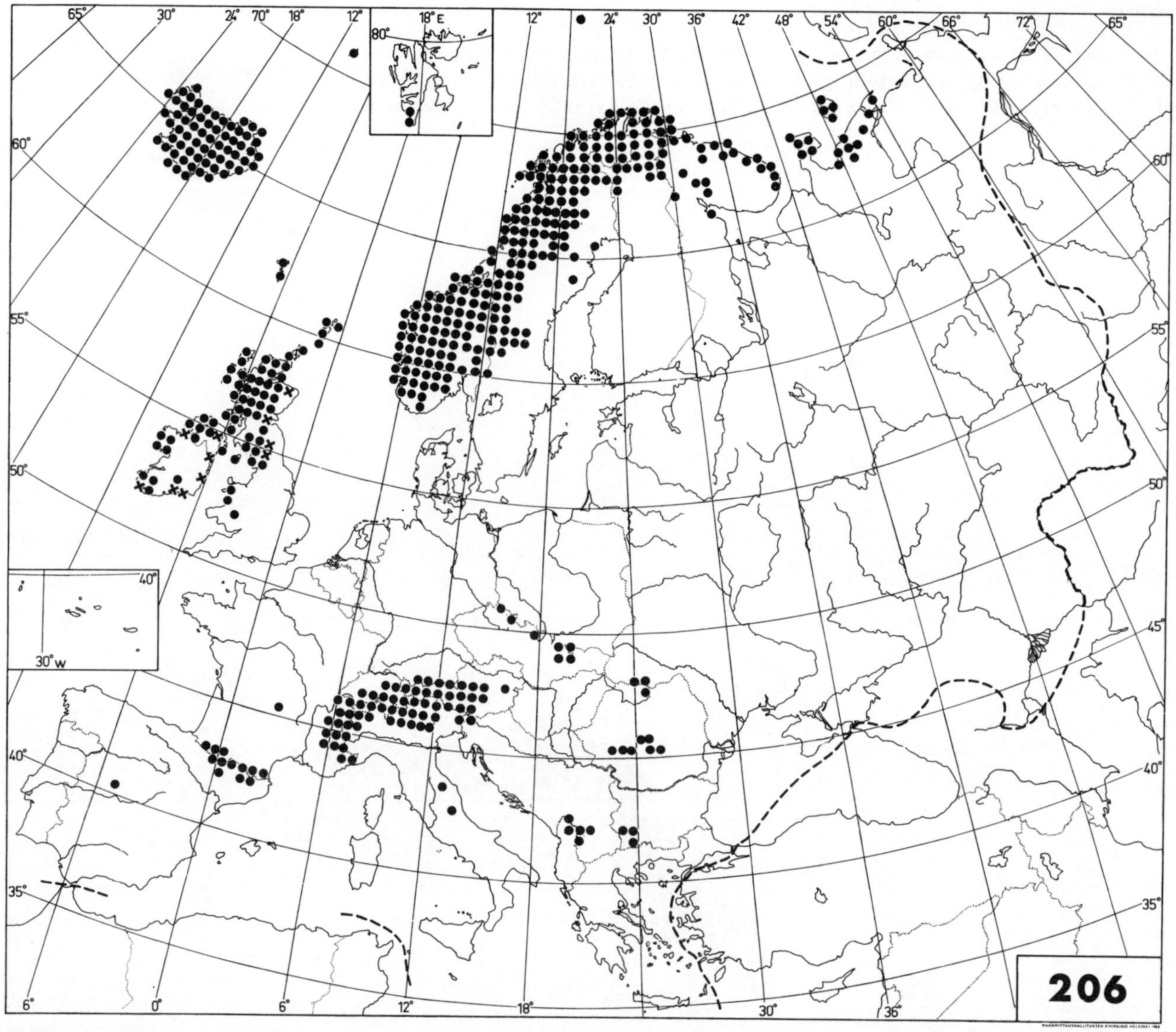

206

Salix herbacea

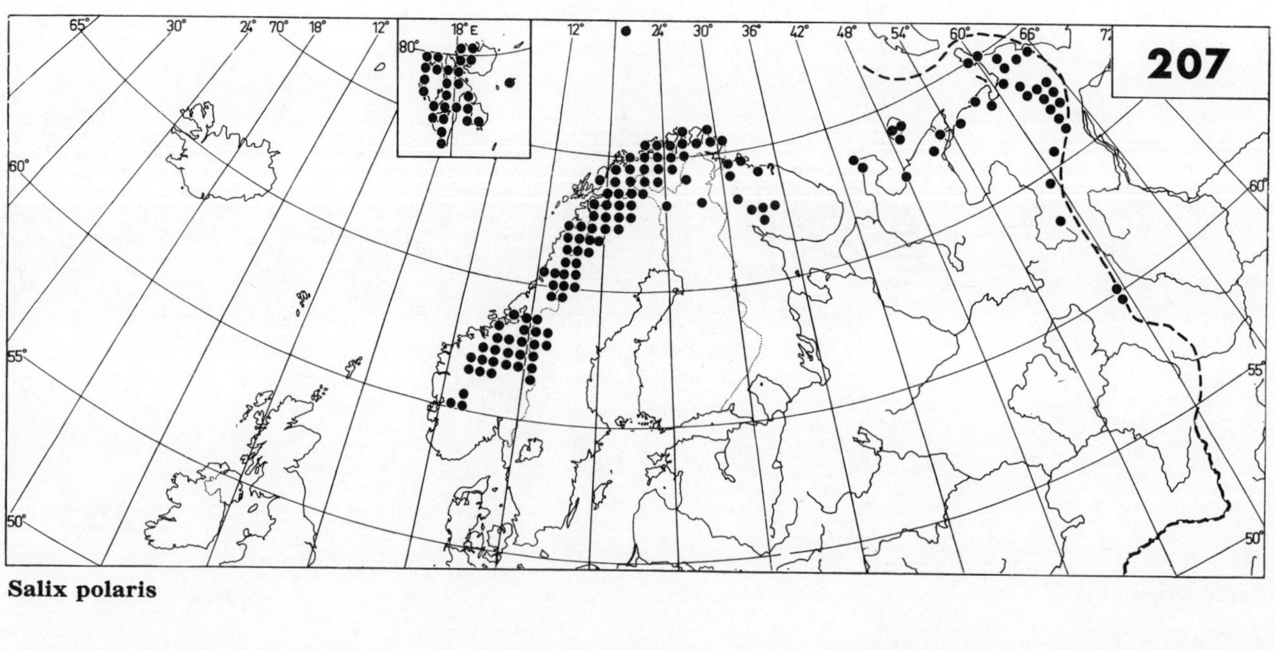

207

Salix polaris

Salix herbacea L. — Map 206.

Notes. Al added (not given in Fl. Eur.).

Total range. Hultén AA 1958: map 24; E. Hadač, Preslia 32: 248 (1960); MJW 1965: map 114a; A. Boratyński & K. Browicz, Atlas Distrib. Trees Shrubs Poland 17: map 35 (1976).

Salix polaris Wahlenb. — Map 207.

Notes. Rs(C) added (not given in Fl. Eur.).

Total range. E. Hadač, Preslia 32: 250 (1960); Skvortsov 1968: 121.

Salix retusa L. – Map 208.

Notes. Cz added (not given in Fl. Eur.).

Total range. Endemic to Europe.

Salix kitaibeliana Willd. — Map 209.

Taxonomy. Included in *Salix retusa* in Skvortsov 1968: 123.

Total range. Endemic to Europe.

Salix serpillifolia Scop. — Map 209.

Notes. Ga and Ge added (not given in Fl. Eur.).

Total range. Endemic to Europe.

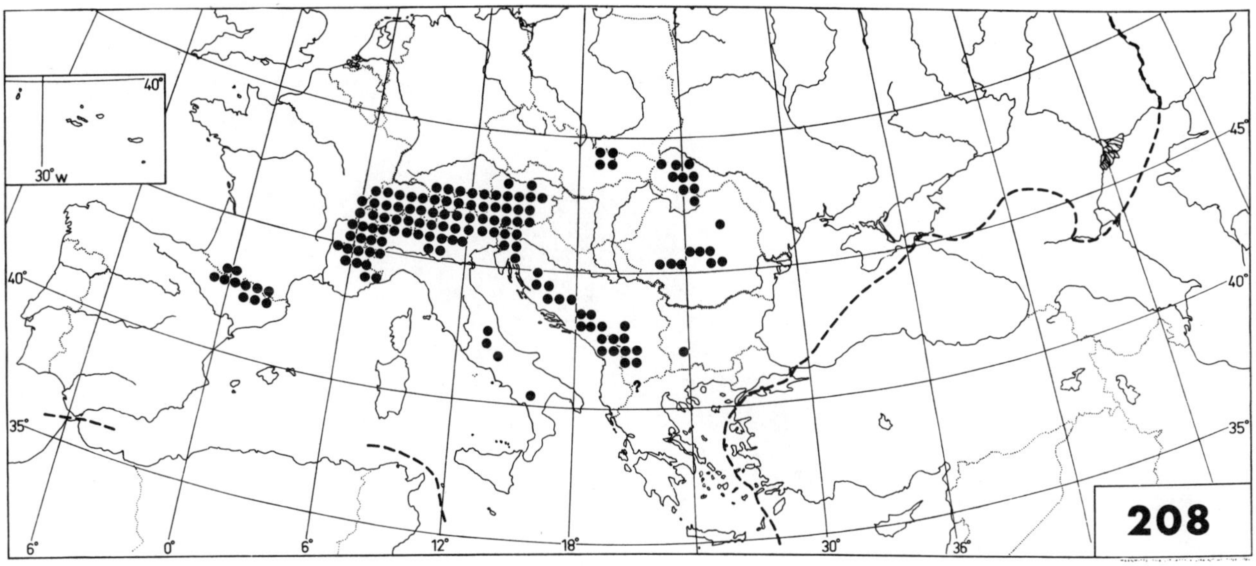

Salix retusa

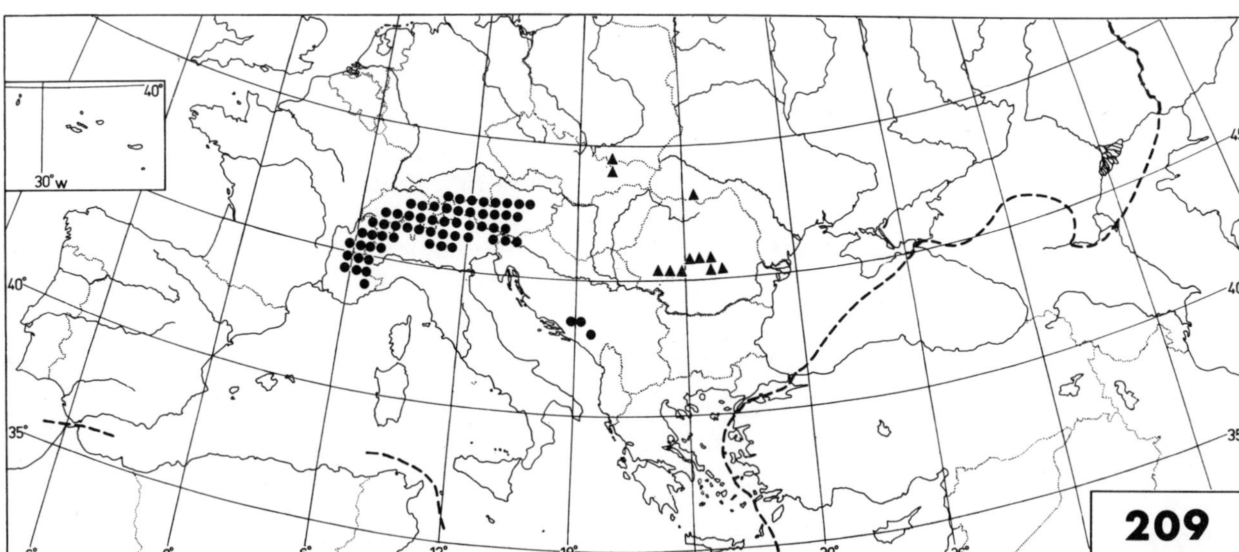

• = **Salix serpillifolia** ▲ = **S. kitaibeliana**

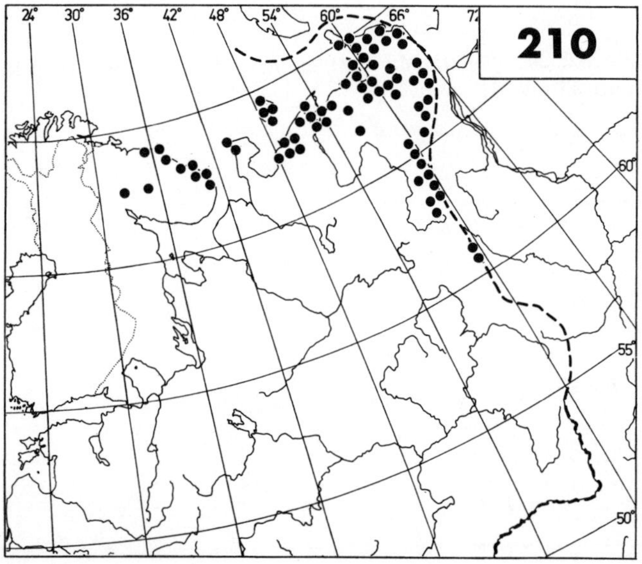

Salix nummularia

Salix nummularia N. J. Andersson — Map 210.

Salix rotundifolia auct. non Trautv.; S. tundricola Schljakov

Taxonomy. R. N. Schljakov, Not. Syst. (Leningrad) 16: 67 (1954). Contrary to the treatment in Fl. Eur., *Salix rotundifolia* Trautv. is considered to represent a separate taxon confined to the Beringian area; see B. Floderus, Svensk Bot. Tidskr. 35: 351—352 (1941); Skvortsov 1968: 122—123, 146; G. W. Argus, Nat. Mus. Nat. Sci. Publ. Bot. 2: 53—61 (1973).

Total range. Skvortsov 1968: 124.

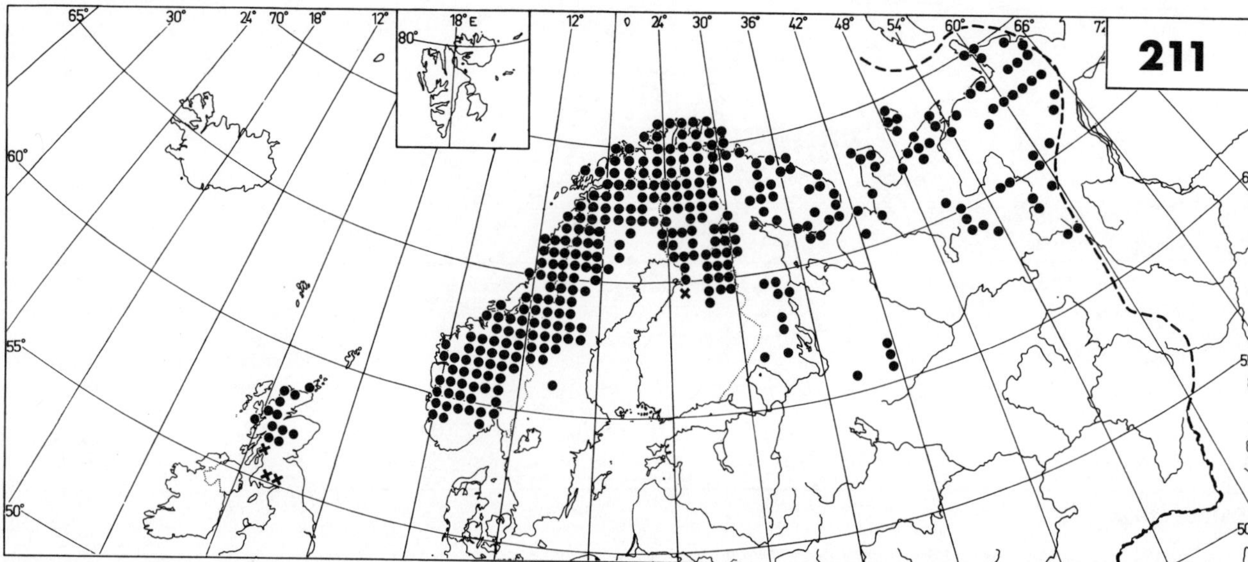

Salix myrsinites

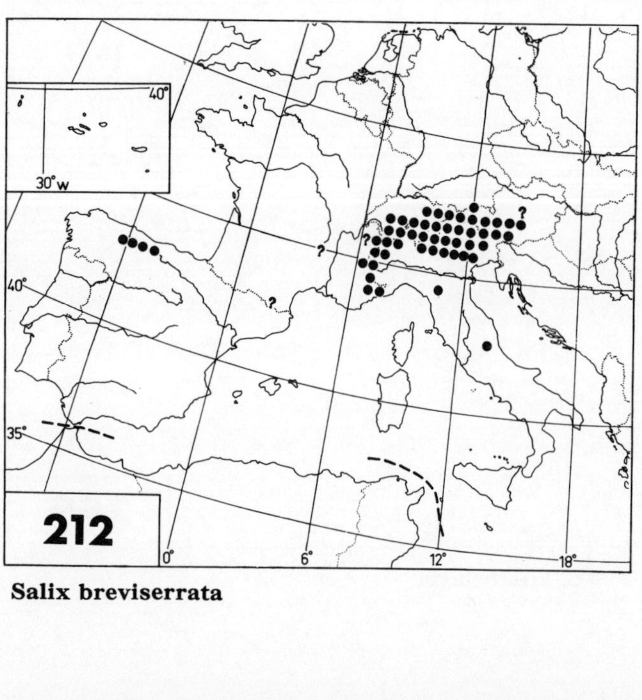

Salix breviserrata

Salix myrsinites L. — Map 211.

Notes. Sb omitted (given in Fl. Eur.).

Total range. Skvortsov 1968: 140.

Salix breviserrata B. Flod. — Map 212.

Salix myrsinites subsp. serrata (Neilr.) Schinz & Thell.

Notes. Ge and He added (not given in Fl. Eur.), Ju omitted (given in Fl. Eur.).

Total range. Endemic to Europe.

Salix alpina Scop. — Map 213.

Salix jacquinii Host

Total range. Endemic to Europe.

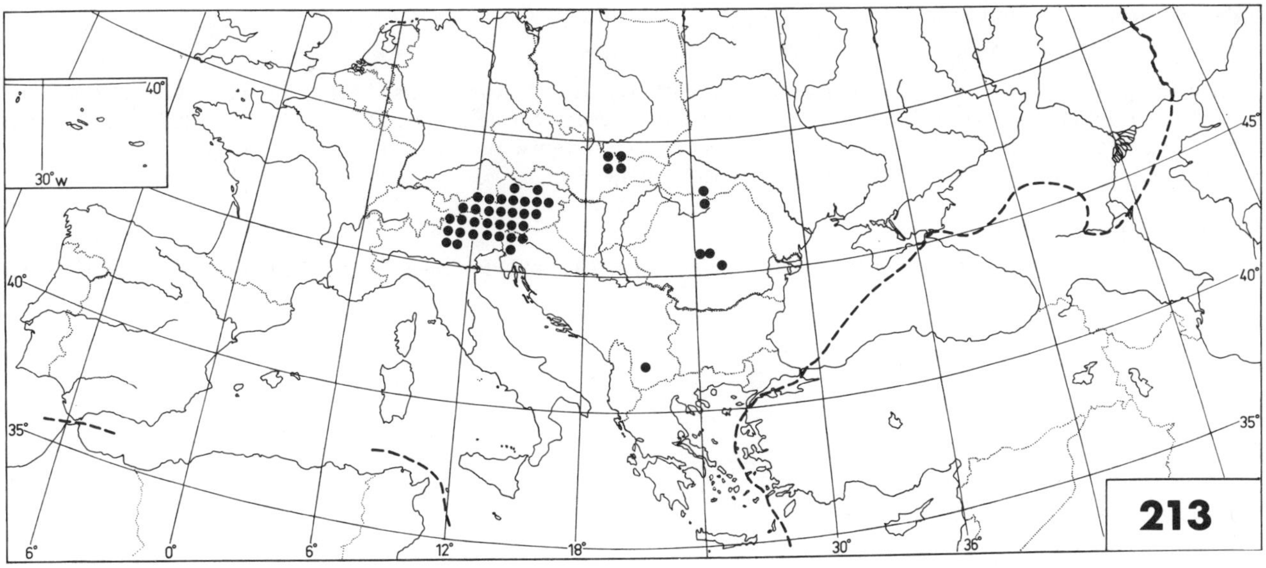

Salix alpina

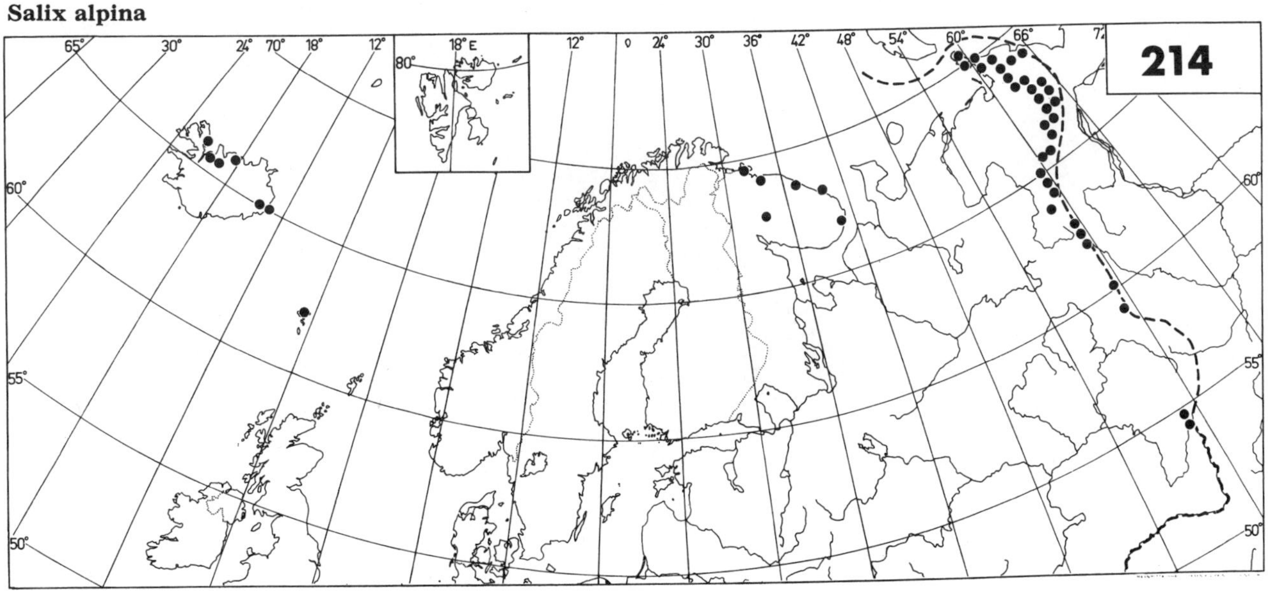

Salix arctica

Salix arctica Pallas — Map 214.

Notes. Fa, Is and Rs(C) added (not given in Fl. Eur.): A. K. Skvortsov, Ann. Naturh. Mus. (Wien) 75: 227 (1971); Á. Löve, Íslenzk Ferdaflóra, p. 194 (1970).

Total range. A. K. Skvortsov in A. I. Tolmatchev, Fl. Arct. URSS 5: 69 (1966) (arctic parts of the Soviet Union); Skvortsov 1968: 135 (Eurasian part); Hultén CP 1971: map 29.

Salix pulchra Cham. — Map 215.

Salix phylicifolia subsp. pulchra (Cham.) Hultén

Taxonomy. Hultén CP 1971: 136; G. W. Argus, Nat. Mus. Nat. Sci. Publ. Bot. 2: 157—167 (1973).

Total range. Hultén Alaska 1968: 359; Skvortsov 1968: 184 (Eurasian part); Hultén CP 1971: map 127.

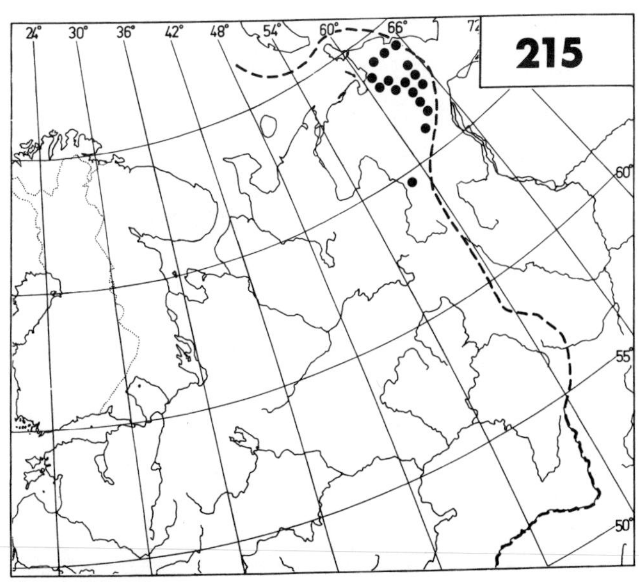

Salix pulchra

Salix glauca L. — Map 216.

Incl. Salix callicarpaea Trautv. and S. stipulifera B. Flod. ex Häyrén

Taxonomy. Hultén CP 1971: 321—323; A. K. Skvortsov, Ann. Naturh. Mus. (Wien) 75: 224—228 (1971).
Notes. Fa omitted (given in Fl. Eur.): A. K. Skvortsov, Ann. Naturh. Mus. (Wien) 75: 227—228 (1971). Rs(C) and Sb added (not given in Fl. Eur.); for Sb, see E. Hultén, Svensk Bot. Tidskr. 58: 350 (1964).
Total range. Skvortsov 1968: 132; Hultén CP 1971: map 46.

S. glauca subsp. **glauca**

Notes. In the map not separated from subsp. *stipulifera*. »Iceland, Faeröer, N. & W. Fennoscandia, arctic Russia» (Fl. Eur. for *S. glauca* s.str.).

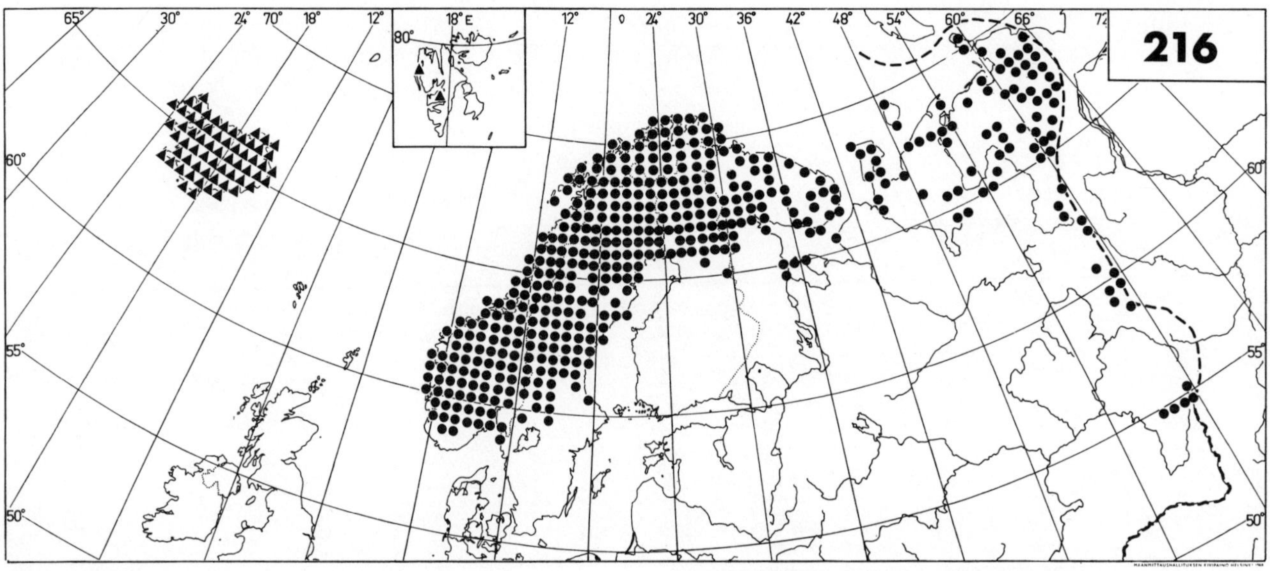

Salix glauca

• = subsp. **glauca** + subsp. **stipulifera**
▲ = subsp. **callicarpaea**

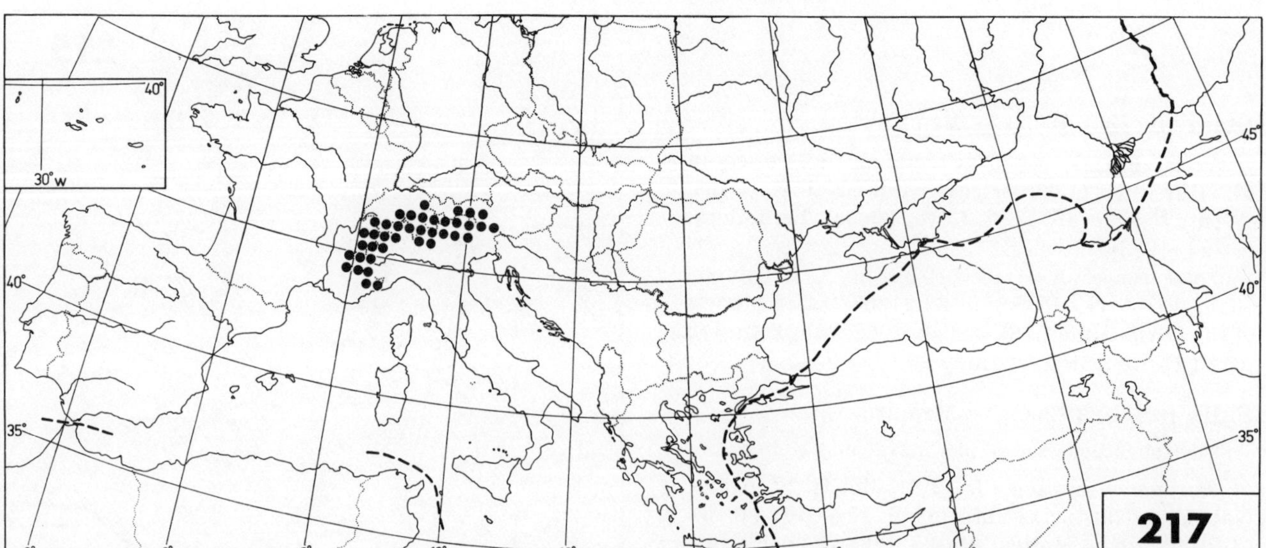

Salix glaucosericea

S. glauca subsp. **stipulifera** (B. Flod. ex Häyrén) Hiitonen

Salix glauca var. appendiculata Wahlenb.; S. stipulifera B. Flod. ex Häyrén

Taxonomy. Treatment as an independent species seems unrealistic. Most of the recent students of the group recognize it only as a minor variant of *Salix glauca*, some even treating it as a synonym. See e.g. N. Hylander, Nord. Kärlväxtfl. 2: 264 (1966); A. K. Skvortsov in A. I. Tolmatchev, Fl. Arct. URSS 5: 70—71 (1966); Skvortsov 1968: 130; Hultén CP 1971: 322.

Notes. Only within the area of subsp. *glauca*, and mapped together with it. »N. Fennoscandia and arctic Russia» (Fl. Eur.).

S. glauca subsp. **callicarpaea** (Trautv.) Böcher — Map 216.

Salix callicarpaea Trautv.; S. cordifolia Pursh subsp. callicarpaea (Trautv.) Löve; S. cordifolia Pursh var. callicarpaea (Trautv.) Fern.

Taxonomy. A. K. Skvortsov, Ann. Naturh. Mus. (Wien) 75: 224—228 (1971).

Notes. Fa omitted (given in Fl. Eur.). Sb added (not given in Fl. Eur.): E. Hultén, Svensk Bot. Tidskr. 58: 350 (1964).

Total range. Hultén AA 1958: map 167 (not up to date as to details); Hultén CP 1971: map 46.

Salix glaucosericea B. Flod. — Map 217.

Taxonomy. Included, without formal recognition, in *Salix glauca* by Skvortsov 1968: 130.

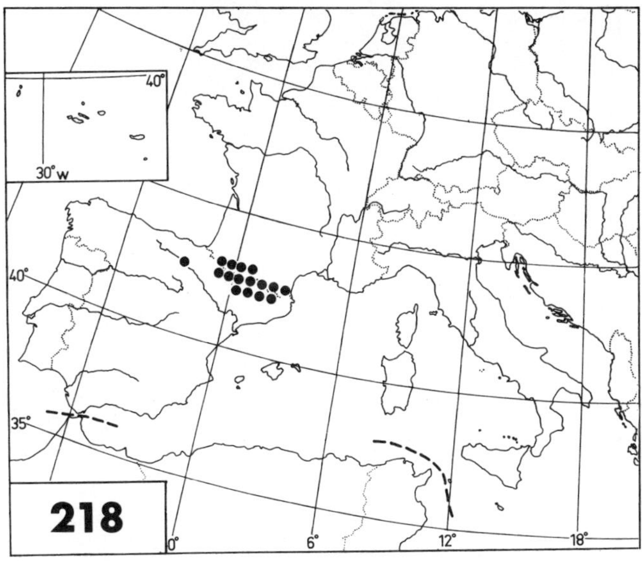

Salix pyrenaica

Salix pyrenaica Gouan — Map 218.

Total range. Endemic to Europe.

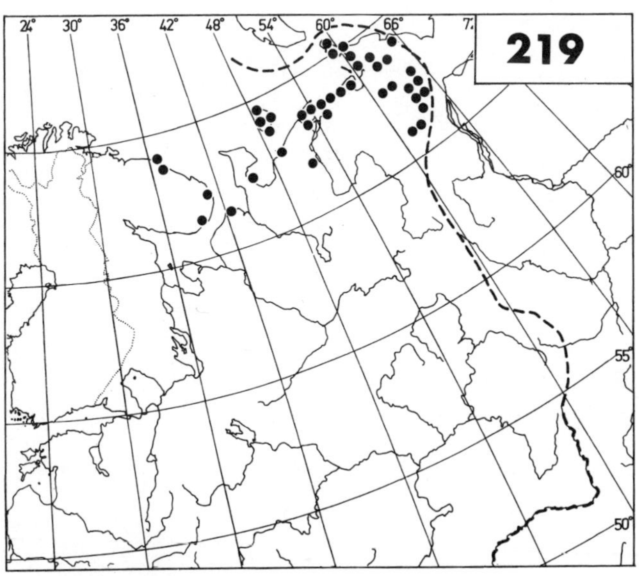

Salix reptans

Salix reptans Rupr. — Map 219.

Total range. A. K. Skvortsov in A. I. Tolmatchev, Fl. Arct. URSS 5: 69 (1966); Skvortsov 1968: 132; B. A. Yurtsev, Problems of Phytogeogr. of North-eastern Asia, p. 126 (Leningrad 1974).

Salix lanata L. — Map 220.

Notes. Fa omitted (given in Fl. Eur.), Rs(C) added (not given in Fl. Eur.).
Total range. Skvortsov 1968: 210 (Eurasian part; incl. *Salix glandulifera* B. Flod.); Hultén CP 1971: map 31.

Salix glandulifera B. Flod.

Salix lanata subsp. glandulifera (B. Flod.) Hiitonen

Taxonomy. Included in *Salix lanata* in Skvortsov 1968: 208, and A. K. Skvortsov, Ann. Naturh. Mus. (Wien) 75: 230 (1971).
Notes. Not mapped, because only partially separated from *Salix lanata*, with which it is fully sympatric in Fennoscandia and N. Russia. »Arctic Europe» (Fl. Eur.).

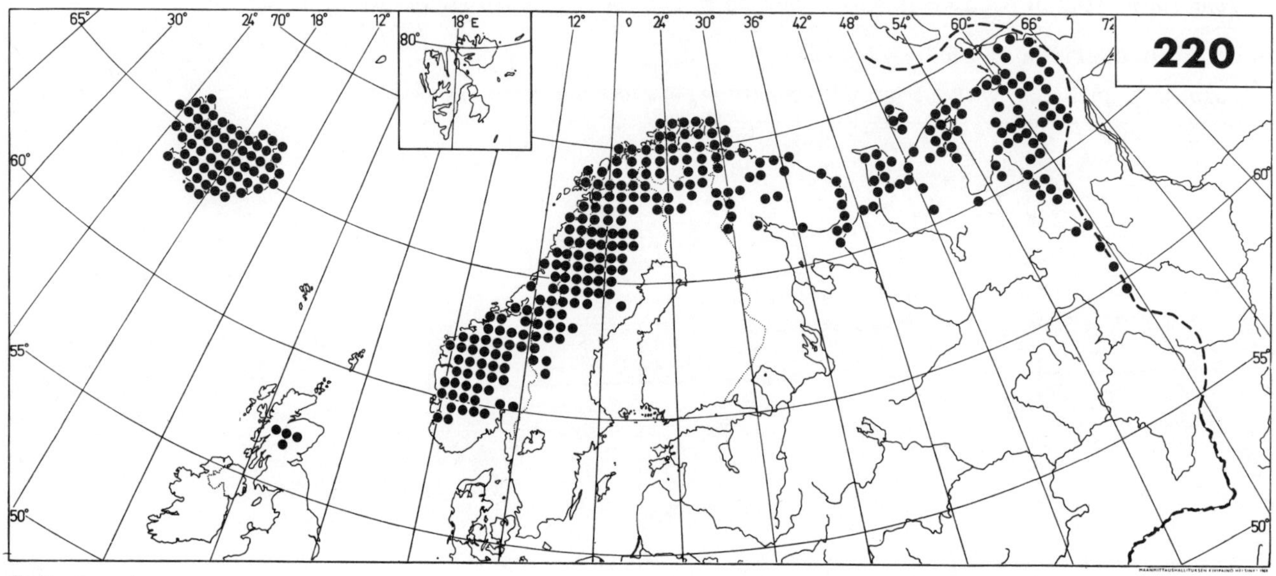

Salix lanata

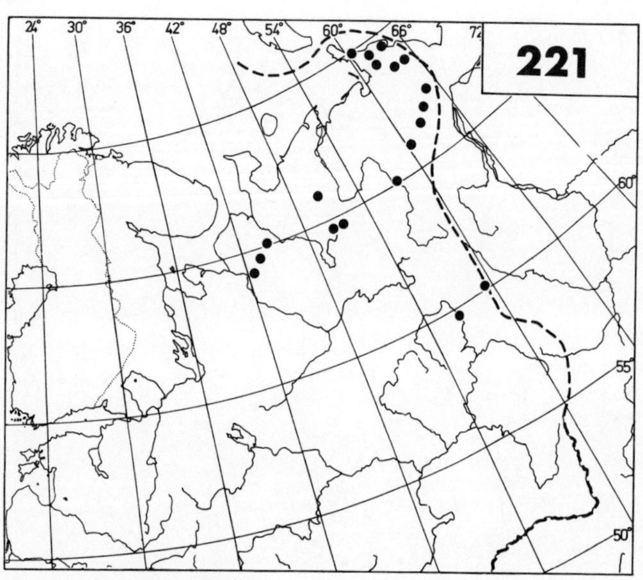

Salix recurvigemmis

Salix recurvigemmis A. Skvortsov — Map 221.

Salix »recurvigemmata» A. Skvortsov

Notes. Rs(N, C). »E. arctic Russia» (Fl. Eur.).
Total range. A. K. Skvortsov, Not. Syst. (Leningrad) 38: 40 (1957); Skvortsov 1968: 210.

Salix phylicifolia L. — Map 222.

Taxonomy. According to Skvortsov 1968: 180—183, *Salix bicolor* Willd. and *S. hegetschweileri* Heer are included in *S. phylicifolia*, as subsp. *rhaetica* (N. J. Andersson) A. Skvortsov, except for plants of the Pyrenees and the French Massif Central, which are given specific rank as *S. basaltica* Coste.
Notes. Da omitted (given in Fl. Eur.); Rs(C) added (not given in Fl. Eur.). Also present in Bu (Rila Mountains), according to information in a letter from J. Chmelař (Brno).
Total range. A. K. Skvortsov, Bull. Biol. Soc. Nat. Moskva 66: 32 (1961); Skvortsov 1968: 181 (incl. *Salix bicolor* and *S. hegetschweileri*, excl. *S. basaltica*); Hultén Alaska 1968: 359; Hultén CP 1971: map 127 (incl. *S. hegetschweileri*).

Salix bicolor Willd. — Map 222.

Salix basaltica Coste; S. schraderiana Willd.; S. weigeliana Willd.

Taxonomy. S. Grapengiesser, Bot. Not. 108: 326 (1955); Neumann & Polatschek 1972: 630. See above under *Salix phylicifolia.*

Notes. Ju omitted (given in Fl. Eur.).

Total range. According to B. Floderus, Ark. Bot. 29 A (18): 8—9 (1939), endemic to Europe (although not marked as such in Fl. Eur.).

Salix hegetschweileri Heer — Map 222.

Salix rhaetica N. J. Andersson; S. bicolor subsp. rhaetica (N. J. Andersson) B. Flod.; S. phylicifolia subsp. rhaetica (N. J. Andersson) A. Skvortsov

Taxonomy. Neumann & Polatschek 1972: 628.

Notes. Ga confirmed (?Ga in Fl. Eur.).

Salix hibernica Rech. fil. — Map 222.

Taxonomy and nomenclature. K. H. Rechinger, Österr. Bot. Zeitschr. 110: 340—341 (1963); A. W. Stelfox, Irish Naturalist's Journ. 15: 25—29 (1965). Male plant unknown.

Total range. Endemic to Ireland.

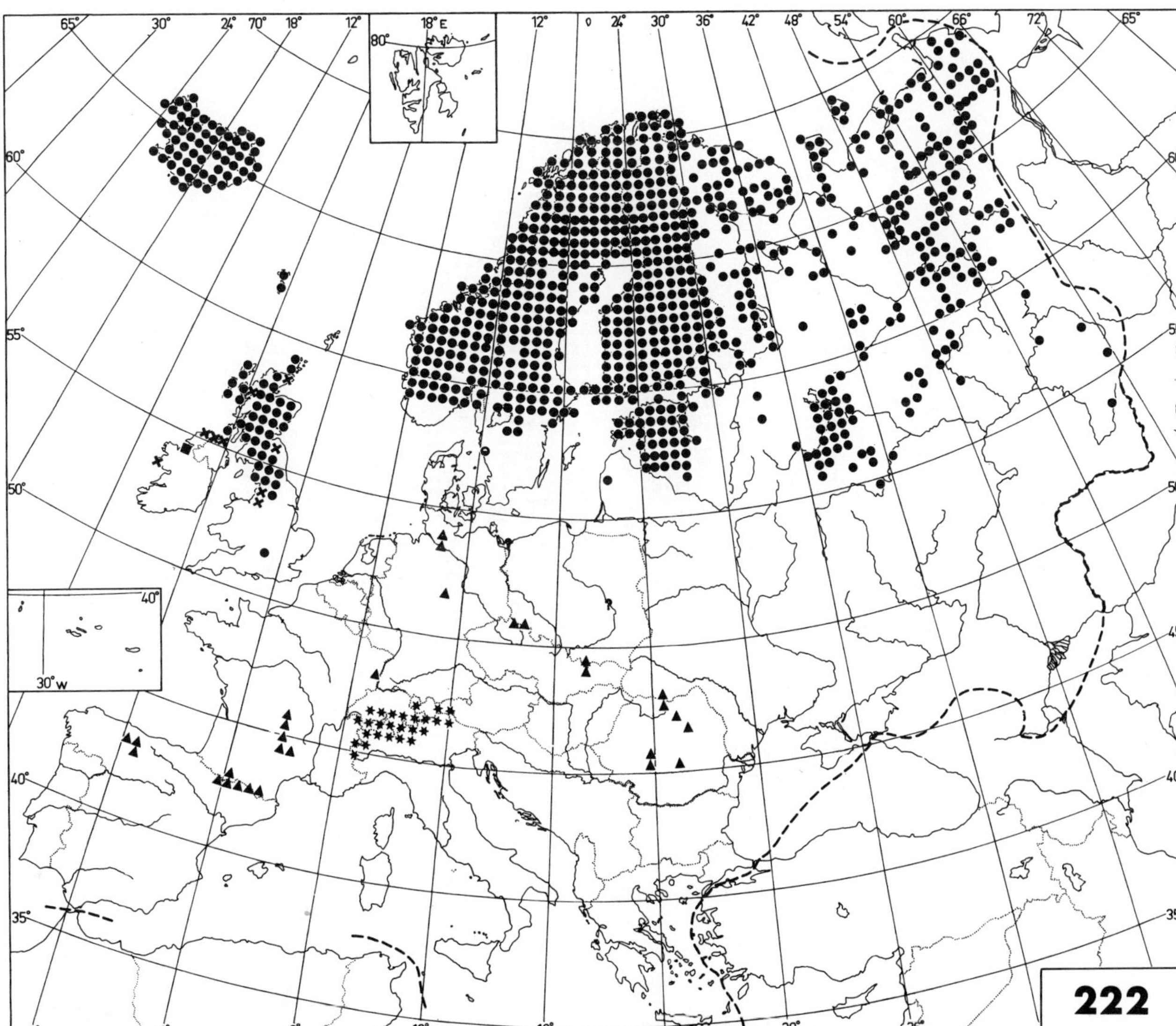

• = **Salix phylicifolia**

▲ = **S. bicolor**

★ = **S. hegetschweileri**

■ = **S. hibernica**

Salix myrsinifolia Salisb. — Map 223.

Salix nigricans auct. non. (?) Sm. Incl. S. borealis (Fries) Nasarov, see below.

Nomenclature. Corrected according to e.g. S. Grapengiesser, Bot. Not. 108: 327 (1955), who also points out that *Salix nigricans* Sm. was originally described as having hairy capsules, a feature excluded by Fl. Eur. from the concept of the species in question. See also S. Rauschert, Feddes Repert. 79: 416 (1969).

Notes. Bu and ?Co added (not given in Fl. Eur.), extinct in Hu (not given in Fl. Eur.). Native occurrence doubtful in Da (given as native in Fl. Eur.): S. Ødum, Bot. Tidsskr. 64: 17 (1968). It has been suggested by J. Chmelař (Brno) that the Pyrenean plants possibly represent *Salix cantabrica,* whilst the data from Corse may belong to *S. apennina,* and those from Bulgaria to *S. silesiaca.*

Total range. Skvortsov 1968: 156 (incl. *Salix borealis*).

Salix borealis (Fries) Nasarov

Salix kolaënsis Schljakov; S. myrsinifolia subsp. borealis (Fries) Hyl.; S. nigricans subsp. borealis (Fries) B. Flod.

Taxonomy. Considered a subspecies of *Salix myrsinifolia* by N. Hylander, Nord. Kärlväxtfl. 2: 268 and 387 (1966), and Skvortsov 1968: 154.

Nomenclature. The citation of the authors of *Salix borealis* is corrected according to the above sources.

Notes. The species is recorded from Scotland, Br (not given in Fl. Eur.), by Neumann & Polatschek 1972: 629. However, judging from the text, this must result from lack of familiarity with the Fennoscandian plant here called *Salix borealis.* The data available do not allow its mapping as a separate unit. »N. Fennoscandia and N. Russia. Fe No Rs(N) Su» (Fl. Eur.).

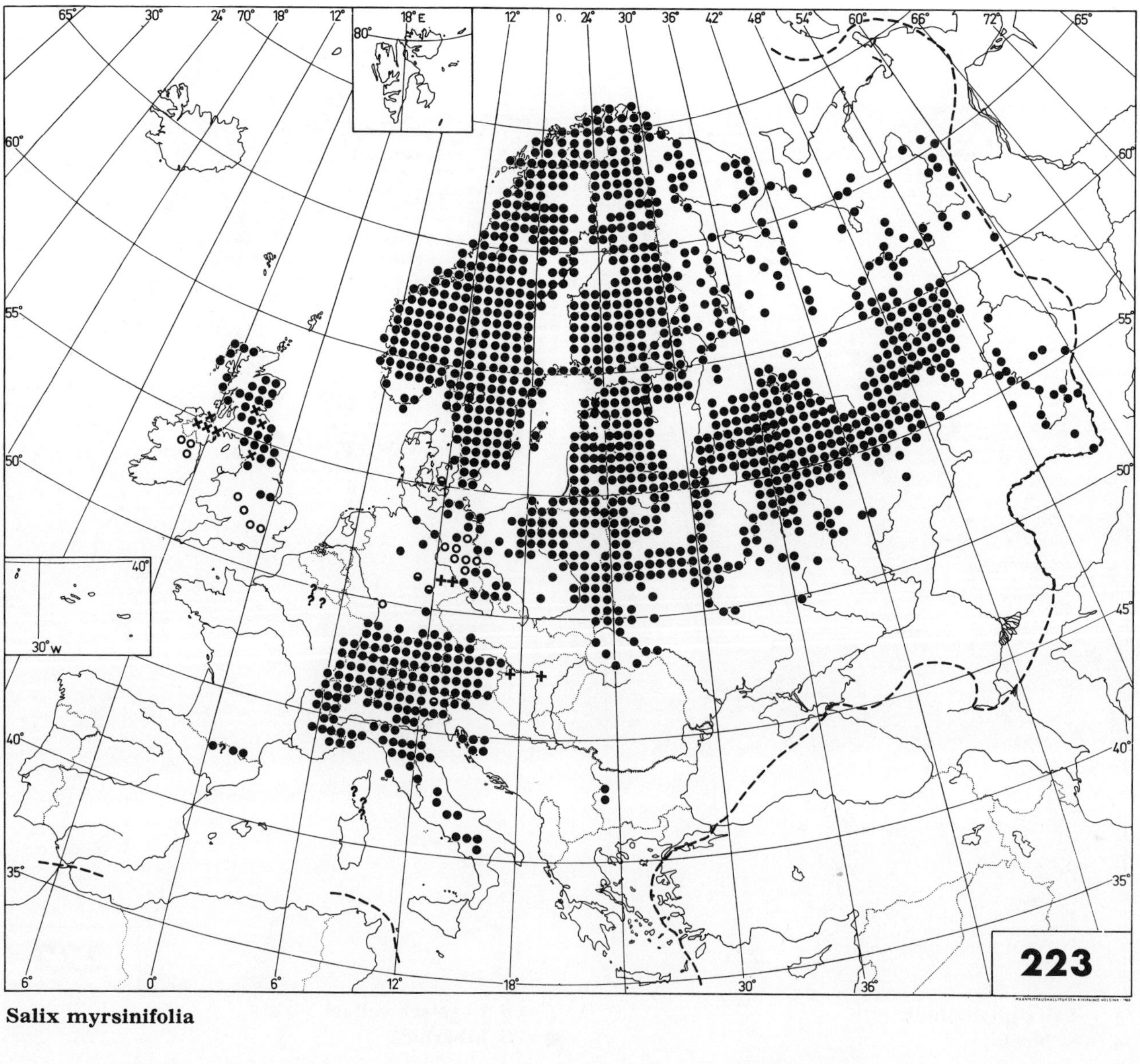

Salix myrsinifolia

Salix apennina A. Skvortsov — Map 224.

Taxonomy. Description in A. K. Skvortsov, Nov. Syst. Pl. Vasc. (Leningrad) 1965: 90—97; Neumann & Polatschek 1972: 628.

Notes. It, Si (the species not mentioned in Fl. Eur.). For Co, see under *Salix myrsinifolia.*

Total range. Endemic to Europe.

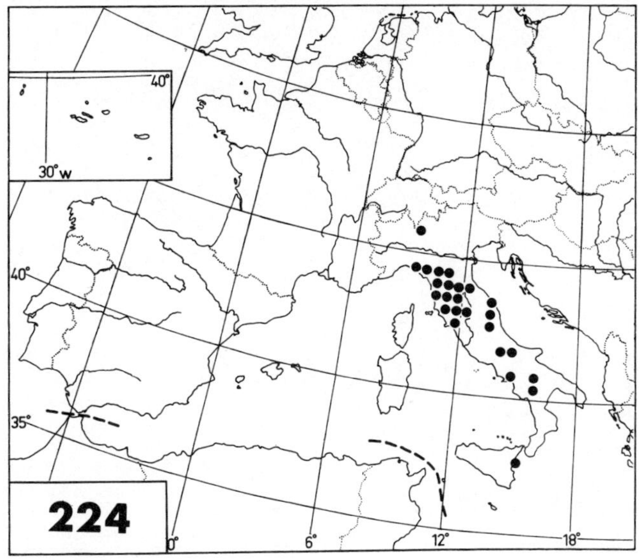

Salix apennina

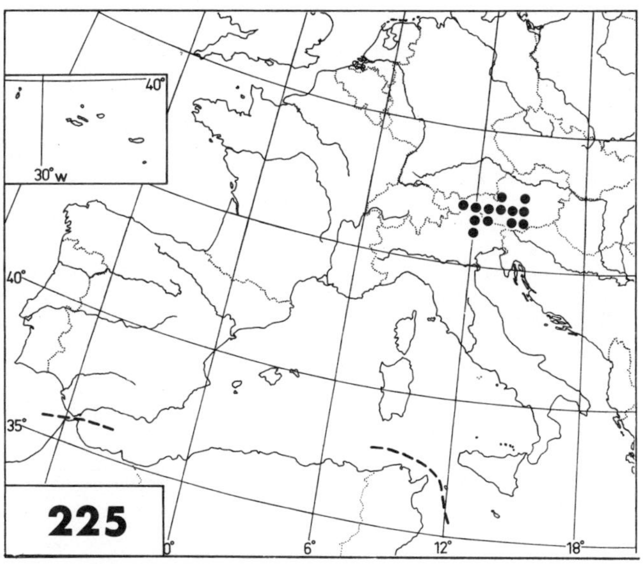

Salix mielichhoferi

Salix mielichhoferi Sauter — Map 225.

Notes. He omitted (?He in Fl. Eur.), It confirmed (?It in Fl. Eur.): Skvortsov 1968: 157.

Total range. Endemic to Europe (although not marked as such in Fl. Eur.).

Salix jenisseensis (Friedrich Schmidt Petrop.) B. Flod. — Map 226.

Salix nigricans var. jenisseensis Friedrich Schmidt Petrop.

Taxonomy. A. K. Skvortsov, Not. Syst. (Leningrad) 19: 83—88 (1959).

Notes. Rs(N) (the species not mentioned in Fl. Eur.).

Total range. A. K. Skvortsov, Not. Syst. (Leningrad) 19: 87 (1959); MJW 1965: map 115d; Skvortsov 1968: 86 and 152.

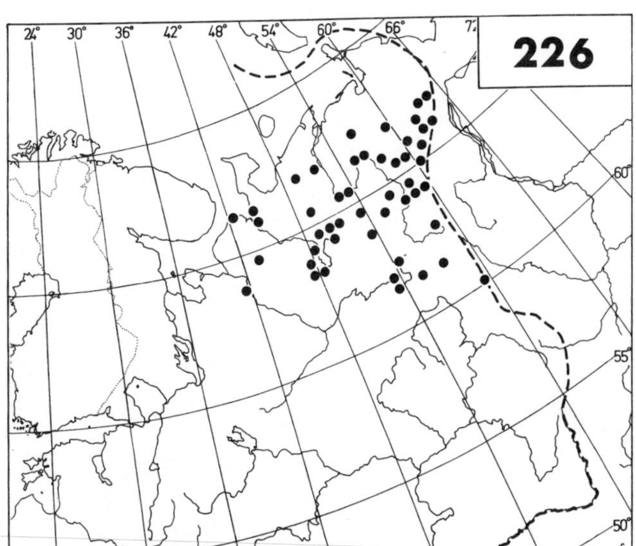

Salix jenisseensis

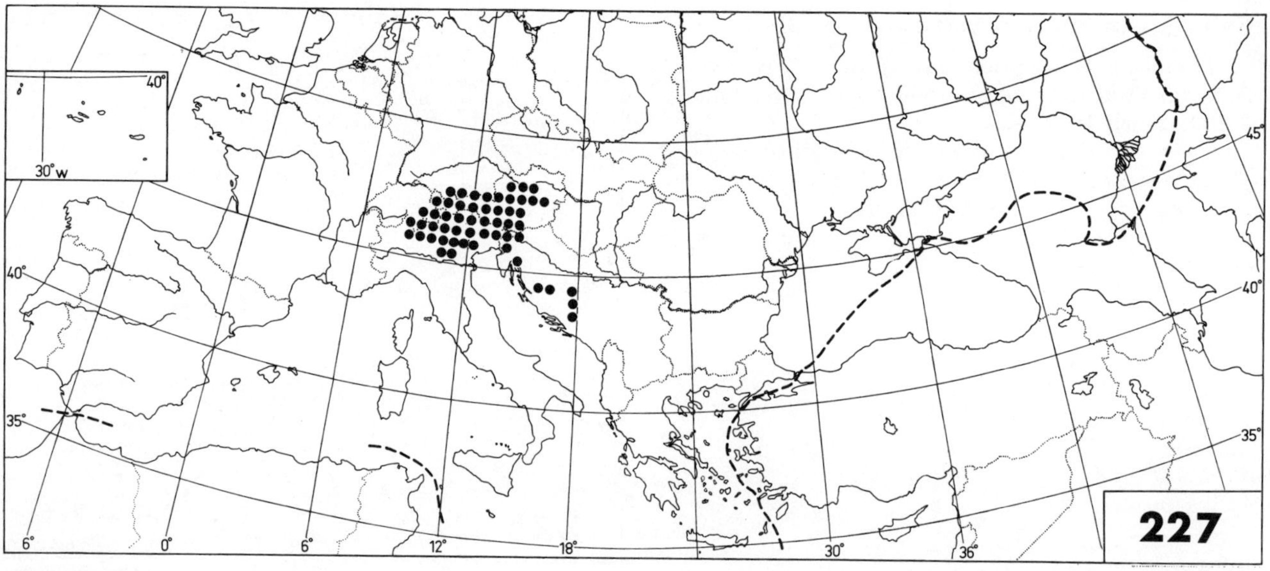

Salix glabra

Salix glabra Scop. — Map 227.

Total range. Endemic to Europe (although not marked as such in Fl. Eur.).

Salix crataegifolia Bertol. — Map 228.

Total range. Endemic to Europe.

Salix pedicellata Desf. — Map 229.

Salix nigricans var. pedicellata (Desf.) Bertol.

Notes. Doubtful in Co (given in Fl. Eur.): cf. J. Vivant, Monde Pl. 61 (351): 14 (1966), Gr omitted (given in Fl. Eur.), doubtful in It (given in Fl. Eur.), Tu not confirmed (?Tu in Fl. Eur.).

Total range. Schmucker 1942: map 130; Skvortsov 1968: 164.

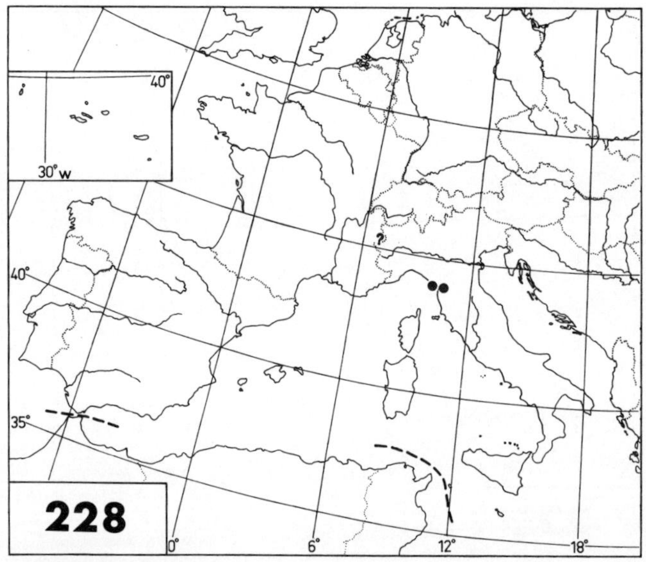

Salix crataegifolia

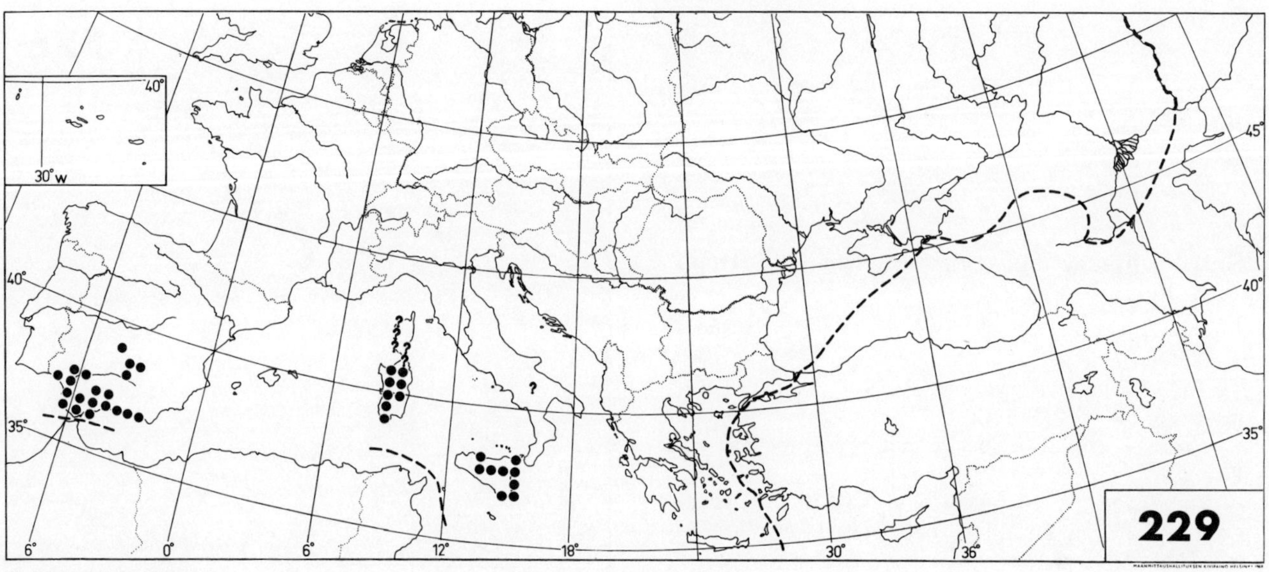

Salix pedicellata

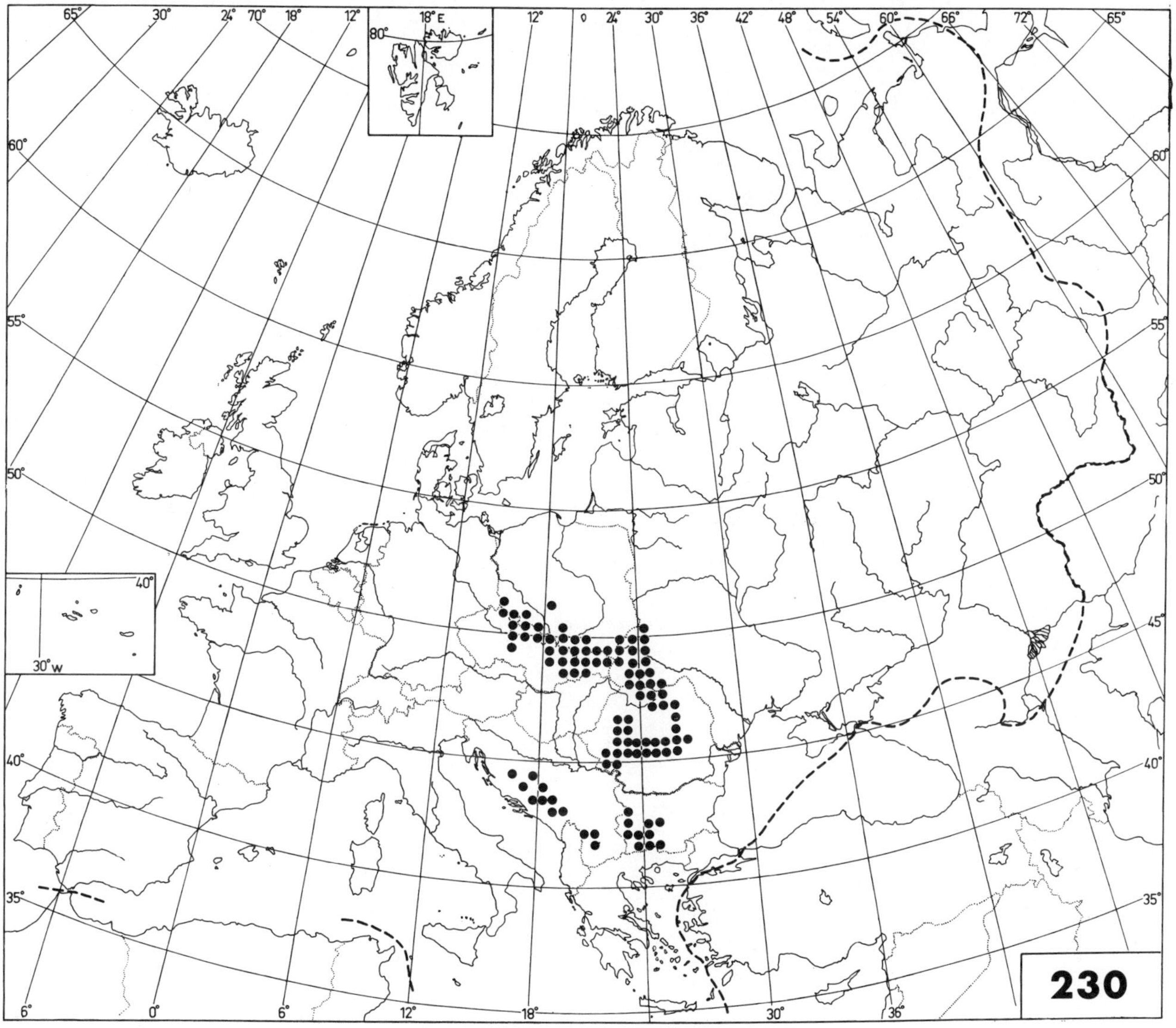

Salix silesiaca

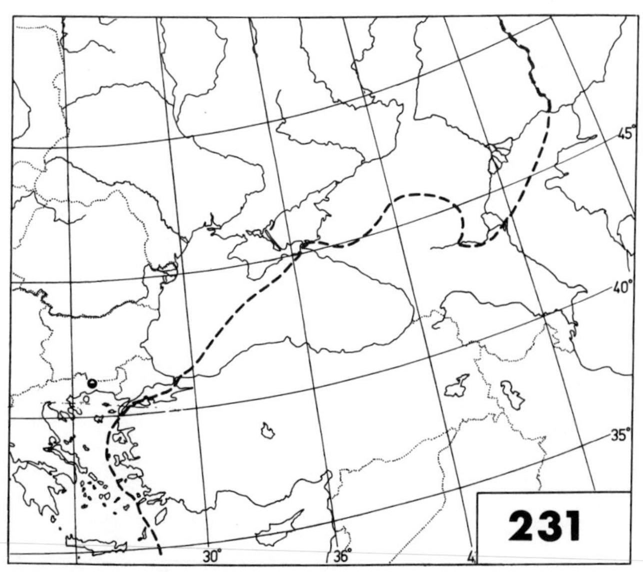

Salix aegyptiaca

Salix silesiaca Willd. — Map 230.

Notes. Ge omitted (given in Fl. Eur.).

Total range. Apparently endemic to Europe (although not marked as such in Fl. Eur.).

Salix aegyptiaca L. – Map 231.

Salix medemii Boiss.

Taxonomy. There is still some taxonomical confusion about *Salix aegyptiaca*: B. Floderus, Ark. Bot. 25 A (11): 1—44 (1933); Skvortsov 1968: 167—171; Neumann & Polatschek 1972: 626—627.

Notes. Doubtfully native in Gr (given as native in Fl. Eur.), Rs(W, K, E) omitted (given in Fl. Eur.): Skvortsov 1968: 168.

Total range. Skvortsov 1968: 167.

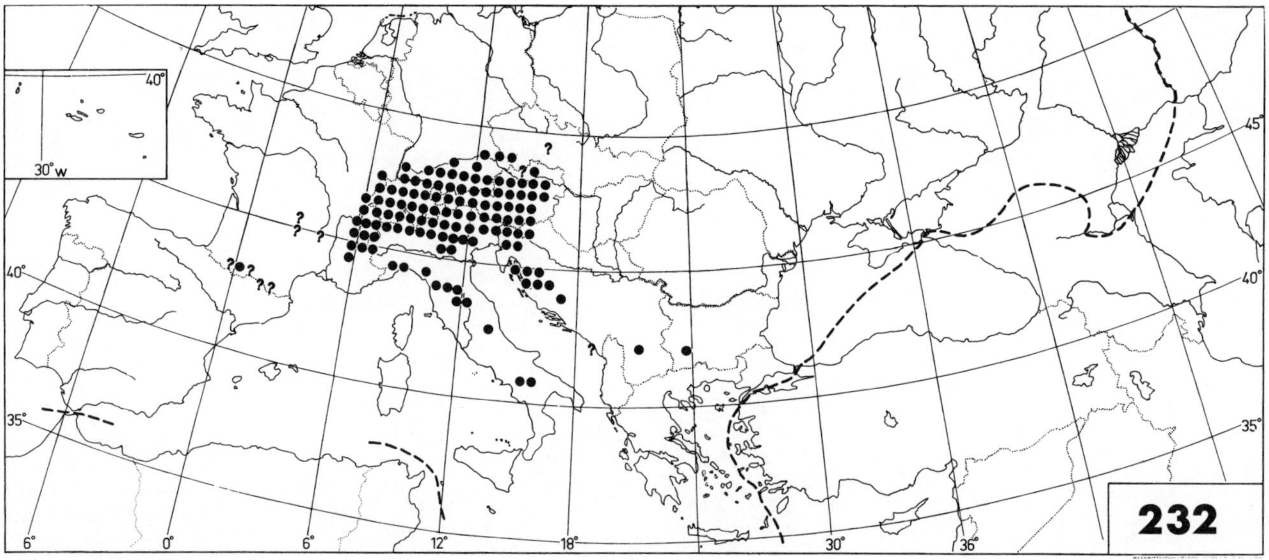

Salix appendiculata

Salix appendiculata Vill. — Map 232.

Salix grandifolia Sér.; S. aurita var. grandifolia (Sér.) Fiori

Notes. Bu added (not given in Fl. Eur.). Cz added (not given in Fl. Eur.): J. Holub, Preslia 38: 79 (1966).

Total range. Endemic to Europe (although not marked as such in Fl. Eur.).

Salix laggeri Wimmer — Map 233.

Salix albicans Bonjean; S. pubescens Schleicher ex Sér.

Taxonomy. K. H. Rechinger, Österr. Bot. Zeitschr. 110: 339—340 (1963); Neumann & Polatschek 1972: 627—628.

Notes. It added (not given in Fl. Eur.).

Total range. Apparently endemic to Europe (although not marked as such in Fl. Eur.).

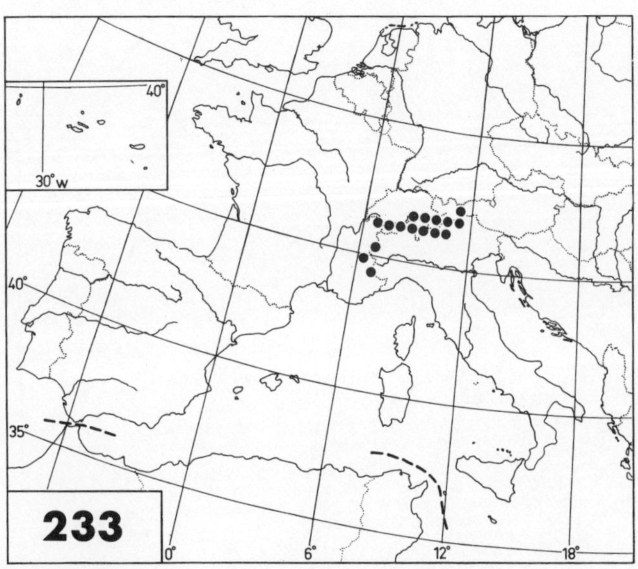

Salix laggeri

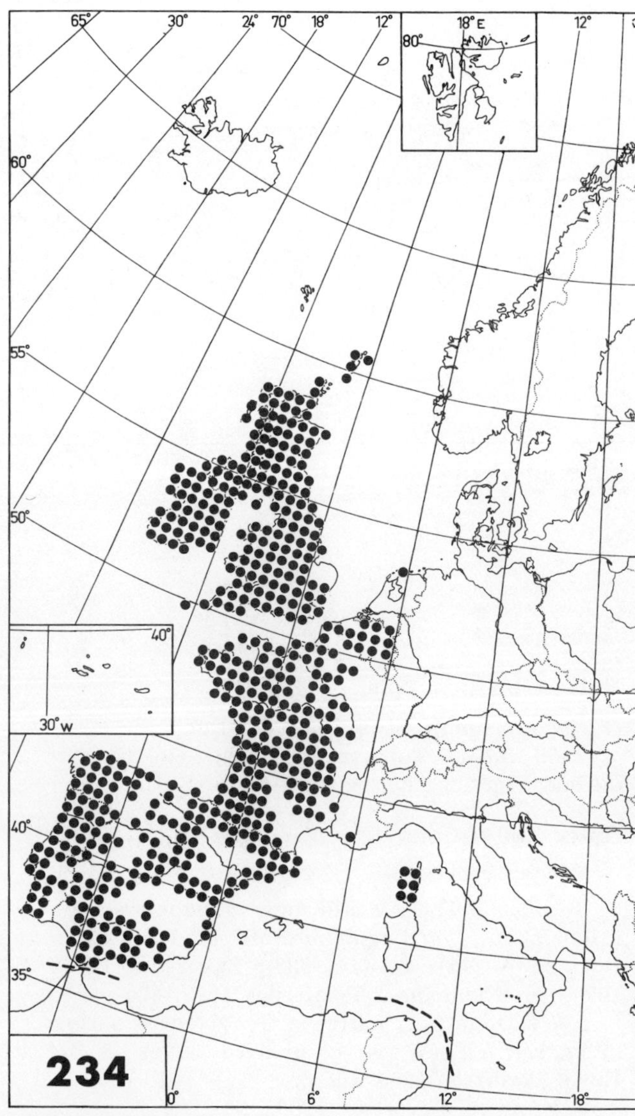

Salix atrocinerea

Salix atrocinerea Brot. — Map 234.

Salix acuminata Thuill. non Sm.; S. cinerascens Link; S. incerta Lapeyr.; S. oleifolia Sm. non Vill.; S. rufinervis DC. Incl. S. atrocinerea subsp. catalaunica R. Görz

Taxonomy. Neumann & Polatschek 1972: 625—626.

Notes. Co and Ho added (not given in Fl. Eur.); for Ho, see J. Mennema & S. J. van Ooststroom, Gorteria 7: 144 (1975). Owing to confusion with *Salix cinerea* (see above), the area is not very accurately known.

Total range. Skvortsov 1968: 170.

Salix cinerea L. — Map 235.

Notes. Co and Sa omitted (given in Fl. Eur.), being considered to be misidentifications of *Salix atrocinerea* and *S. pedicellata*, respectively; Skvortsov 1968: 169. In Ga the western limit of the species is approximative, owing to frequent confusion and/or hybridization with *S. atrocinerea.*

Total range. Skvortsov 1968: 170.

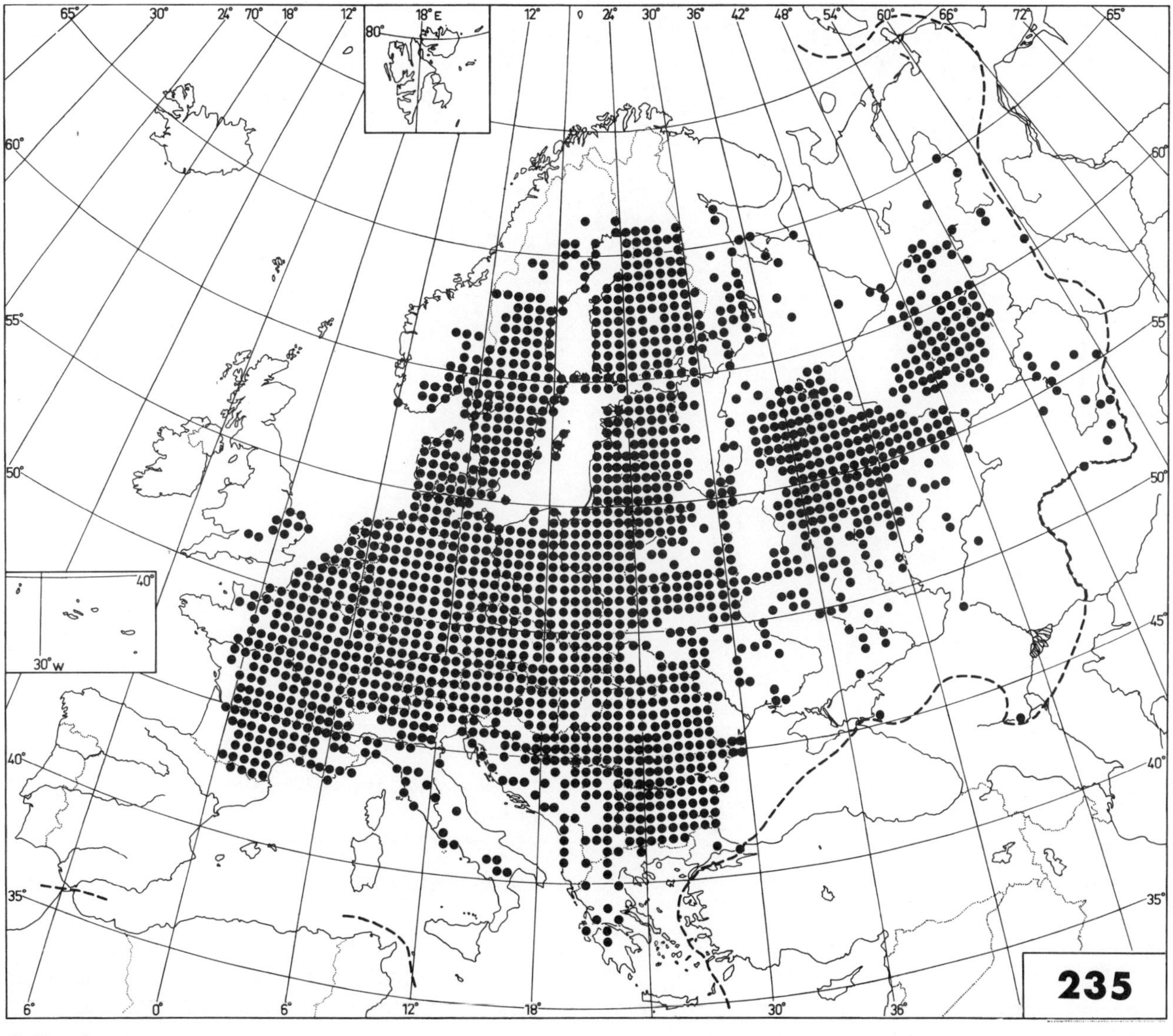

Salix cinerea

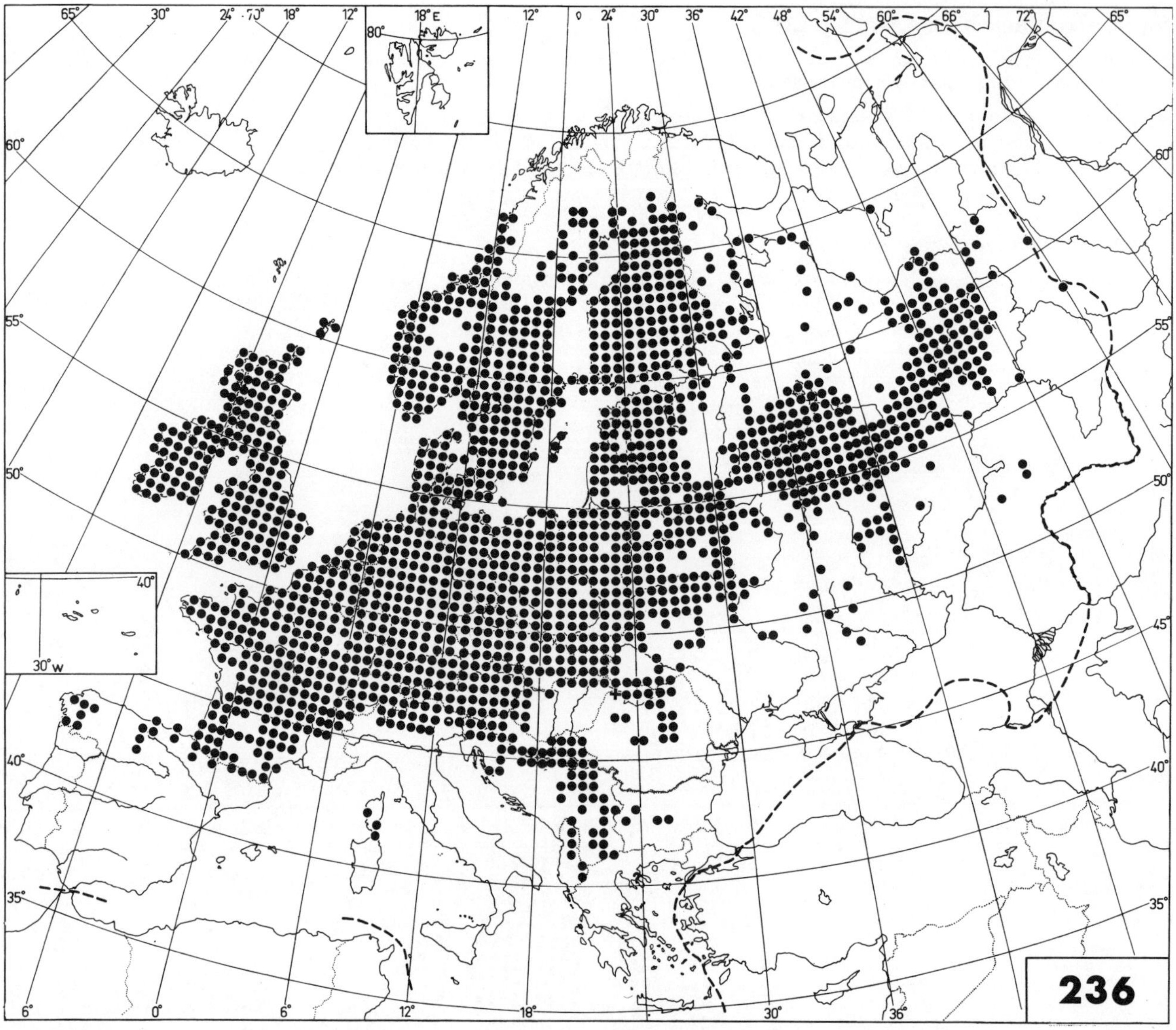

Salix aurita

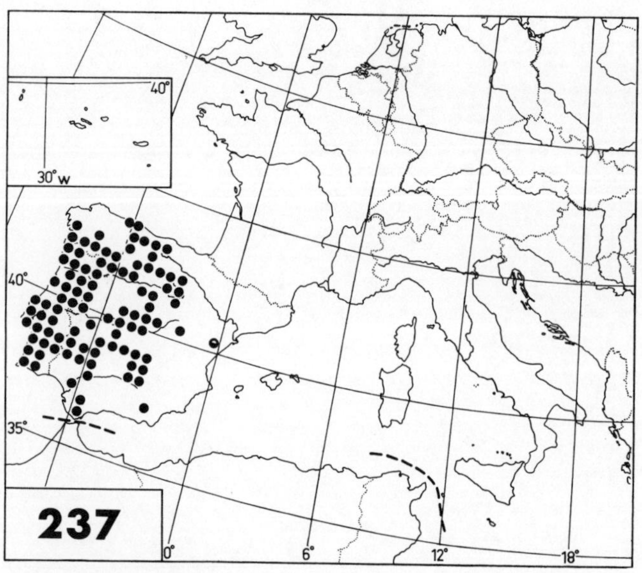

Salix salviifolia

Salix aurita L. — Map 236.

Notes. Al, Bu and ?Gr added (not given in Fl. Eur.).

Total range. Skvortsov 1968: 172.

Salix salviifolia Brot. — Map 237.

Salix oleifolia auct. hisp. non Vill.

Taxonomy. Neumann & Polatschek 1972: 625.

Total range. As far as is known, endemic to Europe (not marked as such in Fl. Eur.).

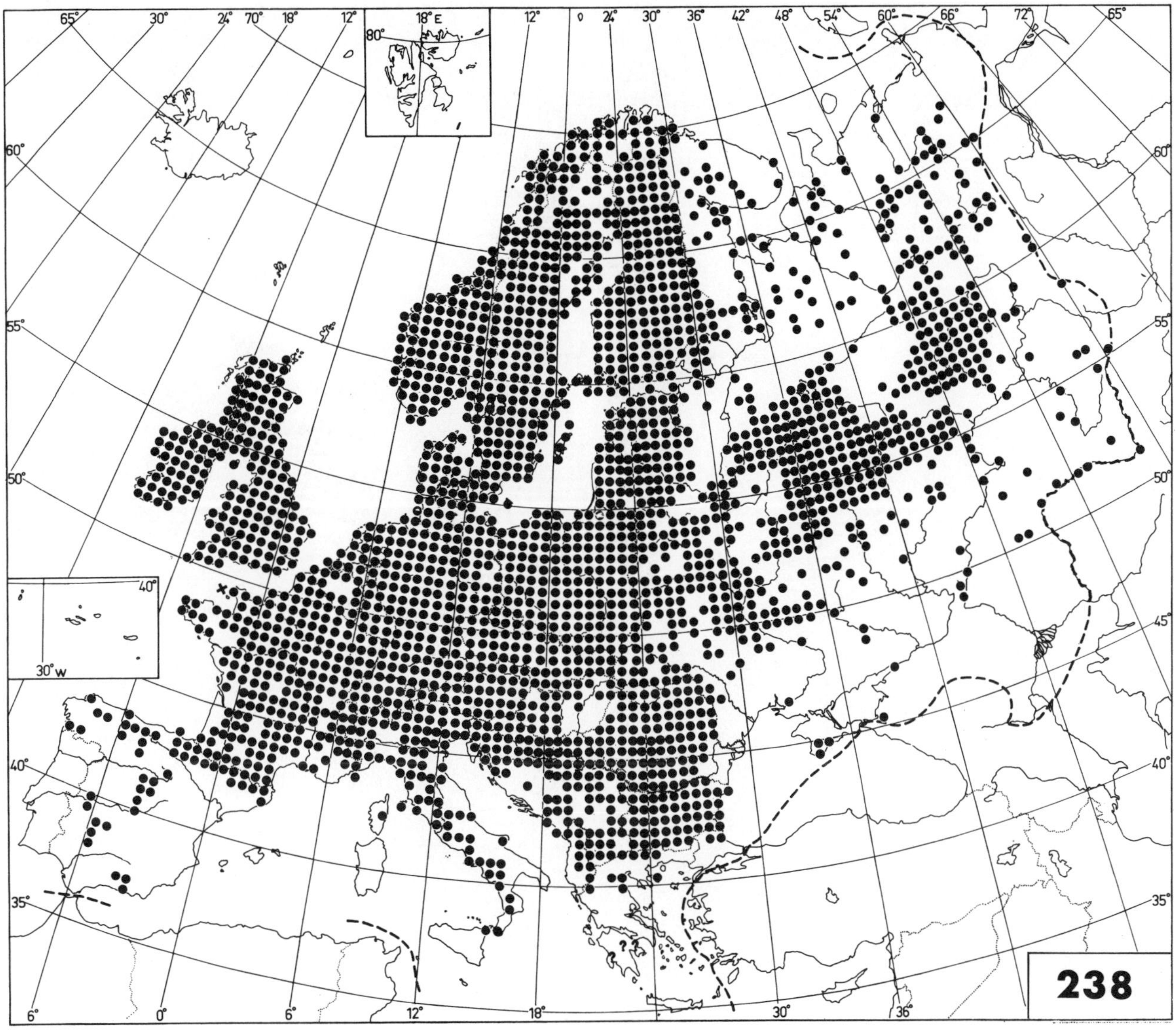

Salix caprea

Salix caprea L. — Map 238.

Incl. Salix coaetanea (Hartman) B. Flod.; S. caprea subsp. sericea (N. J. Andersson) B. Flod. in Lindm. 1926; S. caprea subsp. coaetanea (Hartman) Hiitonen 1933.

Taxonomy. There are widely divergent opinions concerning the appropriate rank of *Salix coaetanea*.

Notes. The continuous series of morphological intermediates connecting at least the Fennoscandian populations identified as *Salix caprea* and *S. coaetanea* does not allow their separate mapping. Fl. Eur. reports *S. coaetanea* from Br, Fe, No, Rs(N) and Su.

Total range. MJW 1965: map 114d; Fenaroli 1967: 83; Skvortsov 1968: 85 and 167 (incl. *Salix coaetanea*).

Salix cantabrica Rech. fil. — Map 239.

Taxonomy. Neumann & Polatschek 1972: 629.

Notes. For possible occurrences in the Pyrenees, see under *Salix myrsinifolia*.

Total range. Endemic to Europe.

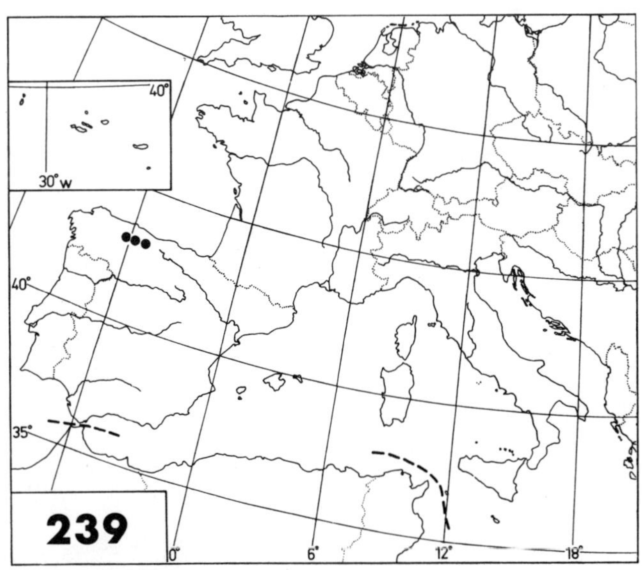

Salix cantabrica

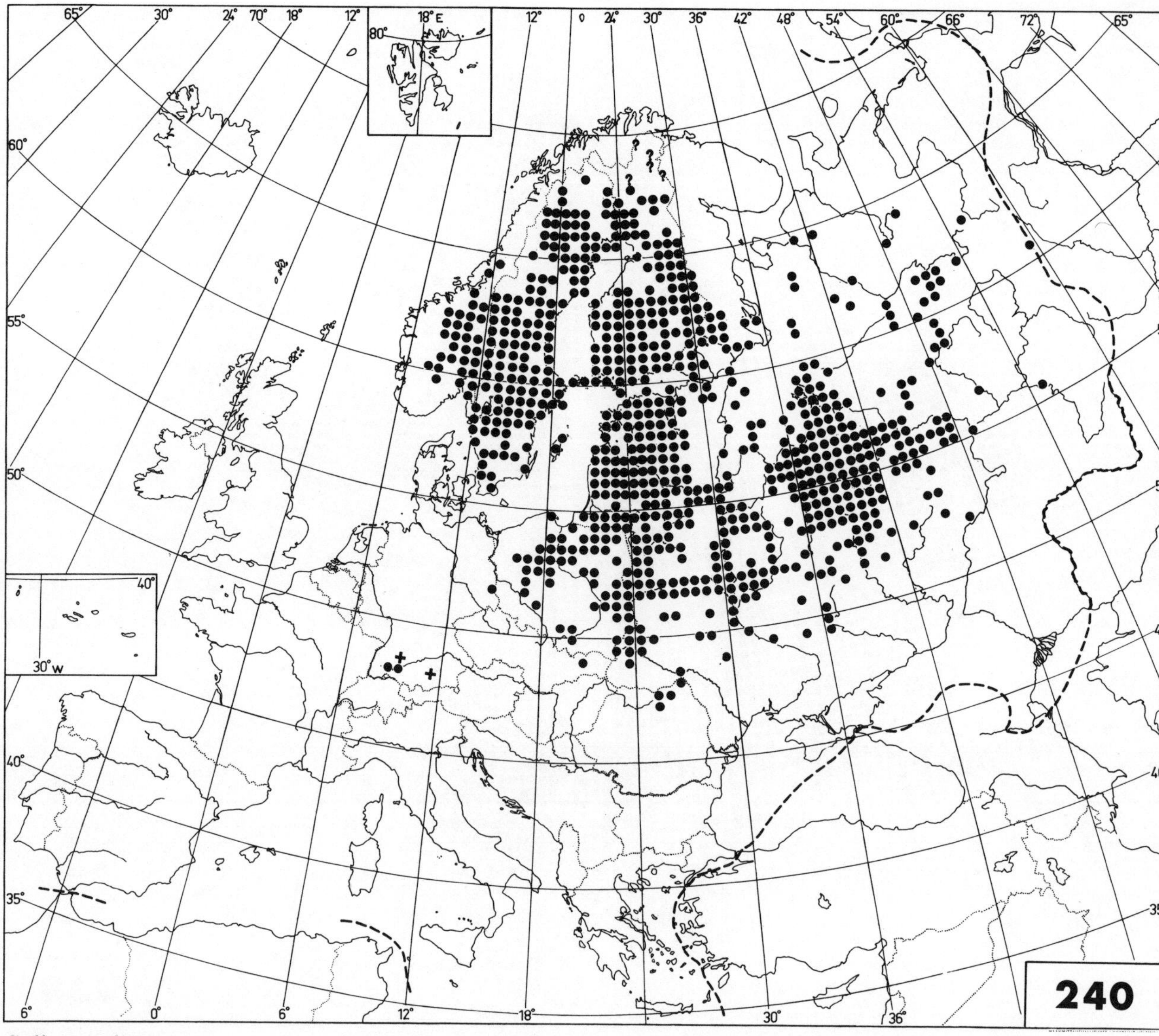

Salix starkeana

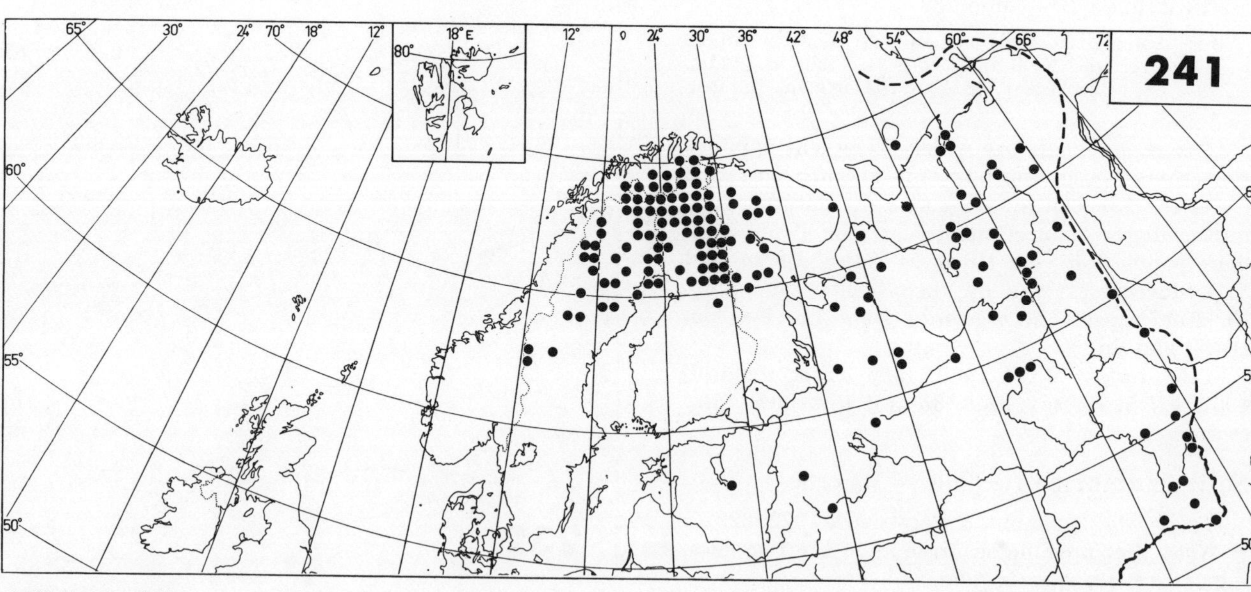

Salix xerophila

Salix starkeana Willd. — Map 240.

Salix depressa auct. non L.; S. livida Wahlenb.; S. vagans N. J. Andersson

Taxonomy. See under *Salix xerophila.*
Notes. He omitted (given in Fl. Eur.).
Total range. Hultén Alaska 1968: 358; Skvortsov 1968: 173; Hultén CP 1971: map 125 (incl. *Salix xerophila*).

Salix xerophila B. Flod. — Map 241.

Salix bebbiana Sarg., ampl. A. Skvortsov; S. cinerascens (Wahlenb.) B. Flod. non Willd.; S. depressa subsp. cinerascens (Wahlenb.) O. Dahl; S. livida var. cinerascens Wahlenb.; S. starkeana subsp. cinerascens (Wahlenb.) Hultén

Taxonomy and nomenclature. I. Hiitonen, Mem. Soc. Fauna Fl. Fenn. 25: 82—83 (1950); A. Cronquist, Rhodora 73: 558—559 (1971); E. Hultén, Bot. Not. 126: 476 (1973).
Notes. Rs(C) added (not given in Fl. Eur.).
Total range. Skvortsov 1968: 173.

Salix myrtilloides L. — Map 242.

Notes. Au added (not given in Fl. Eur.). Possibly extinct in Cz (Cz in Fl. Eur.).
Total range. MJW 1965: map 115a; Skvortsov 1968: 126.

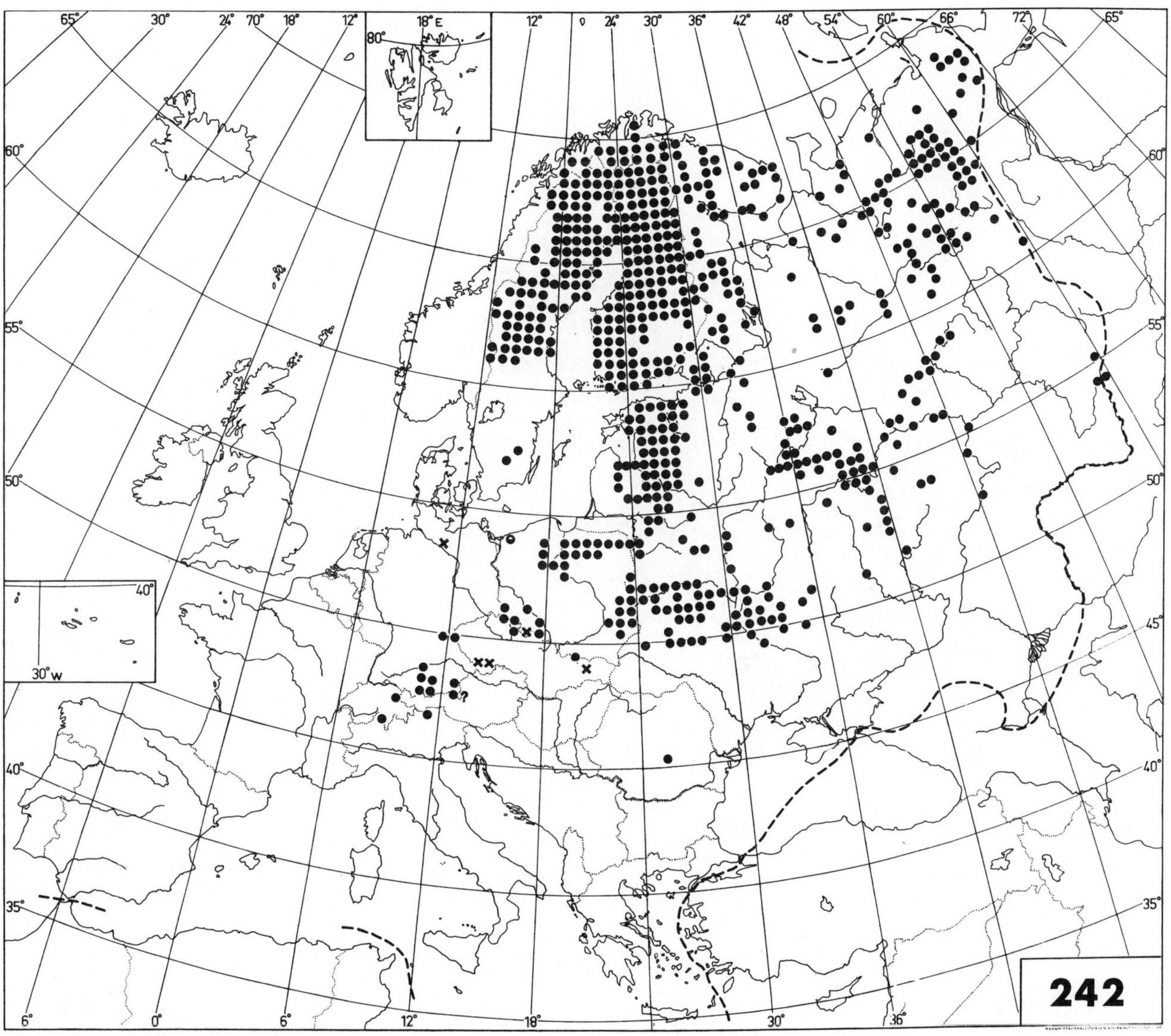

Salix myrtilloides

Salix repens L. — Map 243.

Incl. Salix arenaria L. and S. leiocarpa Mang

Taxonomy. S. Grapengiesser, Svensk Bot. Tidskr. 37: 224—227 (1943); F. Mang, Mitt. Arbeitsgem. Flor. Schleswig-Holstein 10: 1—79 (1962).

Notes. Rs(C) added (not given in Fl. Eur.), Rs(W) omitted (given in Fl. Eur.). The map is provisional, especially as regards the eastern limit of the range.

Total range. According to Skvortsov 1968: 217, endemic to Europe (not marked as such in Fl. Eur.).

Salix arenaria L.

Salix repens subsp. arenaria (L.) Hiitonen; S. repens var. nitida (Sér.) Wenderoth

Notes. Not kept apart from *Salix repens*; mapped together with it as one entity. »Atlantic coasts of Europe extending to Poland. Be Br Da Ga Ge Hb Ho Lu No Po Su» (Fl. Eur.).

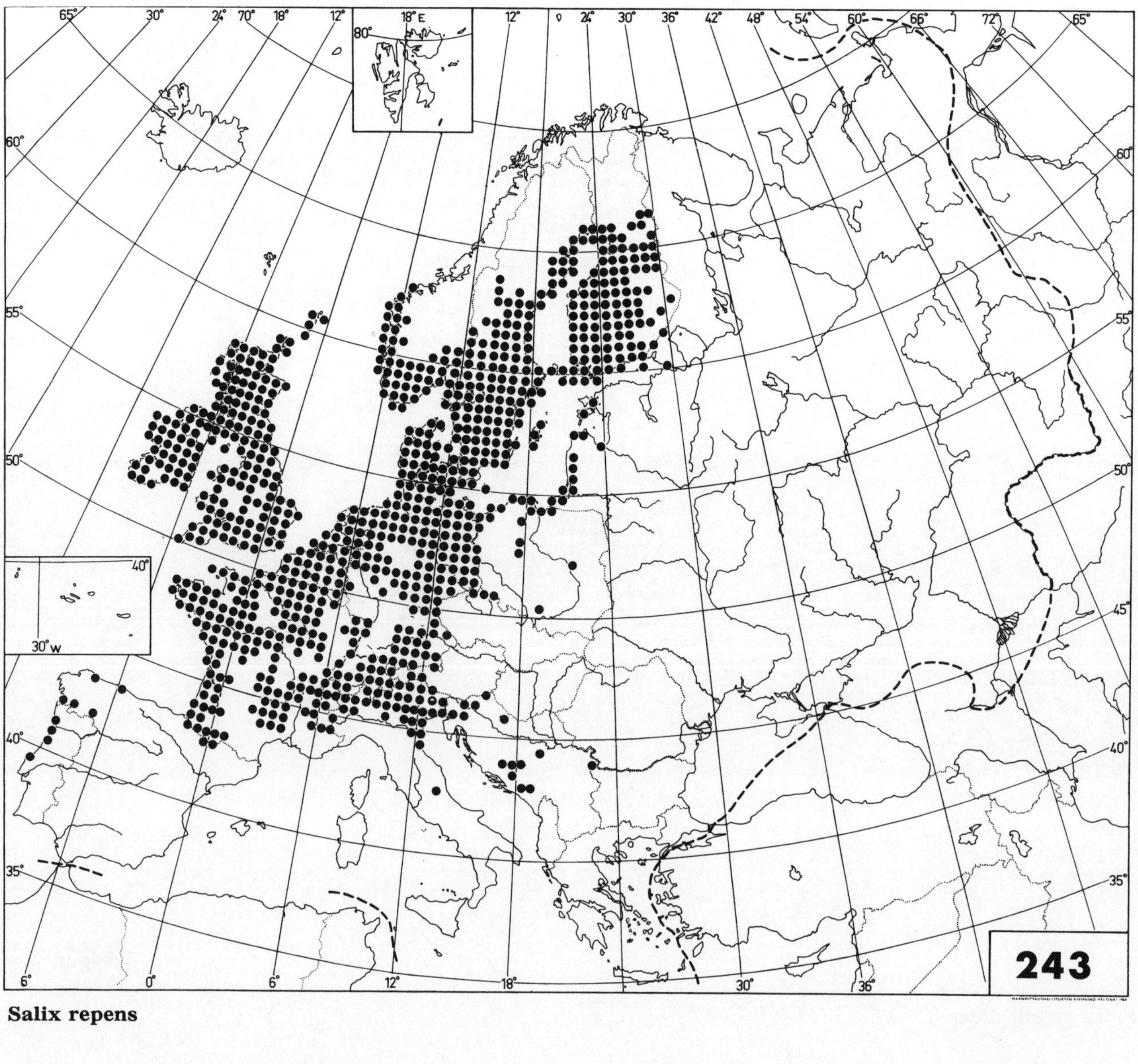

Salix repens

Salix rosmarinifolia L. — Map 244.

Salix repens subsp. rosmarinifolia (L.) Čelak.

Notes. Be and He omitted (given in Fl. Eur.), extinct in Bu (Bu in Fl. Eur.), Fe and Ga added (not given in Fl. Eur.). In several countries, especially in Au and Cz, mapping has proved difficult, if not impossible, because of the frequent occurrence of transitional plants between this taxon and *Salix repens*.

Total range. Skvortsov 1968: 217.

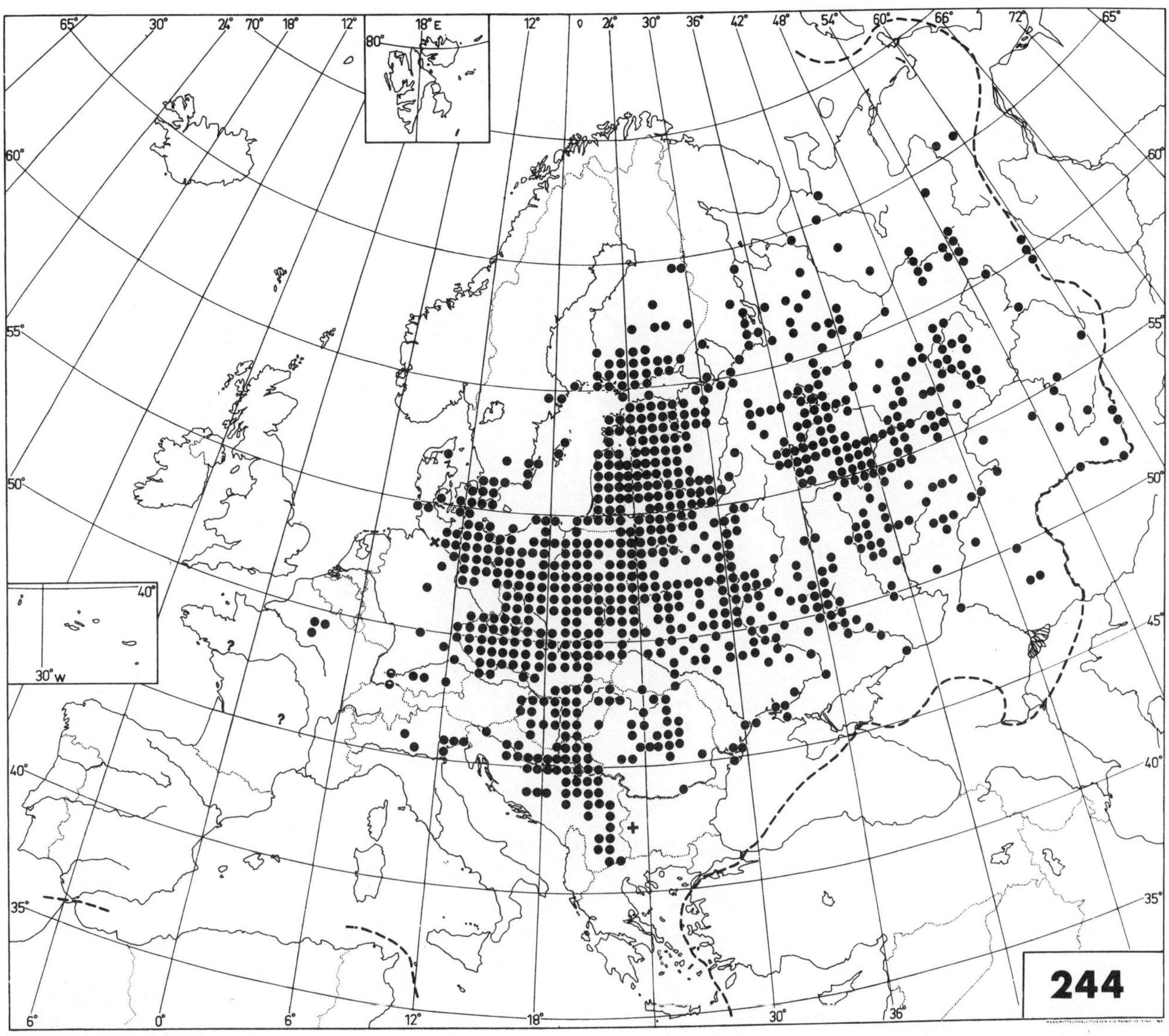

Salix rosmarinifolia

Salix repens L., coll. — Map 245.

Incl. Salix arenaria L. and S. rosmarinifolia L.

Notes. Because of frequent occurrence of intermediates between the three taxa recognized in Fl. Eur., a collective map is given besides those for *Salix repens* (incl. *S. arenaria*) and *S. rosmarinifolia*.

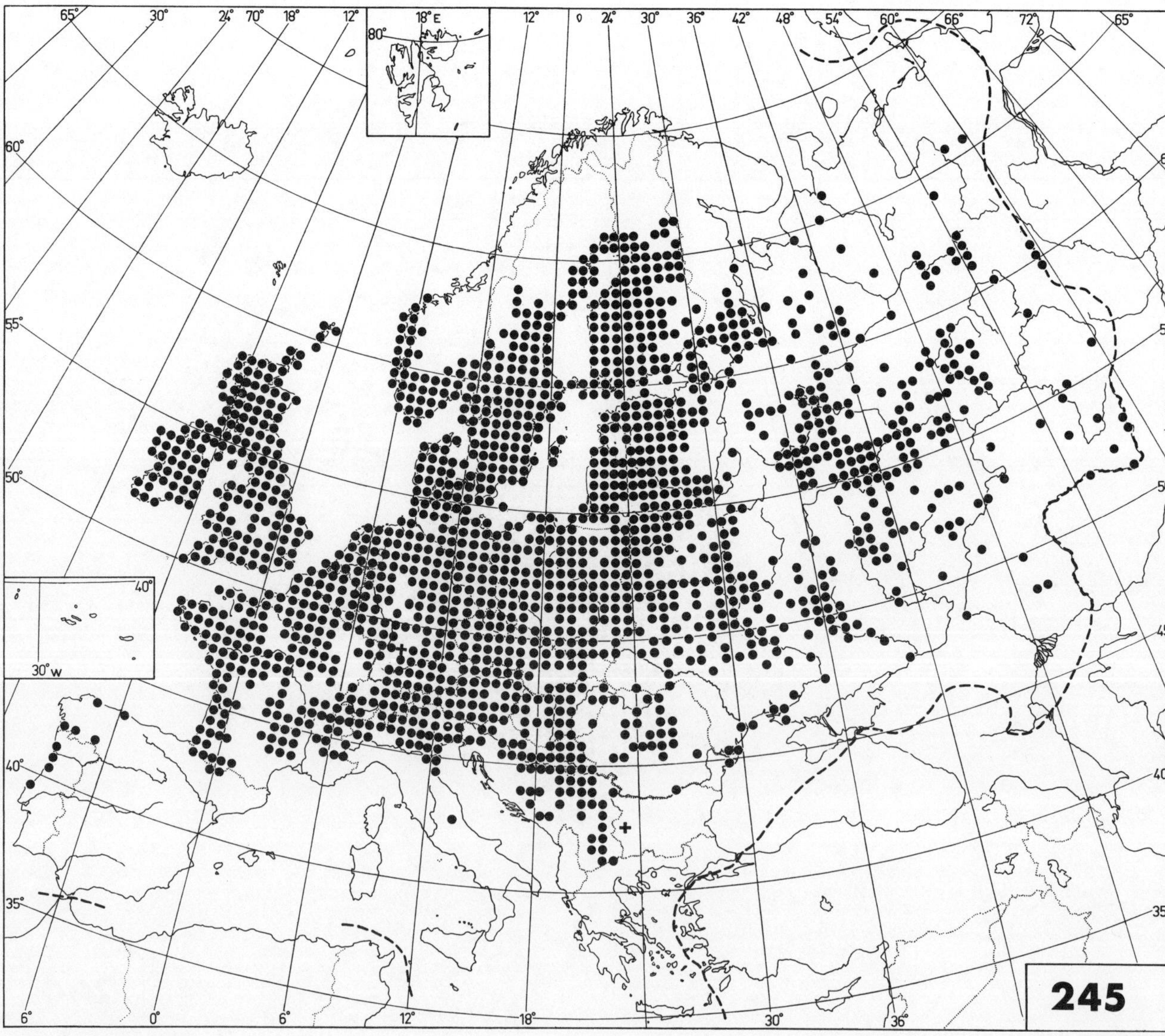

Salix repens coll.

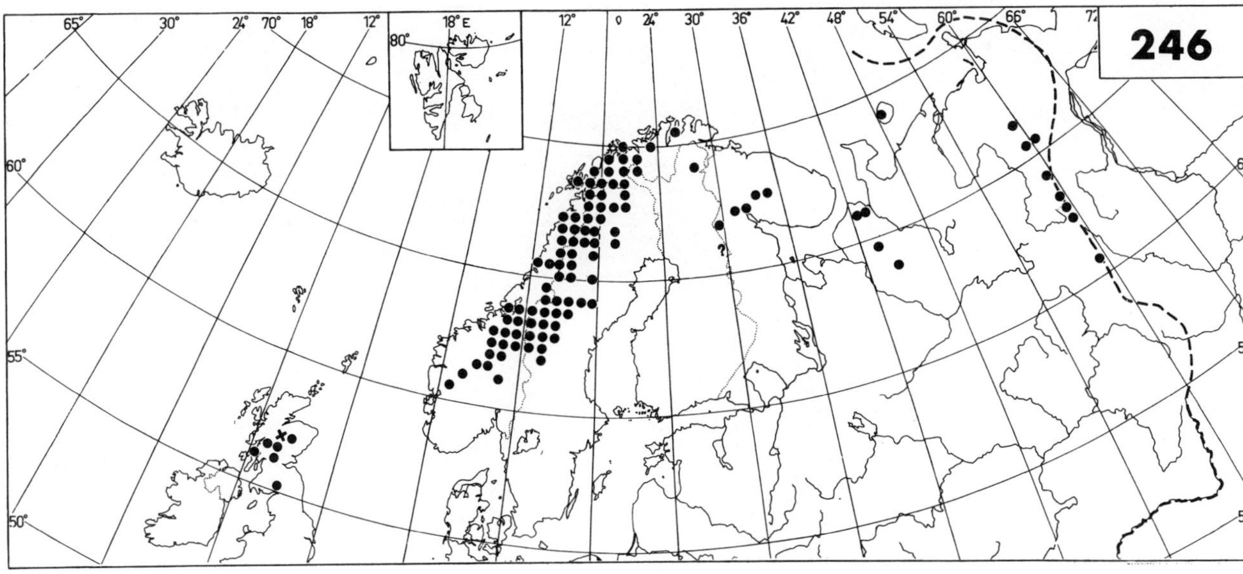

Salix arbuscula

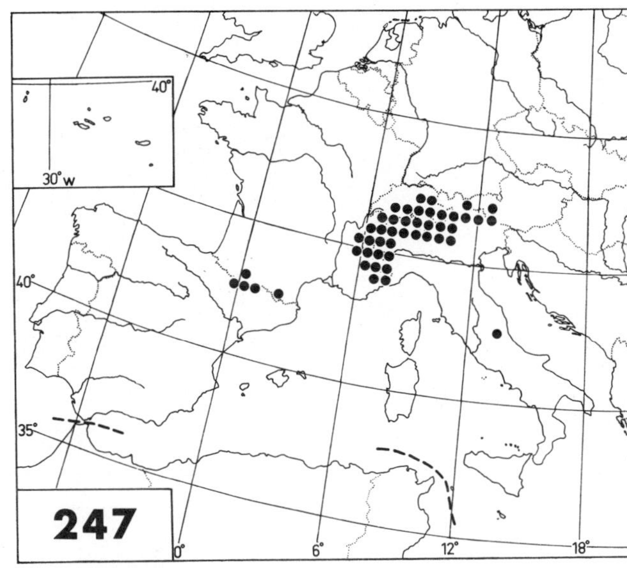

Salix foetida

Salix arbuscula L. — Map 246.

Total range. A. K. Skvortsov, Bull. Biol. Soc. Nat. Moskva 66: 32 (1961); Skvortsov 1968: 86 and 187.

Salix foetida Schleicher ex DC. — Map 247.

Total range. Endemic to Europe.

Salix waldsteiniana Willd. — Map 248.

Notes. Bu and He confirmed (?Bu and ?He in Fl. Eur.).

Total range. Endemic to Europe (although not marked as such in Fl. Eur.).

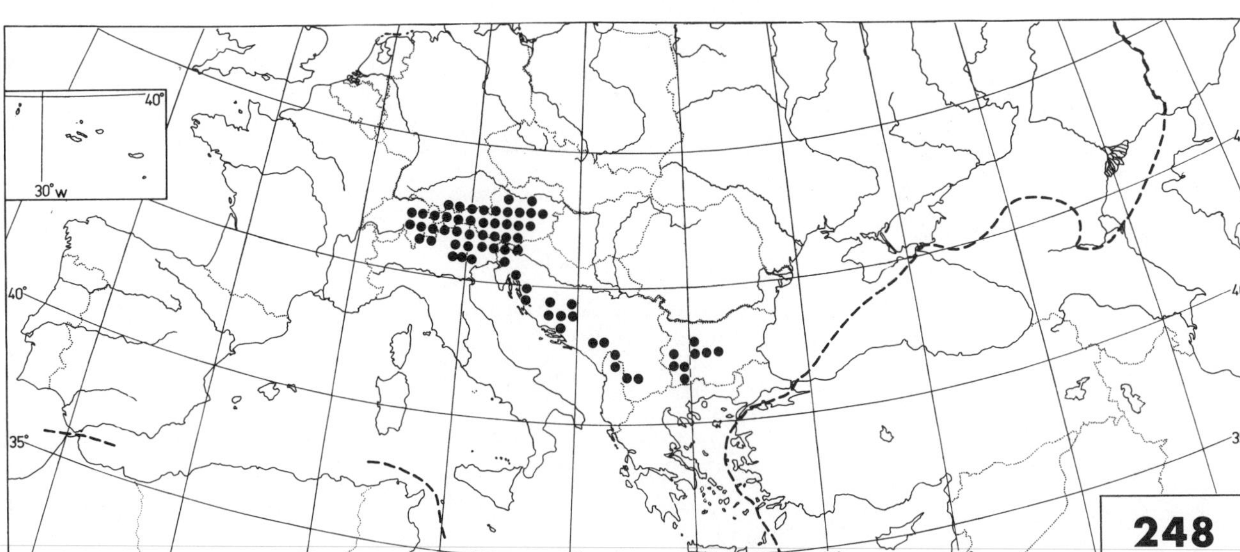

Salix waldsteiniana

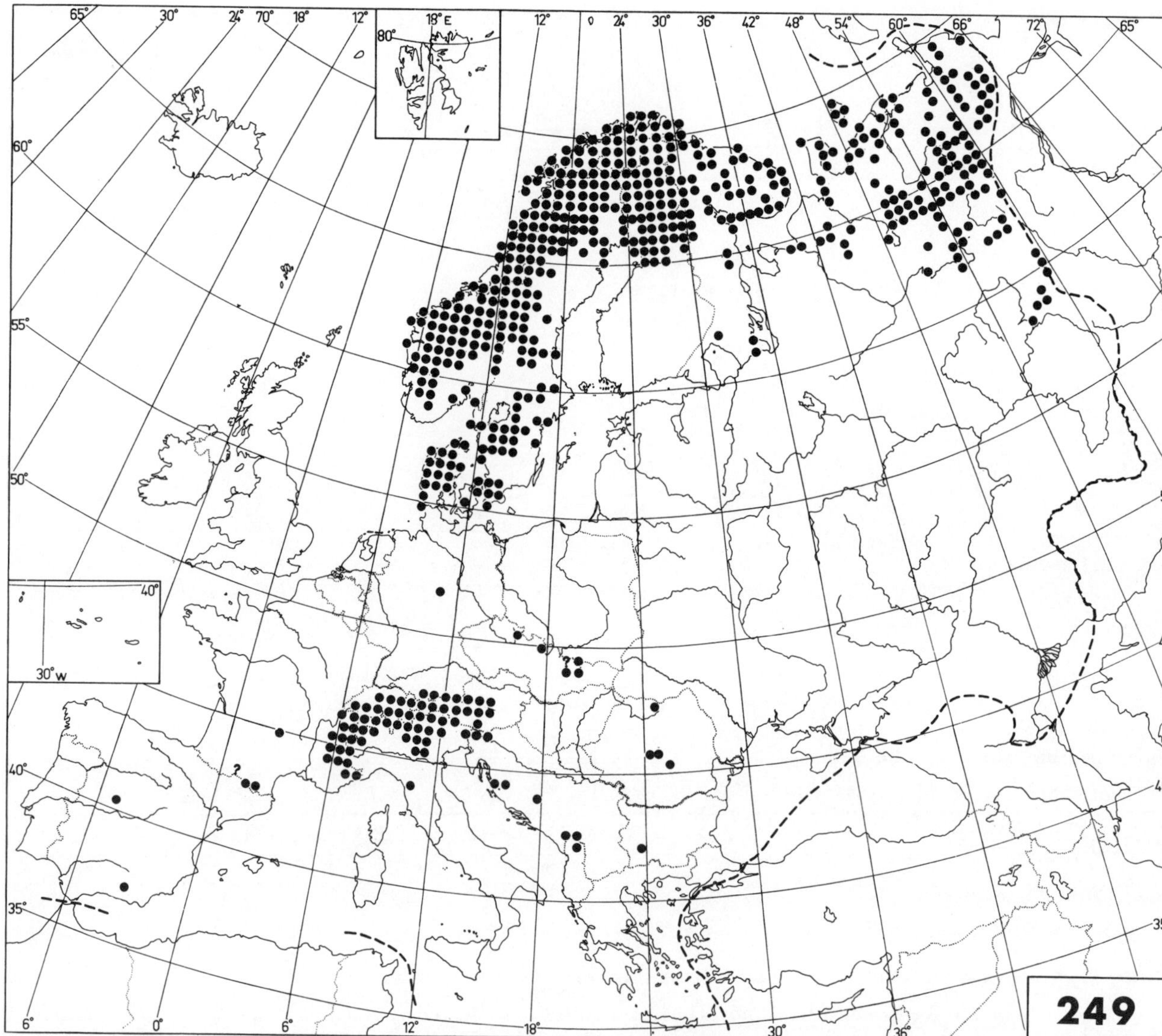

Salix hastata

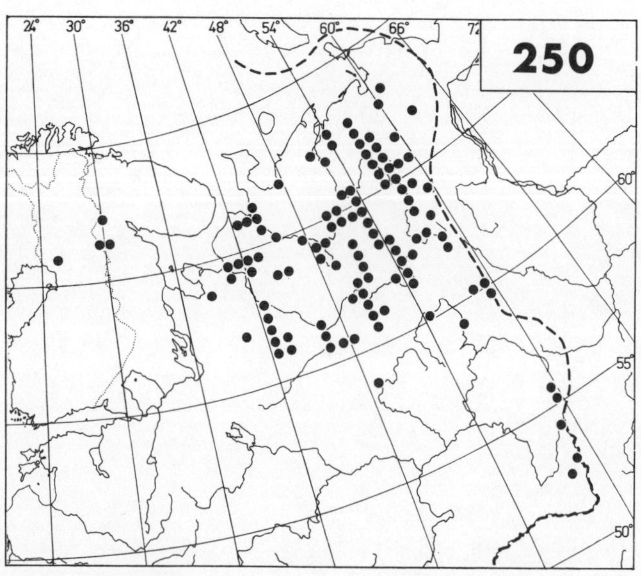

Salix pyrolifolia

Salix hastata L. — Map 249.

Notes. The subspecies recognized in Fl. Eur. (see below) could not be mapped separately. Bu and Rs(C) added (not given in Fl. Eur.), no data received from Rs(W) (given in Fl. Eur.).

Total range. MJW 1965: map 115c; Skvortsov 1968: 149.

S. hastata subsp. **hastata**

»Throughout most of the range of the species» (Fl. Eur.).

S. hastata subsp. **vegeta** N. J. Andersson

»S. Fennoscandia and parts of C. Europe» (Fl. Eur.).

S. hastata subsp. **subintegrifolia** (B. Flod.) B. Flod.

»N. part of the range of the species» (Fl. Eur.).

Salix pyrolifolia Ledeb. — Map 250.

Total range. Skvortsov 1968: 85 and 146.

Salix helvetica Vill. — Map 251.

Notes. Cz and Po added (not given in Fl. Eur.).
Total range. Endemic to Europe.

Salix lapponum L. — Map 251.

Incl. Salix marrubifolia Tausch ex N. J. Andersson in DC.; S. helvetica subsp. marrubifolia (Tausch ex N. J. Andersson) B. Flod.

Taxonomy. Further studies are badly needed on the Central European representatives of the group (*Salix helvetica, S. marrubifolia, S. lapponum*). Among the plants called *S. marrubifolia* in Cz, that of the Sudeten Mts. was recently identified as *S. lapponum*, and the Carpathian plant as *S. helvetica*.

Notes. No data available from Rm (given as *Salix marrubifolia* in Fl. Eur.); Rs(C) added (not given in Fl. Eur.). The species is also mentioned from »Makedonia» in Skvortsov 1968: 205.

Total range. MJW 1965: map 116a; Skvortsov 1968: 205 (incl. *Salix marrubifolia*).

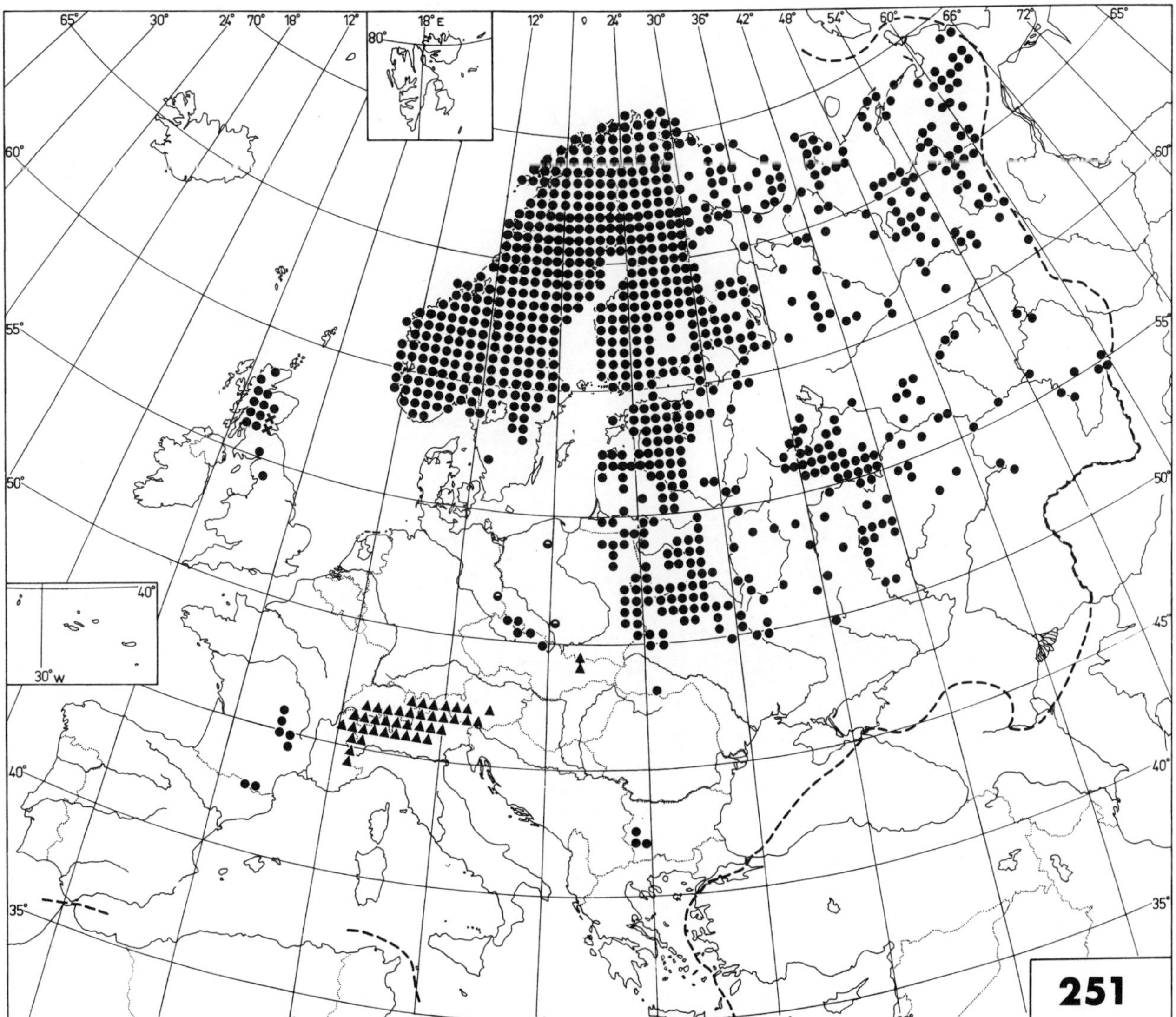

▲ = Salix helvetica • = S. lapponum

Salix viminalis L. — Map 252.

Salix veriviminalis Nasarov. Incl. S. pseudolinearis (Turcz.) Nasarov and S. rossica Nasarov; S. viminalis subsp. rossica (Nasarov) Hyl.

Notes. Introductions not mapped. Owing to the extensive cultivation of the species, with numerous hybrid derivatives, an exact delimitation of the native range is almost impossible. Doubtfully native in Gr and Rm (given as native in Fl. Eur.), not considered native anywhere in Br and Hb (Br, *Hb in Fl. Eur.), *Tu added (not given in Fl. Eur.). Fl. Eur. reports *Salix rossica* from Rs(N, B, C, W, E), with possible overlap with *S. viminalis* s.str. only in Rs(W).

Total range. Schmucker 1942, maps 130 and 172; Čermák 1955: 76; Skvortsov 1968: 194 (incl. *Salix rossica*).

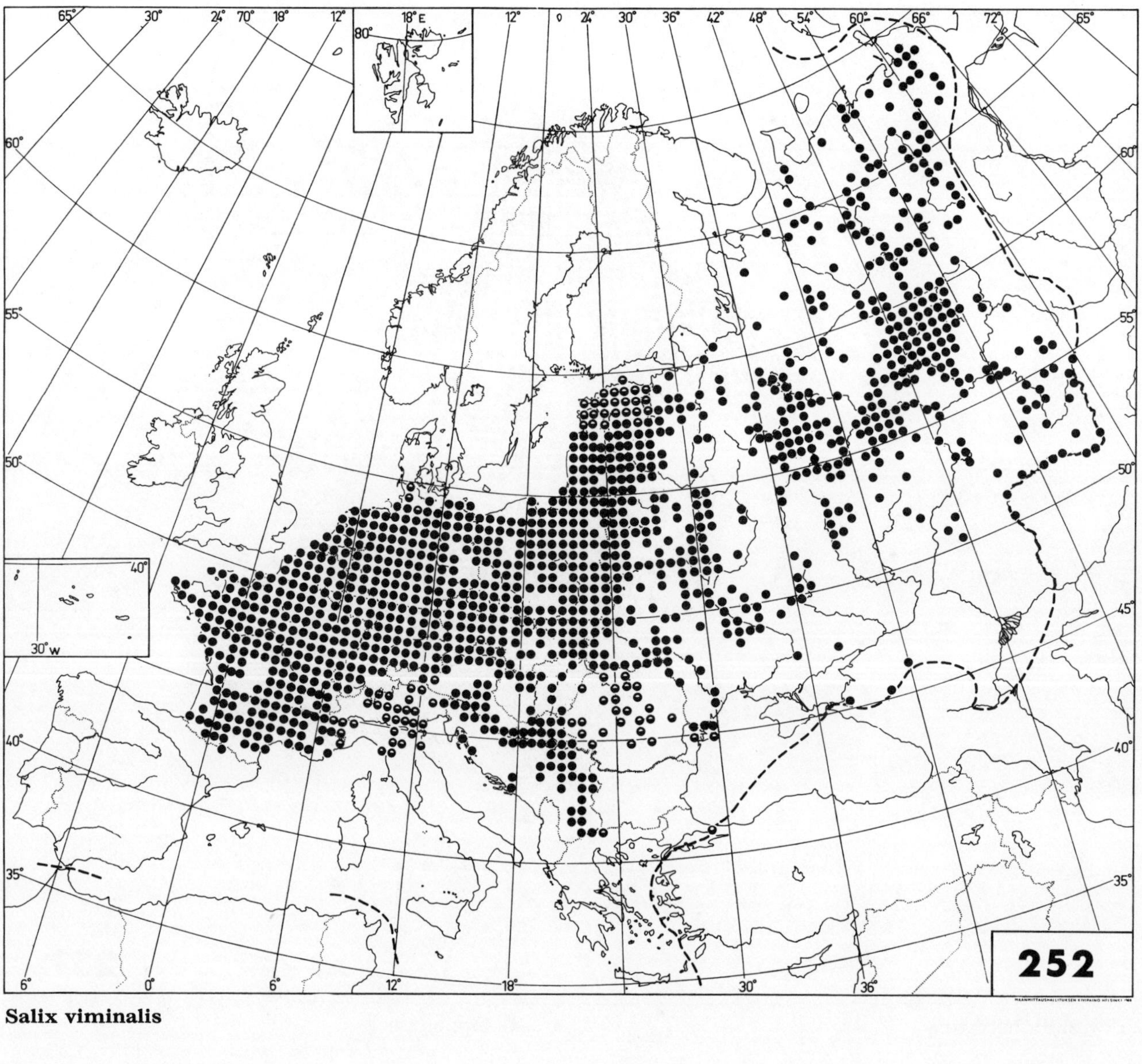

Salix viminalis

Salix burjatica Nasarov — Map 253.

Salix dasyclados auct. non Wimmer

Taxonomy. The hybrid nature (triploid) of *Salix dasyclados* Wimmer (s.str.) was confirmed by Neumann & Polatschek 1972: 623—624, on the basis on non-native material from Niederösterreich, Austria. Skvortsov 1968: 196—199, accords *S. dasyclados* the rank of an independent species, probably owing to the inclusion of a misunderstood eastern taxon, morphologically close to the true *S.* × *dasyclados*, and here premilinarily referred to as *S. burjatica* Nasarov.

Notes. Owing to the taxonomic confusion referred to above, the western limit of the range remains obscure. The data from the Baltic area were given for »*Salix dasyclados* subsp. *baltica* Lacksch.».

Total range. Skvortsov 1968: 197.

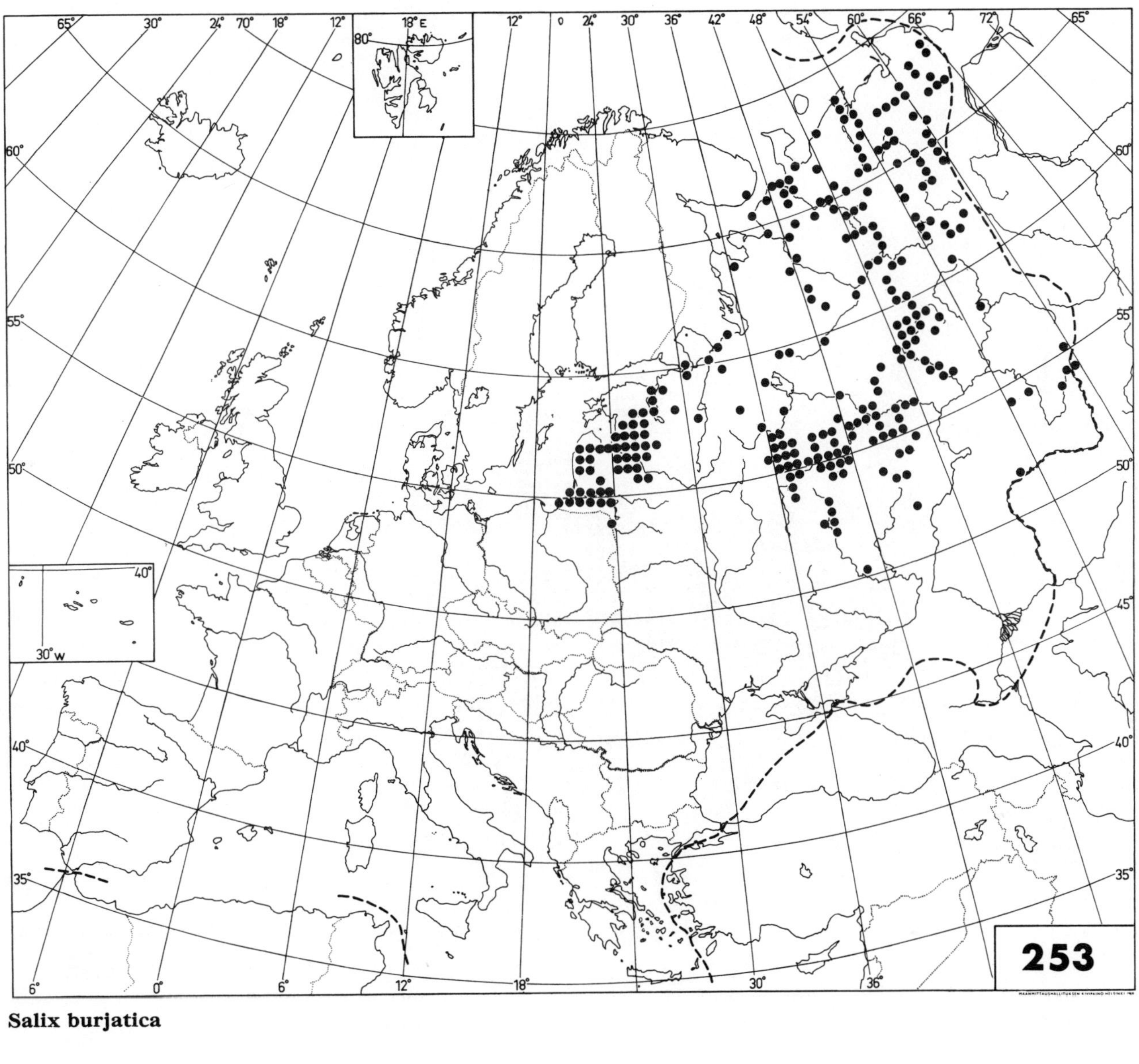

Salix burjatica

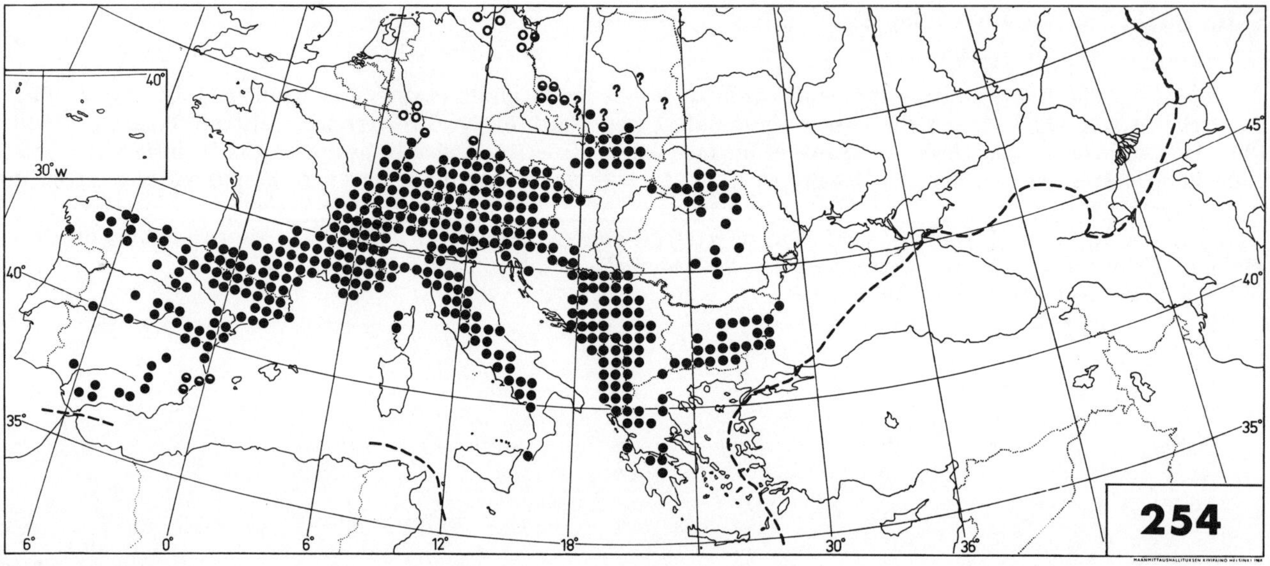

Salix eleagnos

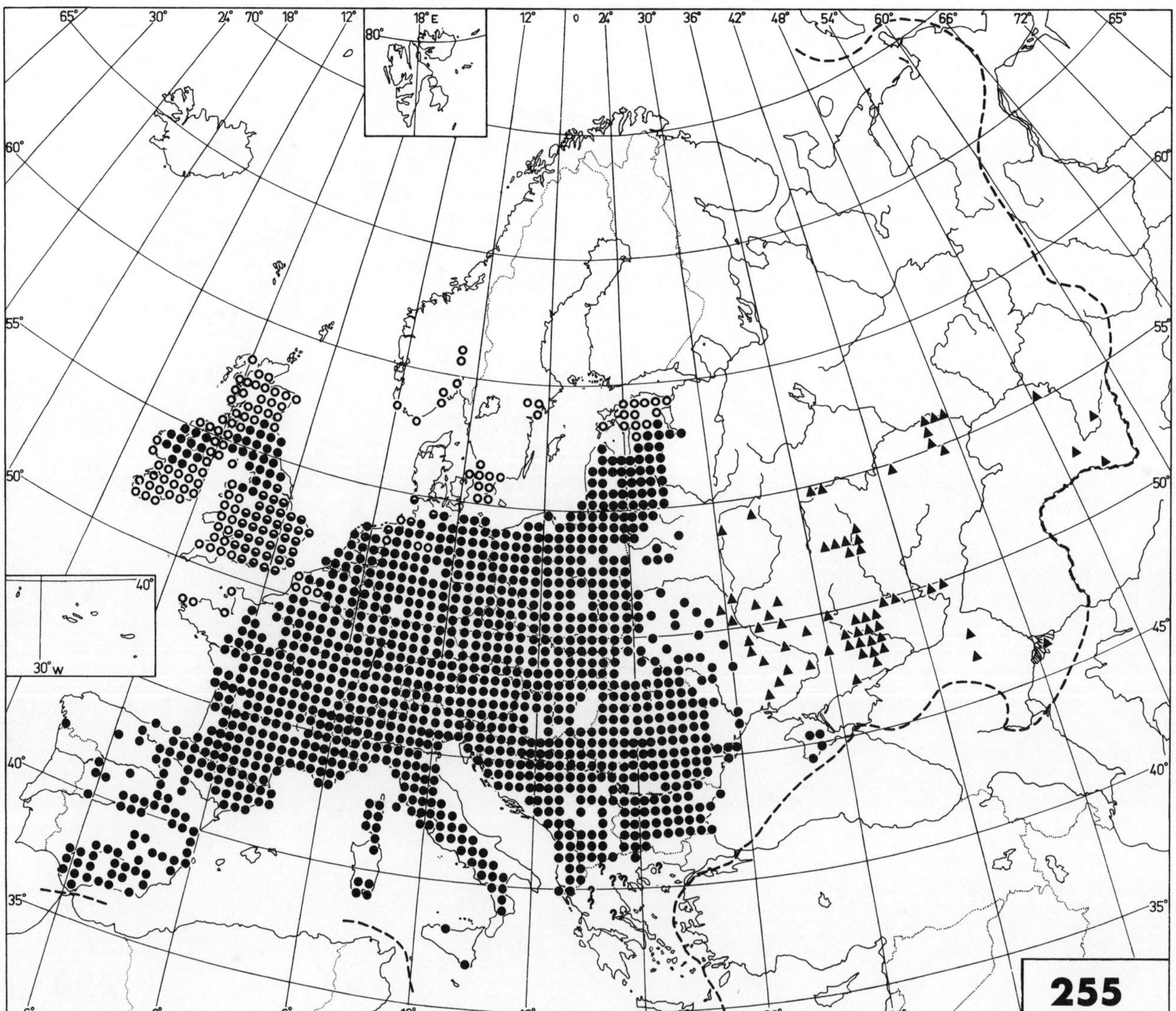

• = **Salix purpurea** ▲ = **S. vinogradovii**

Salix eleagnos Scop. — Map 254.

Salix incana Schrank

Notes. The available data are too incomplete to allow separate mapping of the two subspecies recognized in Fl. Eur.
Total range. Skvortsov 1968: 202.

S. eleagnos subsp. **eleagnos**

»Throughout most of the range of the species» (Fl. Eur.).

S. eleagnos subsp. **angustifolia** (Cariot) Rech. fil.

»Largely replacing (a) [= subsp. *eleagnos*] in S. France and Spain» (Fl. Eur.).

Salix purpurea L. — Map 255.

Notes. The two subspecies could not be mapped separately. Al added (not given in Fl. Eur.), Lu, Rs(N) and Tu omitted (given in Fl. Eur.), Rs(E) omitted (the records representing *S. vinogradovii*; see below) (given in Fl. Eur.).
Total range. MJW 1965: map 116c; A. K. Skvortsov, Nov. Syst. Pl. Vasc. (Leningrad) 1966: 61; Skvortsov 1968: 227.

S. purpurea subsp. **purpurea**

»Widespread, particularly in mountain regions» (Fl. Eur.).

S. purpurea subsp. **lambertiana** (Sm.) A. Neumann ex Rech. fil.

»Widespread, mainly in the lowlands» (Fl. Eur.).

Salix vinogradovii A. Skvortsov — Map 255.

Salix purpurea auct. ross. pro parte, non L.

Taxonomy. A. K. Skvortsov, Nov. Syst. Pl. Vasc. (Leningrad) 1966: 55—58.
Notes. Rs(W, C, E) (the species not mentioned in Fl. Eur.).
Total range. A. K. Skvortsov, Nov. Syst. Pl. Vasc. (Leningrad) 1966: 61; Skvortsov 1968: 227.

Salix amplexicaulis Bory — Map 256.

Notes. It and Tu added (not given in Fl. Eur.).
Total range. A. K. Skvortsov, Nov. Syst. Pl. Vasc. (Leningrad) 1966: 64; Skvortsov 1968: 229.

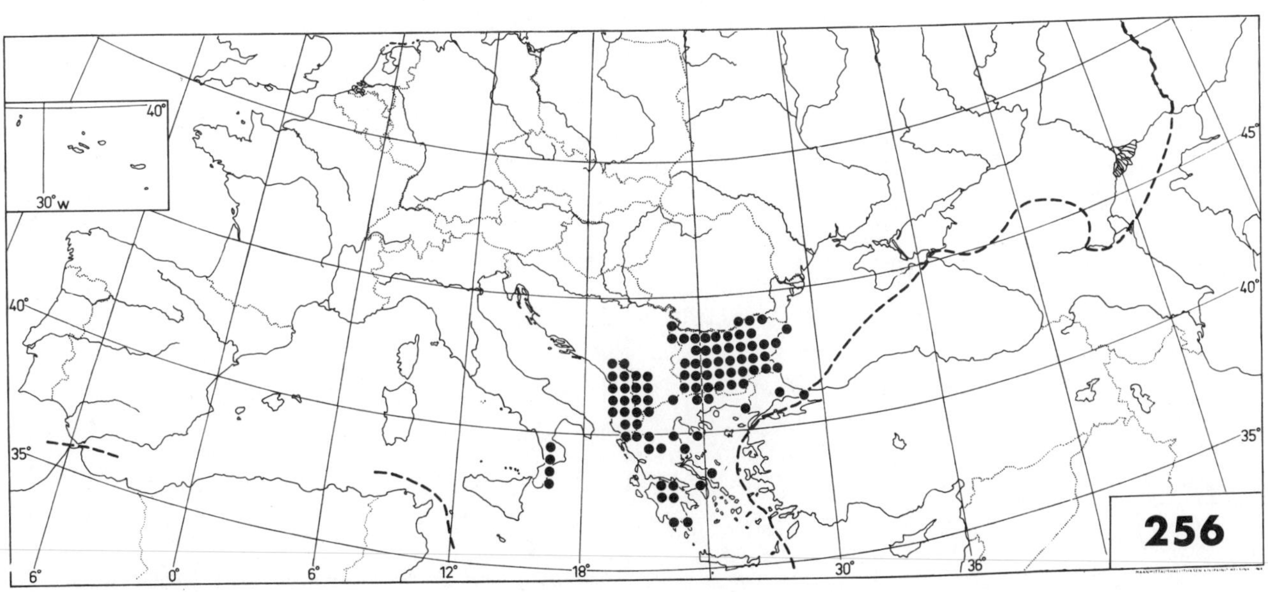

Salix amplexicaulis

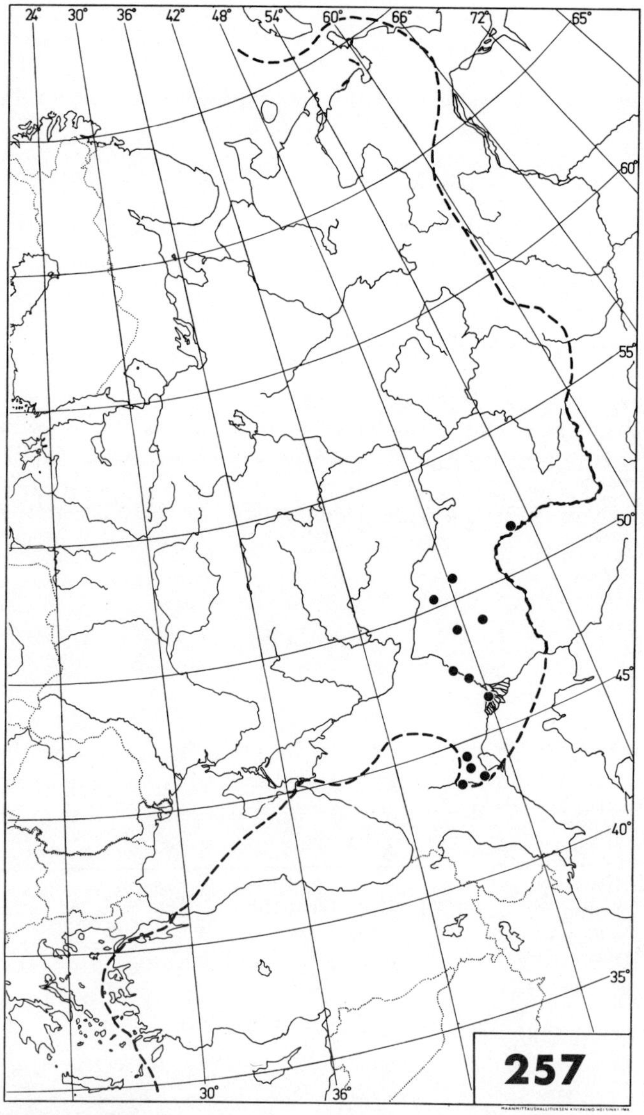

Salix caspica

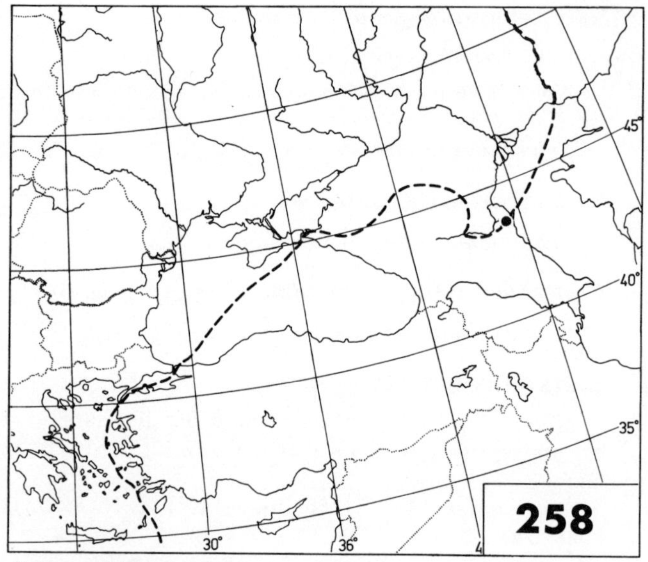

Salix wilhelmsiana

Salix caspica Pallas — Map 257.

Salix volgensis N. J. Andersson (fide Fl. Eur., non Skvortsov)

Total range. MJW 1965: map 116c; Skvortsov 1968: 235.

Salix wilhelmsiana Bieb. — Map 258.

Total range. Skvortsov 1968: 238.

Salix caesia Vill. — Map 259.

Taxonomy. Neumann & Polatschek 1972: 624.
Notes. Rs(C) omitted (given in Fl. Eur.).
Total range. Skvortsov 1968: 225.

Salix tarraconensis Pau — Map 260.

Taxonomy. Neumann & Polatschek 1972: 624—625.

Total range. Endemic to Europe.

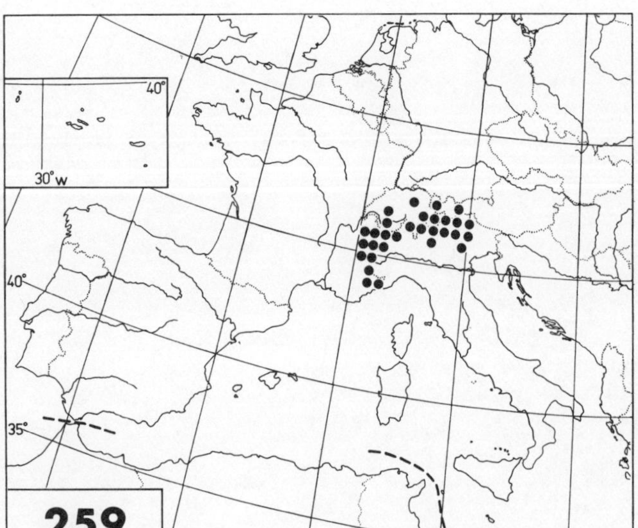

Salix caesia

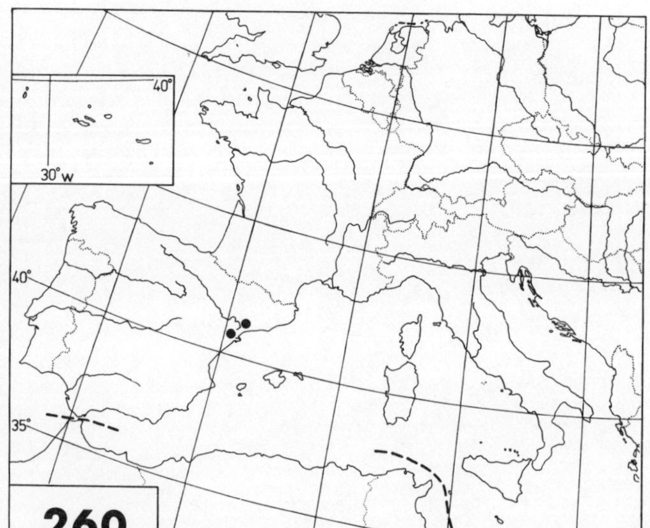

Salix tarraconensis

Salix daphnoides Vill. — Map 261.

Taxonomy. The Scandinavian population (No, Su) approaching *Salix acutifolia* differs from that of Central Europe, and has been recognized as *S. daphnoides* var. *norvegica* Agardh, or even considered to represent the hybrid *S. acutifolia* × *daphnoides*: B. Floderus in Holmberg, Skand. Fl. 1 (b1): 144—148 (1931); N. Hylander, Nord. Kärlväxtfl. 2: 285 (1966).

Notes. Au, Cz, Ga, Ge, He, Hs, It, Ju, No, *Po, Rm, Rs (N, B, C), Su. (Fl. Eur.: No, Su [Au, Ga, Ge, He]). Skvortsov 1968: 212—213.

Total range. Endemic to Europe (although not marked as such in Fl. Eur.).

Salix acutifolia Willd. — Map 261.

Salix daphnoides Will. subsp. acutifolia (Willd.) O. Dahl

Total range. Skvortsov 1968: 213.

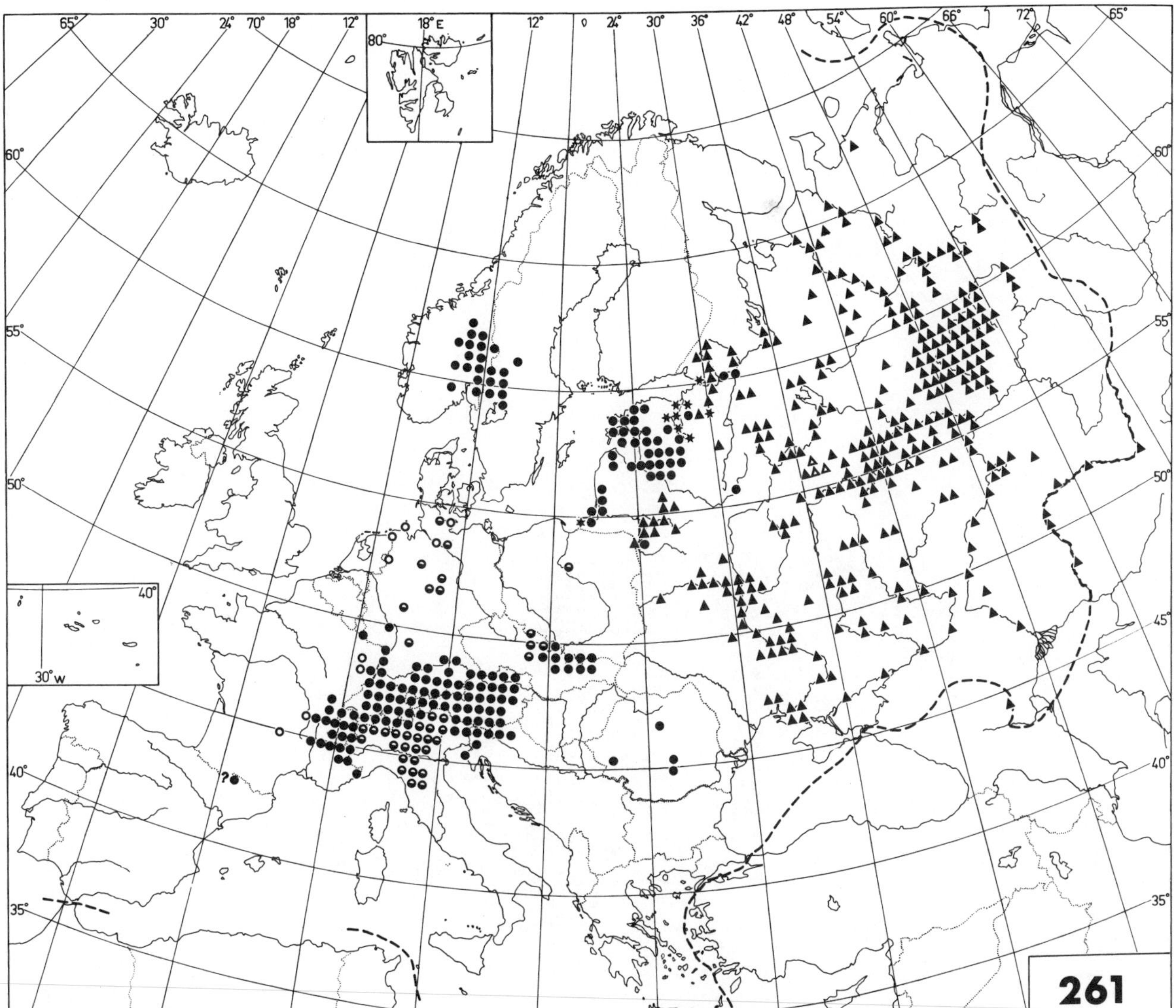

• = **Salix daphnoides**

▲ = **S. acutifolia**

★ = both species

Populus alba L. — Map 262.

Populus bolleana Lauche; P. nivea A. Wesmael

Notes. Considered native in Co, Ga and Hs ([Co, Ga, Hs] in Fl. Eur.), Rs(K) omitted (given in Fl. Eur.), Si added (not given in Fl. Eur.).

Total range. Schmucker 1942: map 142; Čermák 1955: 76; Fenaroli 1967: 96; U. Tamm in Eesti NSV Fl. 8: 520 (1971).

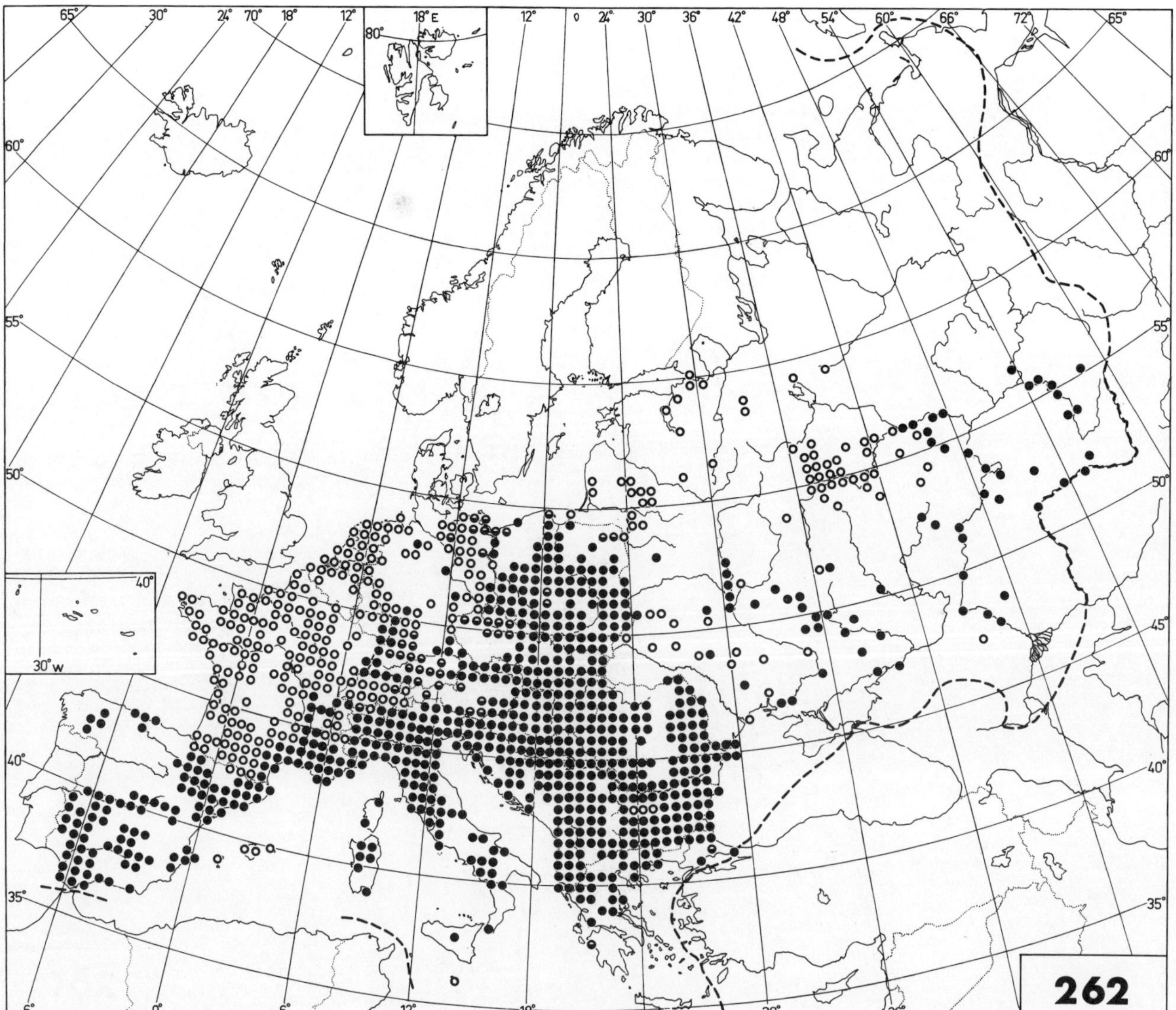

Populus alba

Populus canescens (Aiton) Sm. — Map 263.

Populus alba-tremula E. H. L. Krause; P. bachofenii Wierzb. ex Rochel; P. hybrida Reichenb. non Bieb.

Taxonomy. Considered the hybrid *Populus alba* × *tremula* by e.g. K. Browicz, Atlas Distrib. Trees Shrubs Poland 10: 21—22 (1971).

Notes. Al, Cz, Ju and Rs(E) added (not given in Fl. Eur.), not considered native in Be and Hb (given as native in Fl. Eur.).

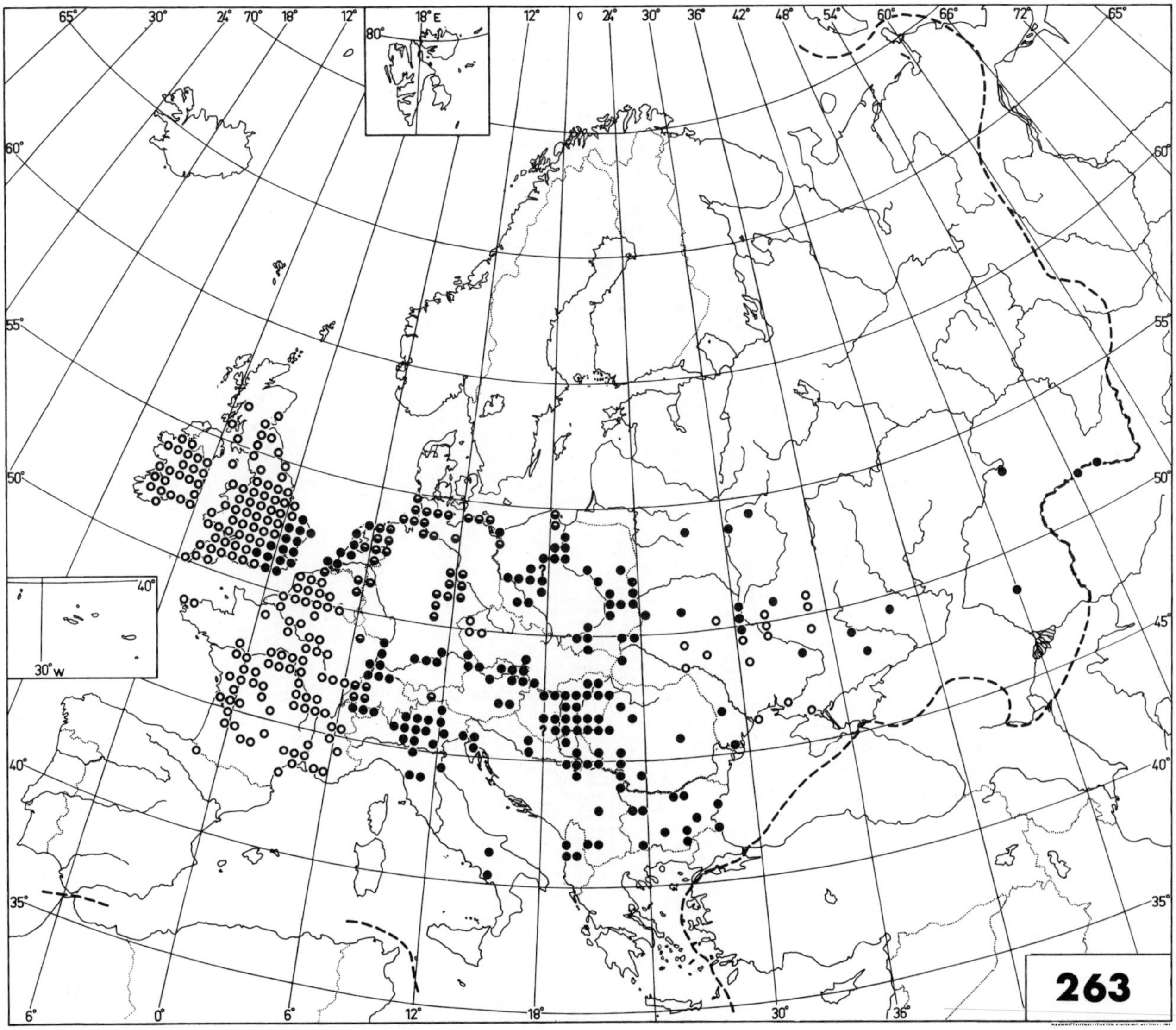

Populus canescens

Populus tremula L. — Map 264.

Total range. Schmucker 1942: map 142; Čermák 1955: 76; MJW 1965: map 112d; Fenaroli 1967: 98; Hultén Alaska 1968: 332; Hultén CP 1971: map 172.

Populus tremula

Populus nigra L. — Map 265.

Populus thevestina Dode

Taxonomy. According to W. Bugała, Arboretum Kórnickie 12: 45—219 (1967), three subspecies can be recognized in Europe, the widespread subsp. *nigra* (with glabrous leaves and shoots) being replaced in Br and Ga by subsp. *betulifolia* (Pursh) W. Wettst. (with petioles, inflorescences and short shoots caducously pubescent), and in S. Europe by subsp. *caudina* (Ten.) Bugała (with a persistent woolly indumentum on the shoots, petioles and inflorescence axes). However, the last-mentioned combination is antedated by *P. nigra* subsp. *neapolitana* (Ten.) Maire, given by Bugała in the synonymy.

Notes. The species is mapped here as one collective taxon. Considered native in Br, Co, Hs, Sa and Si ([Br, Co, Hs, Sa, Si] in Fl. Eur.), Cr omitted (given in Fl. Eur.), Ga added (not given in Fl. Eur.).

Total range. Schmucker 1942: map 142; Čermák 1955: 76; MJW 1965: map 112c; W. Bugała. Arboretum Kórnickie 12 (1967): 75 (the species) and 118 (subsp. *betulifolia* and subsp. *caudina*); Fenaroli 1967: 104.

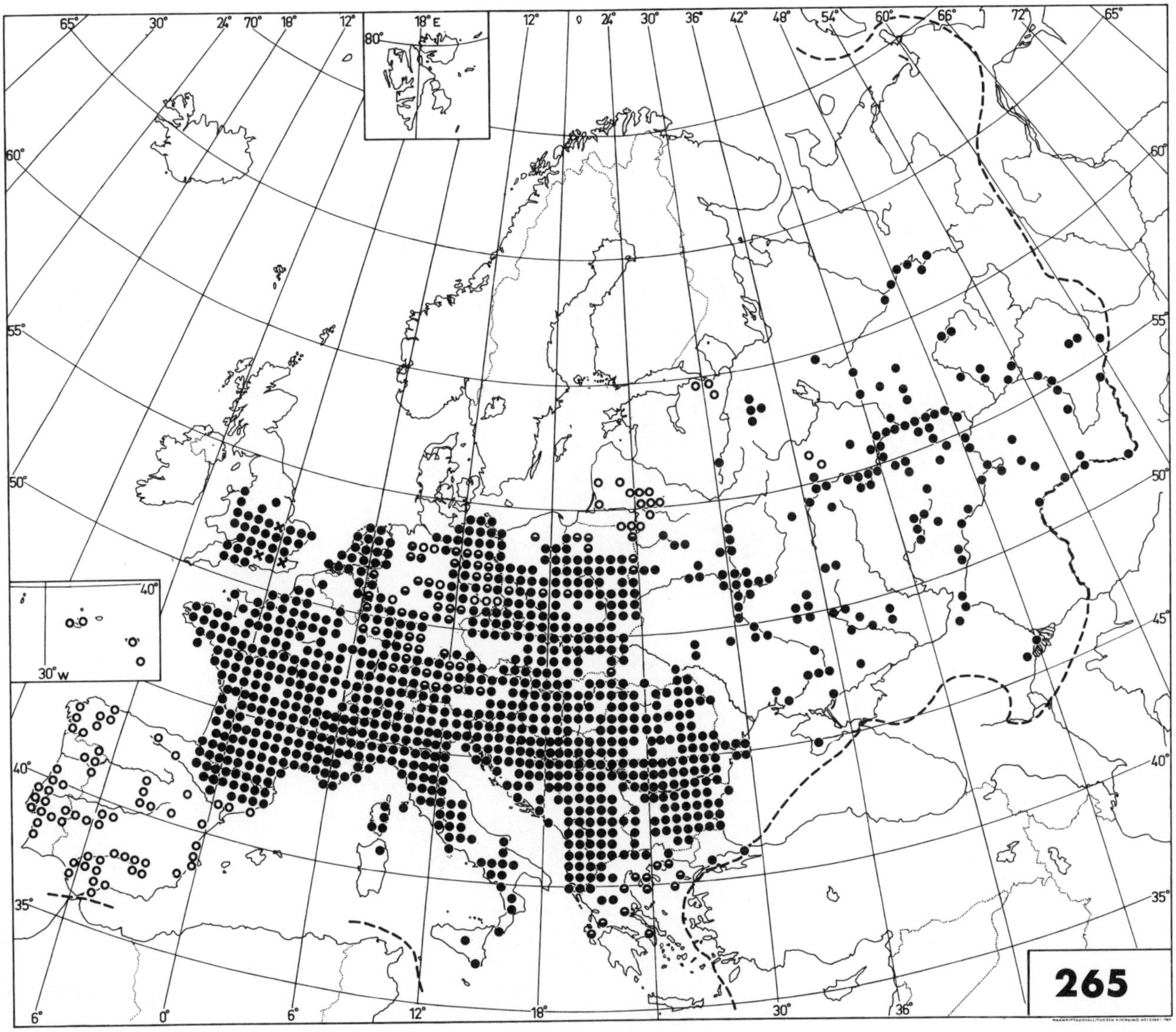

Populus nigra

MYRICACEAE

Myrica faya Aiton — Map 266.

Notes. In Lu naturalized since before 1900 (*Lu in Fl. Eur.).

Myrica caroliniensis Miller

Myrica pennsylvanica Loisel.

Notes. »Naturalized in S. England and the Netherlands» (Fl. Eur.). Not mapped.

Native of E. North America.

Myrica conifera Burm. fil.

Notes. Recently recorded as a fully established alien from [Az] (the species not mentioned in Fl. Eur.): A. Hansen, Bol. Soc. Brot. 46: 233 (1972). Not mapped.

Native of S. Africa.

Myrica gale L. — Map 267.

Gale palustris Chevall.

Notes. Su added (not given in Fl. Eur.).

Total range. Hultén AA 1958: map 199; MJW 1965: map 117a.

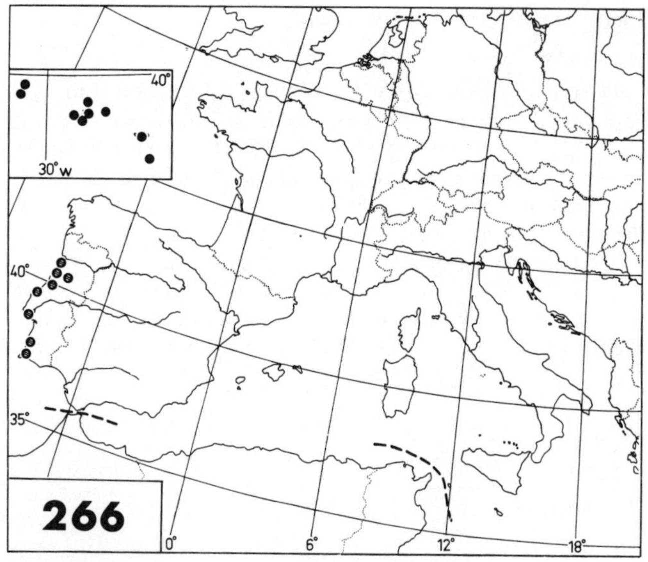

Myrica faya

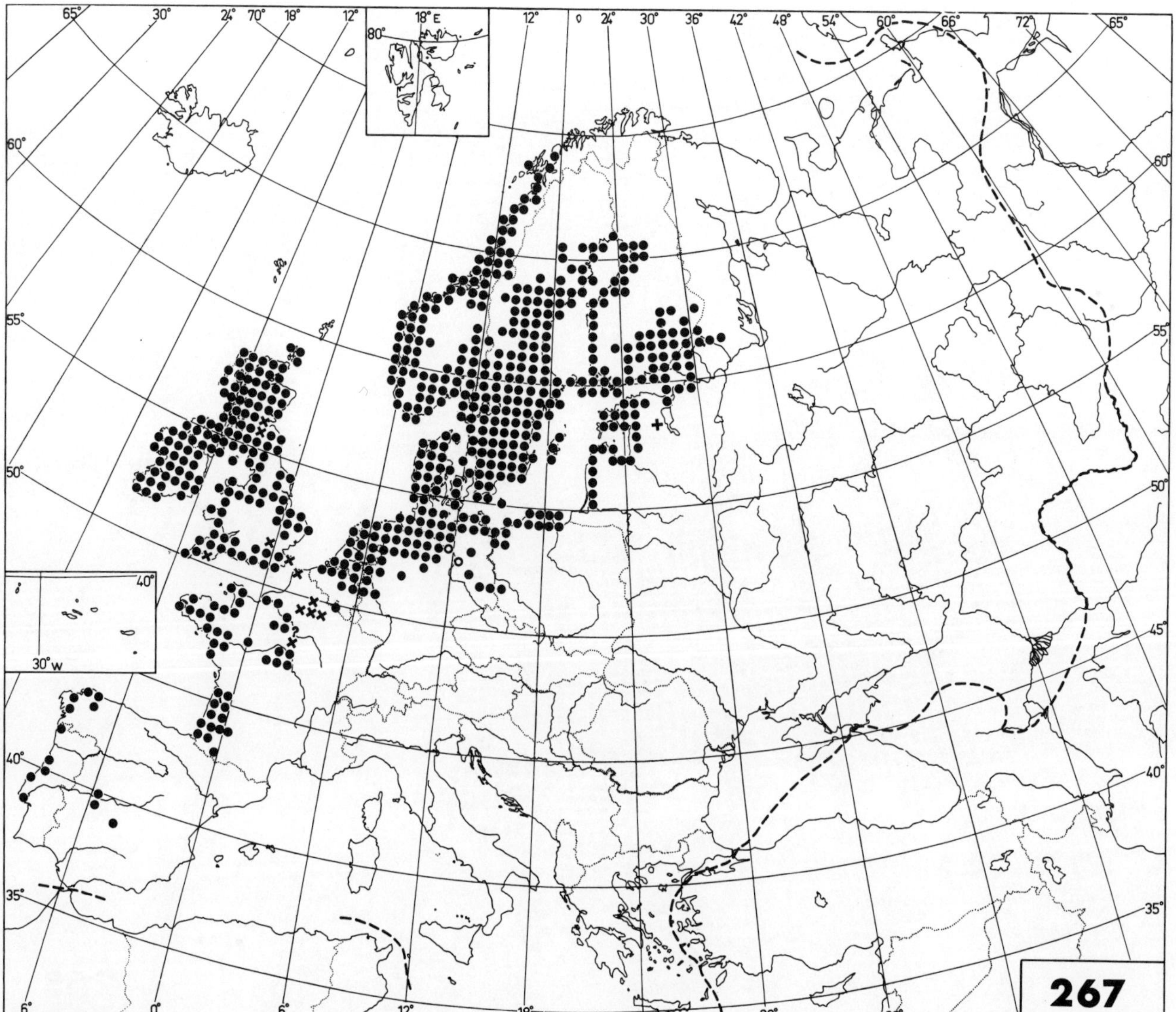

Myrica gale

JUGLANDACEAE

Juglans regia L. — Map 268.

Notes. Al (no data received), *Au, Bu, Gr, *It, Ju, *Si. »Balkan peninsula, Kriti, Romania; widely cultivated elsewhere for its fruits and timber and naturalized in many places» (Fl. Eur.).

Total range. Schmucker 1942: map 122 (not considered native in Europe); Čermák 1955: 82; Fenaroli 1967: 108.

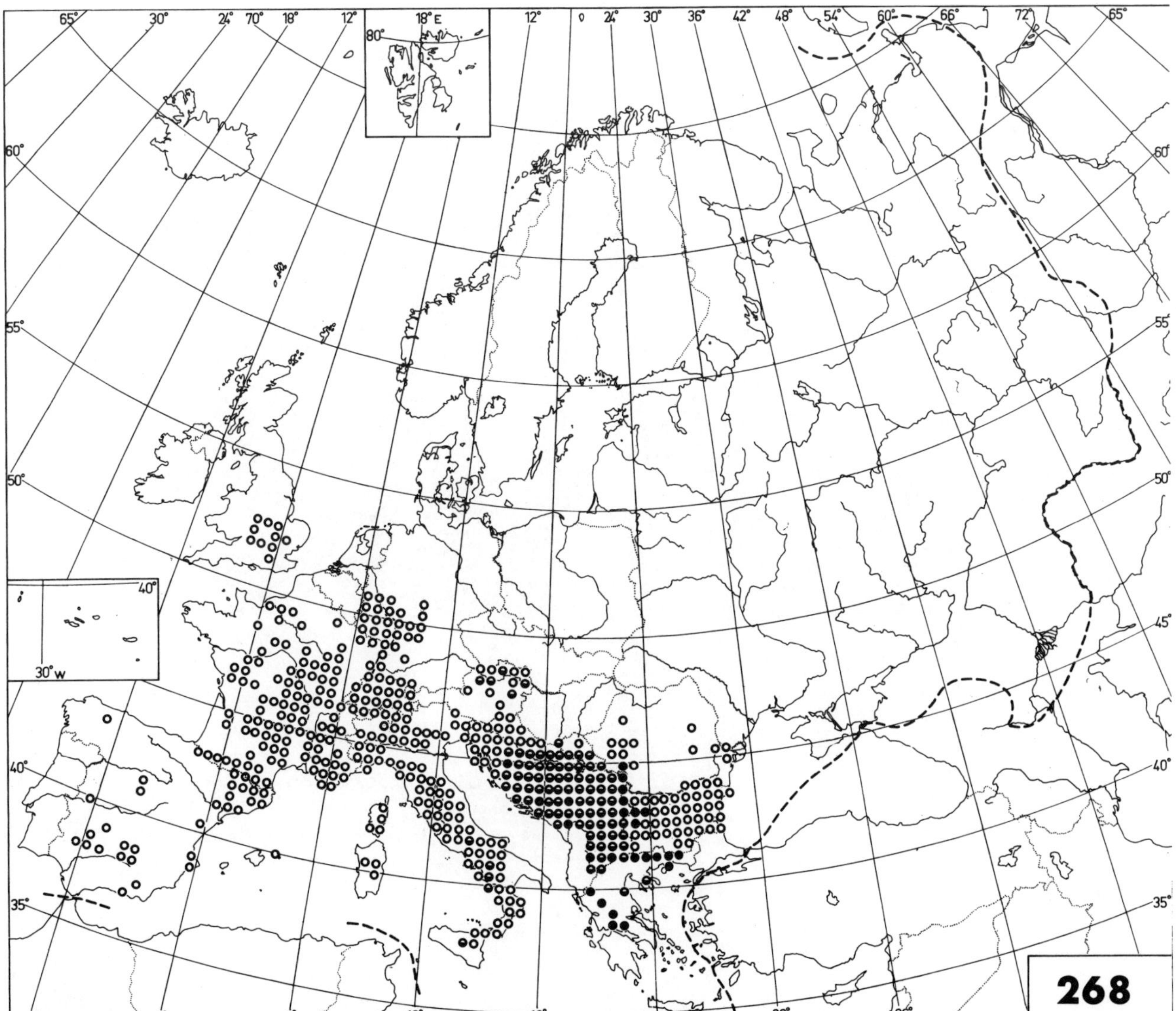

Juglans regia

BETULACEAE

Betula pendula Roth — Map 269.

Betula alba sensu Coste; B. verrucosa Ehrh. Incl. B. aetnensis Rafin. and B. oycoviensis Besser

Taxonomy. The Sicilian population has been referred to *Betula aetnensis* Rafin. (not mentioned in Fl. Eur.).
Notes. Gr and ?Tu added (not given in Fl. Eur.).
Total range. Hegi 1957: 143; MJW 1965: map 118d; Hultén CP 1971: map 81.

Betula pubescens Ehrh. — Map 270.

Incl. Betula callosa Notø; B. celtiberica Rothm. & Vasc.; B. subarctica Orlova

Taxonomy. Hultén CP 1971: 90, 340—343; A. S. Gardiner, Trans. Proc. Bot. Soc. Edinb. 41: 451—459 (1972); P. Kallio & Y. Mäkinen, Rep. Kevo Subarctic Res. Stat. 13 (in press).

Notes. Since the infraspecific taxonomy is still confused, and the records available are inadequate and/or inconsistent, the subspecies recognized in Fl. Eur. could not be mapped separately.

Total range. Schmucker 1942: maps 69—70; Hegi 1957: 154; MJW 1965: map 119a; Hultén CP 1971: maps 81 and 82.

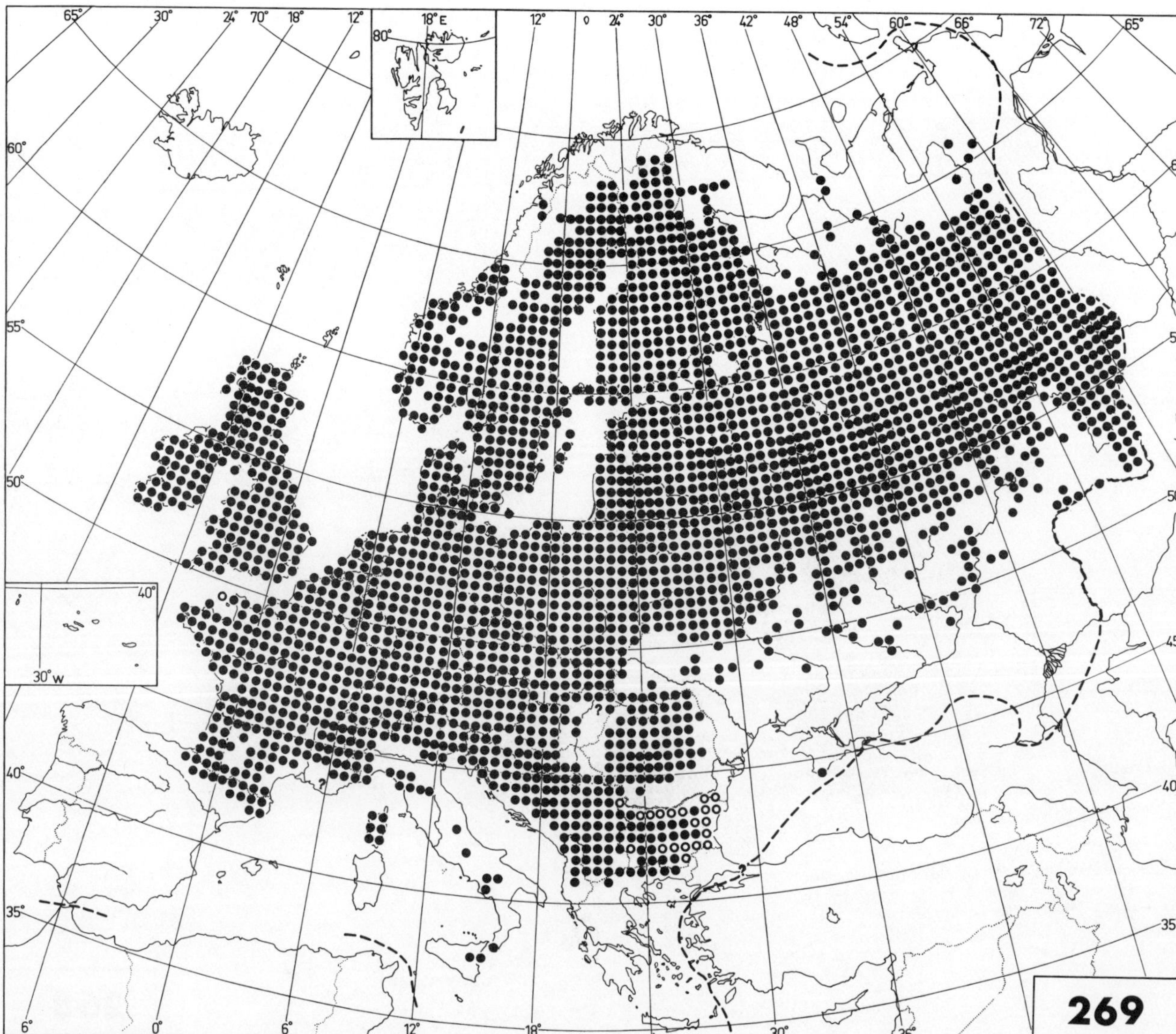

Betula pendula

B. pubescens subsp. **pubescens**

Betula concinna Gunnarsson; B. pubescens subsp. suecica Gunnarsson

Notes. »Usually lowland; northwards to c. 70° N. in Finland and N. Russia» (Fl. Eur.).

Total range. Hultén CP 1971: map 81.

B. pubescens subsp. **carpatica** (Willd.) Ascherson & Graebner

Betula carpatica Waldst. & Kit. ex Willd.; B. coriacea Gunnarsson; B. murithii Gaudin; B. odorata Bechst.; B. pubescens subsp. odorata (Bechst.) E. F. Warburg. Incl. (?) B. borysthenica Klokov

Notes. Adequate records available only from Cz, Ge and Po. Not mapped here. »Arctic Europe, and southwards, mainly in the mountains, to the Pyrenees and Carpathians» (Fl. Eur.).

Total range. Endemic to Europe? (not marked as such in Fl. Eur.). Cf. W. Rothmaler & J. Carvalho e Vasconcellos, Bol. Soc. Brot. Ser. 2, 14: 168 (1940) and K. Browicz, Atlas Distrib. Trees Shrubs Poland 10: 16—17 (1971). However, range certainly wider than that given by Hultén CP 1971: map 82.

B. pubescens subsp. **tortuosa** (Ledeb.) Nyman

Betula tortuosa Ledeb. Incl. B. kusmisscheffii (Regel) Suk.

Taxonomy. P. Kallio & Y. Mäkinen, Rep. Kevo Subarctic Res. Stat. 13 (in press).

Notes. In map 270 those squares have been specially marked from which only subsp. *tortuosa* is recorded. Otherwise not mapped here. Subsp. *tortuosa* is also given in the literature from the mountains of C. Europe. »Arctic Europe; mountains of Scandinavia; Iceland» (Fl. Eur.).

Total range. Schmucker 1942: map 69; Hultén AA 1958: map 47; V. N. Vasil'ev, Mat. Hist. Fl. Veget. USSR 4: 270 (1963) (the Altai population, including the type of the taxon, omitted); Hultén CP 1971: map 82.

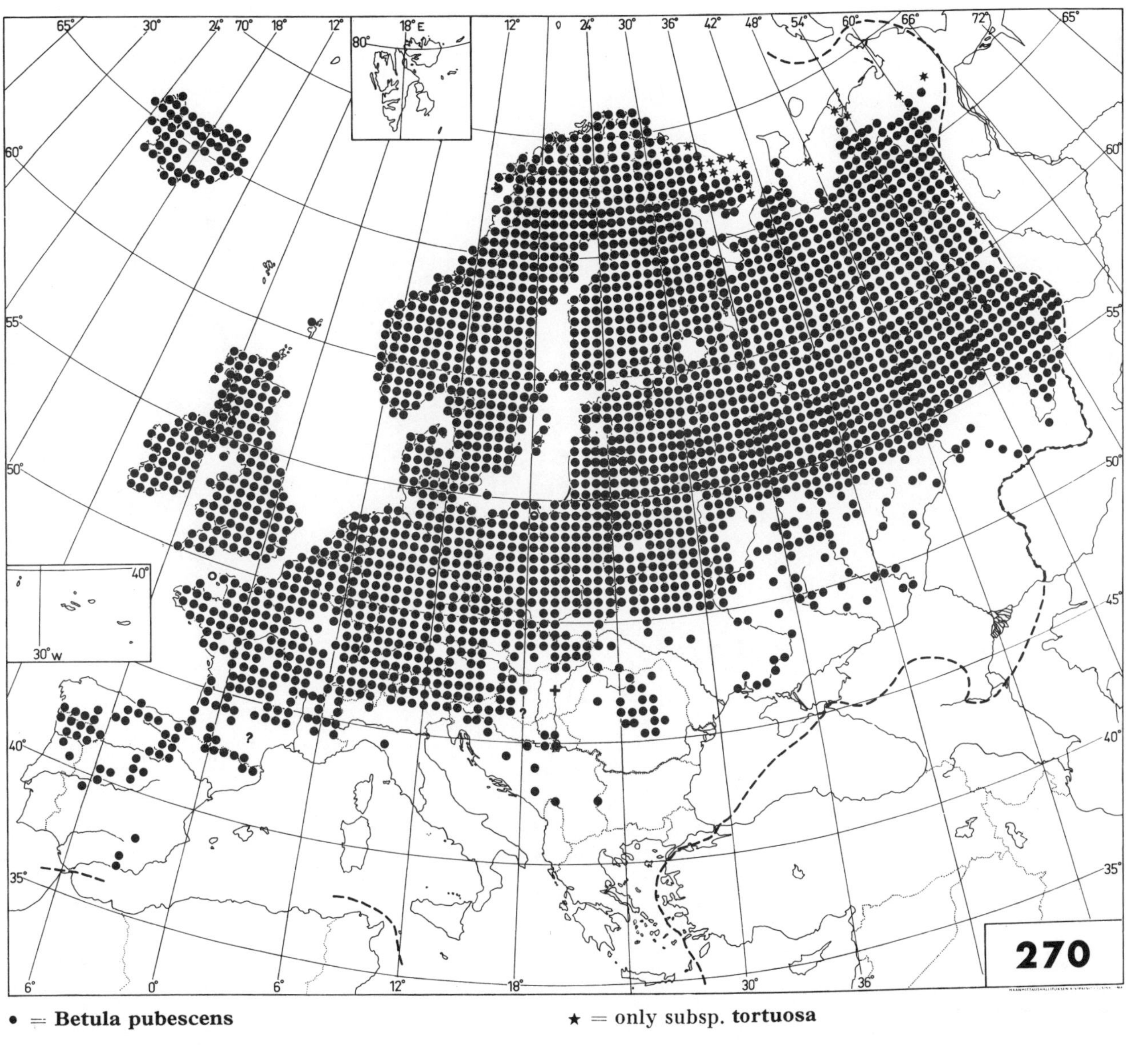

• = Betula pubescens ★ = only subsp. **tortuosa**

Betula humilis Schrank — Map 271.

Notes. Extinct in Cz (given as present in Fl. Eur.): see J. Holub, Preslia 38: 79 (1966). Rs(E) added (not given in Fl. Eur.).

Total range. Schmucker 1942: map 73; Čermák 1955: 79; MJW 1965: map 119c.

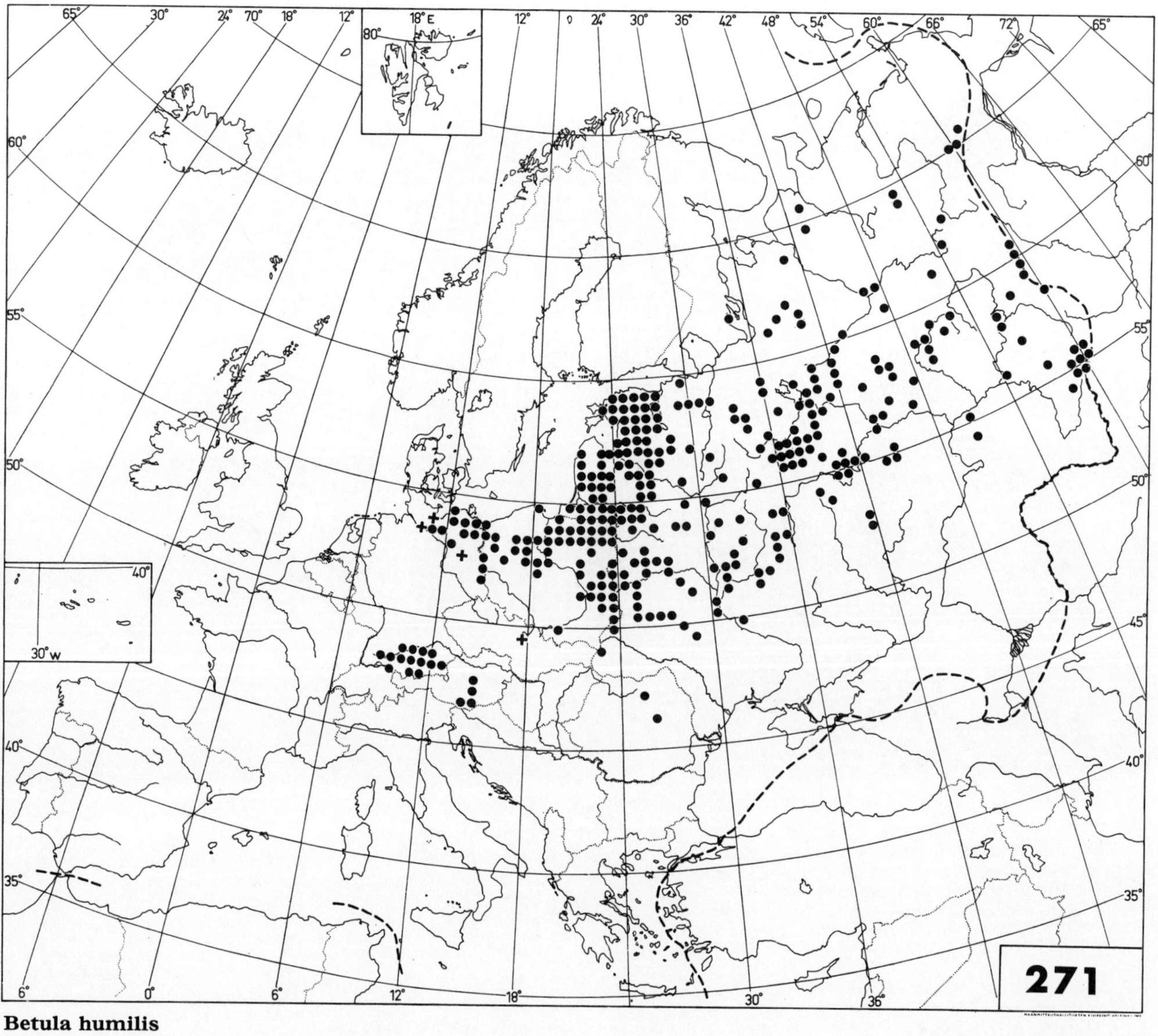

Betula humilis

Betula nana L. — Map 272.

Incl. Betula tundrarum Perf.

Taxonomy. Concerning the taxonomical position of *Betula sukaczewii* Soczava, see A. K. Makhnev & Y. F. Rozhdestvensky, Bot. Žurn. 58: 1107—1117 (1973).

Notes. Extinct in It (the country not given in Fl. Eur.): Nuovo Giorn. Bot. Ital. N.S. 61: 105 (1954).

Total range. Hegi 1957: 161; E. Hadač, Preslia 32: 248 (1960); MJW 1965: map 119b; B. A. Yurtsev, Komarovskie Čtenija 19: 41 (1966); Hultén Alaska 1968: 365; Hultén CP 1971: map 128.

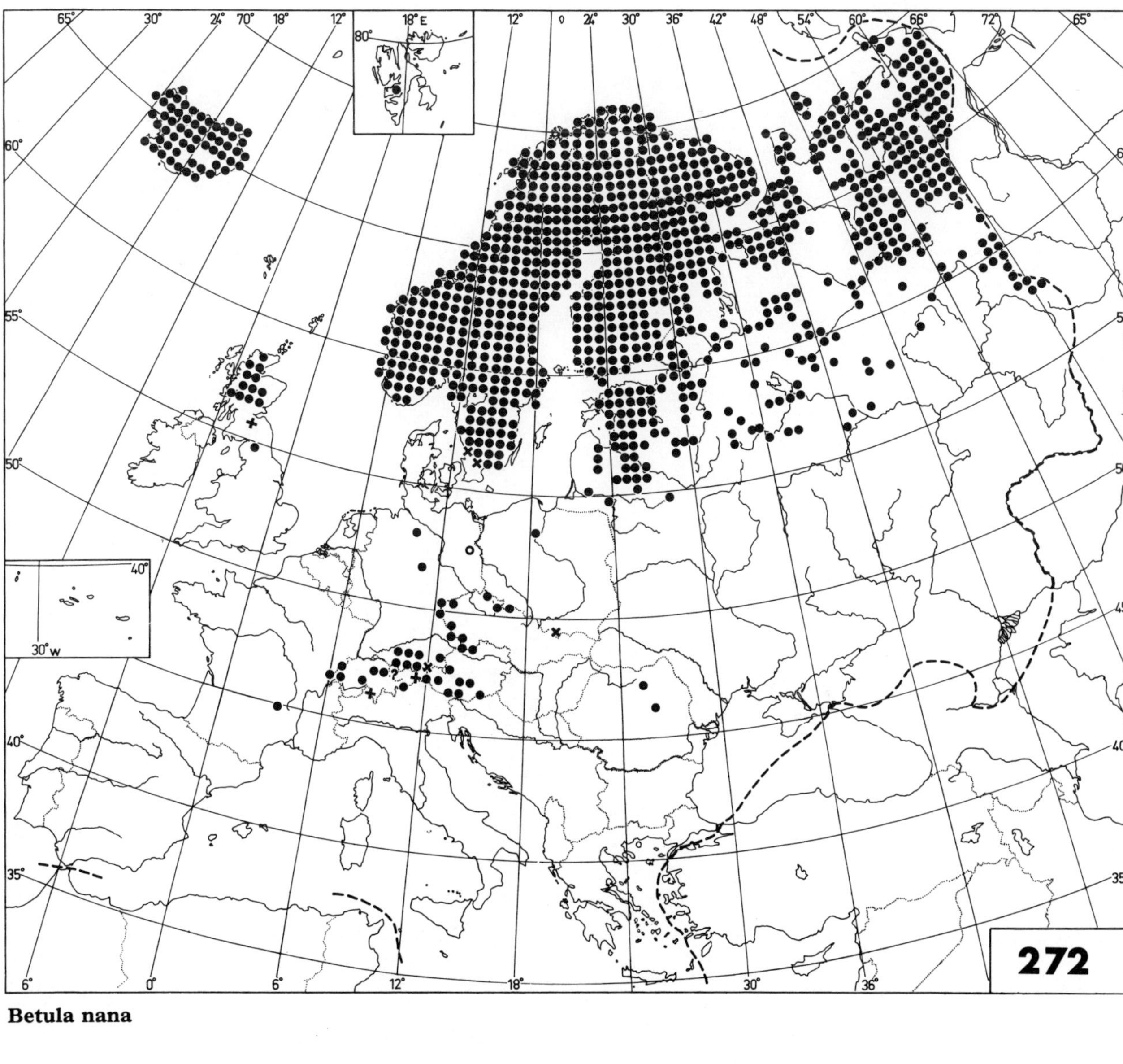

Betula nana

Alnus viridis (Chaix) DC. — Map 273.

Alnus alnobetula (Ehrh.) Hartig; Duschekia viridis (Chaix) Opiz

Taxonomy. For the generic delimitation and nomenclature, see J. Holub & Z. Pouzar, Folia Geobot. Phytotax. (Praha) 2: 406 (1967), and J. Holub, Folia Geobot. Phytotax. (Praha) 2: 421 (1967).

Notes. Rs(C) added (not given in Fl. Eur.).

Total range. Schmucker 1942: map 63; L. Richard, Doc. Carte Veget. Alpes 5: 85 (1967).

A. viridis subsp. **viridis** — Map 273.

Notes. Au, Bu, Cz, Ga, Ge, He, Hu, It, Ju, Po, Rm, Rs (W). The lower limit of native occurrences in C. Europe remains obscure. »Mountains of C. Europe and Balkan Peninsula» (Fl. Eur.).

Total range. Endemic to Europe.

A. viridis subsp. **fruticosa** (Rupr.) Nyman — Map 273.

Alnus crispa (Aiton) Pursh; A. fruticosa Rupr.; Alnaster fruticosa (Rupr.) Ledeb.; Duschekia fruticosa (Rupr.) Pouzar

Notes. Rs(N, C). »N.E. Russia» (Fl. Eur.).

Total range. Schmucker 1942: map 64 (Eurasian part); S. J. Sokolov *et al.*, Derev'ja i kustarniki SSSR 2: 337, map 93 (Moskva—Leningrad 1951); MJW 1965: map 119d; L. Richard, Doc. Carte Veget. Alpes 5: 85 (1967).

A. viridis subsp. **suaveolens** (Req.) P. W. Ball — Map 273.

Alnus suaveolens Req.; Duschekia viridis subsp. suaveolens (Req.) Holub

Total range. Endemic to Corse.

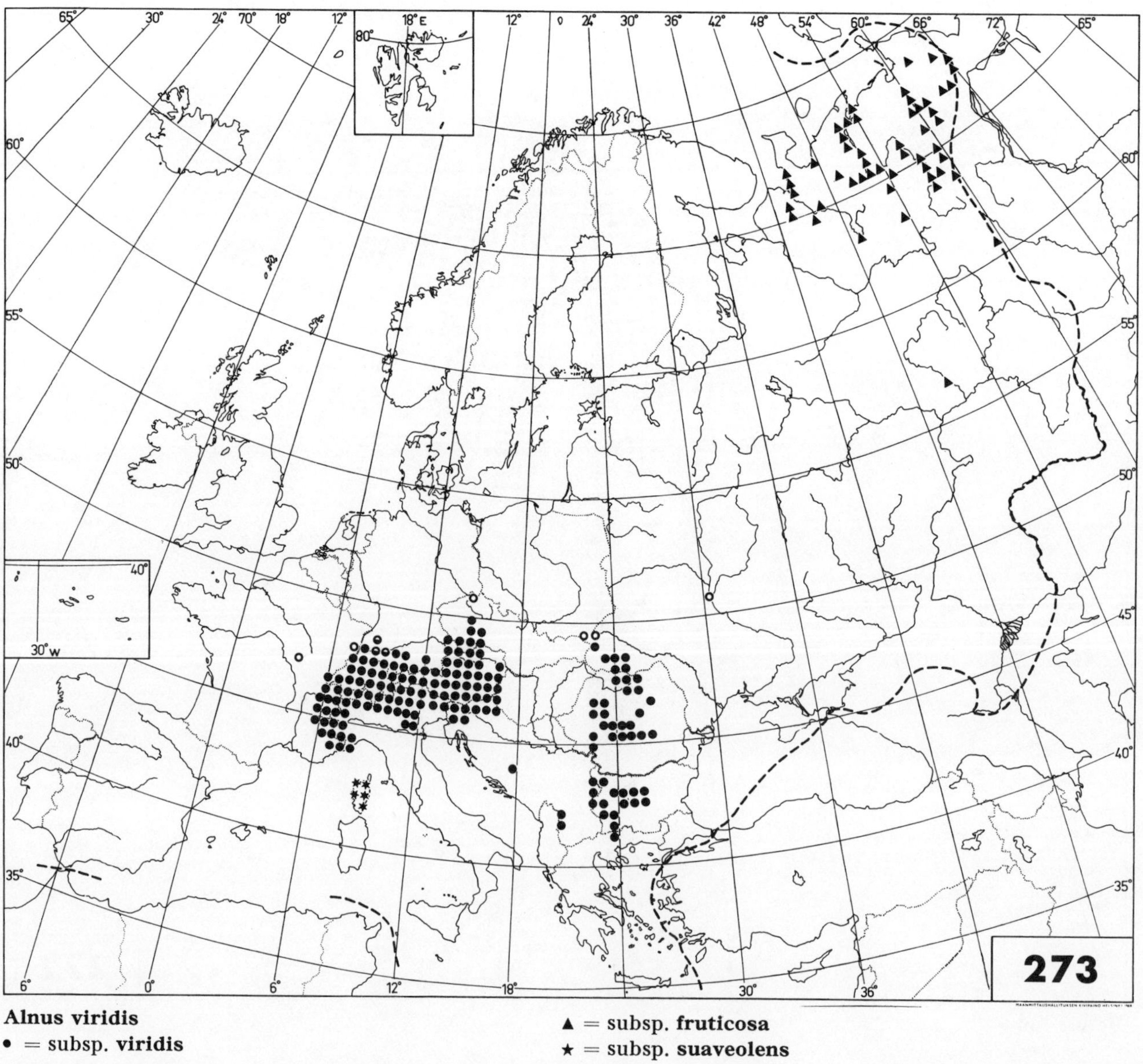

Alnus viridis
• = subsp. **viridis**
▲ = subsp. **fruticosa**
★ = subsp. **suaveolens**

Alnus glutinosa (L.) Gaertner — Map 274.

Taxonomy. F. Yaltirik, Notes Roy. Bot. Gard. Edinb. 28: 15—16 (1967).
Total range. Schmucker 1942: map 62; S. J. Sokolov *et al.*, Derev'ja i kustarniki SSSR 2: 337, map 93 (1951); Hegi 1957: 176; MJW 1965: map 120b; J. Jentys-Szaferowa, Monogr. Bot. 32: 203 (1970).

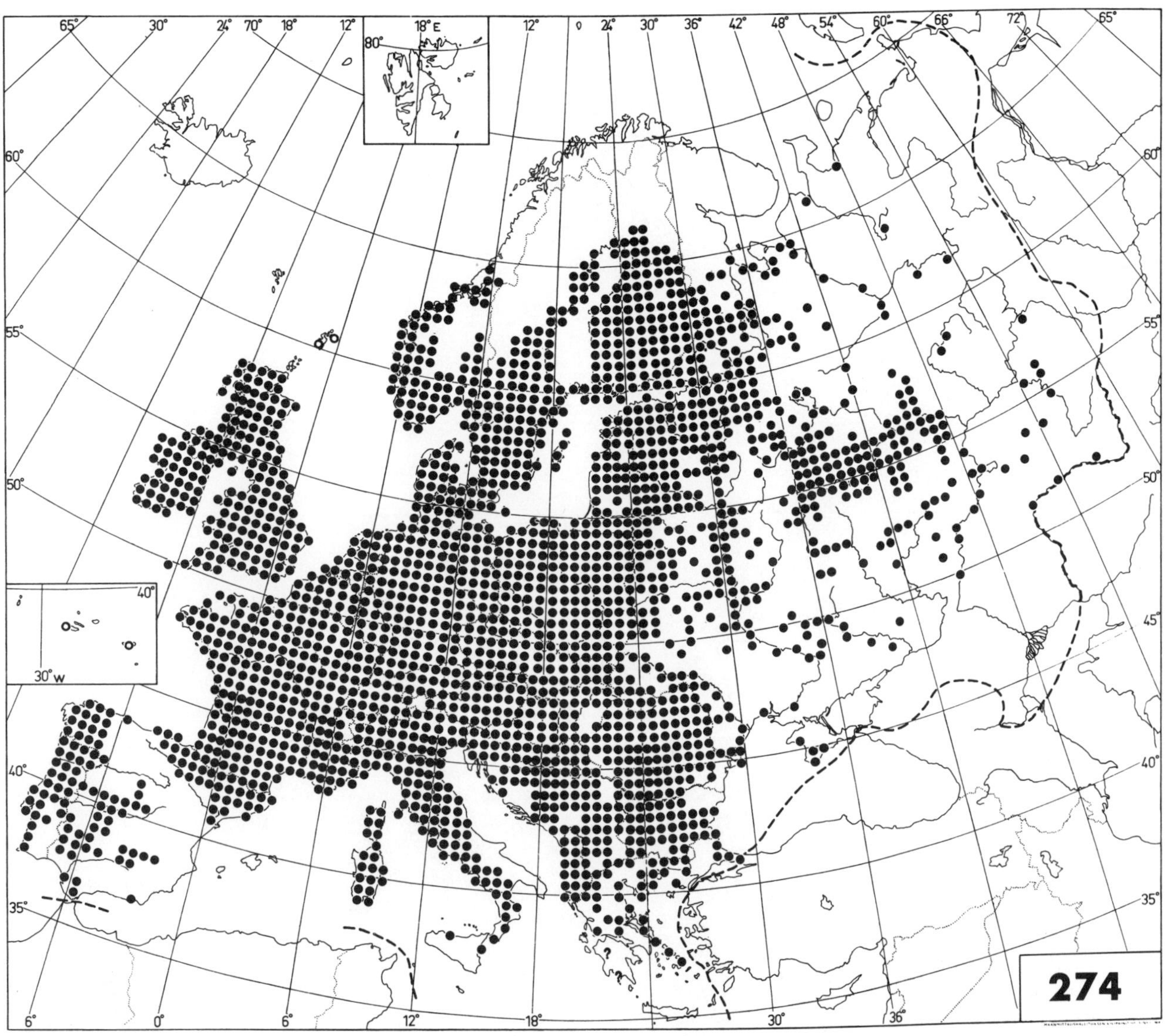

Alnus glutinosa

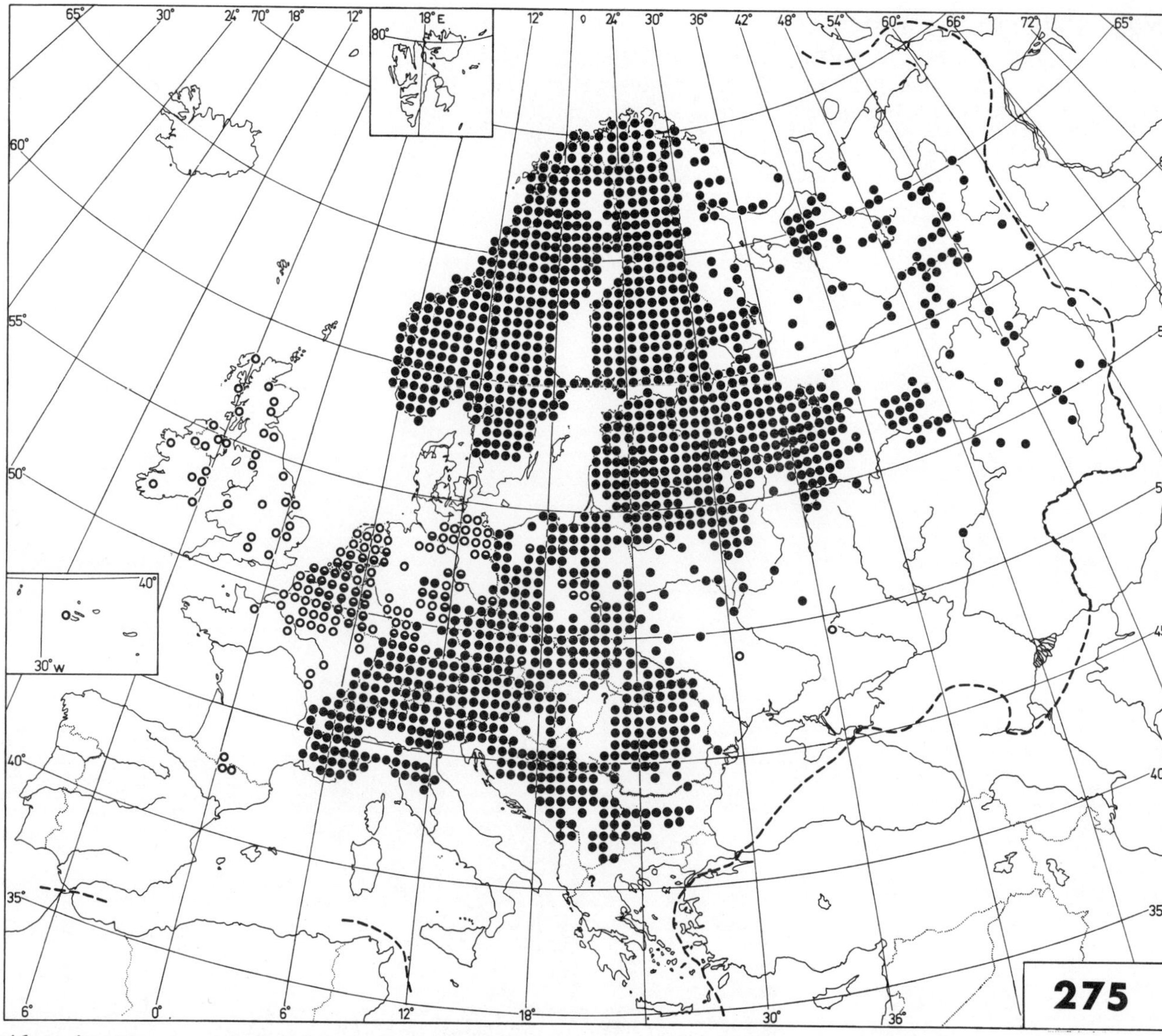

Alnus incana

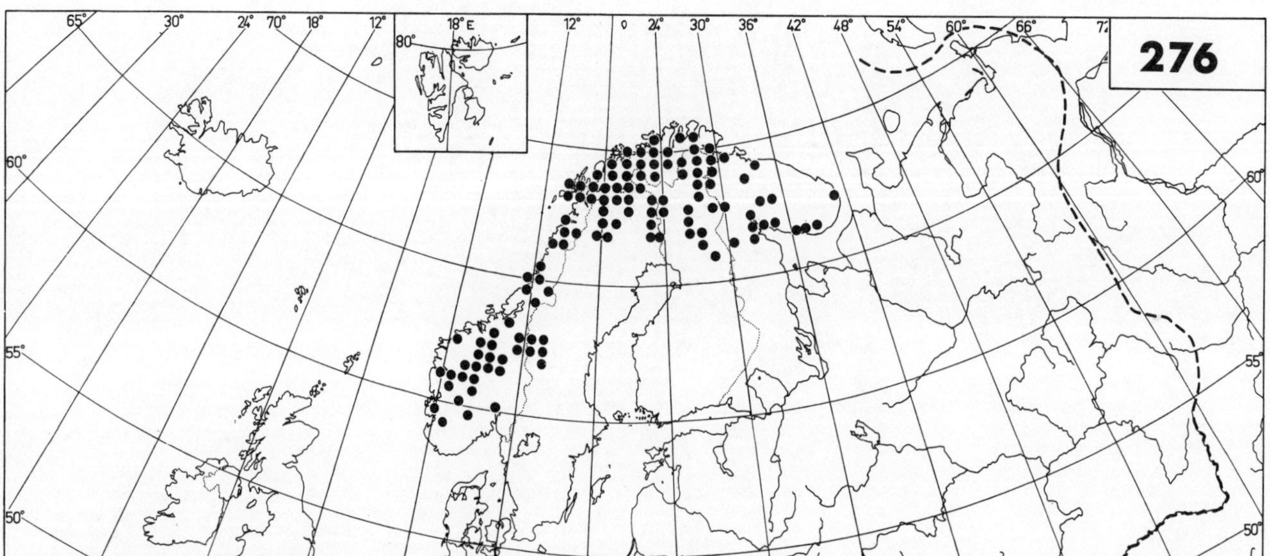

Alnus incana subsp. **kolaënsis**

Alnus incana (L.) Moench — Map 275.

Notes. Besides the map for the collective taxon, a separate map is given to show the approximate area of subsp. *kolaënsis*. *Be, ?Gr, *Ho and Rs(W, E) added (not given in Fl. Eur.).
Total range. Schmucker 1942: map 63; S. J. Sokolov *et al.*, Derev'ja i kustarniki SSSR 2: 337, map 93 (1951); Čermák 1955: 79; Hegi 1957: 176; MJW 1965: map 120a; Fenaroli 1967: 118; Hultén Alaska 1968: 370; Hultén CP 1971: map 126; O. Gjaerevoll, Plantegeografi: 81 (Trondheim — Oslo — Bergen — Tromsø 1973).

A. incana subsp. **incana**

Notes. Not mapped separately. »Throughout the range of the species» (Fl. Eur.).

A. incana subsp. **kolaënsis** (Orlova) Á. & D. Löve — Map 276.

Alnus kolaënsis Orlova; A. incana var. virescens Wahlenb.

Notes. Fe, No, Rs(N), Su. »Arctic Europe» (Fl. Eur.).
Total range. Endemic to Europe (although not marked as such in Fl. Eur.).

Alnus cordata (Loisel.) Loisel. — Map 277.

Alnus cordifolia Ten.

Notes. Al added (not given in Fl. Eur.), Sa omitted (?Sa in Fl. Eur.).
Total range. Endemic to Europe.

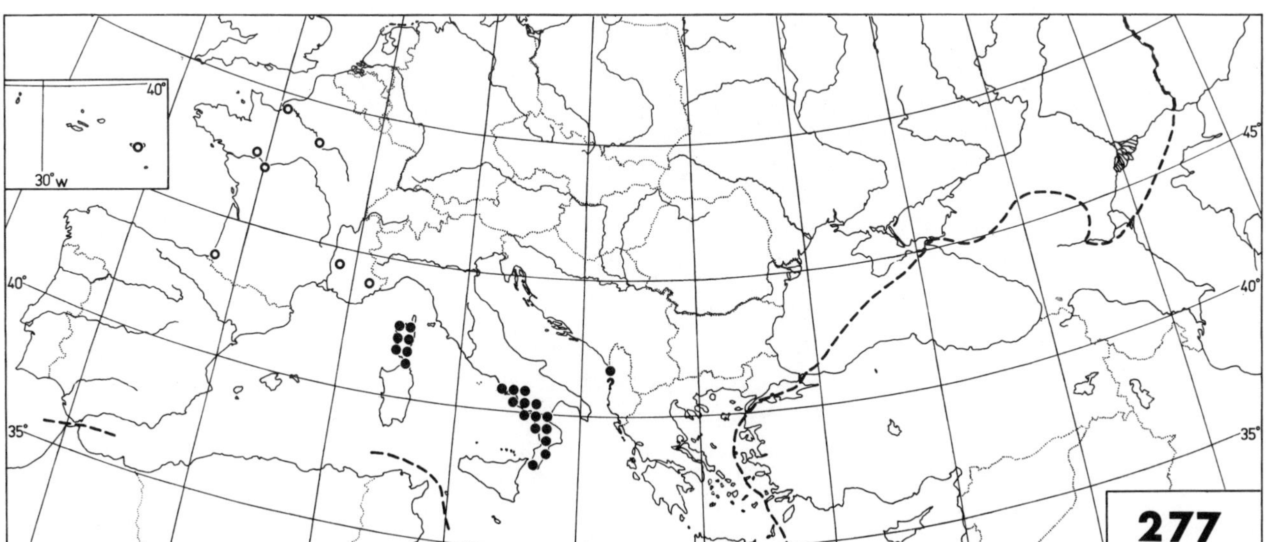

Alnus cordata

CORYLACEAE

Carpinus betulus L. — Map 278.

Notes. Rs(B) added (not given in Fl. Eur.). For the rest of the range within Rs, only the eastern limit is indicated.

Total range. Schmucker 1942: map 80; Hegi 1957: 184; MJW 1965: map 118a; Fenaroli 1967: 124; J. Jentys-Szaferowa, Monogr. Bot. 32: 203 (1970).

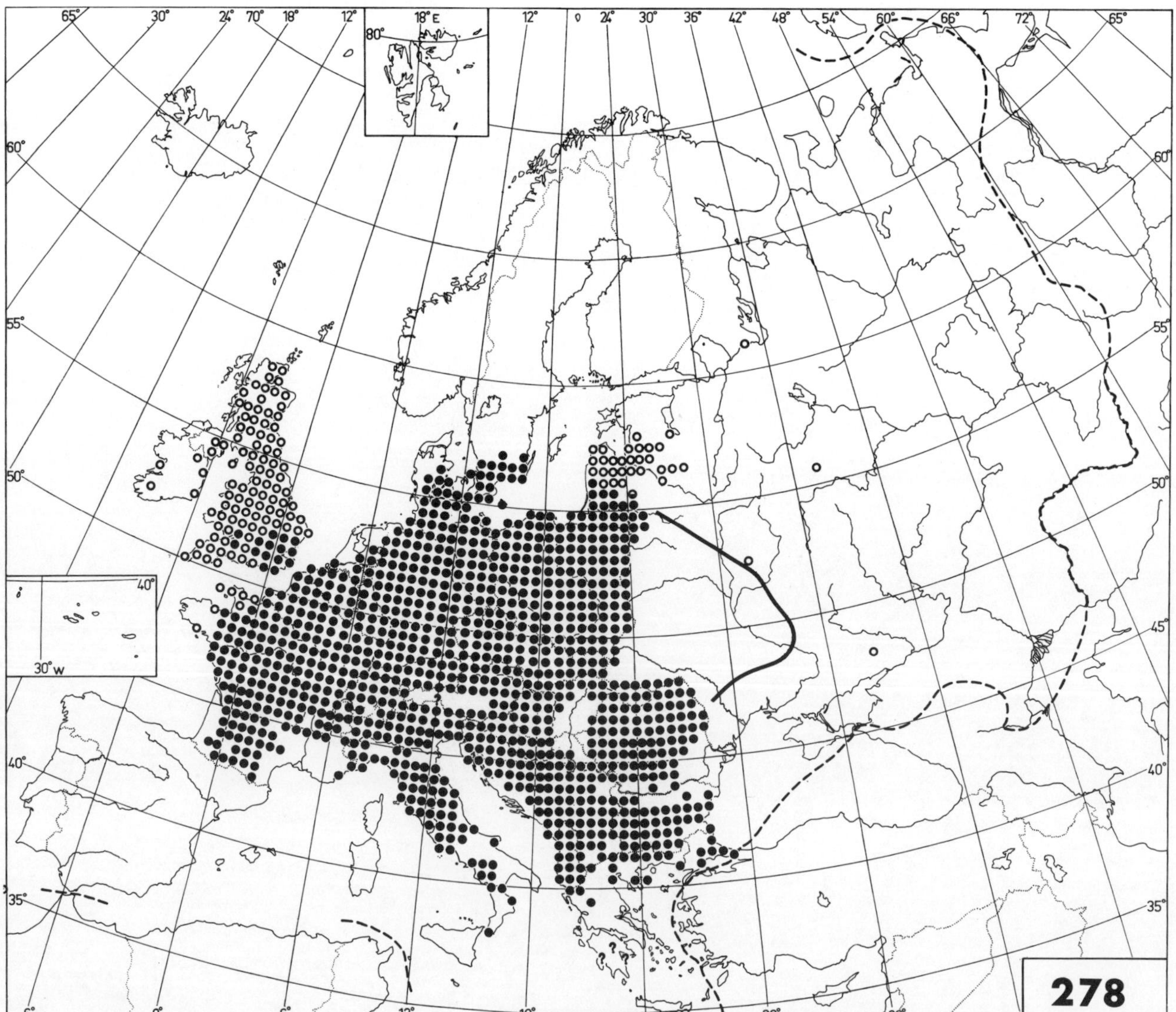

Carpinus betulus Line = E. limit of continuous occurrence

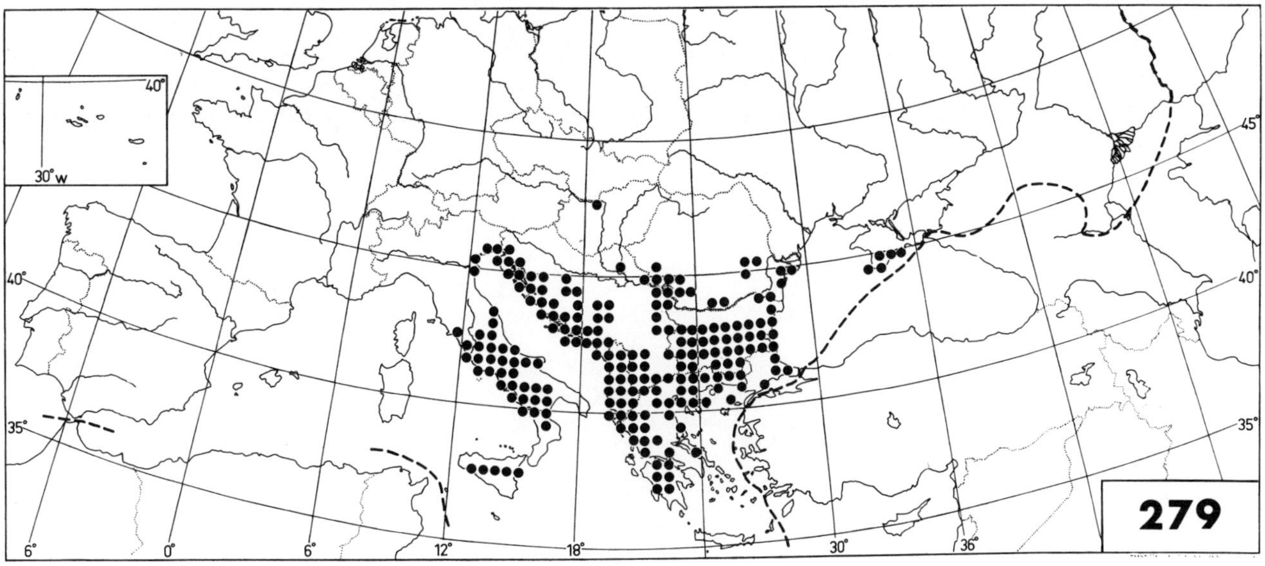

Carpinus orientalis

Carpinus orientalis Miller — Map 279.

Carpinus duinensis Scop.

Total range. Schmucker 1942: map 80; Čermák 1955: 67; Hegi 1957: 184; MJW 1965: map 118a; Fenaroli 1967: 128.

Ostrya carpinifolia Scop. — Map 280.

Ostrya virginiana subsp. carpinifolia (Scop.) Briq.

Notes. Extinct in Hu (?Hu in Fl. Eur.), Tu confirmed (?Tu in Fl. Eur.).
Total range. Schmucker 1942: maps 81—82; Čermák 1955: 67 and 77; Fenaroli 1967: 129.

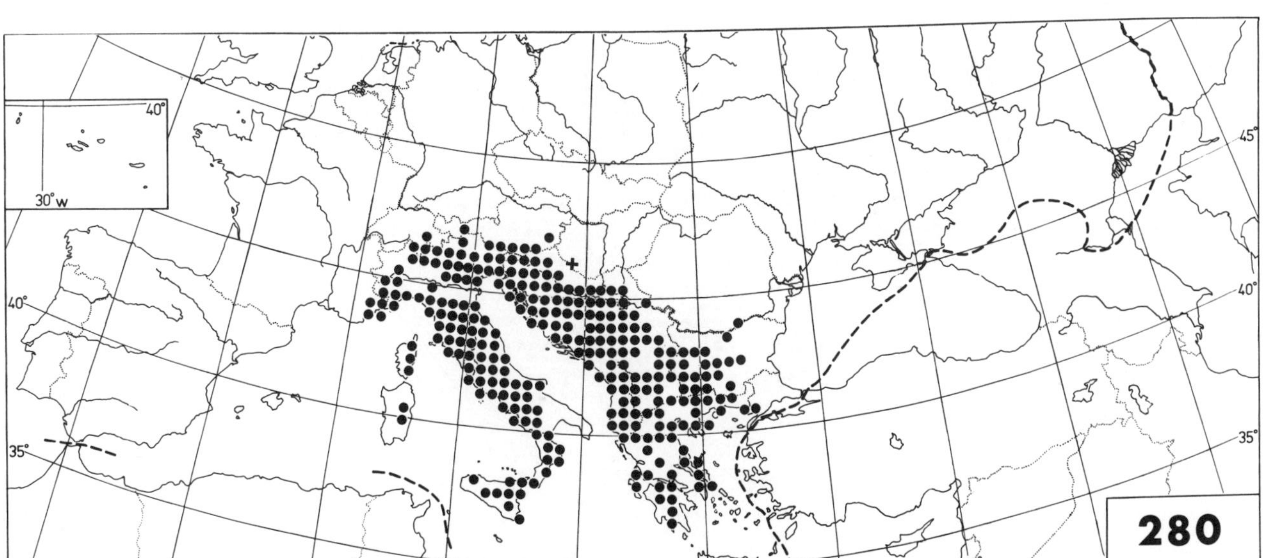

Ostrya carpinifolia

Corylus avellana L. — Map 281.

Notes. Native status doubtful in Cr (?Cr in Fl. Eur.).

Total range. Schmucker 1942: map 86; Čermák 1955: 82; A. Saxer, Beitr. Geobot. Landesaufn. Schweiz 36: 69 (1955); MJW 1965: map 118c; Fenaroli 1967: 134; M. Gostyńska-Jakuszewska & J. Zieliński, Atlas Distrib. Trees Shrubs Poland 18: map 24 (1976).

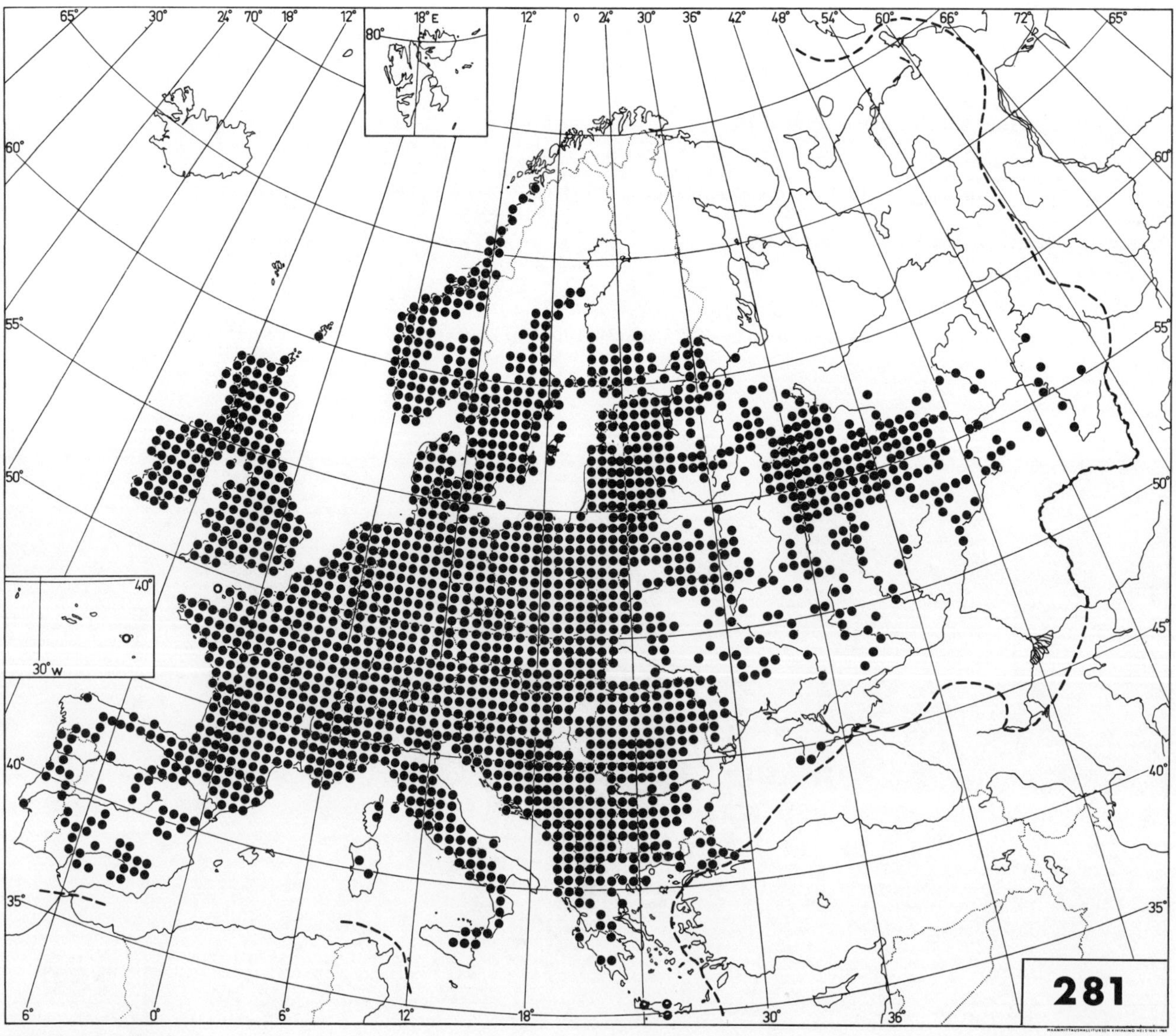

Corylus avellana

Corylus colurna L. — Map 282.

Total range. Schmucker 1942: map 86; Čermák 1955: 82; A. Saxer, Beitr. Geobot. Landesaufn. Schweiz 36: 69 (1955); MJW 1965: map 118c.

Corylus maxima Miller — Map 283.

Corylus tubulosa Willd.

Notes. Tu omitted (?Tu in Fl. Eur.).

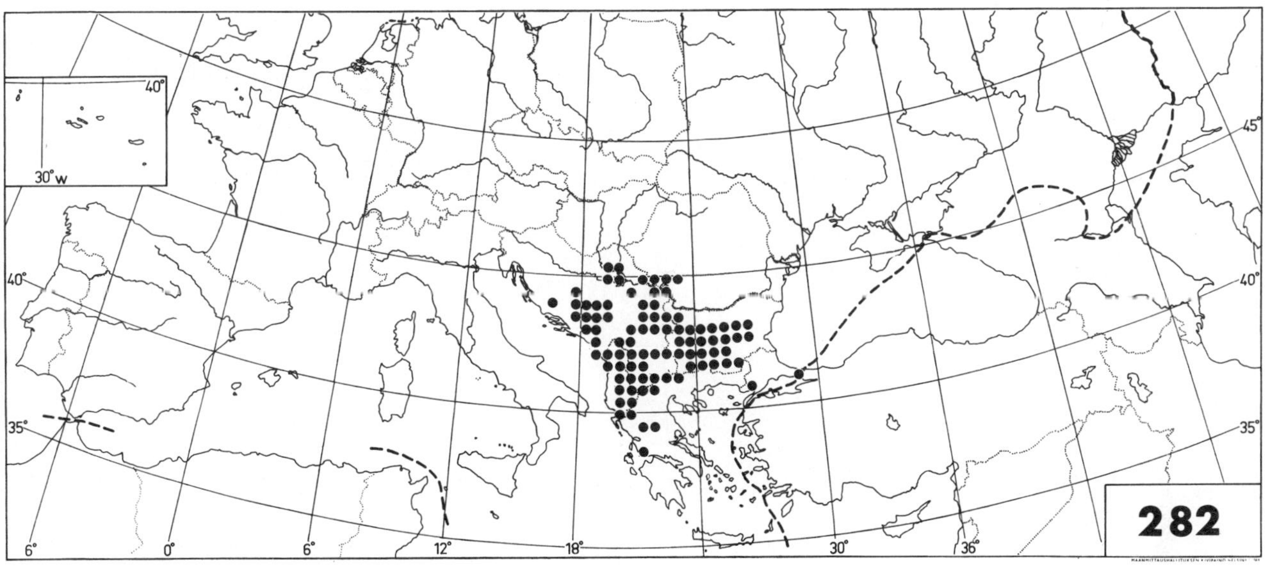

Corylus colurna

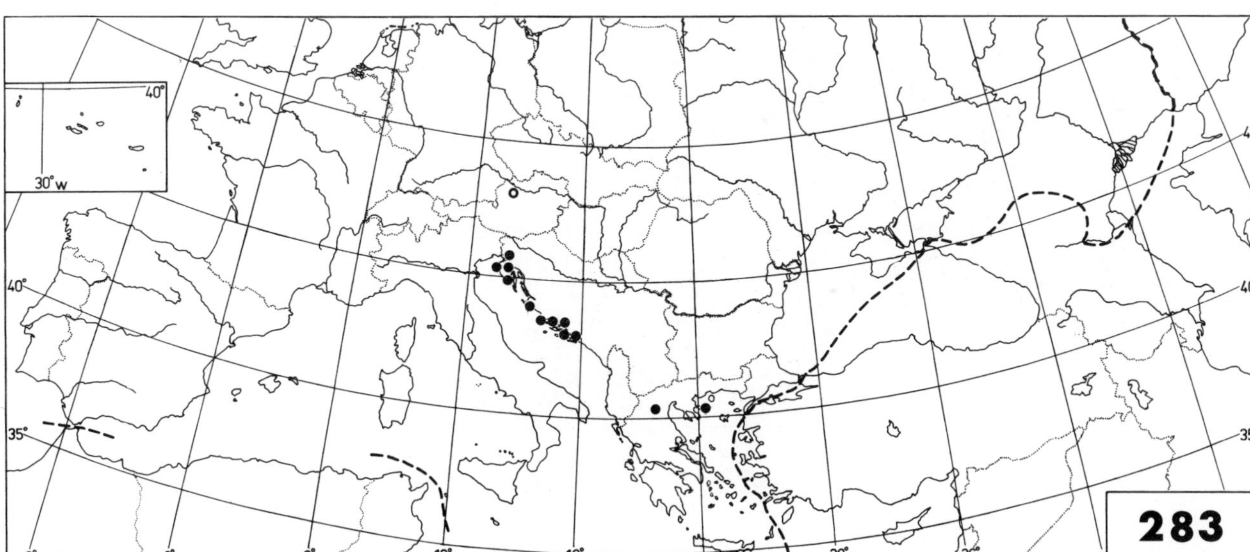

Corylus maxima

FAGACEAE

Fagus sylvatica L. — Map 284.

Taxonomy. Intermediates between *Fagus sylvatica* and *F. orientalis* called *F. moesiaca* (K. Maly) Czecz. are given not only for areas where the two species are sympatric, but through the Balkan peninsula. They have not been mapped separately.

Notes. Introductions not mapped. Rs(B) added (not given in Fl. Eur.). The distribution of the two species in Gr is according to Ch. Moulopoulos »The beech woods of Greece. I. The beech species and their distribution in Greece.» (Thessaloniki 1965).

Total range. Endemic to Europe.

Fagus orientalis Lipsky — Map 285.

Incl. Fagus taurica Popl.

Taxonomy. See under *Fagus sylvatica.*

Notes. Native status doubtful in Rm (given as native in Fl. Eur.), Rs(E) added (not given in Fl. Eur.).

Total range. L. Lämmermeyr, Die Pflanzenareale 1 (2): map 18 (1926); Schmucker 1942: maps 87—88; Čermák 1955: 67; A. Saxer, Beitr. Geobot. Landesaufn. Schweiz 36: 75 (1955); Hegi 1957: 203; MJW 1965: maps 120d and 121a.

Castanea sativa Miller — Map 286.

Castanea vulgaris Lam.

Notes. Al, Au, Bu, *Co, *Cr, *Cz, *Ga, Gr, *Hs, Hu, It, Ju, Sa, Si, Tu. Concerning the age of the occurrences in SW. Germany, see W. Lang, Pollichia 16—18 (1969—71). »S. Europe, from Italy eastwards, and extending northwards to Hungary; extensively planted for its nuts and timber and naturalized in parts of W., C. & N. Europe» (Fl. Eur.).

Total range. Schmucker 142: map 118; Čermák 1955: 67 and 77; MJW 1965: map 121b; Fenaroli 1967: 142.

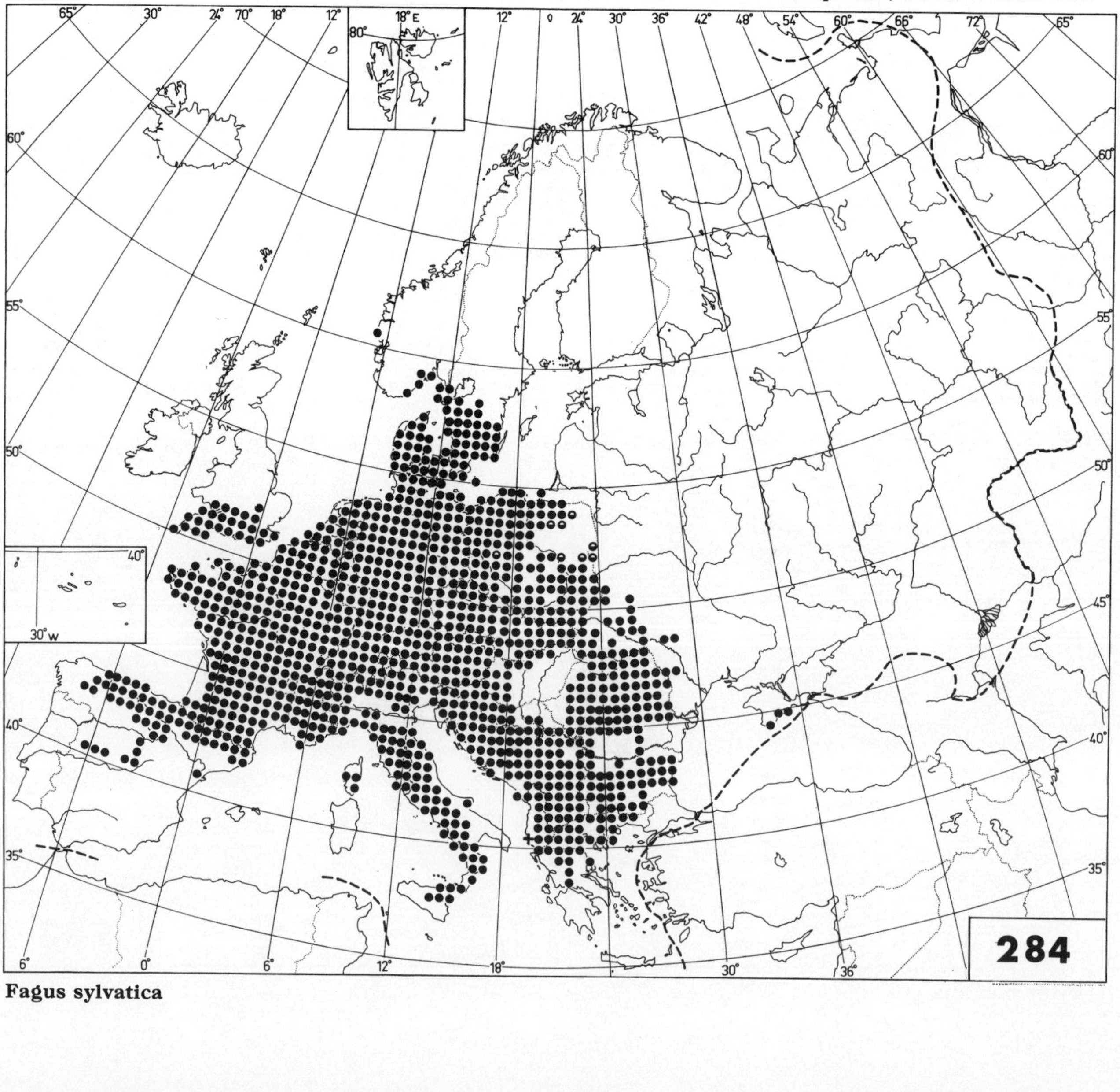

Fagus sylvatica

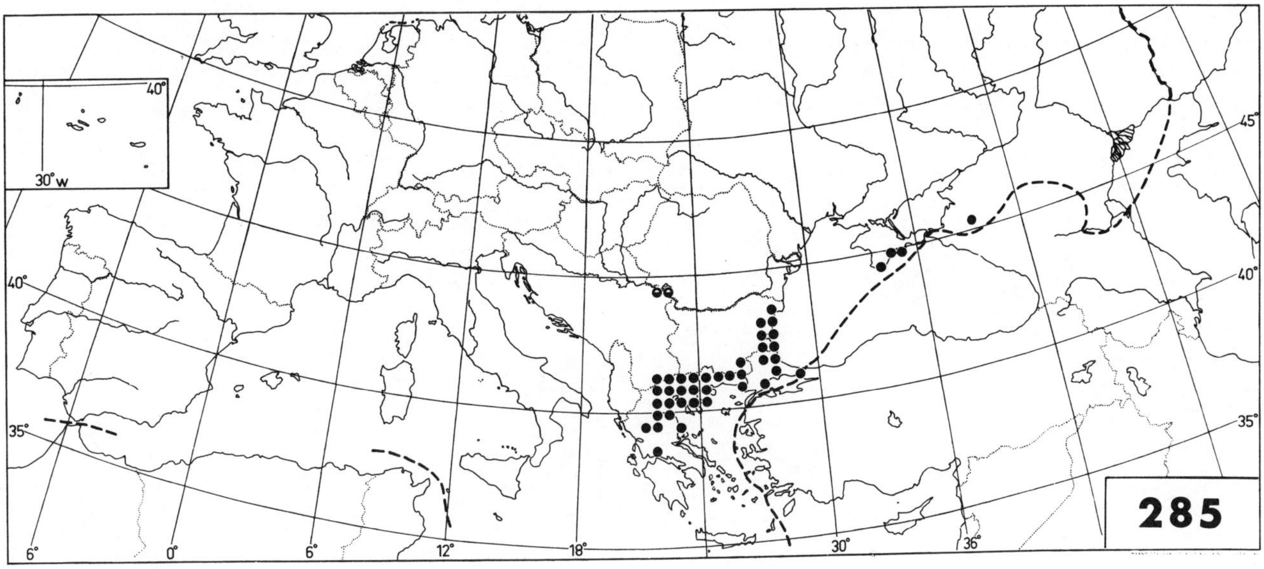

Fagus orientalis

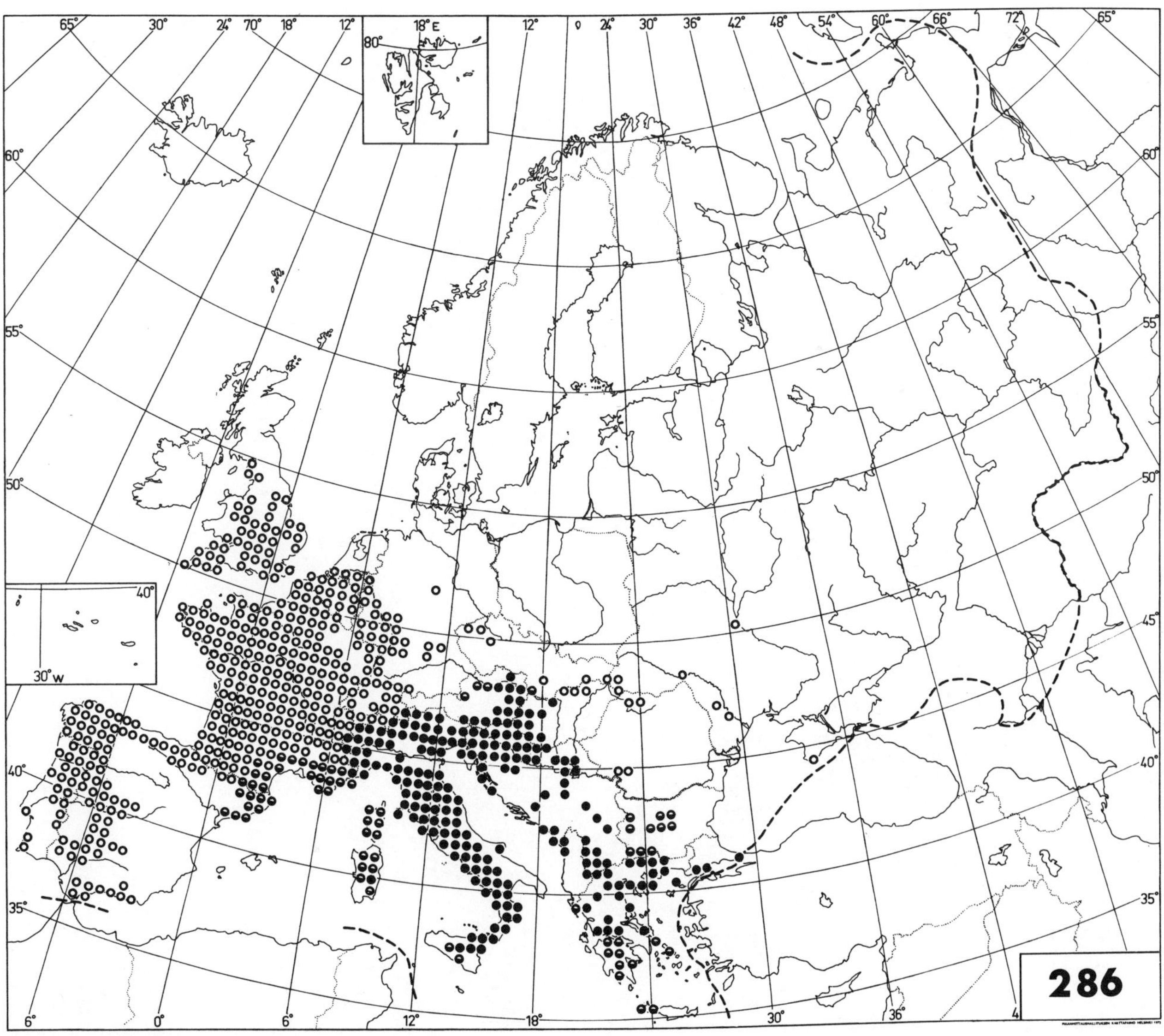

Castanea sativa

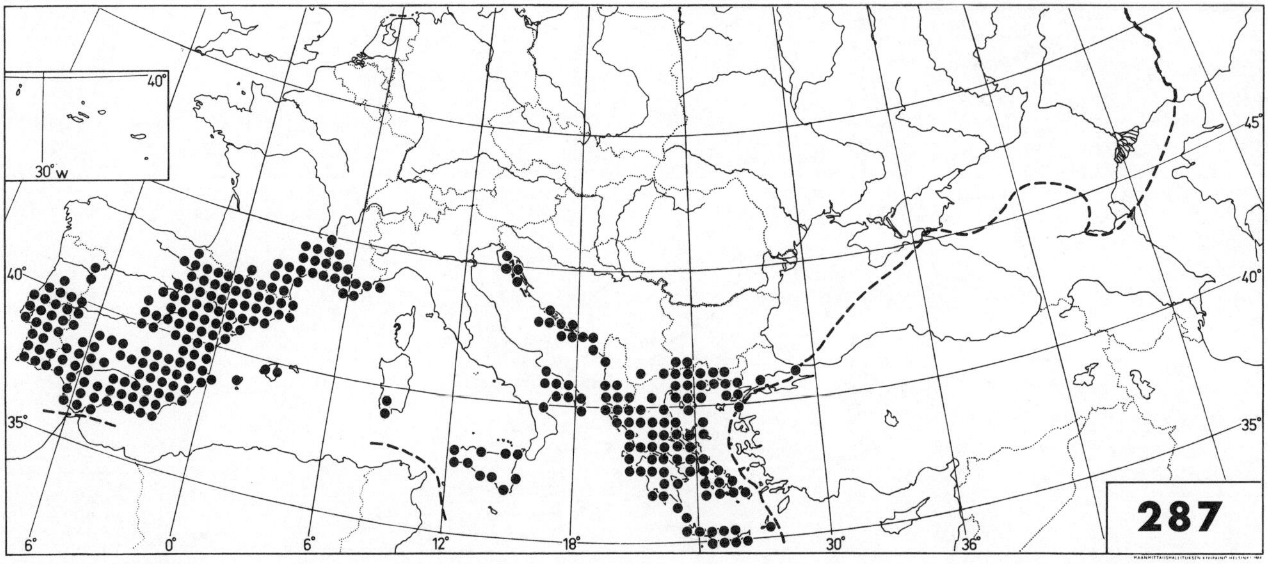

Quercus coccifera

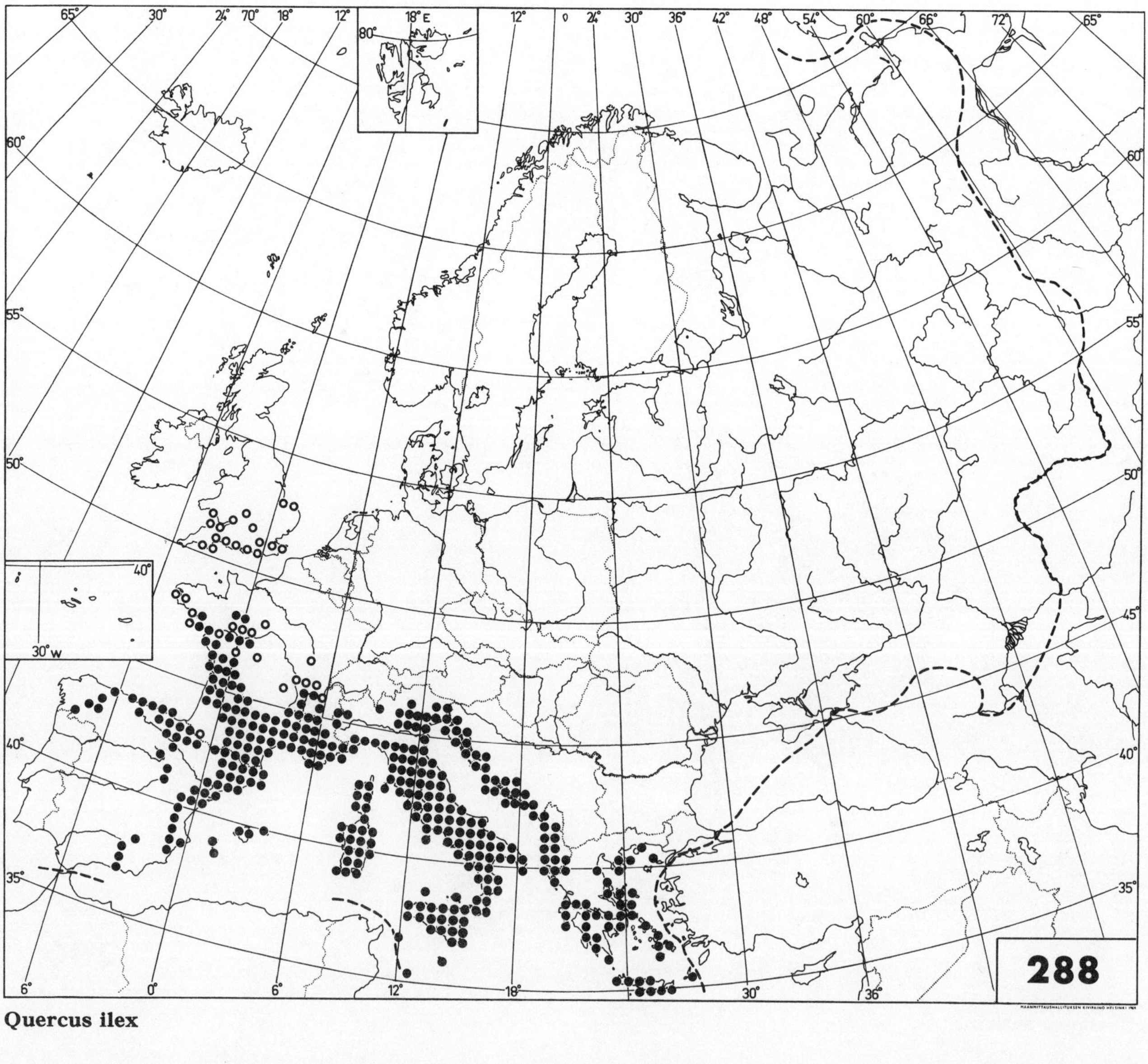

Quercus ilex

Quercus coccifera L. — Map 287.

Quercus calliprinos Webb; Q. mesto Boiss.; Q. pseudococcifera Webb

Notes. Occurrence in Co doubtful (given in Fl. Eur.).

Total range. Schmucker 1942: map 99; Čermák 1955: 78; MJW 1965: map 122d; Fenaroli 1967: 152; S. Sabato, Webbia 27: 518 (1972).

Quercus ilex L. — Map 288.

Quercus avellaeformis Colmeiro & Boutelou; Q. gracilis Lange; Q. smilax L.

Notes. Lu and Tu omitted (given in Fl. Eur.).

Total range. Schmucker 1942: map 94; MJW 1965: map 122c; Fenaroli 1967: 148; H. Meusel & R. Schubert, Flora 160: 170 (1971) (all four incl. *Quercus rotundifolia*); S. Sabato, Webbia 27: 518 (1972). Cf. G. Lang, Feddes Repert. 81: 325 (1970).

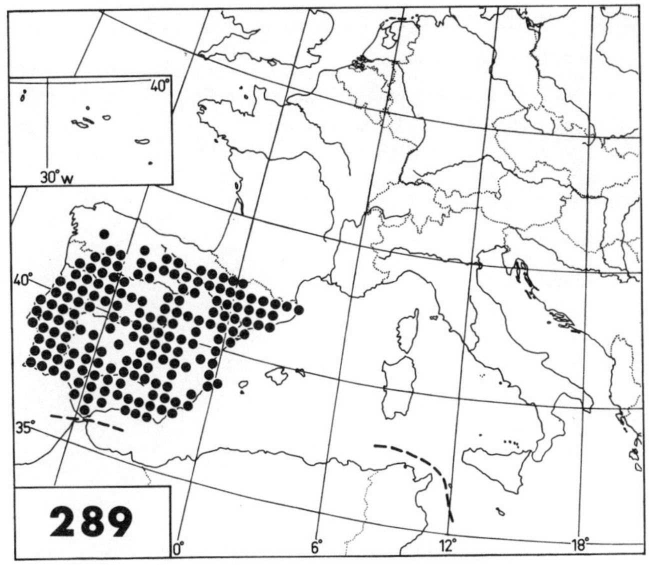

Quercus rotundifolia

Quercus rotundifolia Lam. — 289.

Quercus ballota Desf.; Q. ilex subsp. rotundifolia (Lam.) T. Morais

Taxonomy. C. Sáenz de Rivas, Anal. Inst. Bot. Cavanilles 25: 243—262 (1968).

Notes. Ga omitted (given in Fl. Eur.).

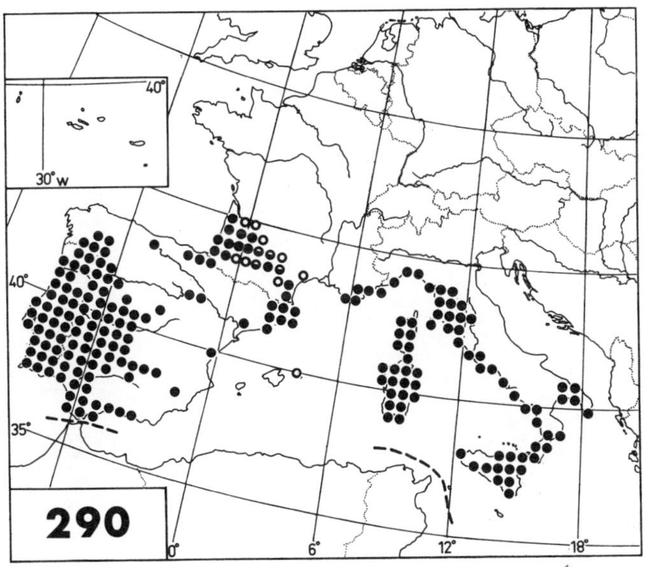

Quercus suber

Quercus suber L. — Map 290.

Incl. Quercus occidentalis Gay

Notes. Ju omitted (given in Fl. Eur.).

Total range. Schmucker 1942: map 97; L. Fenaroli, Fl. Mediterranea I: 60 (Milano 1962); Fenaroli 1967: 151.

Quercus trojana Webb — Map 291.

Quercus macedonica DC.

Total range. Schmucker 1942: map 96 (gives the species as endemic to Europe); Fenaroli 1967: 171; P. Fukarek, Feddes Repert. 81: 166 (1970).

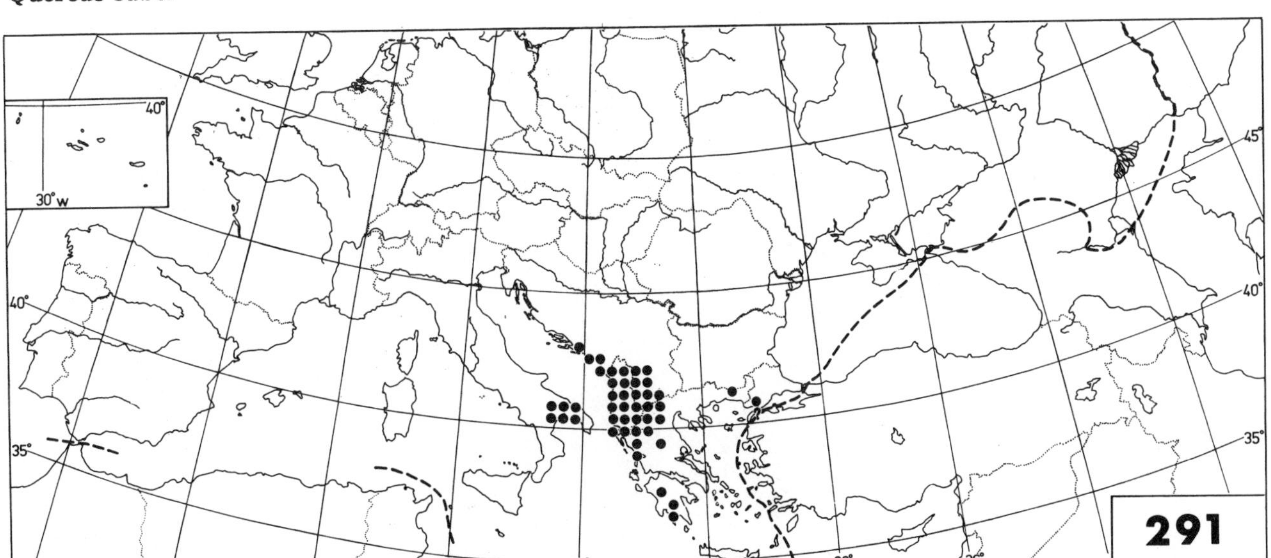

Quercus trojana

Quercus macrolepis Kotschy — Map 292.

Quercus aegilops auct.

Notes. The species was formerly extensively cultivated and the exact boundaries of the true native range remain uncertain. Not likely to be indigenous in Cr or in Ju (Cr, ?Ju in Fl. Eur.).

Total range. Schmucker 1942: map 90; Fenaroli 1967: 169.

Quercus crenata Lam. — Map 293.

Quercus pseudosuber G. Santi

Taxonomy. Although Fl. Eur. gives *Quercus crenata* as a probable hybrid, M. Barbero *et al.*, Bull. Soc. Bot. France 119: 121—125 (1972), wish it to be recognized as a distinct species, this being in accordance with the treatment by K. H. Rechinger in Hegi 1957: 226—227.

Notes. Ga, It, Ju, Si. »S. Europe from France to Albania» (Fl. Eur.).

Total range. Schmucker 1942: map 97.

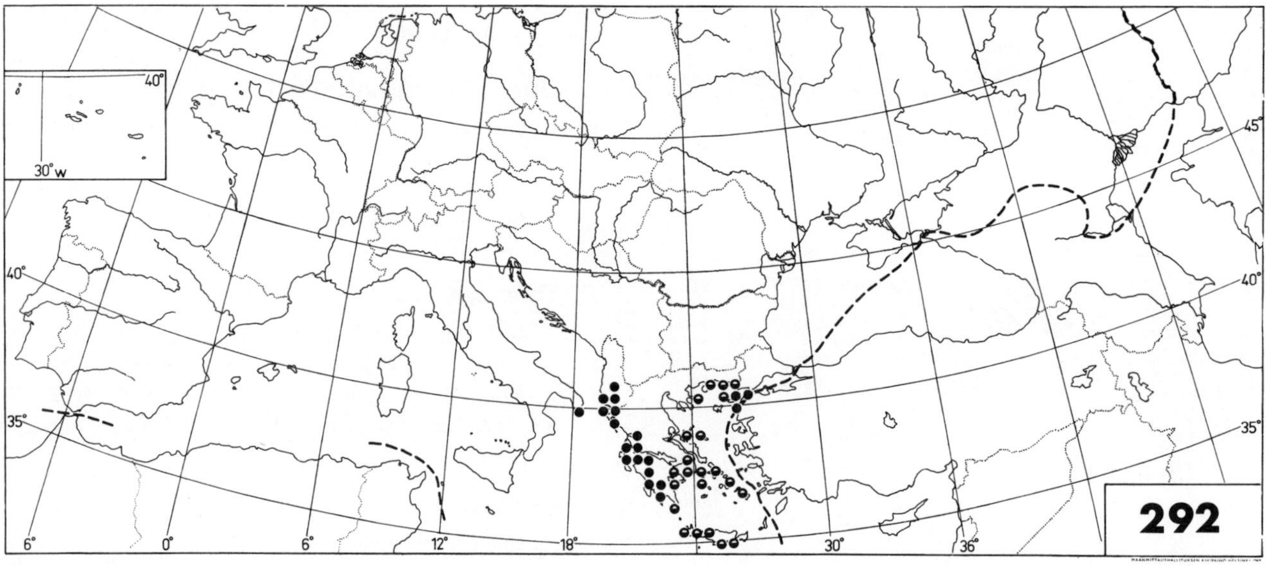

Quercus macrolepis

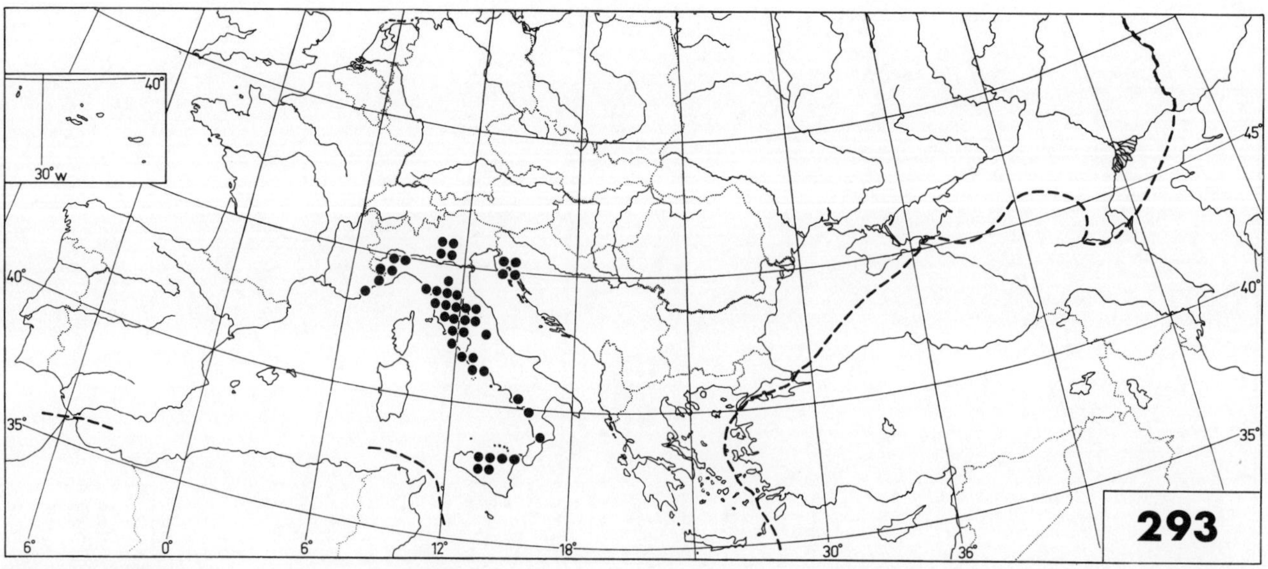

Quercus crenata

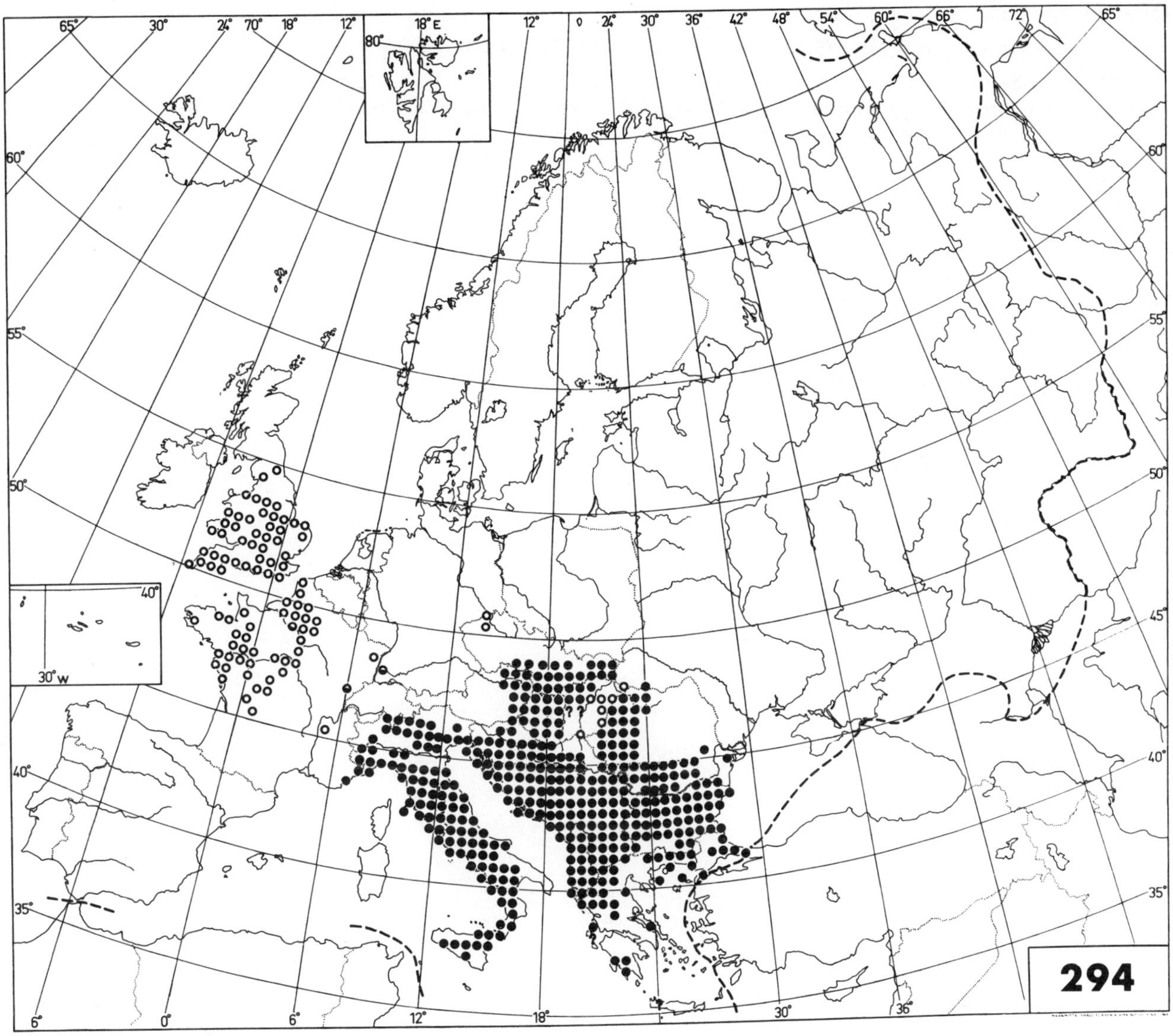

Quercus cerris

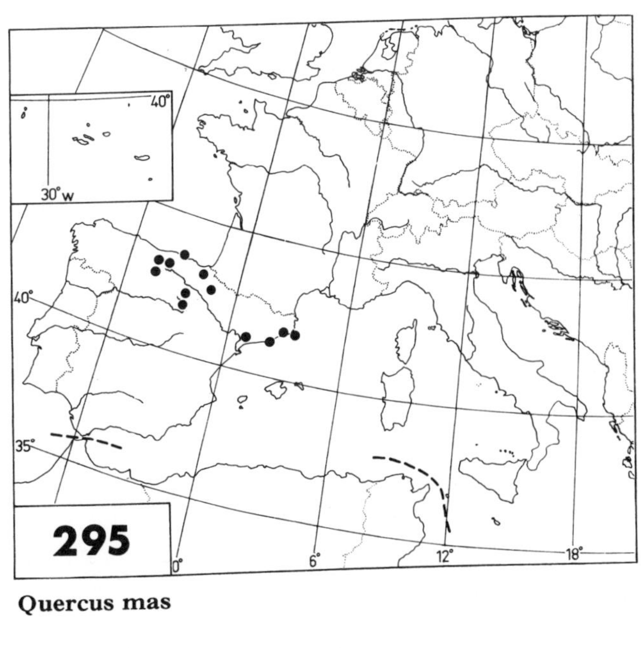

Quercus mas

Quercus cerris L. — Map 294.

Incl. Quercus austriaca Willd.; Q. cerris subsp. austriaca (Willd.) O. Schwarz; Q. thracica Stefanov & Nedjalkov

Taxonomy. Concerning the taxonomical status of *Quercus thracica*, see B. Stefanoff, Acta Bot. Acad. Sci. Hung. 19: 343—357 (1973).

Notes. Doubtful in Cr (given in Fl. Eur.), doubtfully native in Ge (given as native in Fl. Eur.).

Total range. Schmucker 1942: map 98; Čermák 1955: 69; MJW 1965: map 123a; Fenaroli 1967: 166.

Quercus mas Thore — Map 295.

Notes. See under *Quercus polycarpa*. No exact records available from Ga (given in Fl. Eur.).

Total range. Endemic to Europe.

Quercus polycarpa Schur — Map 296.

Quercus petraea subsp. polycarpa (Schur) Soó; Q. dshorochensis auct. non C. Koch

Taxonomy. See under *Quercus petraea.*

Notes. As it proved very difficult to make separate maps for *Quercus mas*, *Q. polycarpa*, *Q. dalechampii* and *Q. petraea* proper, a collective map for *Q. petraea* sensu lato is included. Au and Cz added (not given in Fl. Eur.).

Total range. MJW 1965: map 122a.

Quercus petraea (Mattuschka) Liebl. — Map 297.

Quercus sessiliflora Salisb.; Q. sessilis Ehrh.

Taxonomy. G. L. Menitsky, Nov. Syst. Pl. Vasc. (Leningrad) 1968: 87, gives *Quercus dalechampii* and *Q. polycarpa* in the synonymy of *Q. petraea* subsp. *medwediewii* (A. Camus) Menitsky.

Notes. See also the map for *Quercus petraea* sensu lato. Rs(E) added (not given in Fl. Eur.), Si omitted (given in Fl. Eur.).

Total range. Schmucker 1942: map 93; Čermák 1955: 69; Hegi 1957: 233; MJW 1965: map 122a; Fenaroli 1967: 159.

Quercus dalechampii Ten. — Map 298.

Quercus sessiliflora subsp. glomerata Jav. (non Q. glomerata Lam.)

Taxonomy. See under *Quercus petraea.*

Notes. See under *Quercus polycarpa.* Cz added (not given in Fl. Eur.).

Total range. Schmucker 1942: map 102; L. Fenaroli in Conosci L'Italia II. La Flora, p. 211 (Milano 1958).

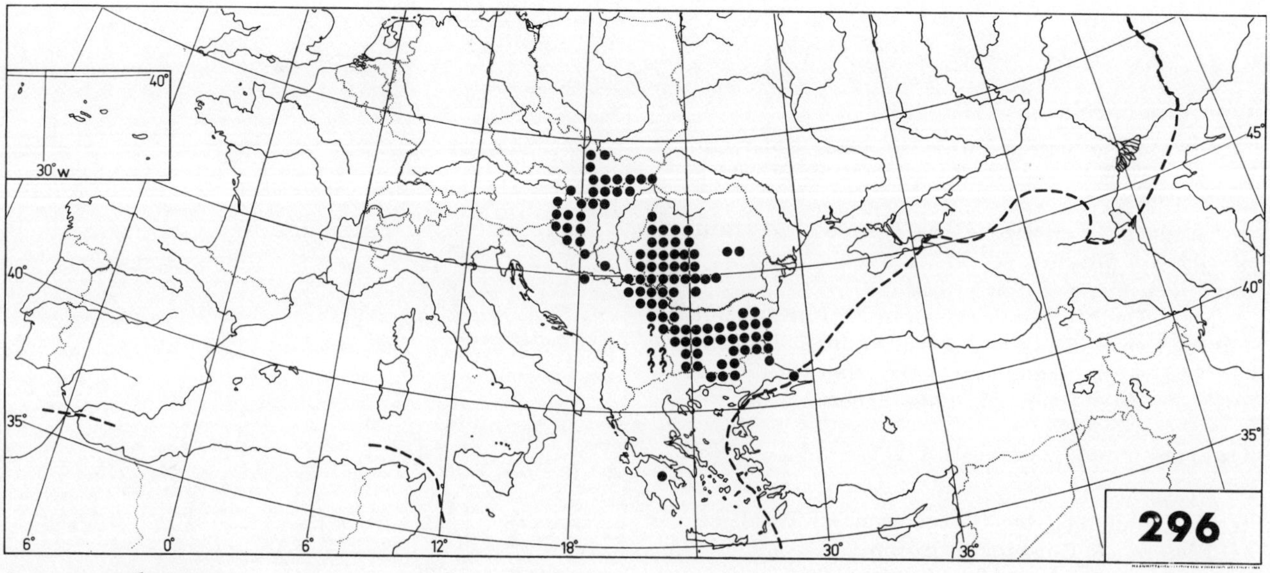

Quercus polycarpa

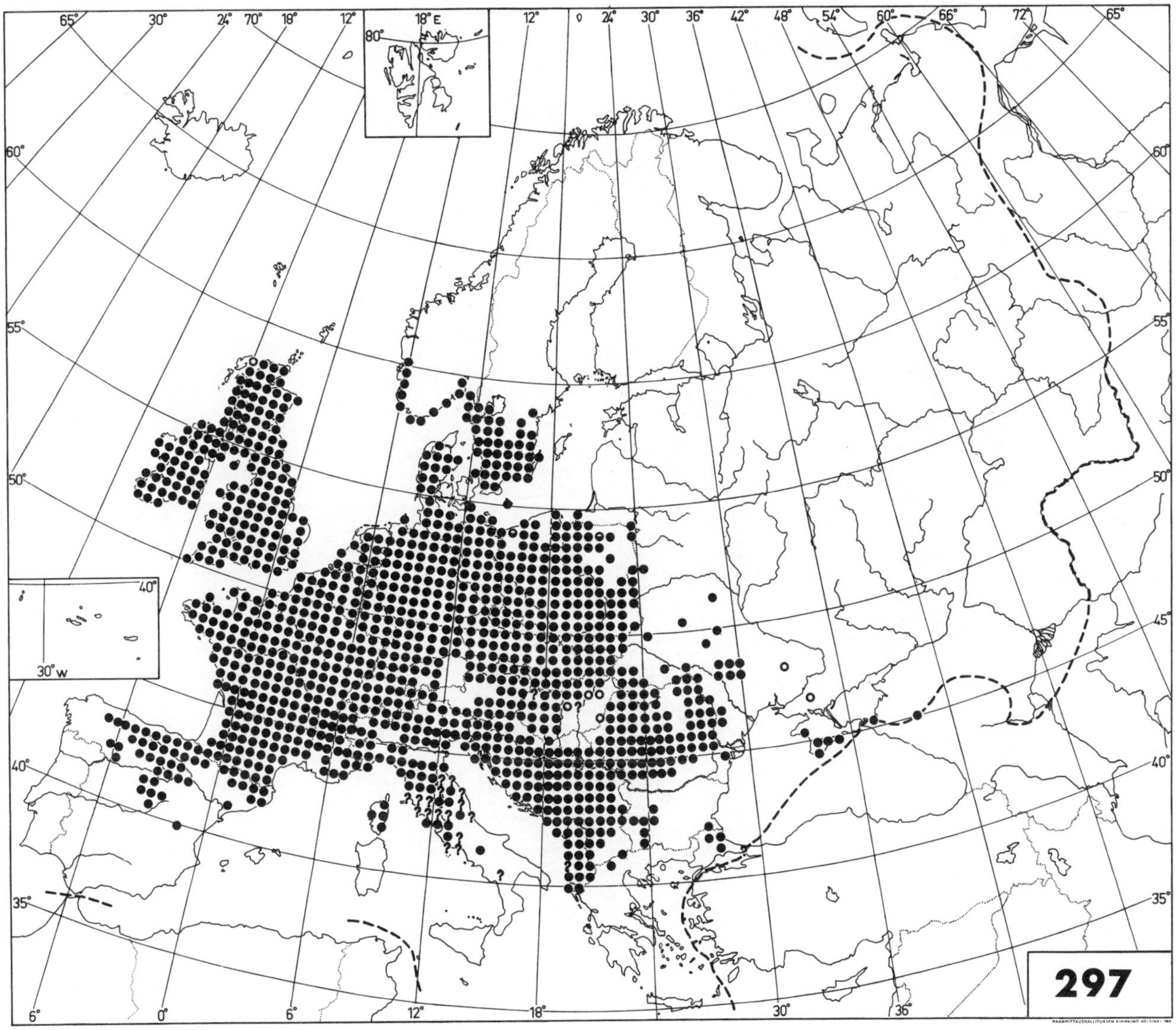

Quercus petraea

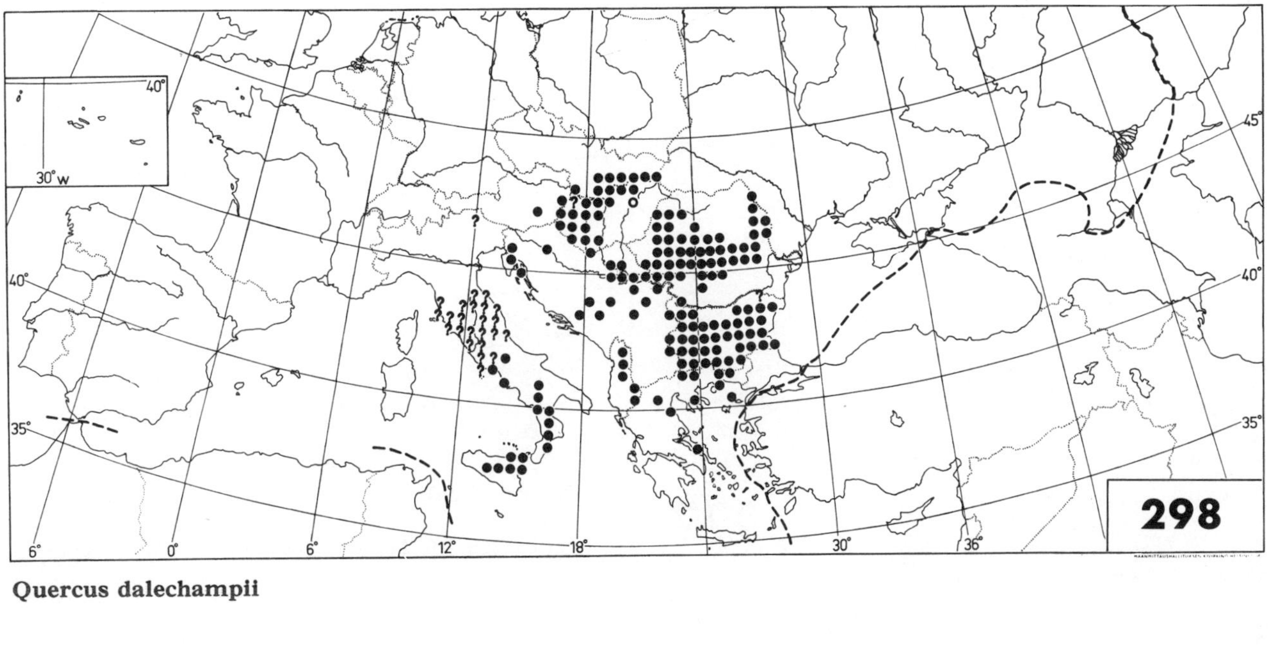

Quercus dalechampii

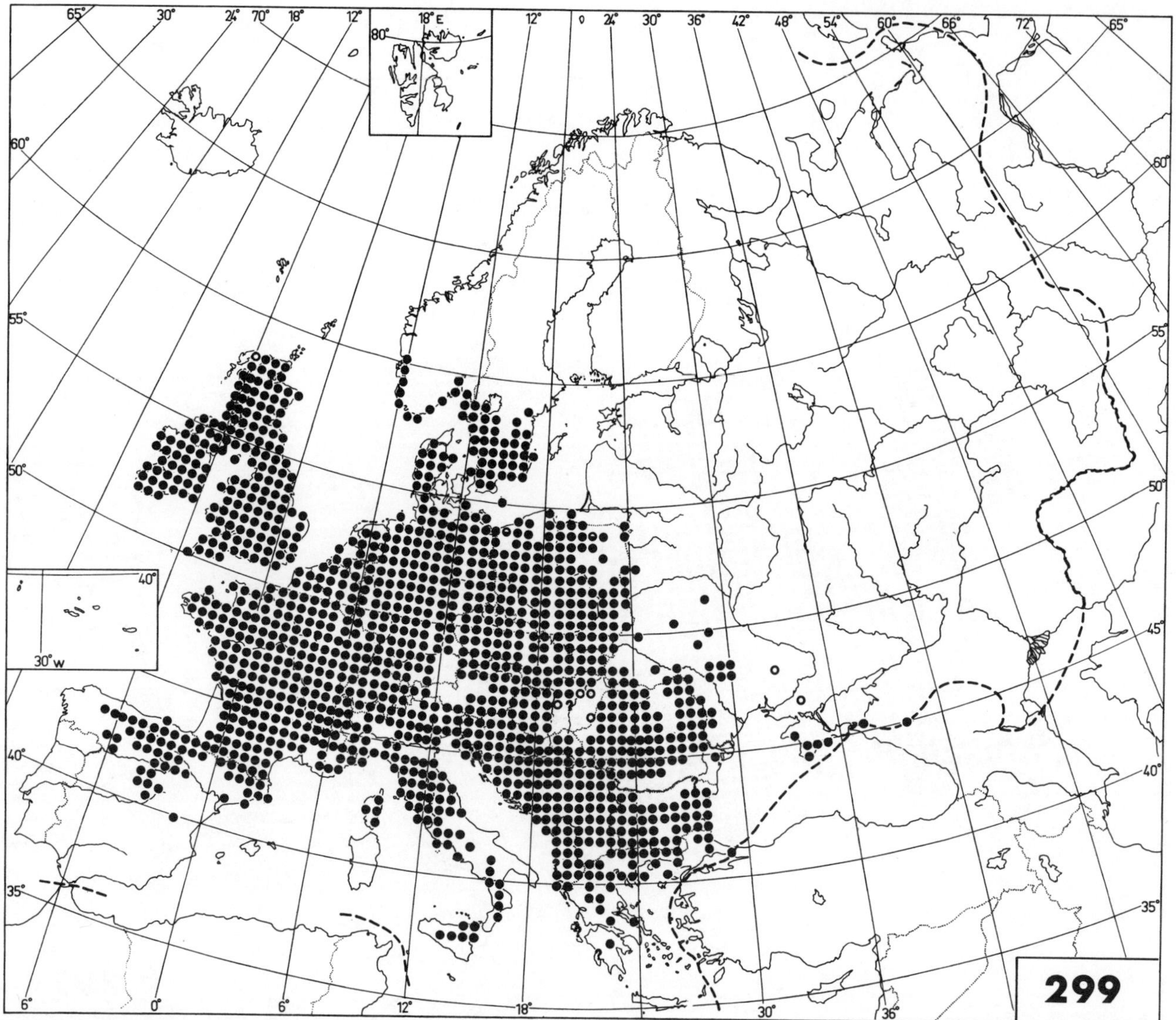

Quercus petraea sensu lato

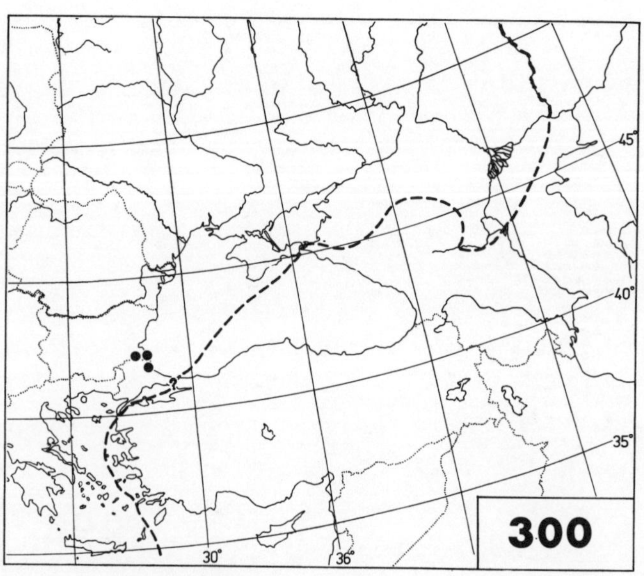

Quercus hartwissiana

Quercus petraea (Mattuschka) Liebl., sensu lato — Map 299.

Quercus dalechampii + Q. mas + Q. petraea s.str. + Q. polycarpa

Total range. J. Jentys-Szaferowa, Monogr. Bot. 32: 203 (1970).

Quercus hartwissiana Steven — Map 300.

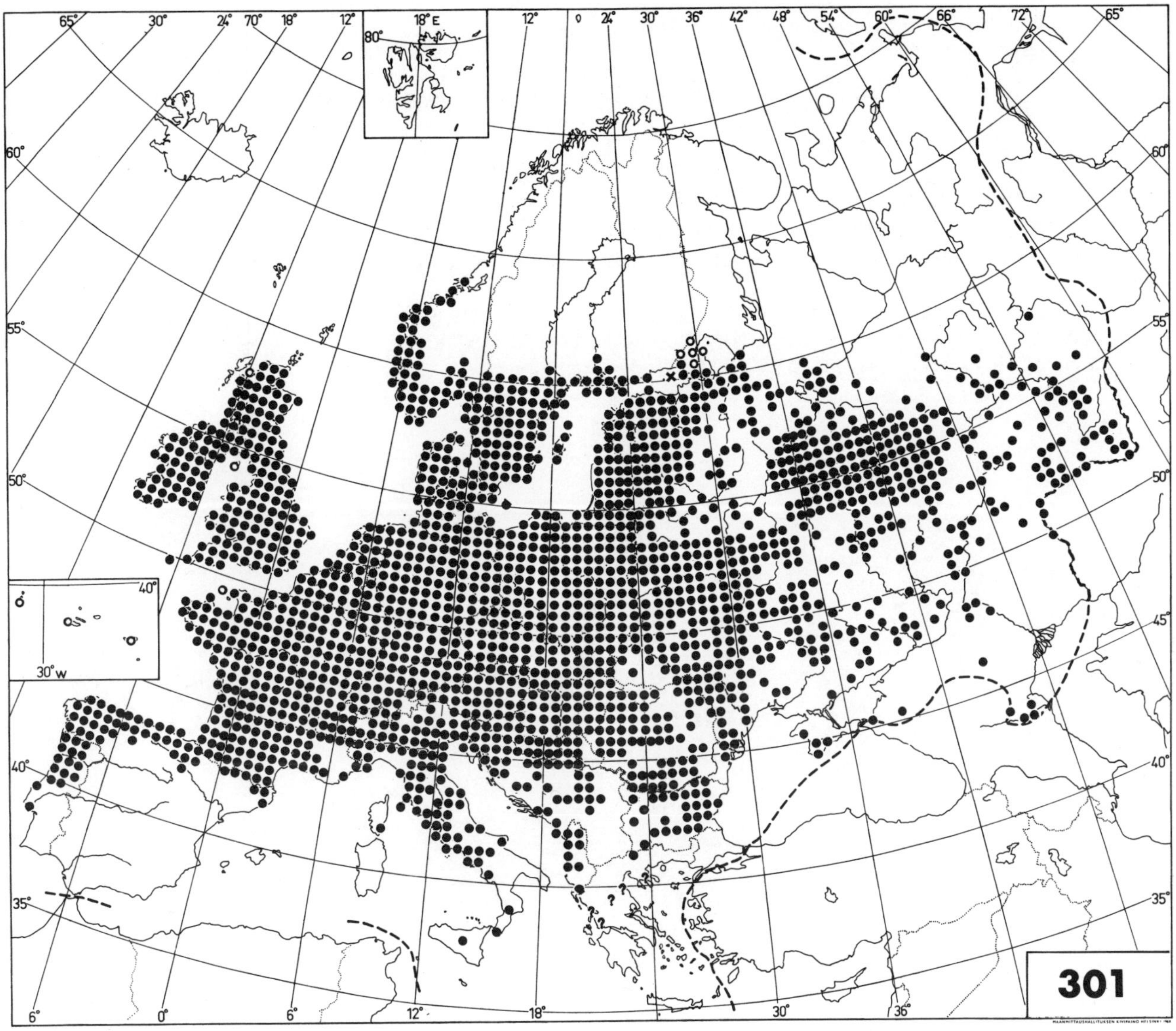

Quercus robur subsp. **robur**

Quercus robur L.

Quercus pedunculata Ehrh. ex Hoffm.

Total range. Schmucker 1942: map 93; Čermák 1955: 69 and 78; Hegi 1957: 233; MJW 1965: map 121d; Fenaroli 1967: 156; J. Jentys-Szaferowa, Monogr. Bot. 32: 203 (1970).

Q. robur subsp. **robur** — Map 301.

Incl. Quercus estremadurensis O. Schwarz and Q. robur subsp. puberula (Lasch) Weimarck

Taxonomy. Extensive introgression between *Quercus petraea* and *Q. robur* has been demonstrated in parts of NW. Europe, e.g. by J. E. Cousens, Watsonia 5: 273–286 (1963).

Notes. Co confirmed (?Co in Fl. Eur.), possibly present in Gr (according to old records of usually reliable authors; not given as present in Fl. Eur.), Sa omitted (given in Fl. Eur.).

Q. robur subsp. **brutia** (Ten.) O. Schwarz — Map 302.

Quercus brutia Ten.

Notes. Al, Gr and Ju added (not given in Fl. Eur.). The additions made in accordance with information from P. Fukarek (Sarajevo).

Total range. Endemic to Europe.

Quercus pedunculiflora C. Koch — Map 303.

Quercus haas auct. non Kotschy; Q. pedunculata subsp. pedunculiflora (C. Koch) Maire & Petitmengin

Taxonomy. G. L. Menitsky, Bot. Žurn. 51: 1245—1265 (1966), Nov. Syst. Pl. Vasc. (Leningrad) 1968: 81—94 (1968).

Notes. Ju added (not given in Fl. Eur.), no data available from Rs(K) (given in Fl. Eur.). Concerning possible occurrence in Cz, see D. Magic, Biológia (Bratislava) 30: 65—74 (1975).

Total range. Schmucker 1942: map 102; MJW 1965: map 121d.

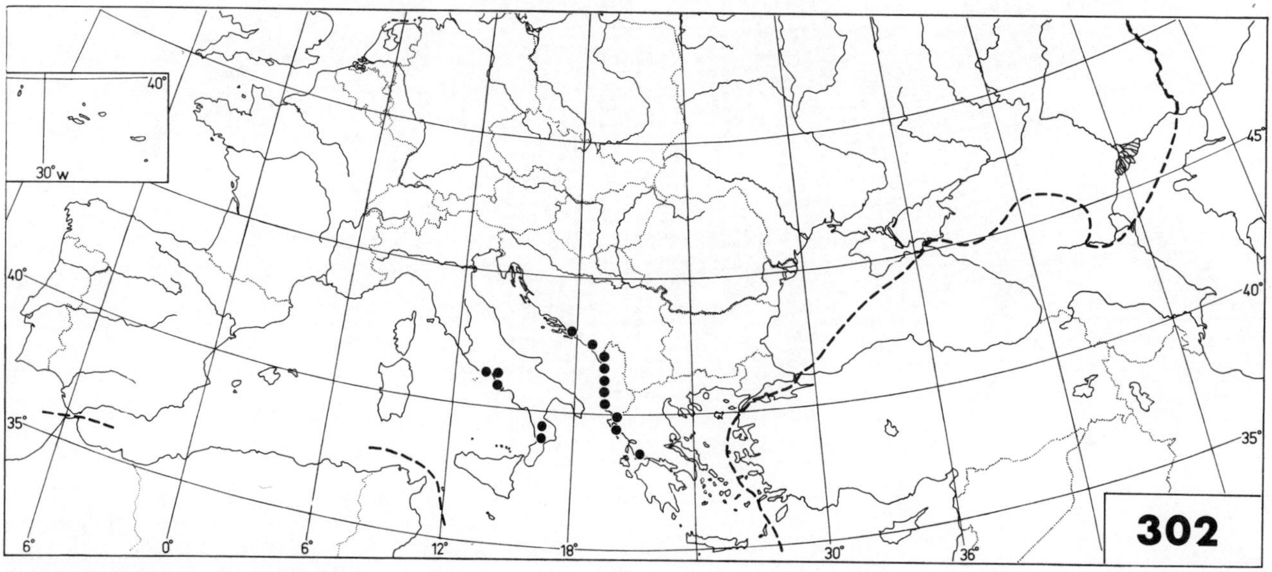

Quercus robur subsp. **brutia**

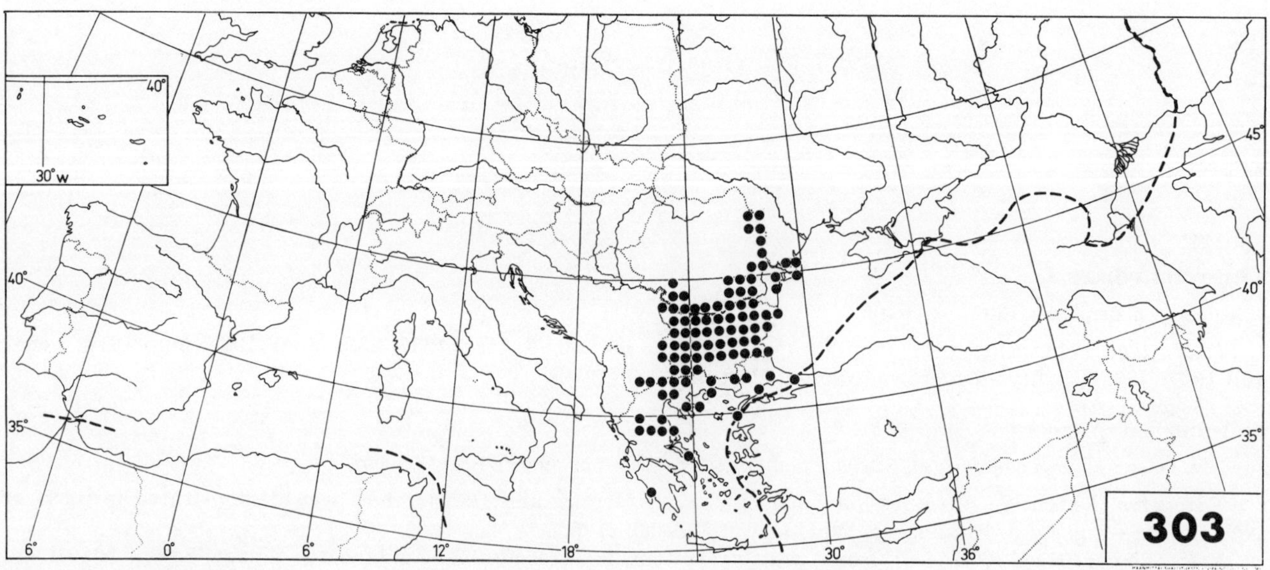

Quercus pedunculiflora

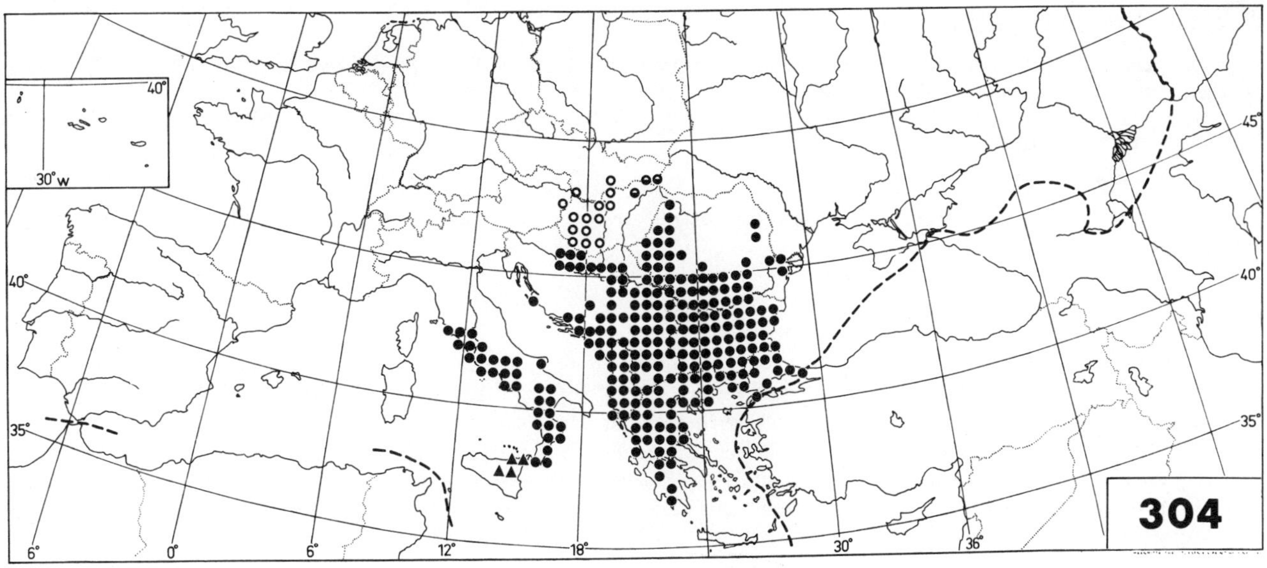

▲ = **Quercus sicula** • = **Q. frainetto**

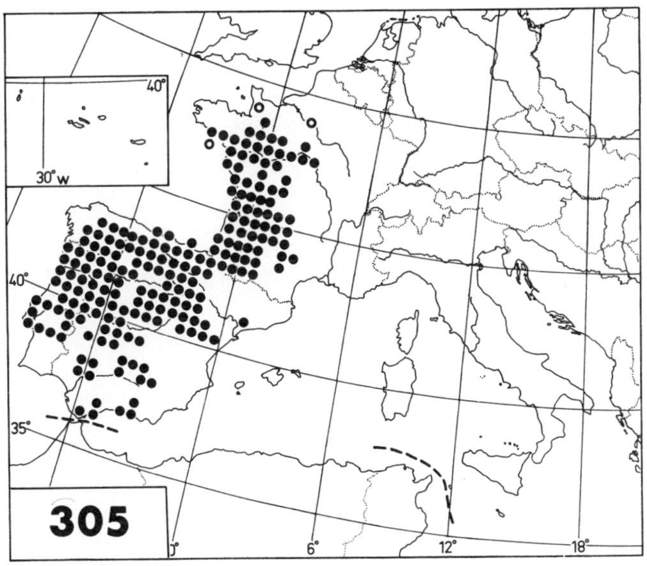

Quercus pyrenaica

Quercus sicula Borzi — Map 304.

Quercus borzii A. Camus

Total range. Endemic to Sicilia.

Quercus frainetto Ten. — Map 304.

Quercus conferta Kit.; Q. farnetto Ten.

Notes. *Cz added (not given in Fl. Eur.), native status doubtful in Hu (given as native in Fl. Eur.).

Total range. Schmucker 1942: maps 96 and 100; Fenaroli 1967: 160; P. Fukarek, Feddes Repert. 81: 166 (1970).

Quercus pyrenaica Willd. — Map 305.

Quercus toza Bast.

Notes. It omitted (given in Fl. Eur.).

Total range. Schmucker 1942: maps 95 and 100; H. Walter, Einführung in die Phytologie III (2): 174 (1954).

Quercus congesta C. Presl — Map 306.

Notes. Ga omitted (given in Fl. Eur.), It added (not given in Fl. Eur.).

Total range. Endemic to Europe.

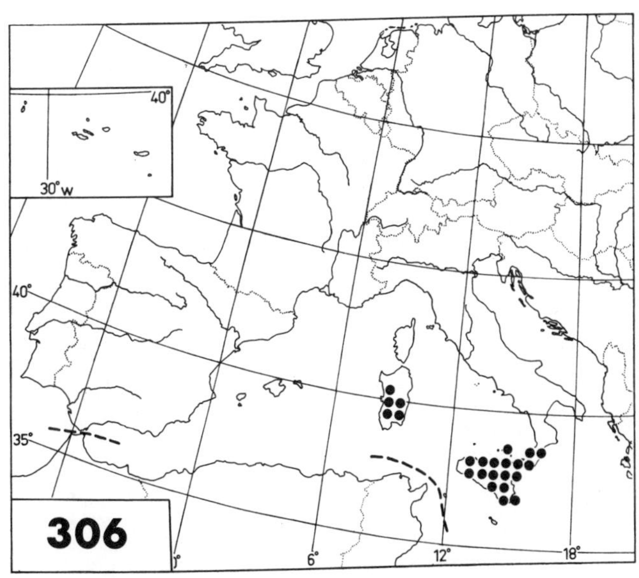

Quercus congesta

Quercus brachyphylla Kotschy

Notes. Here included in *Quercus pubescens*; see under this taxon. »W. & S. Greece; Aegean region» (Fl. Eur.).
Total range. Schmucker 1942: map 100.

Quercus virgiliana (Ten.) Ten.

Quercus apennina auct.

Notes. Not mapped separately; see under *Quercus pubescens.* »S. Europe, from Corse and Sardegna to the Black Sea» (Fl. Eur.).
Total range. Schmucker 1942: map 101.

Quercus pubescens Willd., sensu lato — Map 307.

Quercus apennina auct.; Q. brachyphylla Kotschy; Q. crispata Steven; Q. infectoria Olivier; Q. lanuginosa Thuill.; Q. pubescens subsp. anatolica O. Schwarz; Q. pubescens subsp. palensis (Palassou) O. Schwarz; Q. virgiliana (Ten.) Ten.

Taxonomy. It seems that the correlation of characters used in distinguishing *Quercus brachyphylla*, *Q. virgiliana* and *Q. infectoria* from *Q. pubescens* s.str., and the subspecies of the last-mentioned taxon from each other, corresponds poorly, if at all, to the situation in nature, this being especially true of the Balkan area. Consequently, this group of taxa is treated here as one single complex entity, a solution based on advice from K. H. Rechinger (Vienna) and W. Greuter (Geneva).
Notes. Cr added (given only for *Quercus brachyphylla* in Fl. Eur.).
Total range. Schmucker 1942: map 95; Čermák 1955: 69; MJW 1965: map 122b; Fenaroli 1967: 162; J. Holub & V. Jirásek, Folia Geobot. Phytotax. (Praha) 3: 298 (1968); K. Browicz & J. Zieliński, Atlas Distrib. Trees Shrubs Poland 13: map 28 (1973).

Q. pubescens subsp. **pubescens**

Notes. »Throughout the range of the species, except Spain and Pyrenees» (Fl. Eur.). Not mapped separately.

Q. pubescens subsp. **anatolica** O. Schwarz

Quercus crispata Steven; Q. lanuginosa subsp. crispata (Steven) Nyman

Taxonomy and nomenclature. It may be pointed out that *Quercus pubescens* subsp. *anatolica* O. Schwarz, accepted by Fl. Eur., is antedated by *Q. lanuginosa* subsp. *crispata* (Steven) Nyman. See also G. L. Menitsky, Nov. Syst. Pl. Vasc. (Leningrad) 1968: 90–91.
Notes. »E. part of Balkan peninsula; Krym» (Fl. Eur.). Not mapped separately.

Q. pubescens subsp. **palensis** (Palassou) O. Schwarz

Notes. »Pyrenees; N.E. Spain» (Fl. Eur.). Not mapped separately.
Total range. Endemic to Europe.

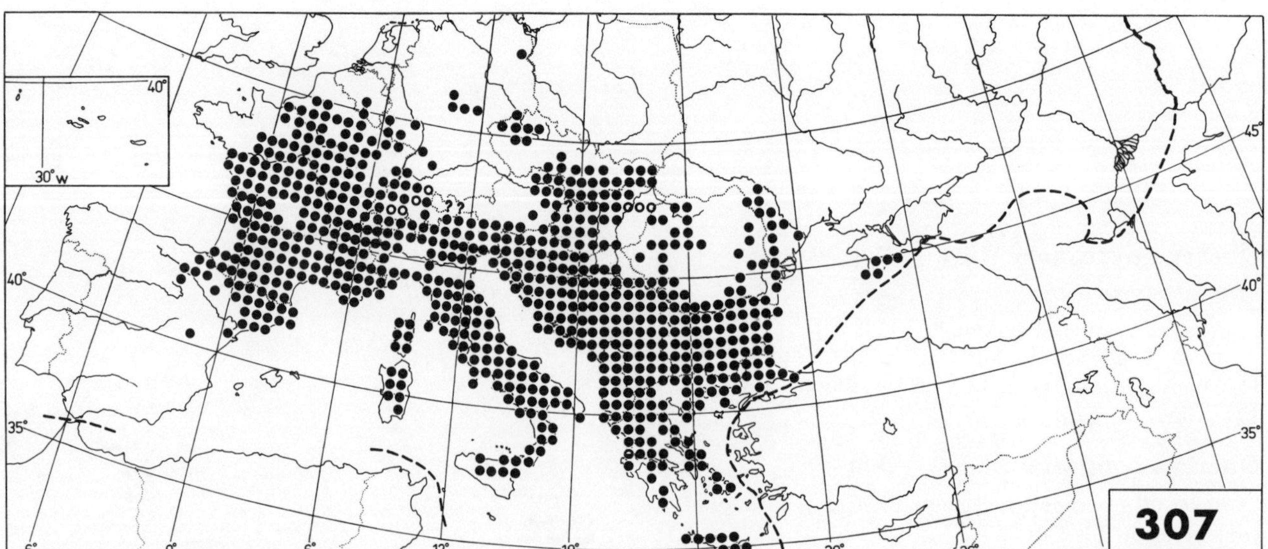

Quercus pubescens sensu lato

Quercus canariensis Willd. — Map 308.

Quercus faginea subsp. baetica (Webb) DC.; Q. salzmanniana (Webb) Coutinho

Taxonomy. C. Sáenz de Rivas, Anal. Inst. Bot. Cavanilles 26: 45—52 (1970); E. F. Galiano & S. Silvestre, Lagascalia 4: 111—112 (1974).

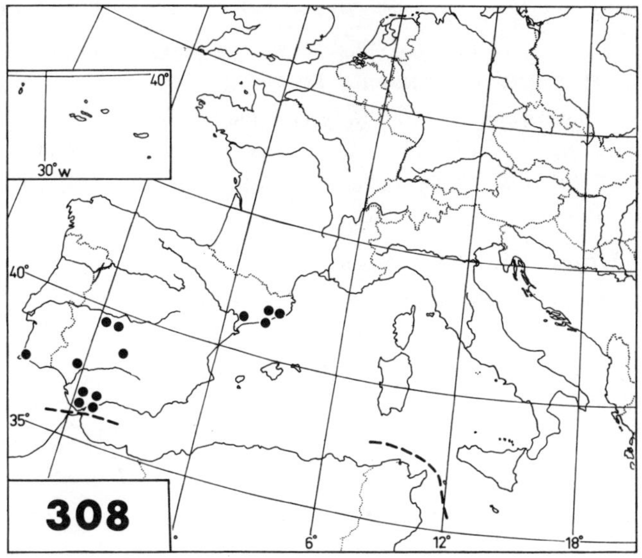

Quercus canariensis

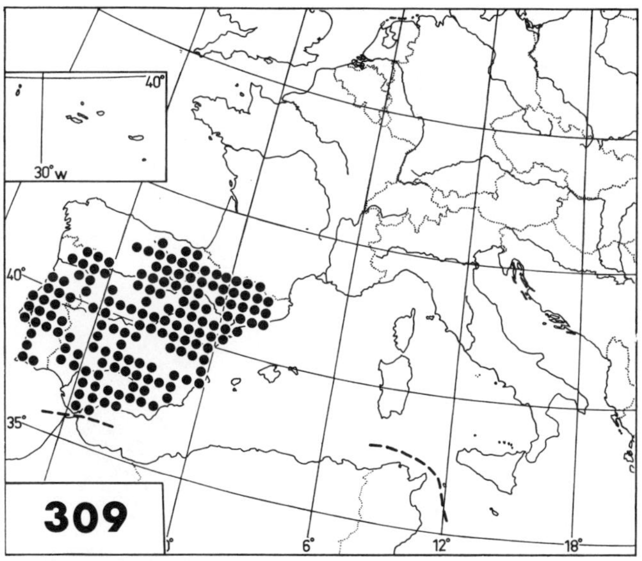

Quercus faginea

Quercus faginea Lam. — Map 309.

Incl. Quercus alpestris Boiss. and Q. valentina Cav.; Quercus lusitanica auct. non Lam.

Taxonomy. Two subspecies are recognized by C. Sáenz de Rivas, V Simp. Fl. Eur. Trab. Comunic., pp. 335—350 (Sevilla 1969): subsp. *faginea* and subsp. *broteri* (Coutinho) A. Camus.

Notes. Bl omitted (given in Fl. Eur.).

Quercus fruticosa Brot. — Map 310.

?Quercus lusitanica Lam., nom. ambig.; Q. humilis Lam.

Total range. Schmucker 1942: maps 94 and 96.

Quercus infectoria Olivier

Notes. Not mapped separately; see under *Quercus pubescens.* »N. Aegean region» (Fl. Eur.).

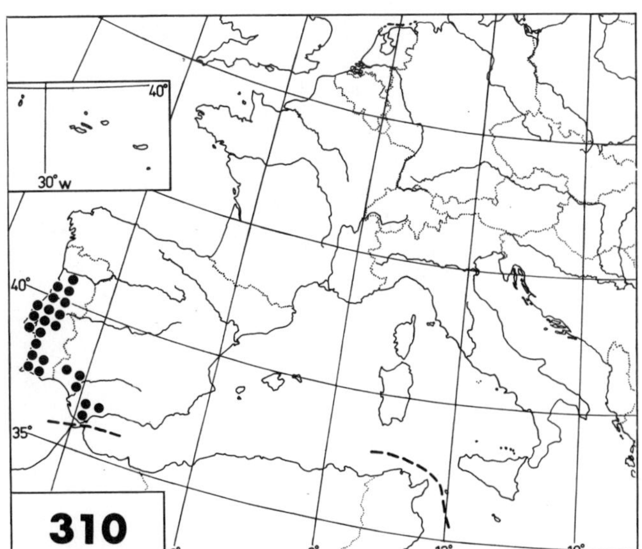

Quercus fruticosa

ULMACEAE

Ulmus glabra Hudson — Map 311.

Ulmus montana With.; U. scabra Miller

Notes. Co added (not given in Fl. Eur.).

Total range. Schmucker 1942: map 154; Čermák 1955: 72; Hegi 1958: 257; MJW 1965: map 123c; Fenaroli 1967: 174.

Ulmus elliptica C. Koch

Notes. Given as native in Rs(K) in Fl. Eur. Not mapped.

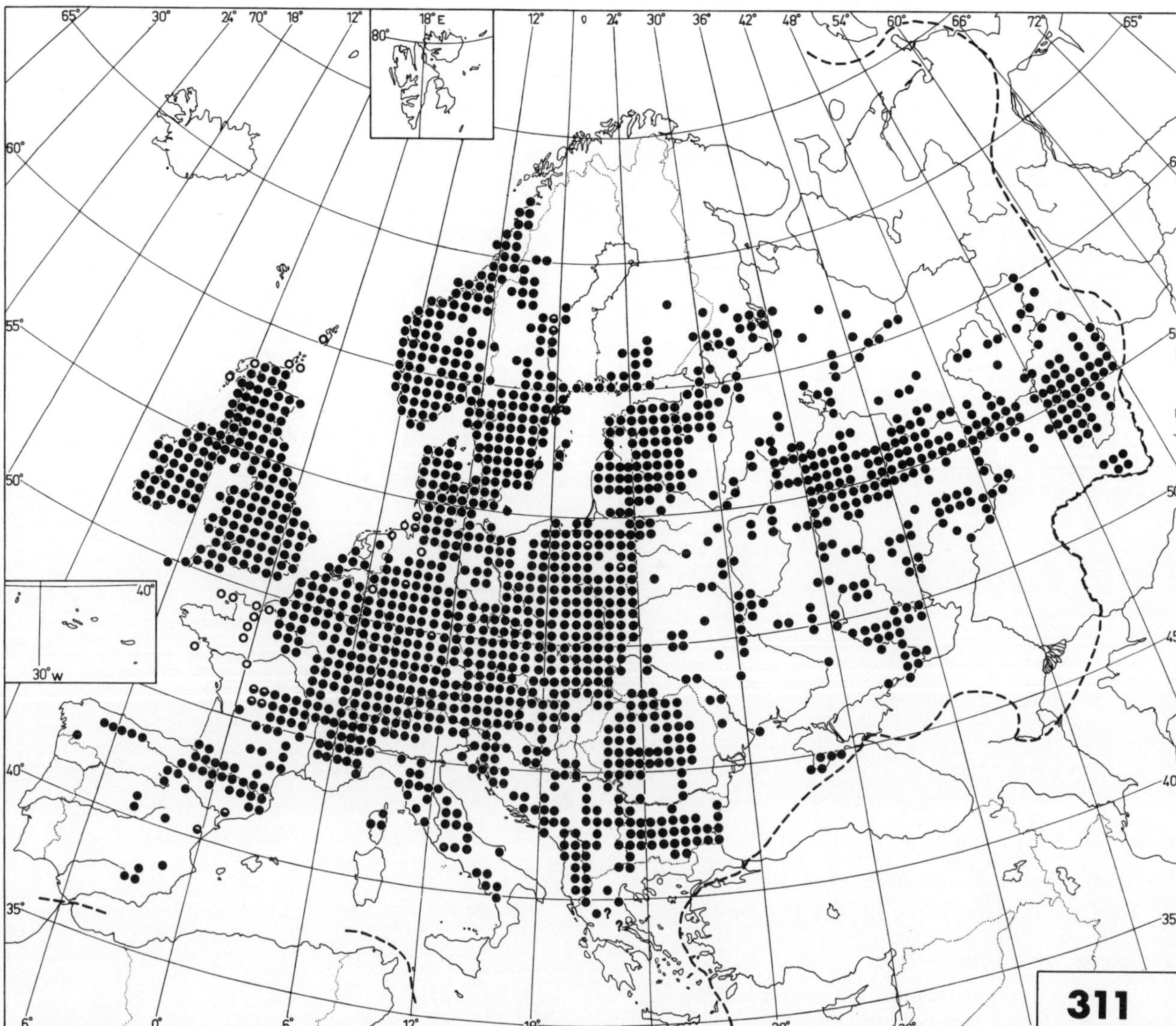

Ulmus glabra

Ulmus minor Miller, coll. — Map 312.

Ulmus campestris auct. non L.; U. carpinifolia G. Suckow; U. diversifolia Melville; U. foliacea sensu Hayek; U. glabra Miller non Hudson; U. stricta (Aiton) Lindley. Incl. U. canescens Melville and U. procera Salisb.

Nomenclature. R. H. Richens, Feddes Repert. (Fl. Eur. Not. Syst. 7) 79: 1—2 (1968).

Notes. Until a comprehensive revision of the group is made, the only practical solution seems to be to map *Ulmus canescens* and *U. procera* under *U. minor*. Not native in Bl (given as native in Fl. Eur.), native status doubtful in Br (given as native in Fl. Eur.), Sa added (as *U. canescens*, not given in Fl. Eur.).

Total range. Schmucker 1942: map 154; Hegi 1958: 257; MJW 1965: map 123d; Fenaroli 1967: 176.

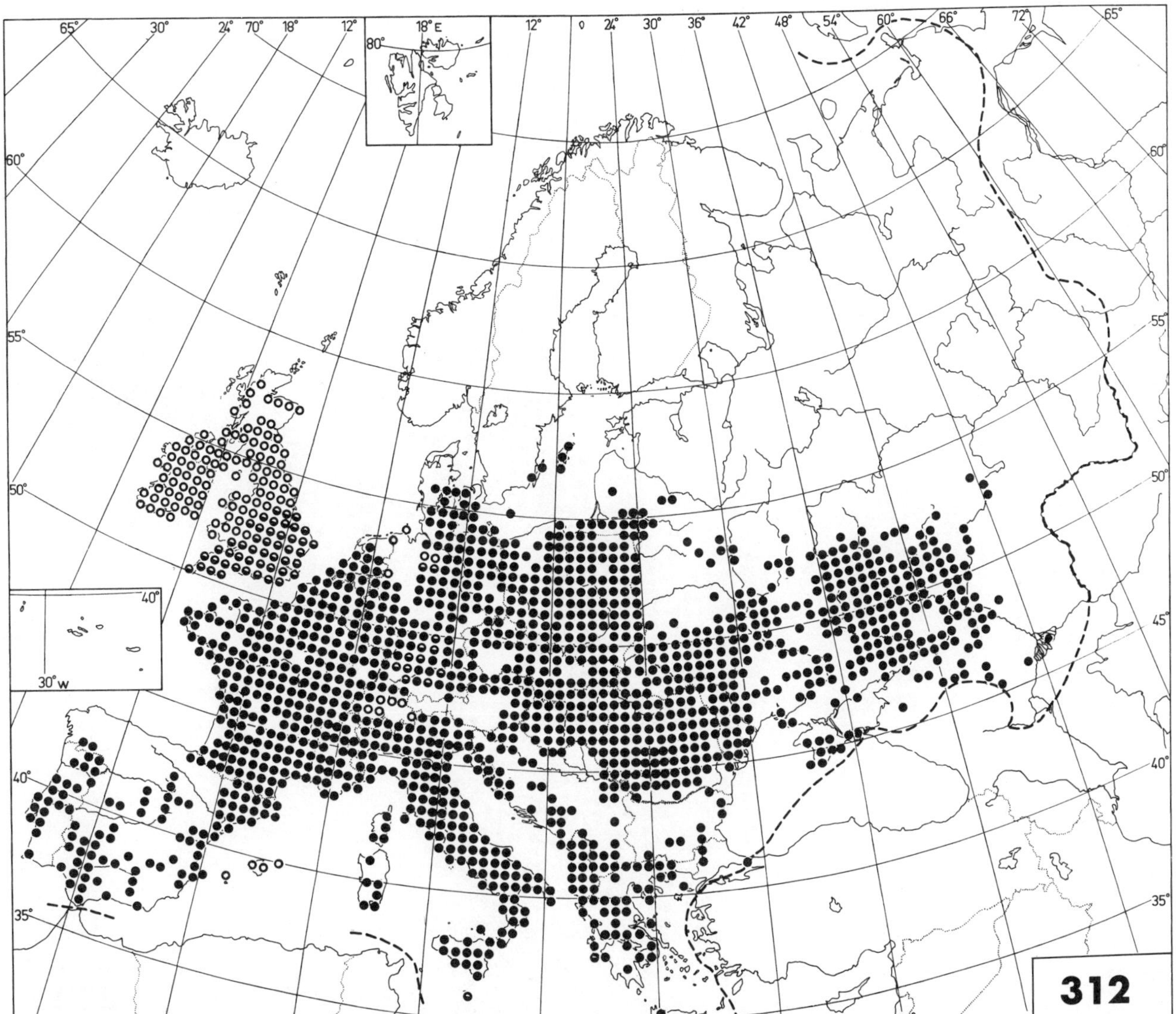

Ulmus minor coll.

Ulmus laevis Pallas — Map 313.

Ulmus effusa Willd.; U. pedunculata Foug.

Notes. Only old and unconfirmed or otherwise unreliable records for Gr (given as native in Fl. Eur.), not native in He (given in Fl. Eur.), Tu added (not given in Fl. Eur.).

Total range. Schmucker 1942: map 154; Čermák 1955: 72; MJW 1965: map 124a; Fenaroli 1967: 178.

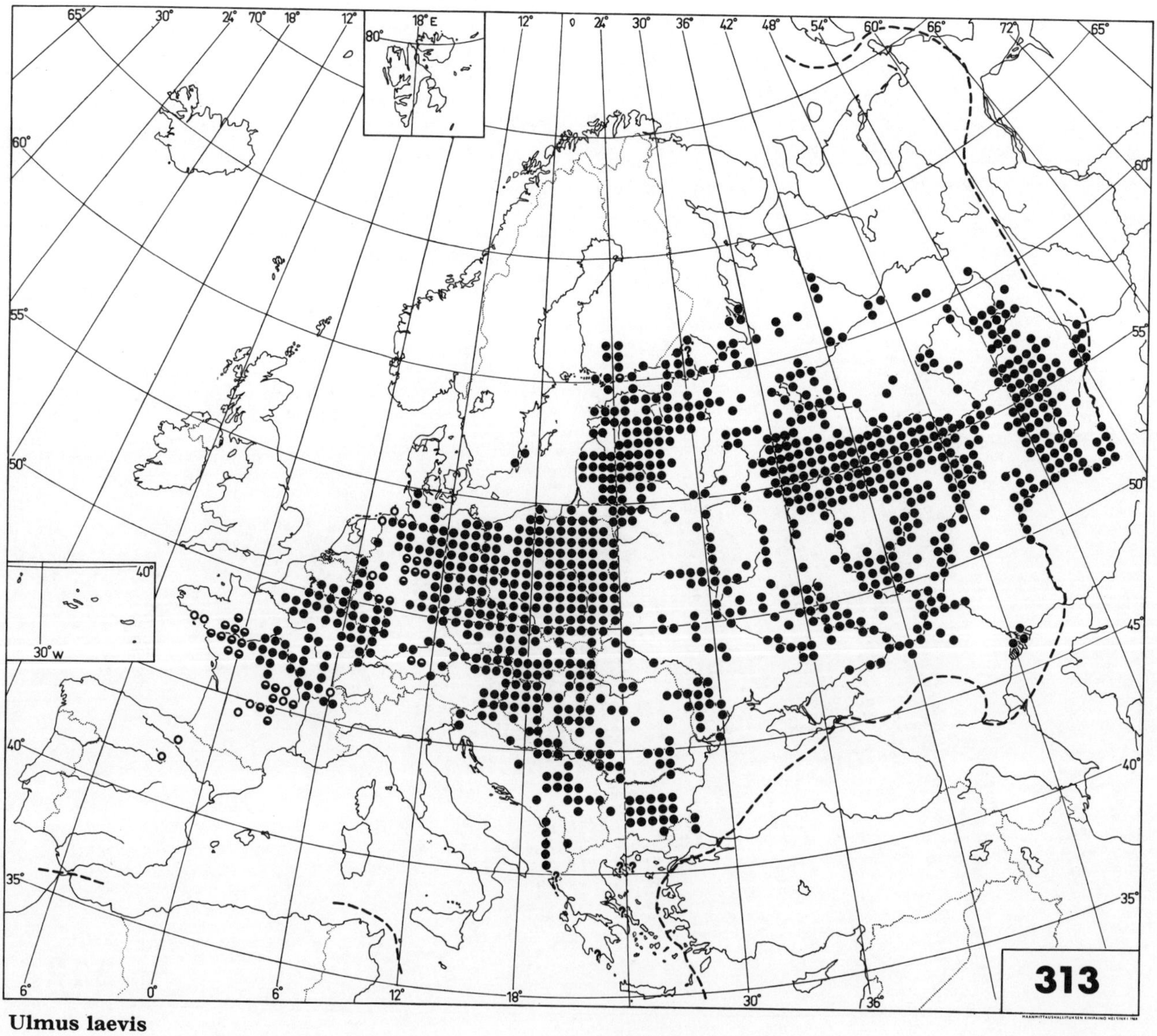

Ulmus laevis

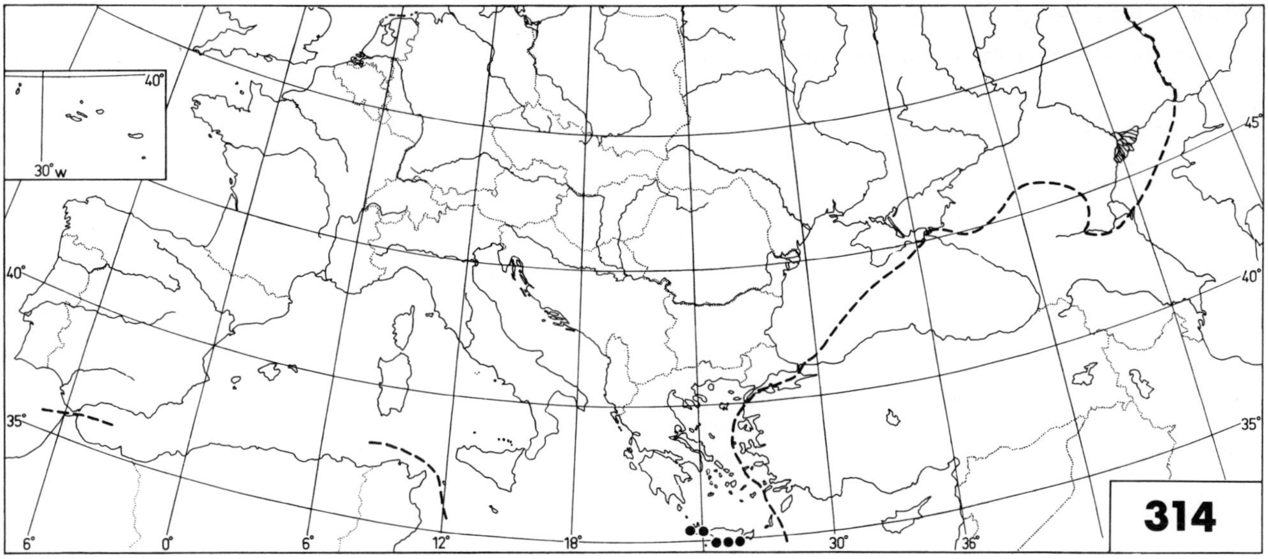

Zelkova abelicea

Zelkova abelicea (Lam.) Boiss. — Map 314.

Zelkova cretica (Sm.) Spach

Celtis australis L. — Map 315.

Notes. Bl and Gr added (not given in Fl. Eur.), Cr omitted (given in Fl. Eur.), not native in He (given as native in Fl. Eur.). Native status perhaps uncertain at least for part of the French and Iberian occurrences.

Total range. Schmucker 1942: map 159 (western part of the range); MJW 1965: 123b.

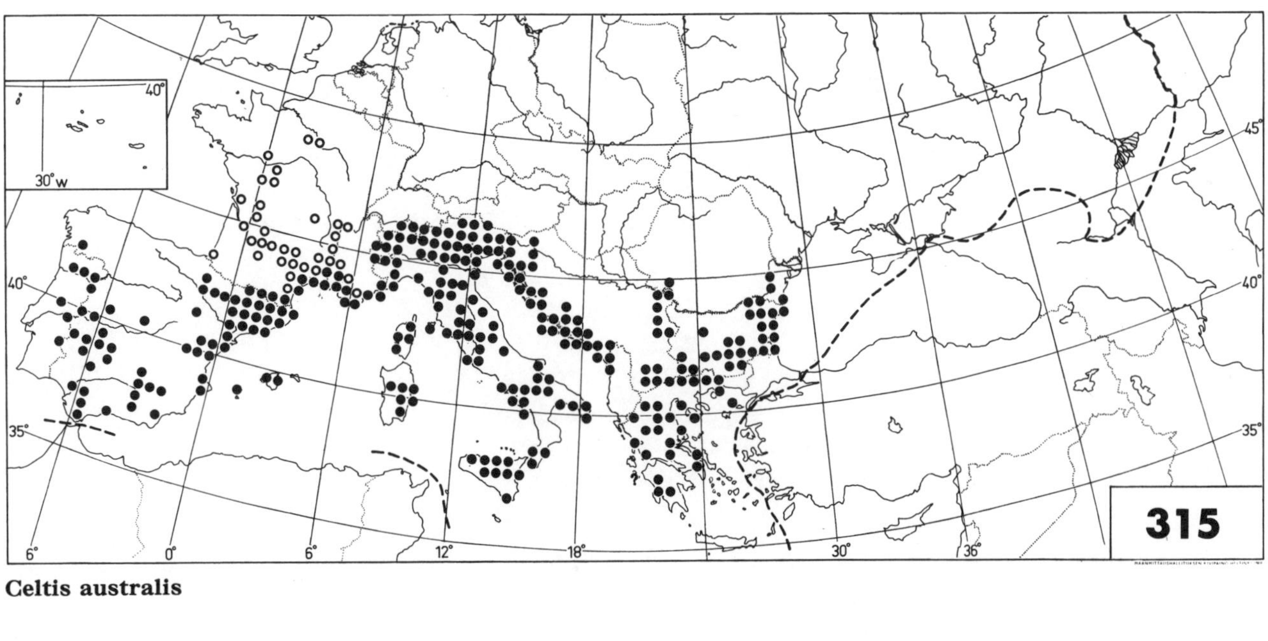

Celtis australis

Celtis caucasica Willd. — Map 316.

Notes. Ju added (»?Macedonia» in Fl. Eur.).
Total range. Schmucker 1942: map 160; MJW 1965: 123b.

Celtis glabrata Steven ex Planchon — Map 316.

Celtis tournefortii Lam. — Map 317.

Celtis aetnensis (Tornab.) Strobl; C. aspera (Ledeb.) Steven

Notes. Bu omitted (given in Fl. Eur.); Cr added (not given in Fl. Eur.): W. Greuter, Ann. Mus. Goulandris 1: 27 (1973).
Total range. Schmucker 1942: map 159.

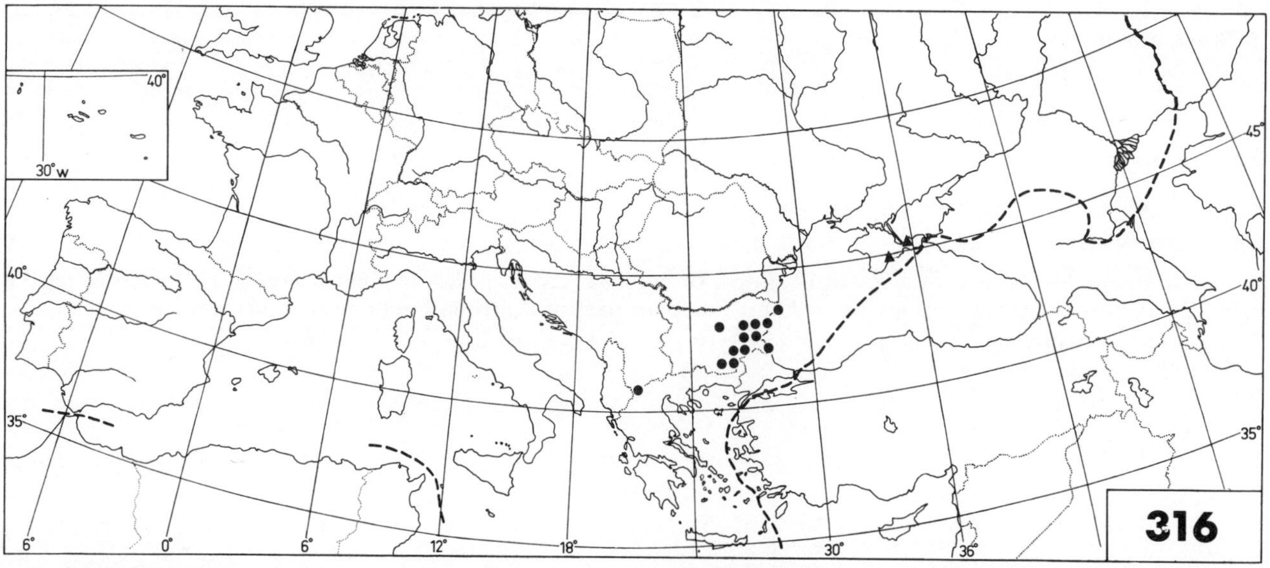

● = **Celtis caucasica** ▲ = **C. glabrata**

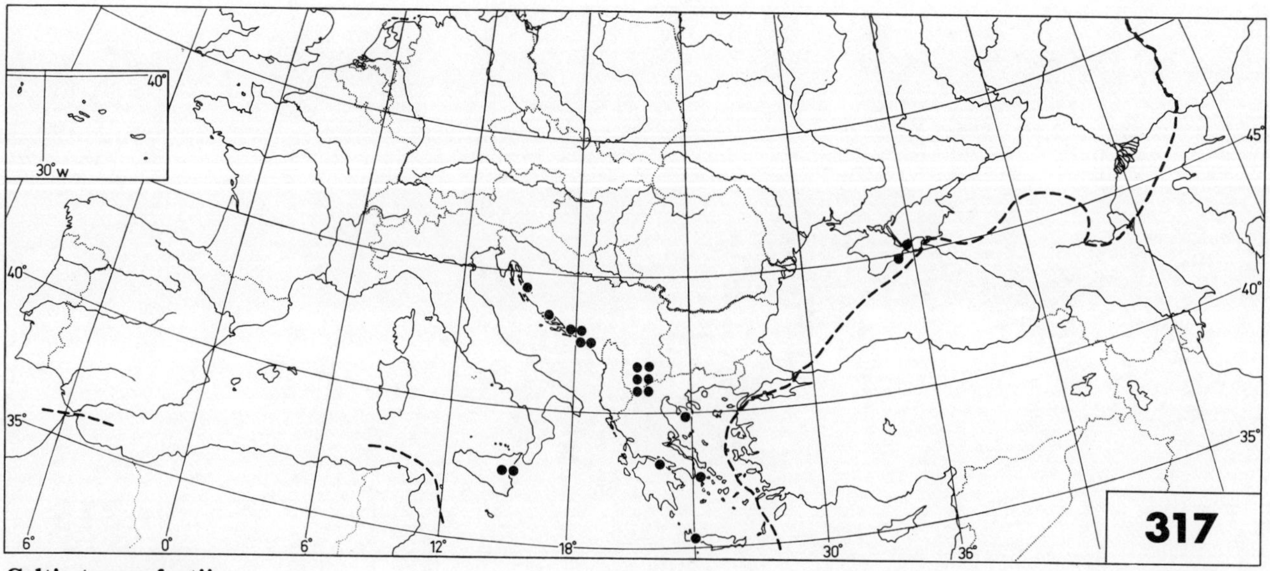

Celtis tournefortii

MORACEAE

Ficus carica L. — Map 318.

Notes. *Al, *Bl, *Co, Cr, *Ga, Gr, *Hs, It, *Ju, Sa, Si, *Tu. »Extensively cultivated and widely naturalized in S. Europe; perhaps native in the southern parts of the Iberian peninsula, Italy, the Balkan peninsula and U.S.S.R.» (Fl. Eur.).

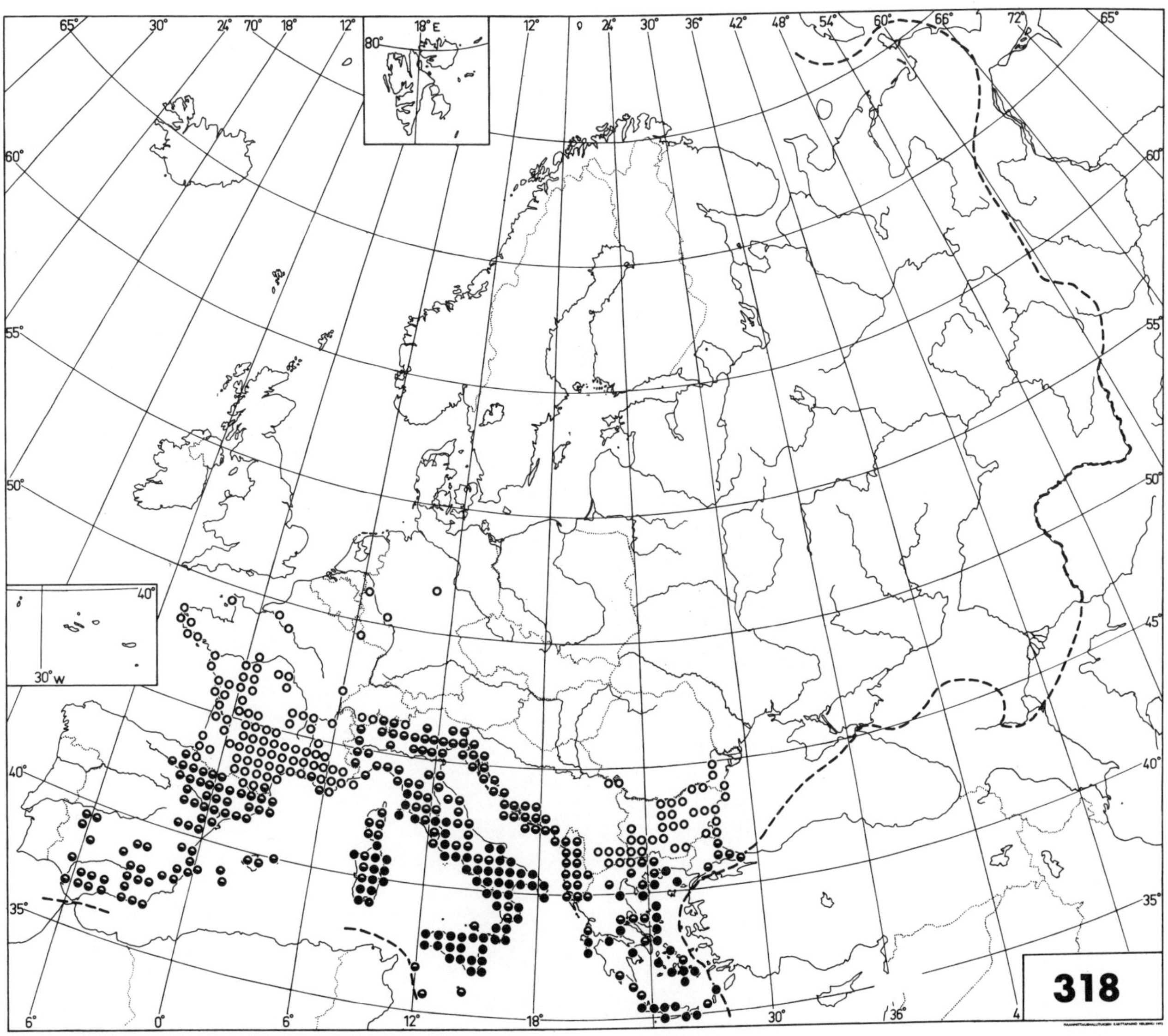

Ficus carica

CANNABACEAE

Humulus lupulus L. — Map 319.

Notes. Not native in Hb (given as native in Fl. Eur.).

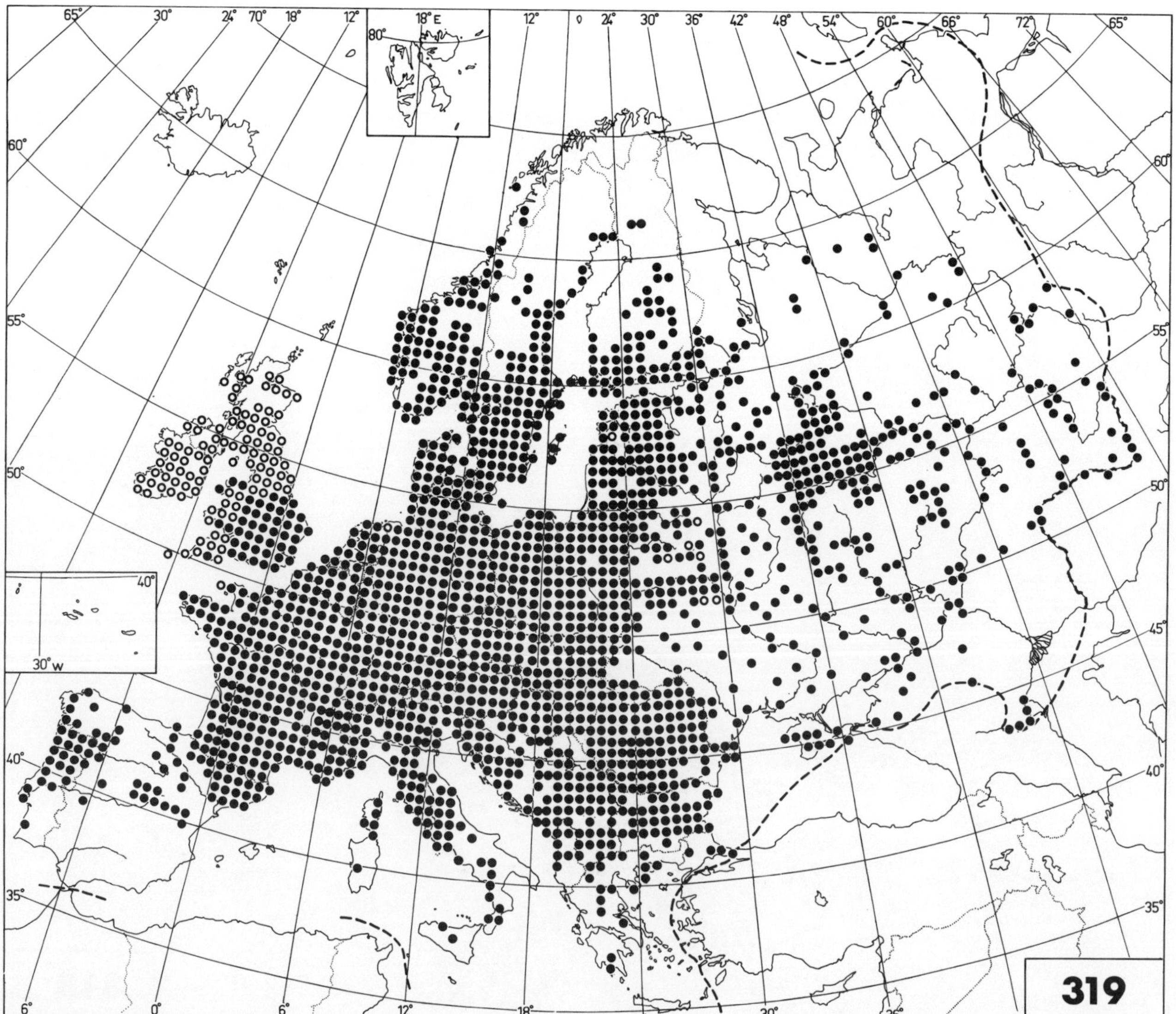

Humulus lupulus

Humulus scandens (Loureiro) Merrill — Map 320.

Humulus japonicus Siebold & Zucc.

Notes. [Hu, It] (the species not mentioned in Fl. Eur.).
Native of E. Asia.

Cannabis sativa L.

Incl. Cannabis ruderalis Janisch.; C. sativa subsp. spontanea Serebr.

Notes. The material available does not allow the production of a satisfactory map. For European records of *Cannabis ruderalis*, with no distinction between established and casual occurrences, see J. Holub, Preslia 38: 79 (1966). »Native on alluvial sands in S.E. Russia; widely cultivated . . . , and naturalized in a large part of Europe; a frequent casual elsewhere» (Fl. Eur.).
Total range. Hultén CP 1971: map 221.

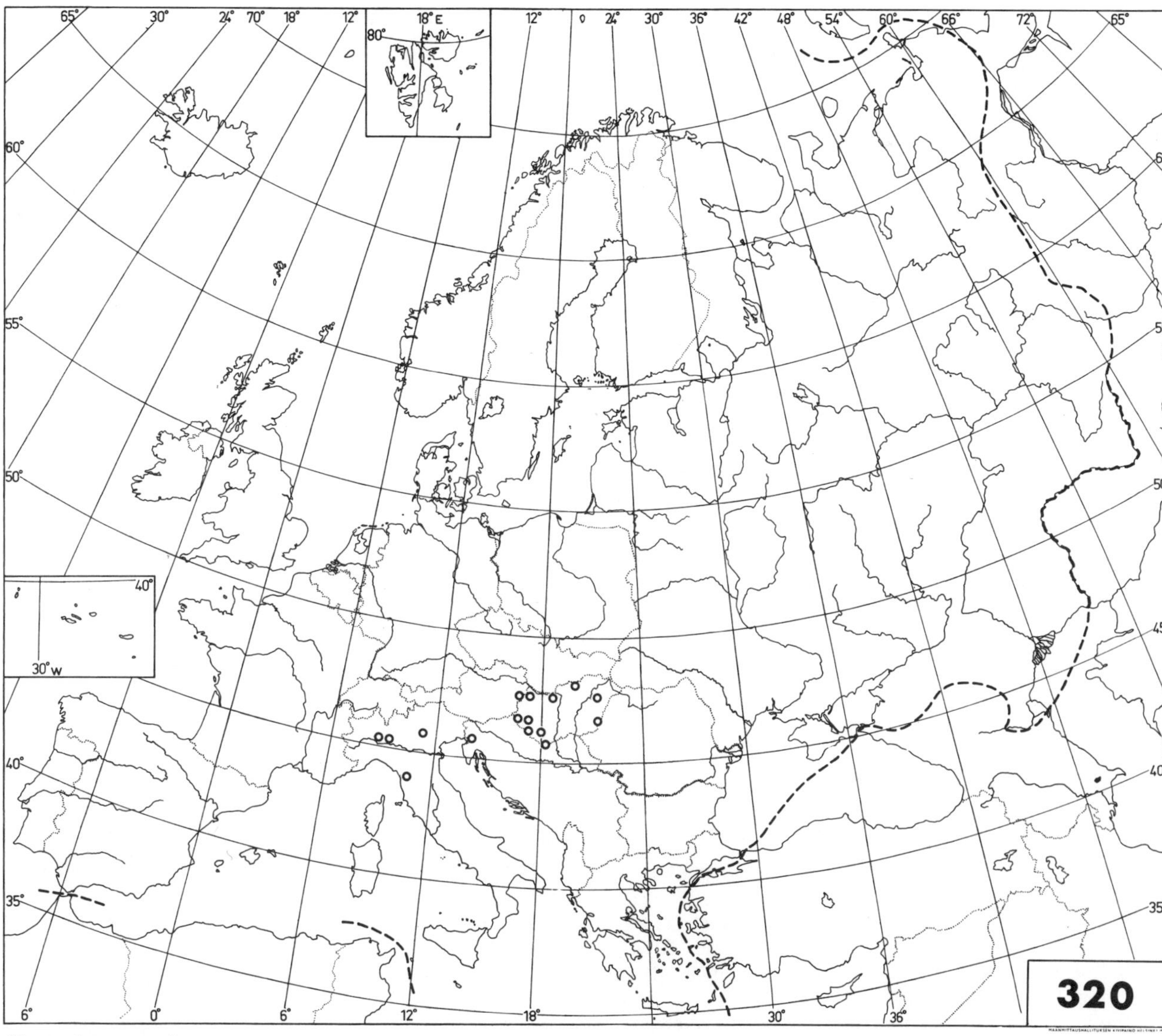

Humulus scandens

URTICACEAE

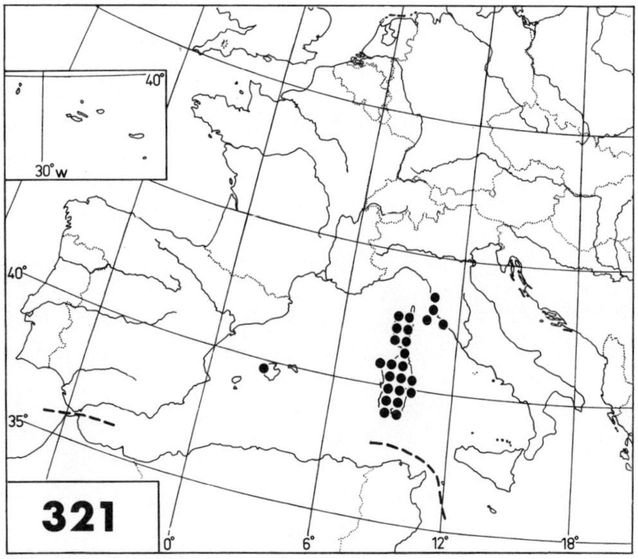

Urtica atrovirens

Urtica atrovirens Req. ex Loisel. — Map 321.

Total range. Endemic to Europe.

Urtica dioica L. — Map 322.

Urtica galeopsifolia Wierzb. ex Opiz; U. gracilis Aiton. Incl. U. pubescens Ledeb.

Notes. Besides a map for the species as a whole, a separate map is given of subsp. *sondenii* (Simmons) Hyl., which, morphologically and ecologically, is certainly one of the most distinct European subentities of the species. Az omitted (given in Fl. Eur.).

Total range. MJW 1965: map 124b; Hultén CP 1971: map 285 (holarctic part).

U. dioica subsp. **dioica**

Taxonomy. Tetraploid with 2n = 52.

Notes. Throughout Europe. Not mapped separately.

U. dioica subsp. **sondenii** (Simmons) Hyl. — Map 323.

Urtica dioica subsp. gracilis (Aiton) Selander, pro parte, non U. gracilis Aiton s. str. orig.; U. dioica var. sondenii Simmons; U. sondenii (Simmons) Avrorin

Taxonomy. S. Selander, Svensk Bot. Tidskr. 41: 264—282 (1947), Hultén CP 1971: 294. Diploid with 2n = 26: I. J. Bassett & C. W. Crompton, Taxon 21: 680 (1972); Á. & D. Löve, Taxon 24: 505 (1975).

Nomenclature. The subspecific combination used was validated by N. Hylander in T. Ahti & L. Hämet-Ahti, Ann. Bot. Fennici 8: 47 (1971).

Notes. Fe, No, Rs (N), Su (the subspecies not recognized by Fl. Eur.).

Total range. Hultén AA 1958: map 202; Hultén CP 1971: map 285.

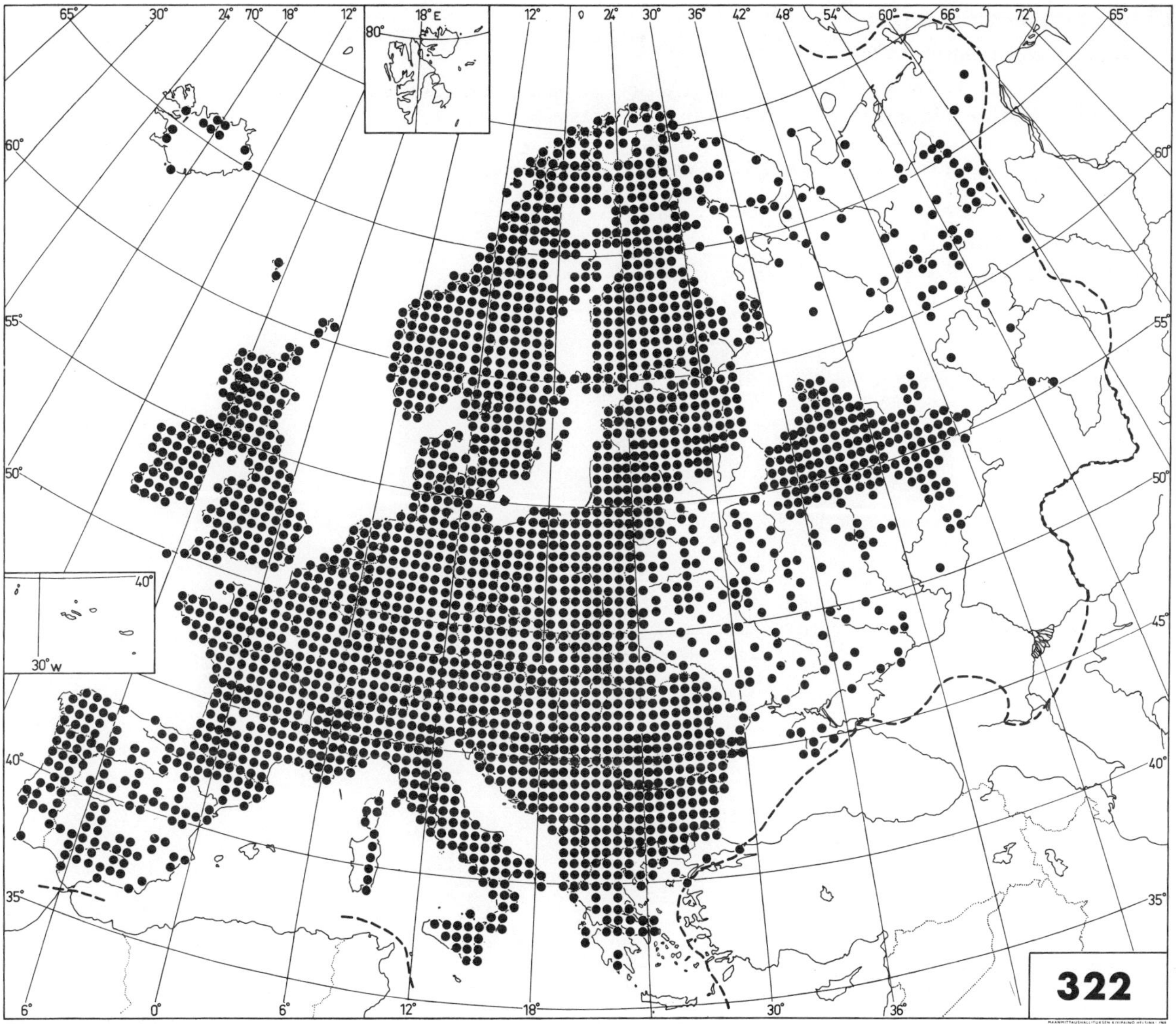

Urtica dioica

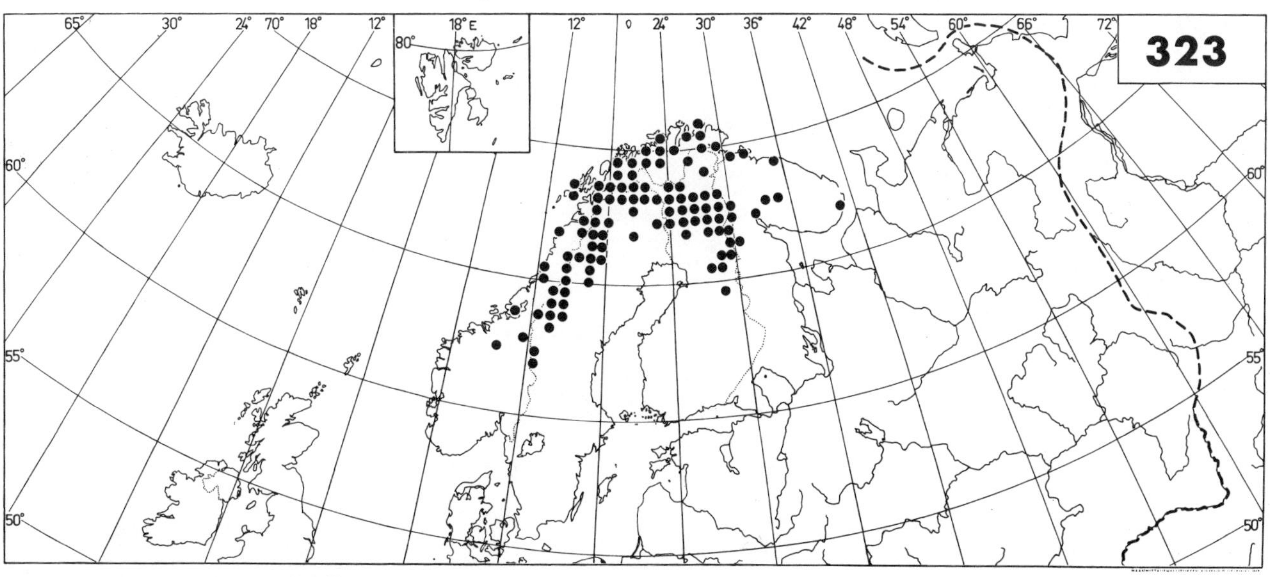

Urtica dioica subsp. **sondenii**

Urtica kioviensis Rogow. — Map 324.

Urtica hulensis Feinbr.

Taxonomy. B. R. Baum & J. M. Gillett, Österr. Bot. Zeitschr. 114: 320—323 (1967).

Notes. Bu and Ge added (not given in Fl. Eur.): P. Konczak *et al.*, Verh. Bot. Ver. Brandenburg 105: 108—116 (1968), P. Gutte *et al.*, Gleditschia 1: 95—97 (1973).

Total range. P. Konczak *et al.*, Verh. Bot. Ver. Brandenburg 105: 109 (1968). Outside Europe, recorded only from Israel.

Urtica cannabina L.

Notes. »Naturalized in U.S.S.R. [Rs(W, E)]» (Fl. Eur.). Not mapped.

Native of C. and S. Asia.

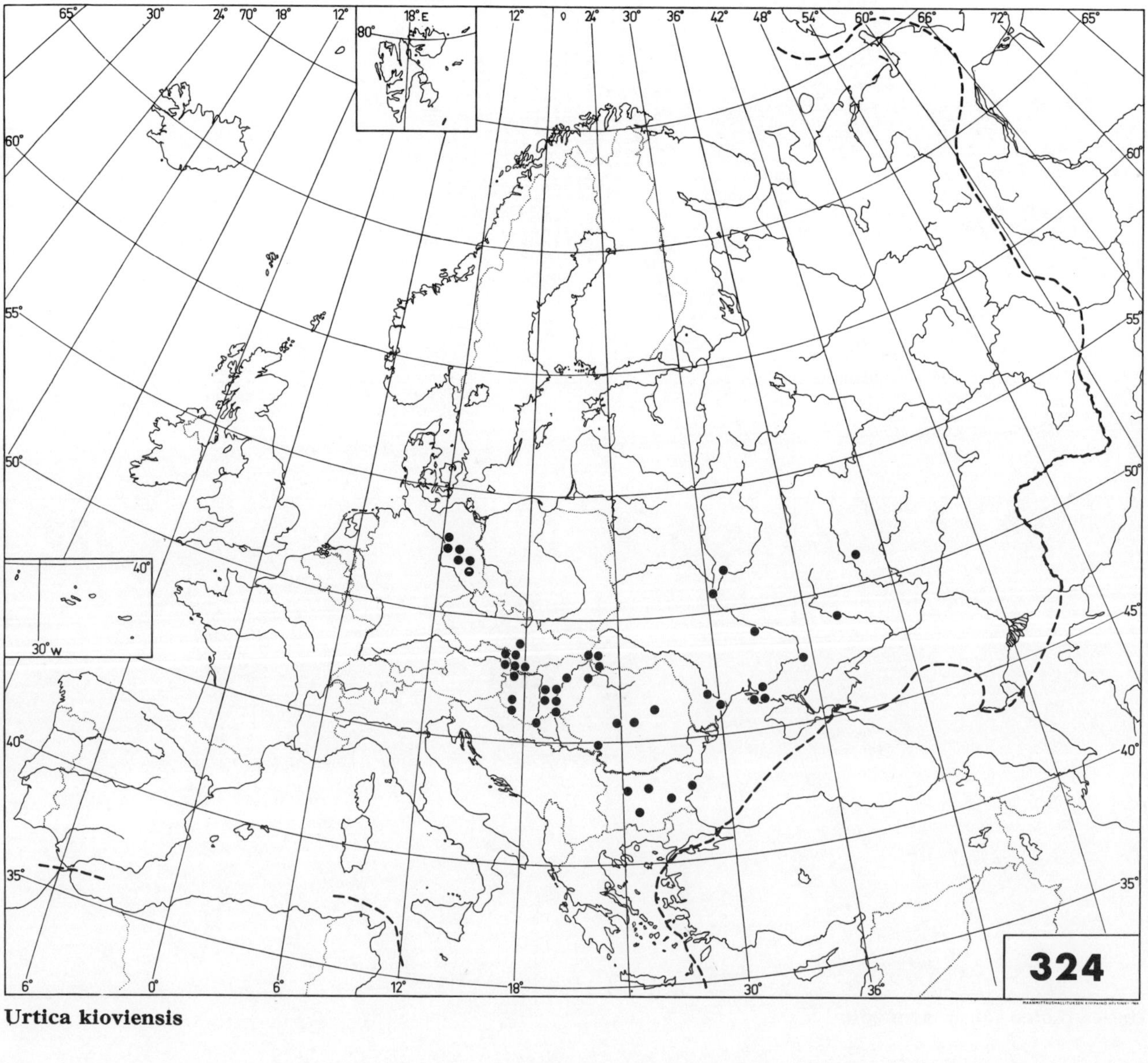

Urtica kioviensis

Urtica rupestris Guss. — Map 325.

Total range. Endemic to Sicilia.

Urtica urens L. — Map 326.

Notes. Fa omitted (given in Fl. Eur.).

Total range. Hultén Alaska 1968: 371; Hultén CP 1971: map 264 (holarctic part).

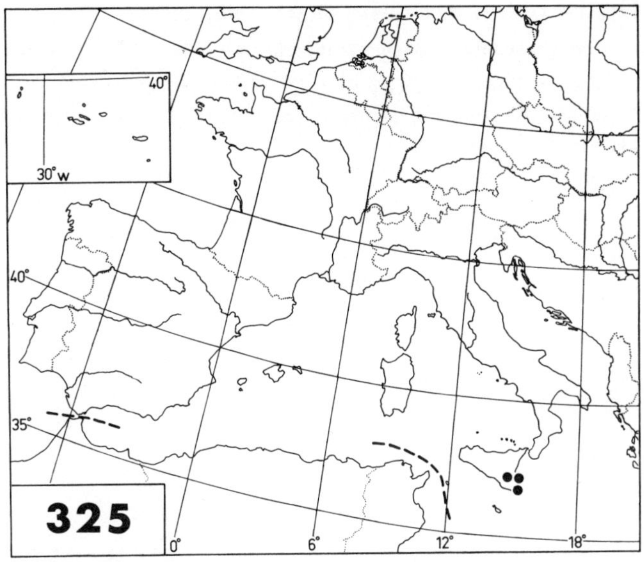

Urtica rupestris

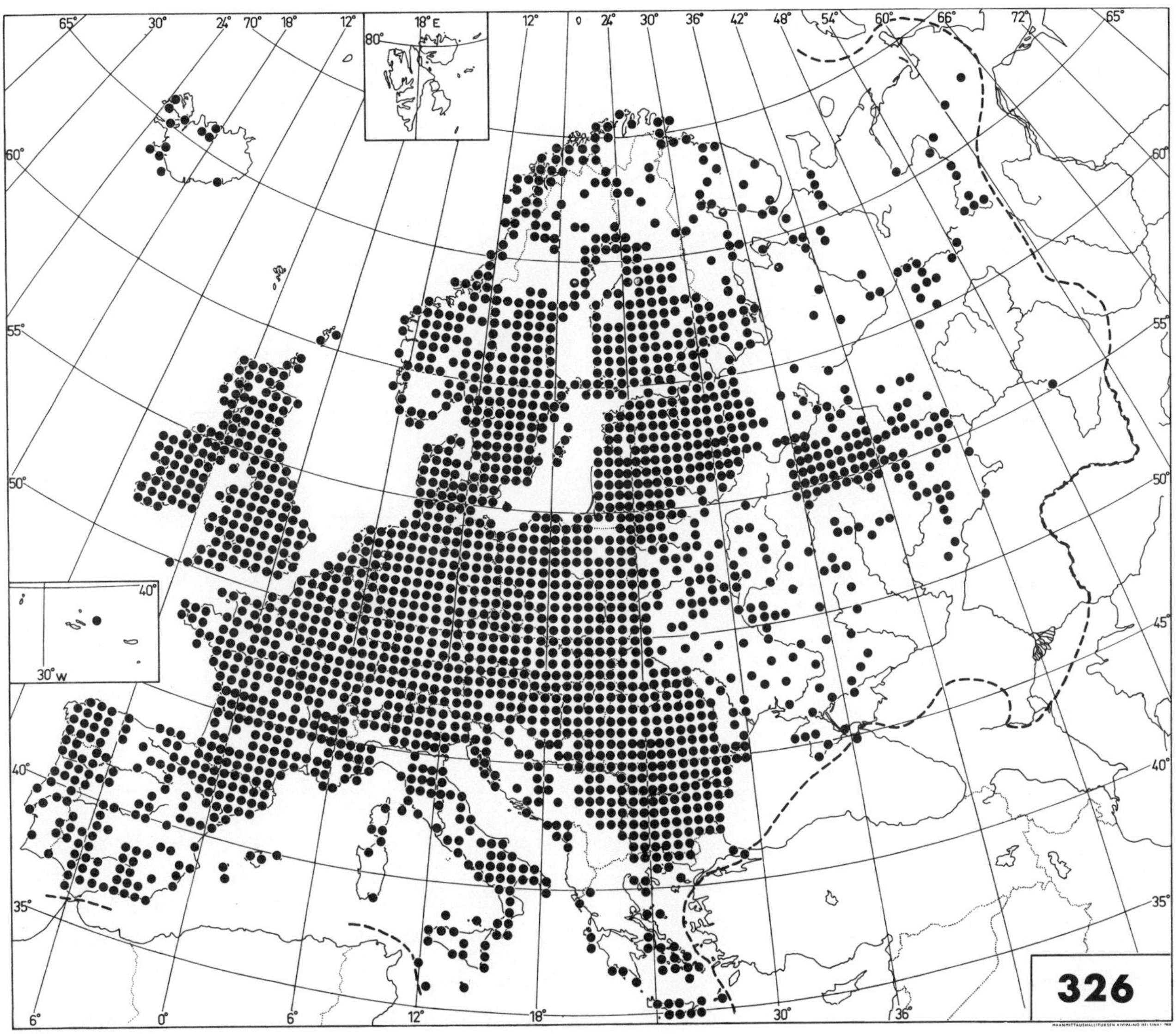

Urtica urens

Urtica membranacea Poiret — Map 327.

Urtica caudata Vahl non Burm. fil.; U. dubia Forskål, nom. inval.

Nomenclature. W. Greuter, Boissiera 13: 32 (1967).

Notes. Al and Ho omitted (given in Fl. Eur.); Hs added (not given in Fl. Eur.): E. F. Galiano & B. Valdés, Lagascalia 4: 114 (1974).

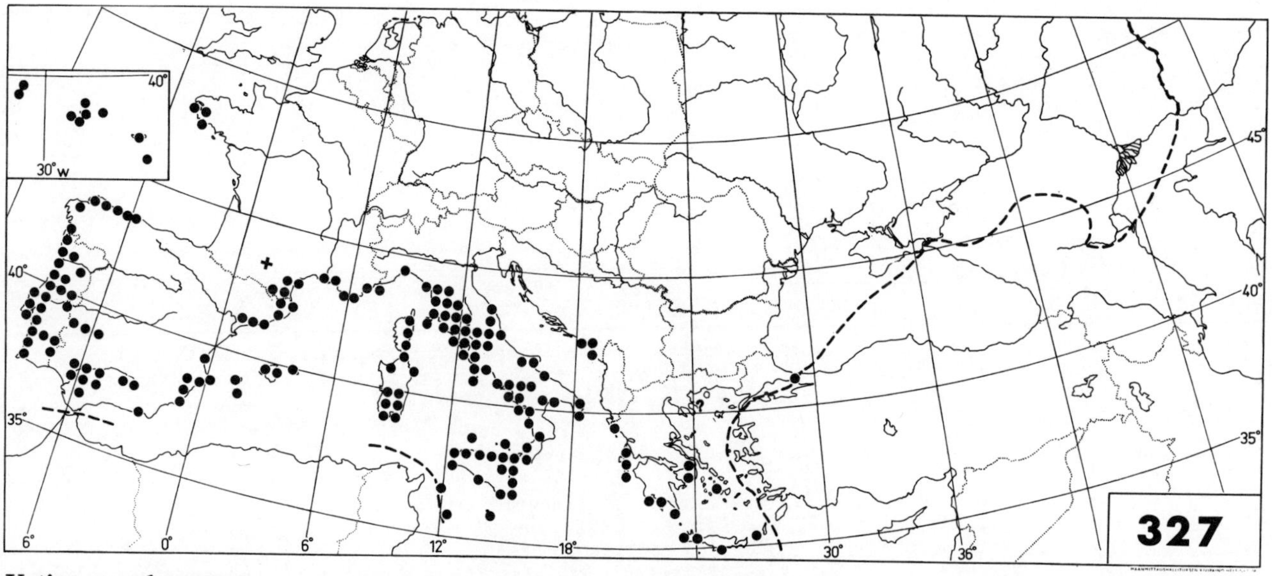

Urtica membranacea

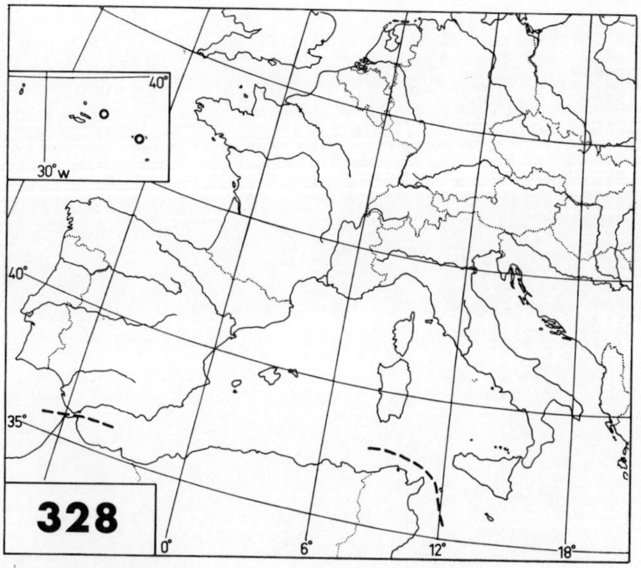

Urtica morifolia

Urtica morifolia Poiret — Map 328.

Notes. [Az] added (the species not mentioned in Fl. Eur.): E. Sjögren, Mem. Soc. Brot. 22: 118—119 (1973).

Native of Madeira and the Canary Islands.

Urtica pilulifera L. — Map 329.

Urtica dodartii L.

Notes. Au, Be, He and Hu omitted ([Au, Be, He, Hu] in Fl. Eur.), extinct in Br ([Br] in Fl. Eur.), probably extinct in Cz ([Cz] in Fl. Eur.), considered native in Ge ([Ge] in Fl. Eur.), Rs(W) added (not given in Fl. Eur.). Many of the extinctions have probably concerned established, not native, occurrences.

Pilea microphylla (L.) Liebm.

Notes. »Naturalized in the Balkan peninsula, [Gr. Ju]» (Fl. Eur.). Not mapped.
Native of tropical America.

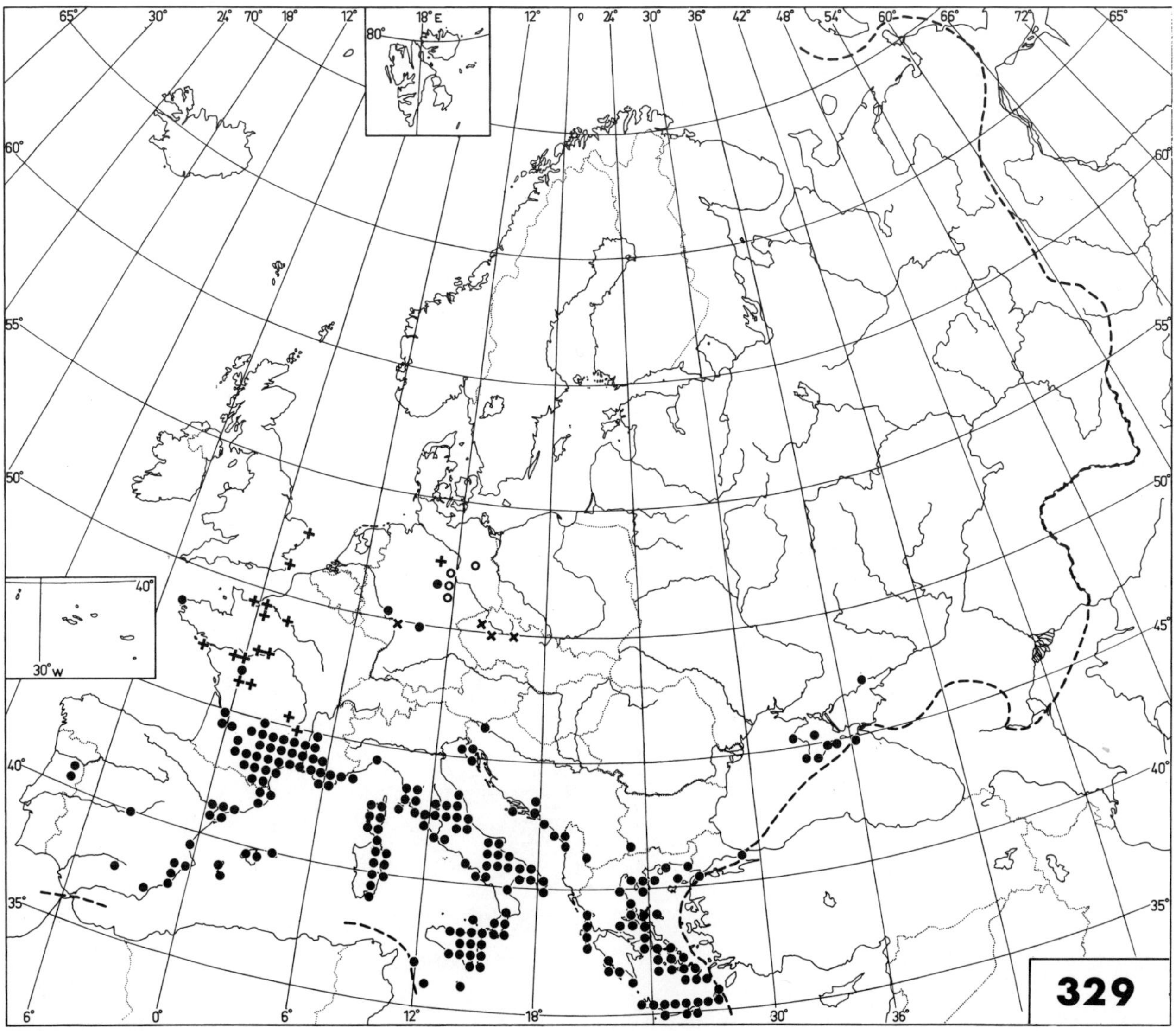

Urtica pilulifera

Parietaria officinalis L. — Map 330.

Parietaria erecta Mert. & Koch

Notes. It is sometimes very difficult or almost impossible to distinguish between native occurrences and old introductions, and the decisions made in the different countries are certainly not completely consistent with each other. In addition, especially in older times, large-leaved plants of *Parietaria judaica* have occasionally been mistaken for this species. Cr, Sa and Si omitted (given in Fl. Eur.), doubtfully native in Ho ([Ho] in Fl. Eur.), [Su] added (not given in Fl. Eur.).

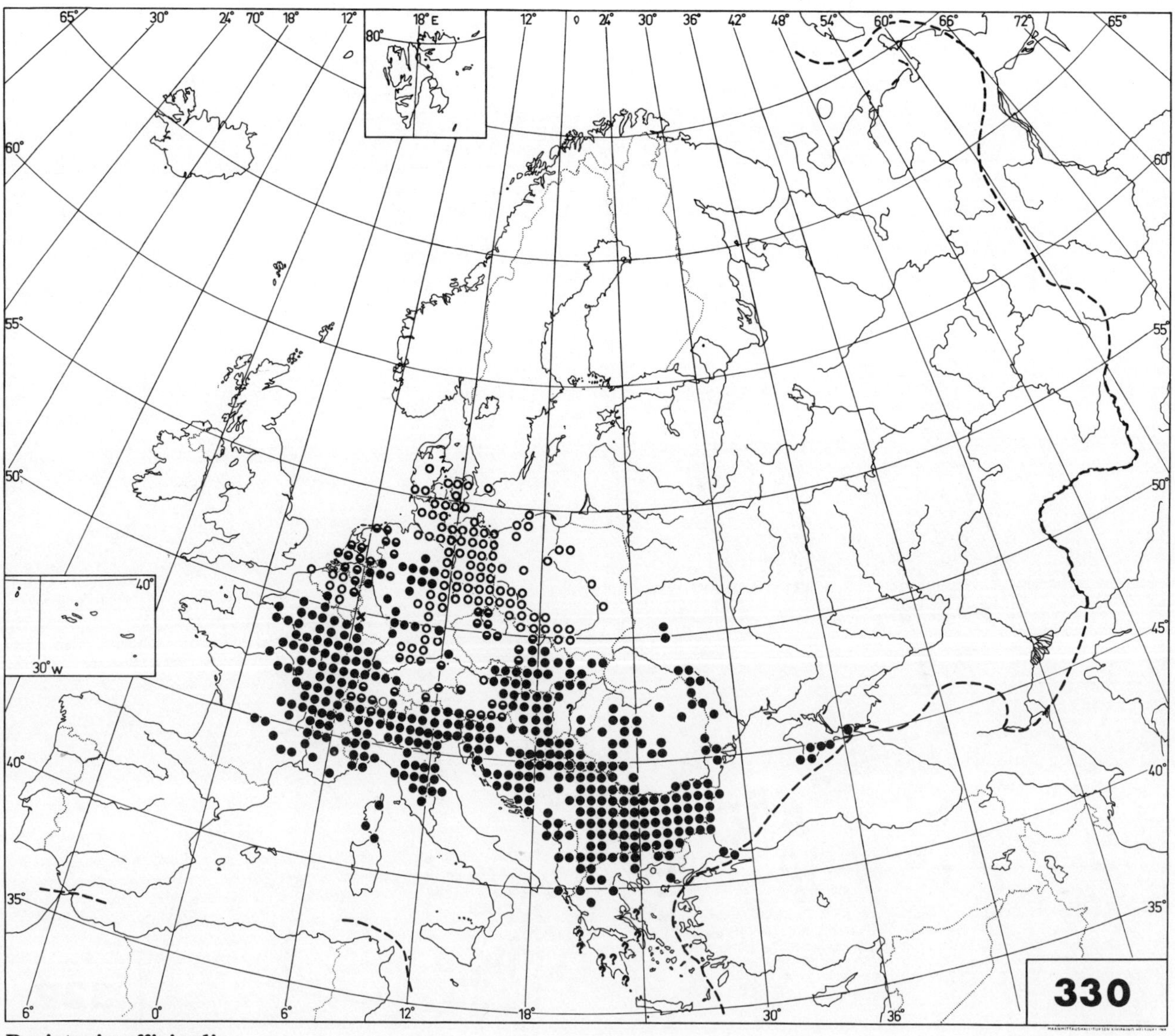

Parietaria officinalis

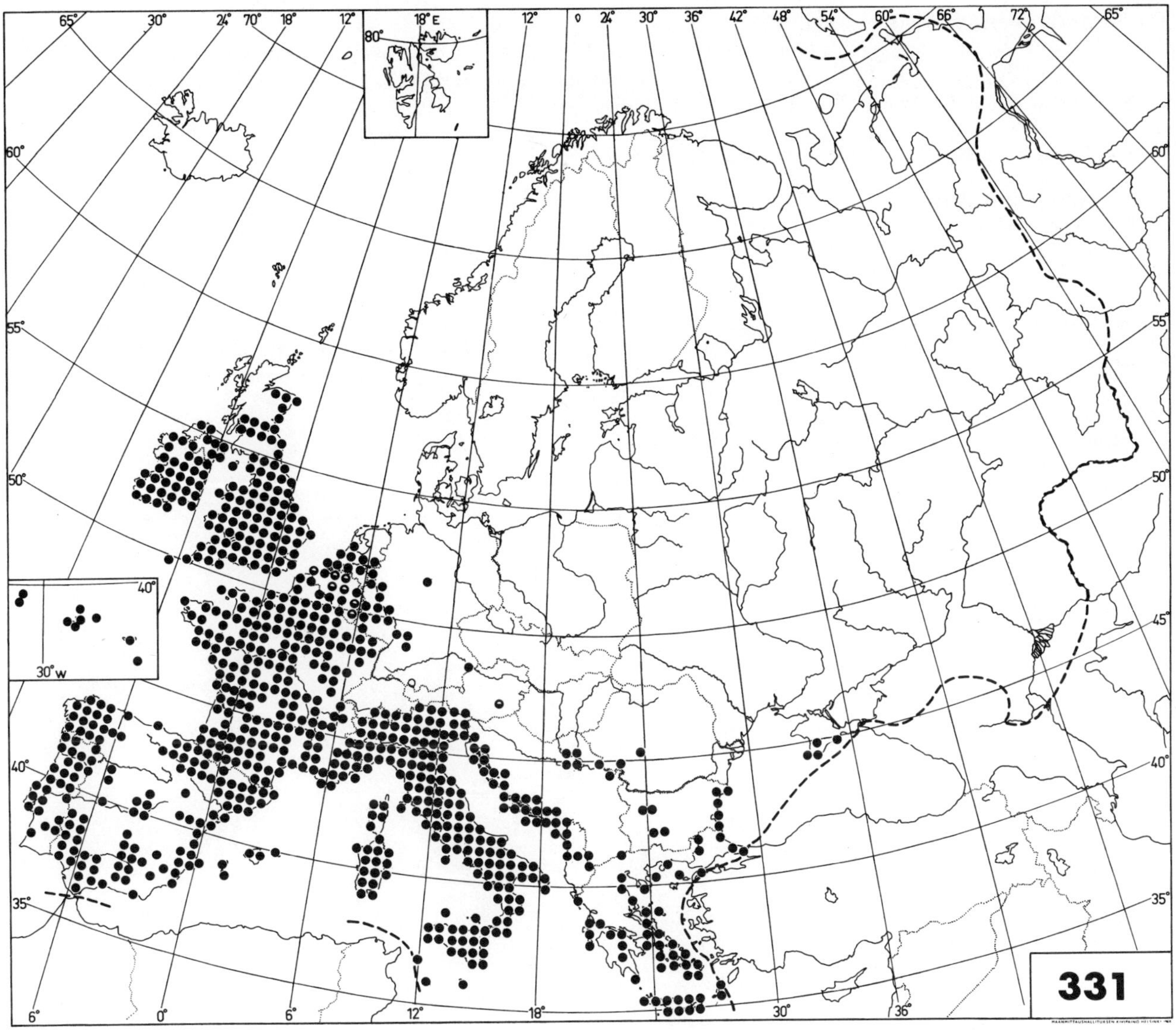

Parietaria judaica

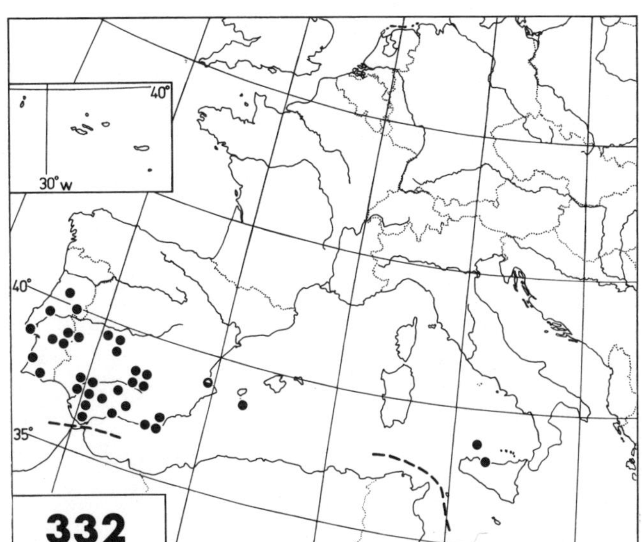

Parietaria mauritanica

Parietaria judaica L. — Map 331.

Parietaria diffusa Mert. & Koch; P. officinalis auct. non L.; P. punctata Willd.; P. ramiflora Moench

Nomenclature. J. Mennema & S. Segal, Gorteria 3: 110—111 (1967); W. Greuter, Boissiera 13: 33 (1967); C. C. Townsend, Watsonia 6: 365—370 (1968); S. Rauschert, Feddes Repert. 83: 645 (1973).

Notes. Of uncertain status in Au ([Au] in Fl. Eur.): rocky places and old walls on the castle hill of Graz; doubtfully native in Be ([Be] in Fl. Eur.); Hu omitted ([Hu] in Fl. Eur.).

Parietaria mauritanica Durieu — Map 332.

Notes. Bl and Si added (not given in Fl. Eur.): M. Lojacono Pojero, Fl. Sicula 2 (2): 351 (1904); J. Chrtek, Folia Geobot. Phytotax. (Praha) 8: 426 (1973).

Parietaria serbica Pančić — Map 333.

Parietaria lusitanica subsp. serbica (Pančić) P. W. Ball, pro parte (excl. P. chersonensis (A. F. Láng) Dörfler)

Taxonomy. J. Chrtek, Folia Geobot. Phytotax. (Praha) 8: 425—426 (1973).

Notes. Bu, Ju: J. Chrtek, Folia Geobot. Phytotax. (Praha) 8: 426 (1973), Rm added (individual countries not given in Fl. Eur.).

Total range. As far as is known, endemic to Europe.

Parietaria lusitanica L.

Excl. Parietaria serbica Pančić

Taxonomy. J. Chrtek, Folia Geobot. Phytotax. (Praha) 8: 425—426 (1973).

Notes. Rm and Rs(E) omitted (given in Fl. Eur.).

P. lusitanica subsp. **lusitanica** — Map 334.

Notes. Hu omitted (given in Fl. Eur.). Bl, Bu, Co, Cr, Ga, Gr, Hs, It, Ju, Lu, Sa, Si. »S. Europe, eastwards to c. 25° E» (Fl. Eur.).

P. lusitanica subsp. **chersonensis** (A. F. Láng) Chrtek — Map 334.

Parietaria chersonensis (A. F. Láng) Dörfler; P. lusitanica var. chersonensis A. F. Láng; P. lusitanica subsp. serbica (Pančić) P. W. Ball, pro parte (typo excl.)

Notes. Rs (W, K).

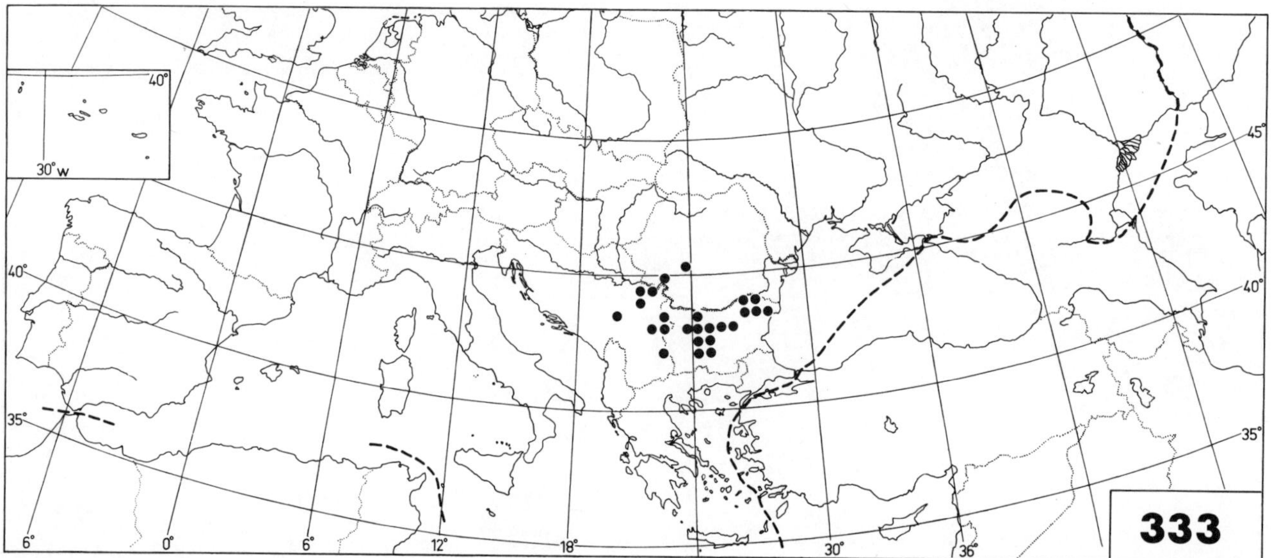

Parietaria serbica

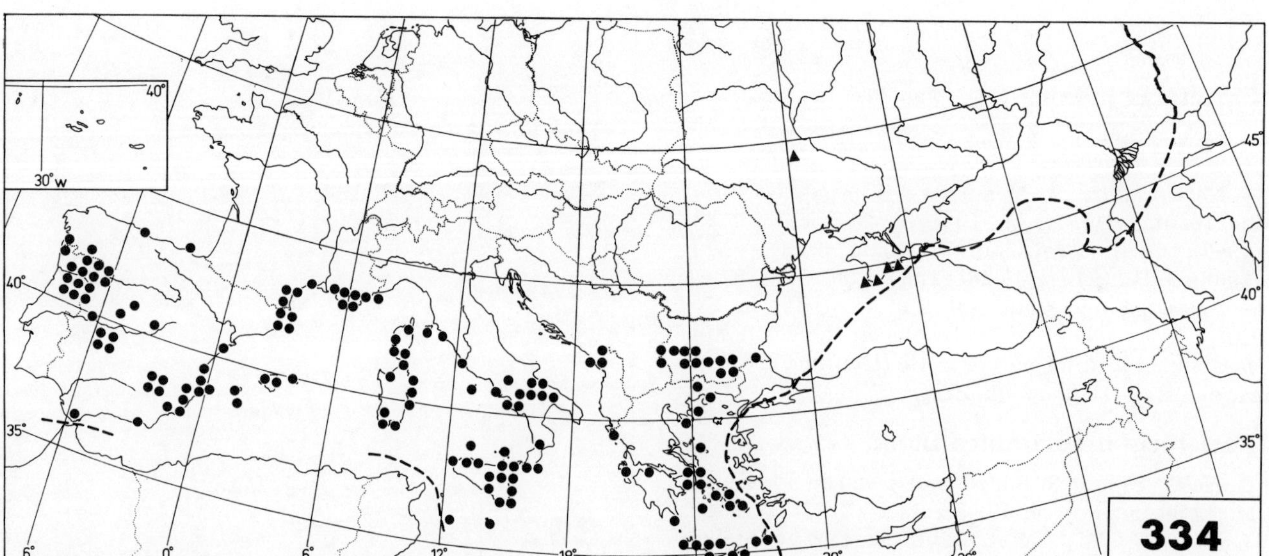

Parietaria lusitanica

● = subsp. **lusitanica**

▲ = subsp. **chersonensis**

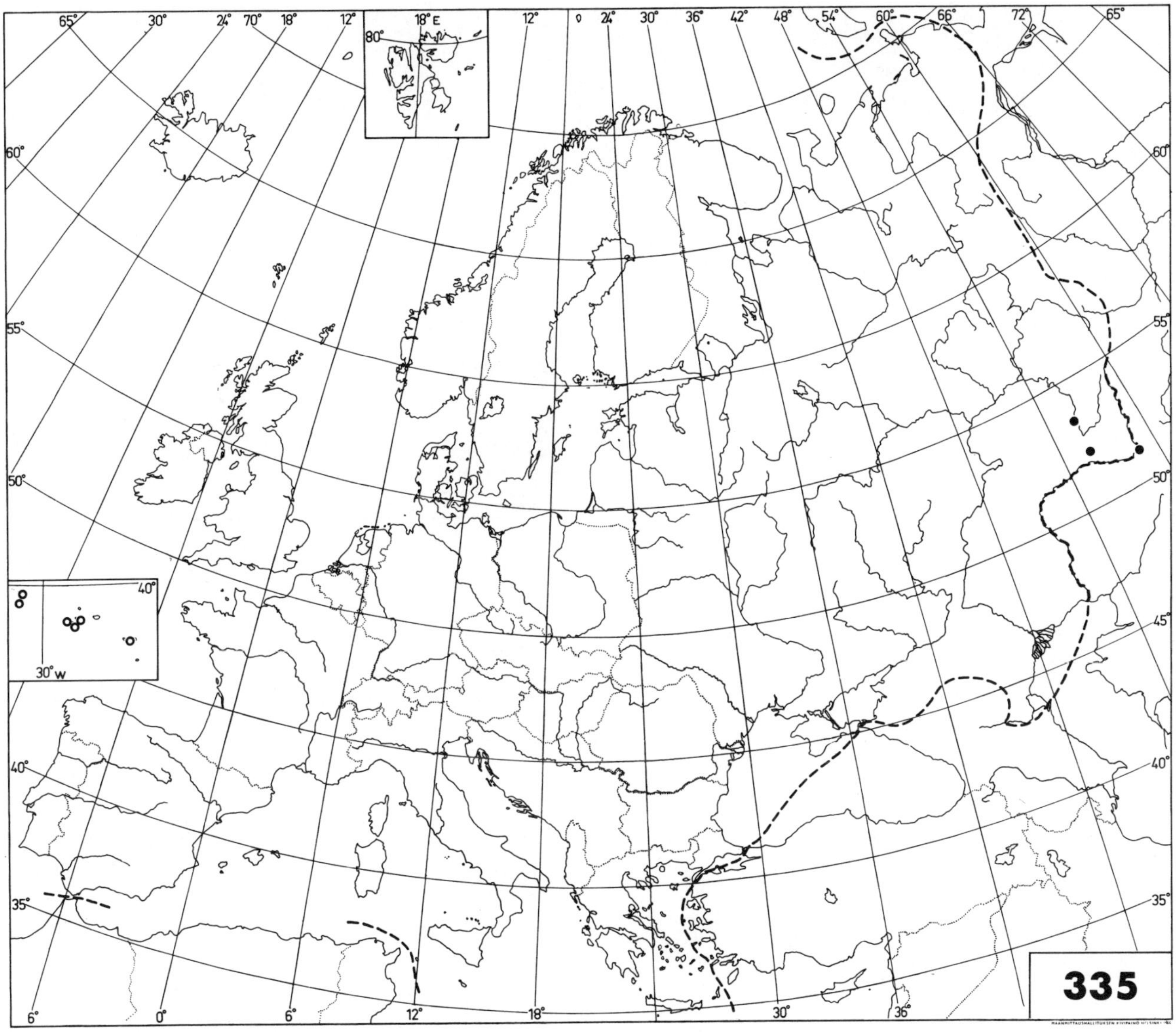

335

Parietaria debilis

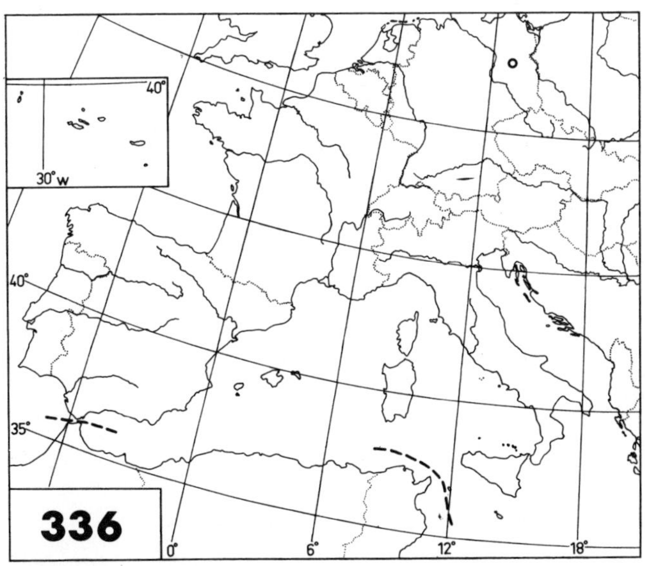

336

Parietaria pensylvanica

Parietaria debilis Forster fil. — Map 335.

Parietaria micrantha Ledeb.

Taxonomy. Parietaria micrantha Ledeb., i.e. the Ural plant, was included as a variety in *P. lusitanica* subsp. *chersonensis* by J. Chrtek, Folia Geobot. Phytotax. (Praha) 8: 426 (1973). As this was done without any mention of *P. debilis* Forster fil., and since the key to the taxonomical solution is clearly outside Europe, the species concept of Fl. Eur. is followed.

Notes. In Az as a recently introduced alien (considered native in Fl. Eur.): E. Sjögren, Mem. Soc. Brot. 22: 119 (1973).

Parietaria pensylvanica Willd. — Map 336.

Notes. [Ge] (the species not mentioned in Fl. Eur.): H. Sukopp & H. Scholz, Ber. Deutsch. Bot. Ges. 77: 419—426 (1965). Concerning the occurrence of the species in Da, Copenhagen, see A. Hansen, Fl. og Fauna 81: 37 (1975).

Native of N. America.

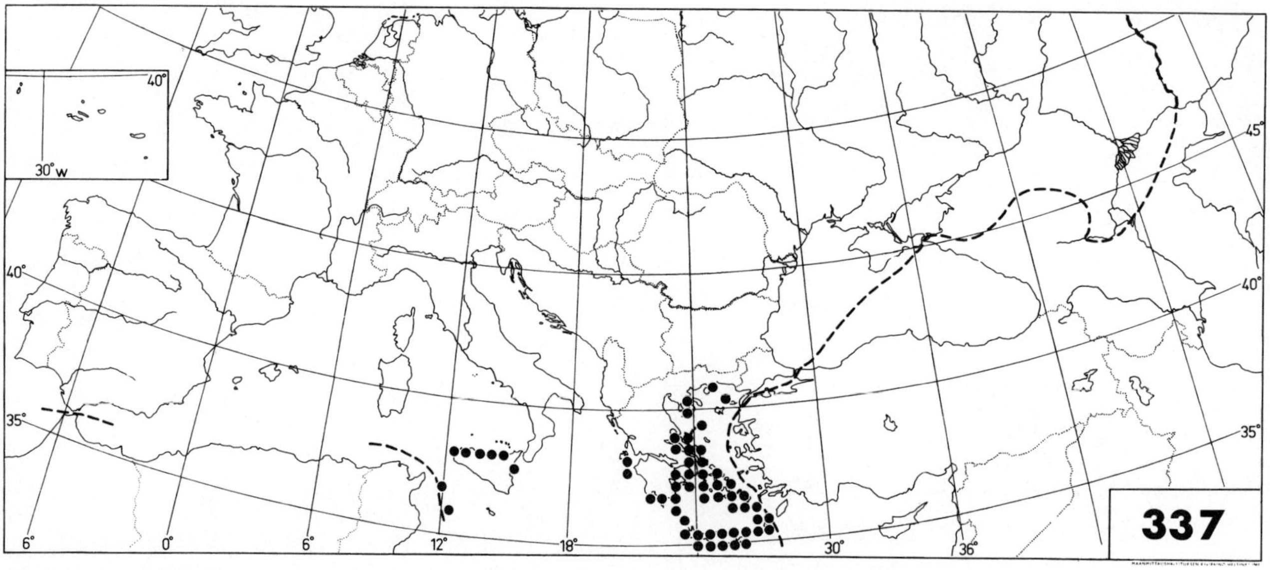

Parietaria cretica

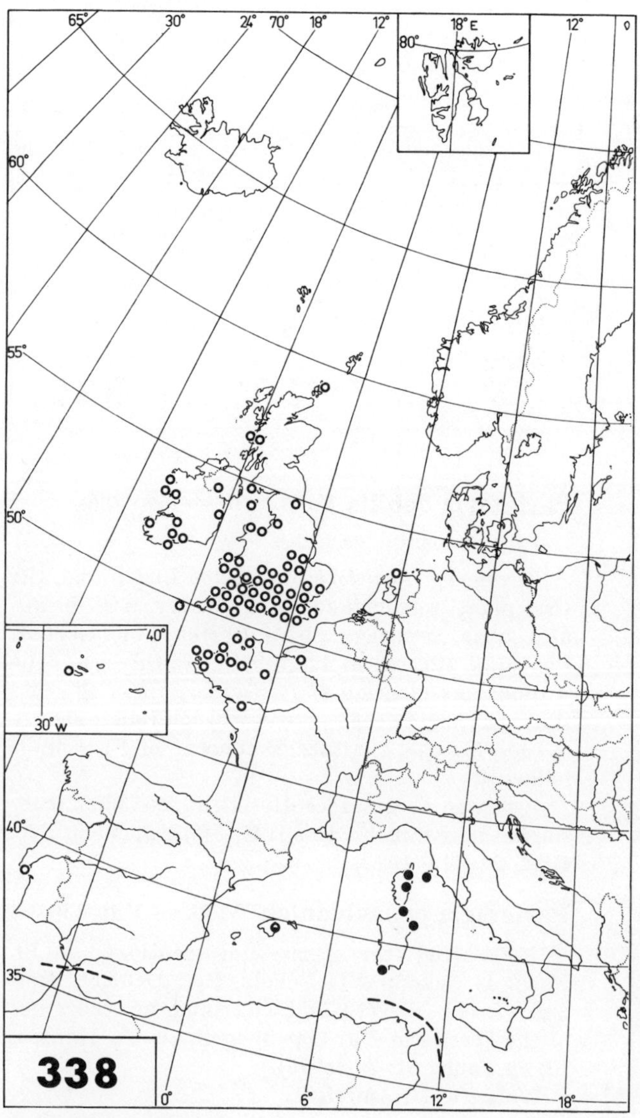

Soleirolia soleirolii

Parietaria cretica L. — Map 337.

Soleirolia soleirolii (Req.) Dandy — Map 338.

Helxine soleirolii Req.; Parietaria soleirolii (Req.) Sprengel

Notes. The map is provisional as to the secondary area. [Az] added (not given in Fl. Eur.), native status doubtful in Bl (given as native in Fl. Eur.), [Lu] added (not given in Fl. Eur.): P. A. Florschütz, Agron. Lusit. 20: 197—198 (1959).

Total range. Endemic to islands of W. Mediterranean.

Forsskalea tenacissima L. — Map 339.

Forsskalea cossoniana Webb

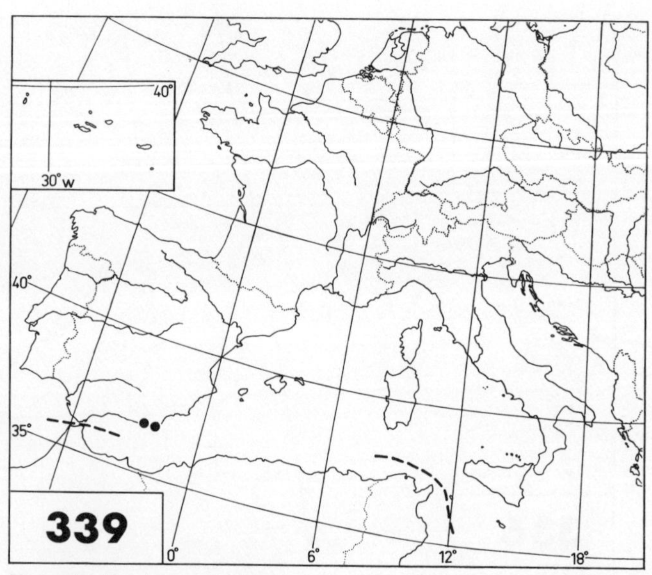

Forsskalea tenacissima

SANTALACEAE

Comandra elegans (Rochel ex Reichenb.) Reichenb. fil. — Map 340.

Thesium elegans Rochel ex Reichenb.

Notes. Al omitted (given in Fl. Eur.), Tu added (not given in Fl. Eur.).

Osyris alba L. — Map 341.

?Osyris mediterranea Bubani

Osyris quadripartita Salzm. ex Decne — Map 342.

Osyris lanceolata Steudel & Hochst. ex A. DC.

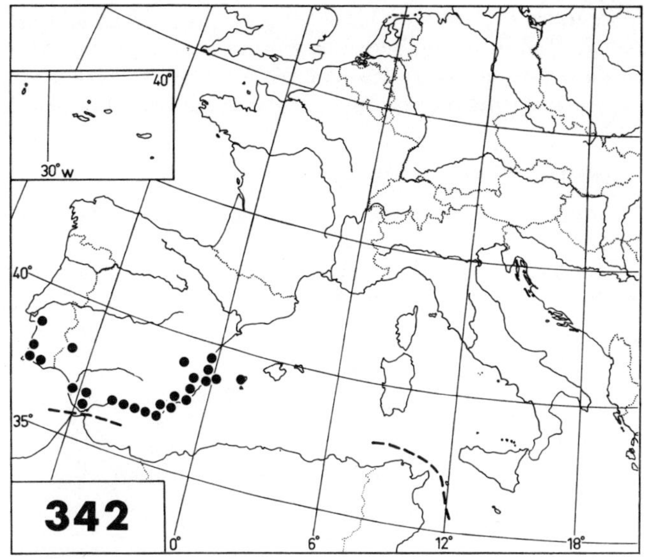

Osyris quadripartita

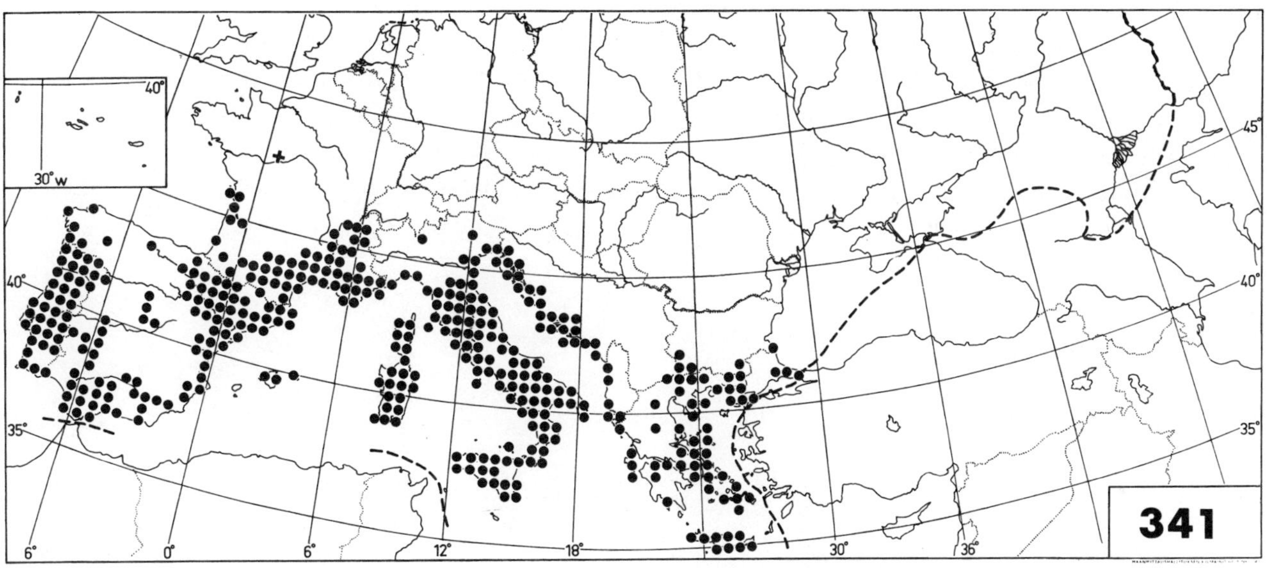

Osyris alba

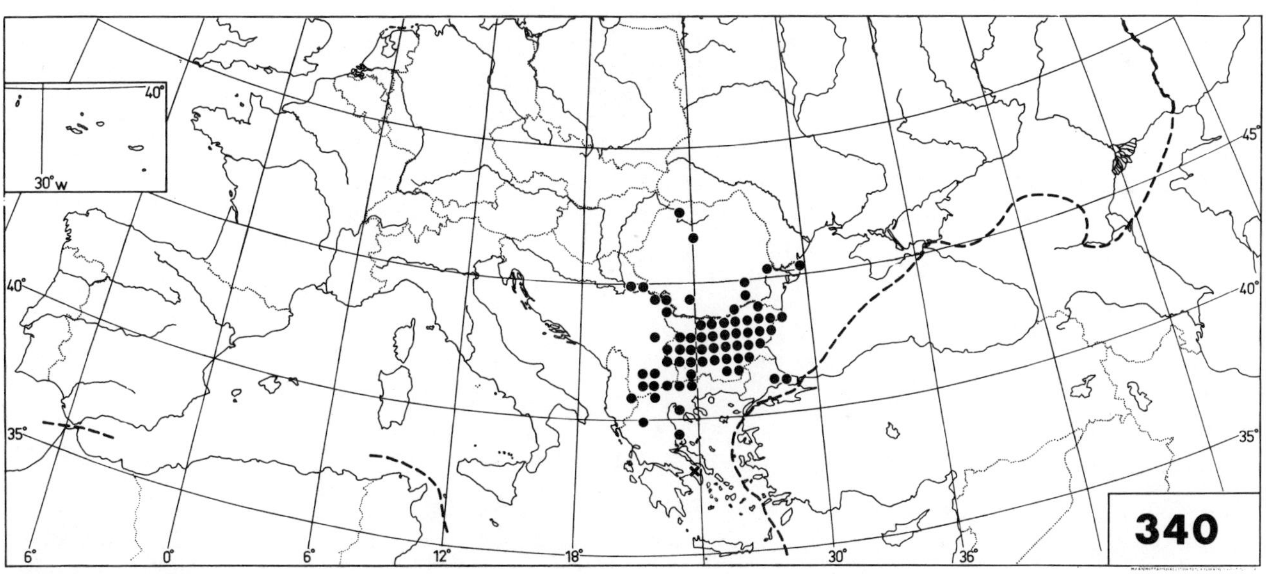

Comandra elegans

A series of monographic studies on the genus *Thesium* L. by R. Hendrych (Praha) has resulted not only in the recognition of a number of new European taxa, but also in the revision of the system affecting the sequence of species. The new arrangement (R. Hendrych, Acta Univ. Carol. (Biol.) 1970: 351 (1972)) is followed here, as it evidently forms an essential part of the advances made in the taxonomy of the genus as a whole.

Thesium ebracteatum Hayne — Map 343.

Notes. ?It added (not given in Fl. Eur.): R. Hendrych, Preslia 41: 235 (1969), Rs(W) added (not given in Fl. Eur.).

Total range. Just a few occurrences in the Ural area, east of the European border. Hendrych 1968: 39; E. Weinert in F. Fukarek & U. Schneider, Natur Naturschutz Mecklenburg 6: 74 (1968); R. Hendrych, Preslia 41: 231 (1969).

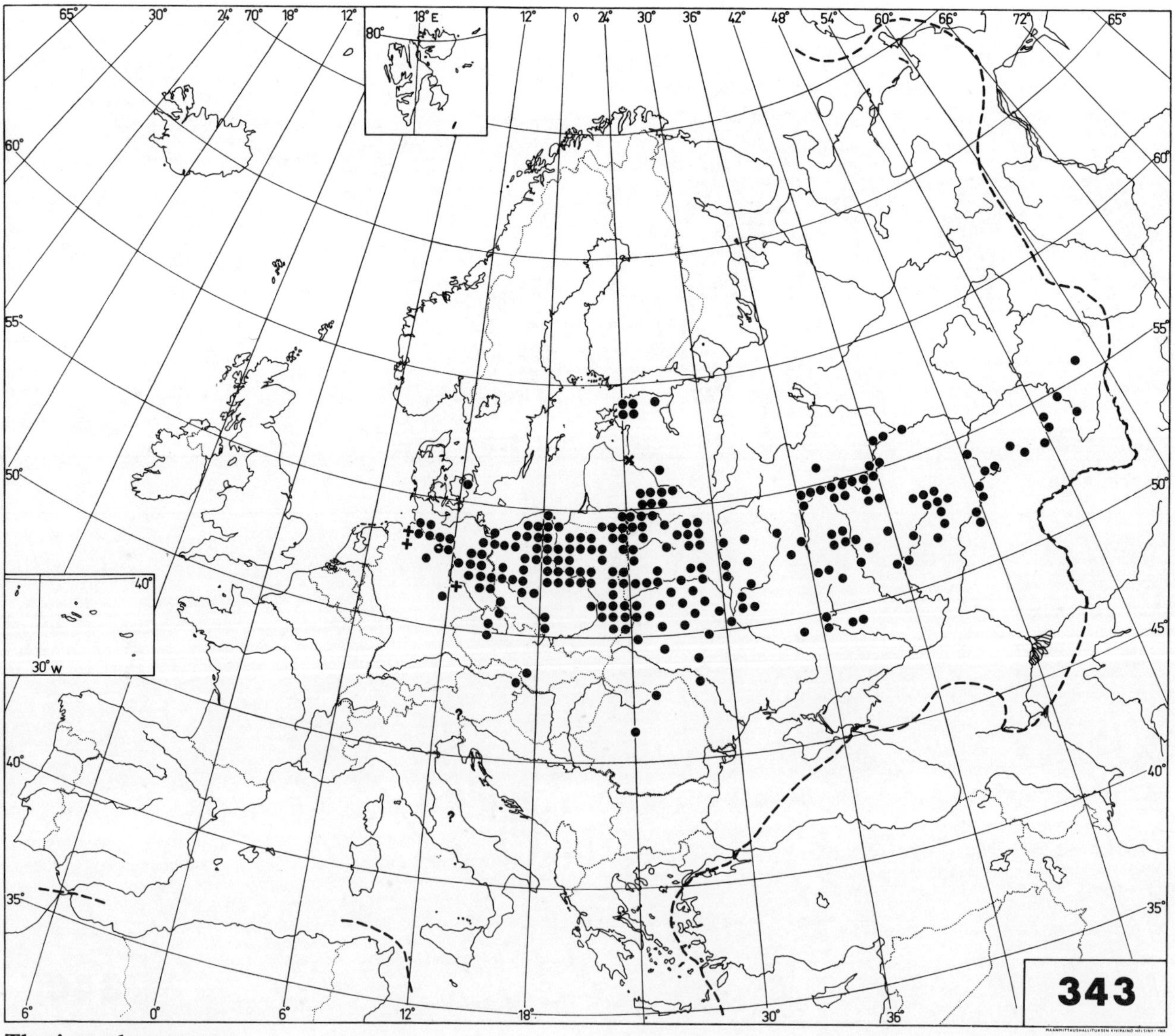

Thesium ebracteatum

Thesium linophyllon L. — Map 344.

Thesium intermedium Schrader; T. linifolium Schrank; T. linophyllon var. linifolium (Schrank) Fiori

Taxonomy. R. Hendrych, Acta Univ. Carol. (Biol.) 1969: 119—170 (1970).
Notes. Rs(B) added (not given in Fl. Eur.).
Total range. Endemic to Europe (although not marked as such in Fl. Eur.).

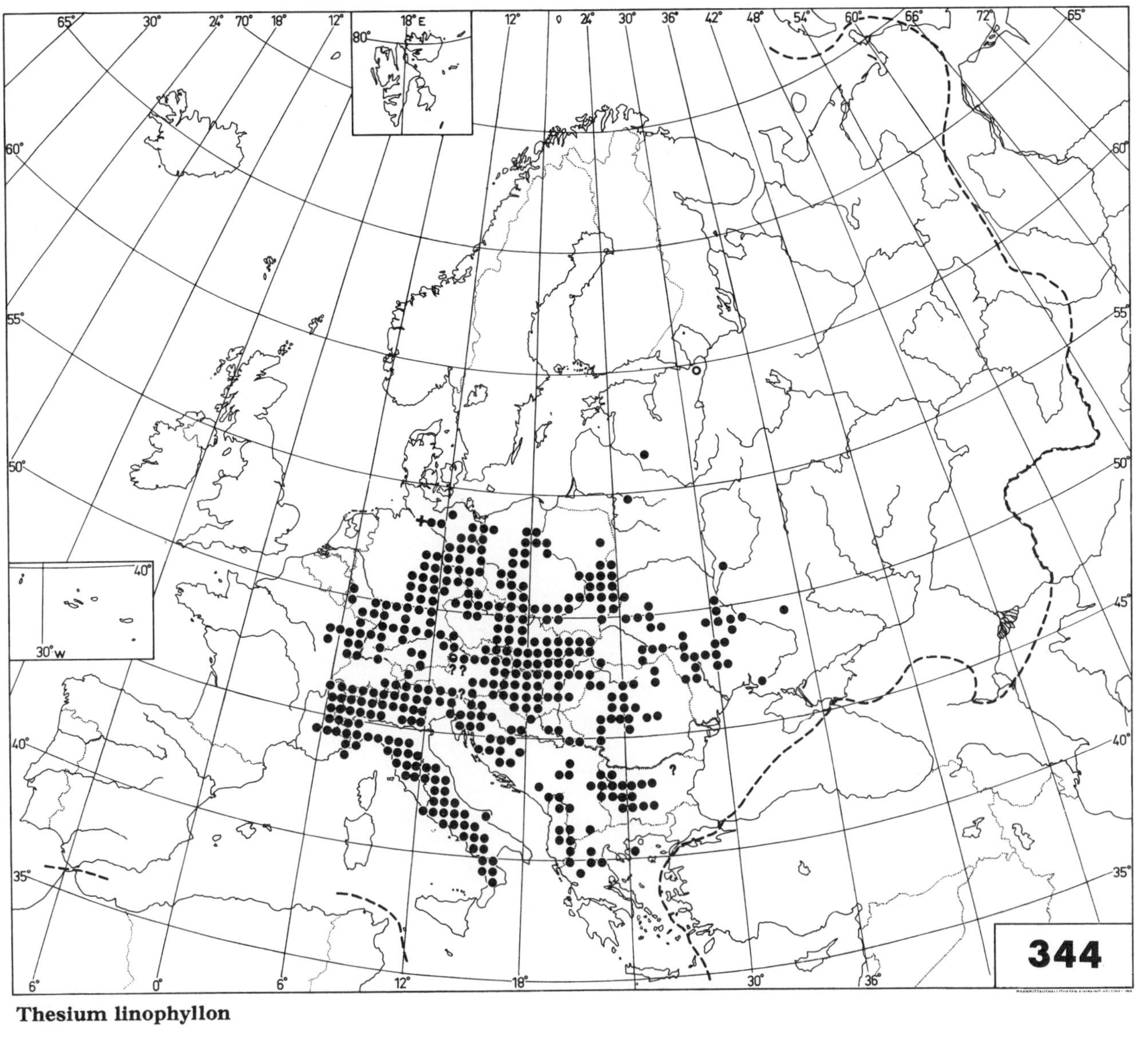

Thesium linophyllon

Thesium bavarum Schrank — Map 345.

Thesium montanum Ehrh. ex Hoffm.; T. linophyllon subsp. montanum (Ehrh. ex Hoffm.) Čelak.; T. linophyllon var. bavarum (Schrank) Fiori

Taxonomy. R. Hendrych, Acta Univ. Carol. (Biol.) 1969: 119—170 (1970).

Notes. Al added (not given in Fl. Eur.): R. Hendrych, Acta Univ. Carol. (Biol.) 1969: 165 (1970). Po omitted (given in Fl. Eur.).

Total range. Outside Europe, one locality known in N. Anatolia. MJW 1965: map 124d; Hendrych 1968: 40; R. Hendrych, Acta Univ. Carol. (Biol.) 1969: 164 (1970).

Thesium procumbens C. A. Meyer — Map 346.

Notes. Rs(E) omitted (given in Fl. Eur.).

Total range. Hendrych 1968: 45.

Thesium bergeri Zucc. — Map 346.

Total range. Hendrych 1968: 45.

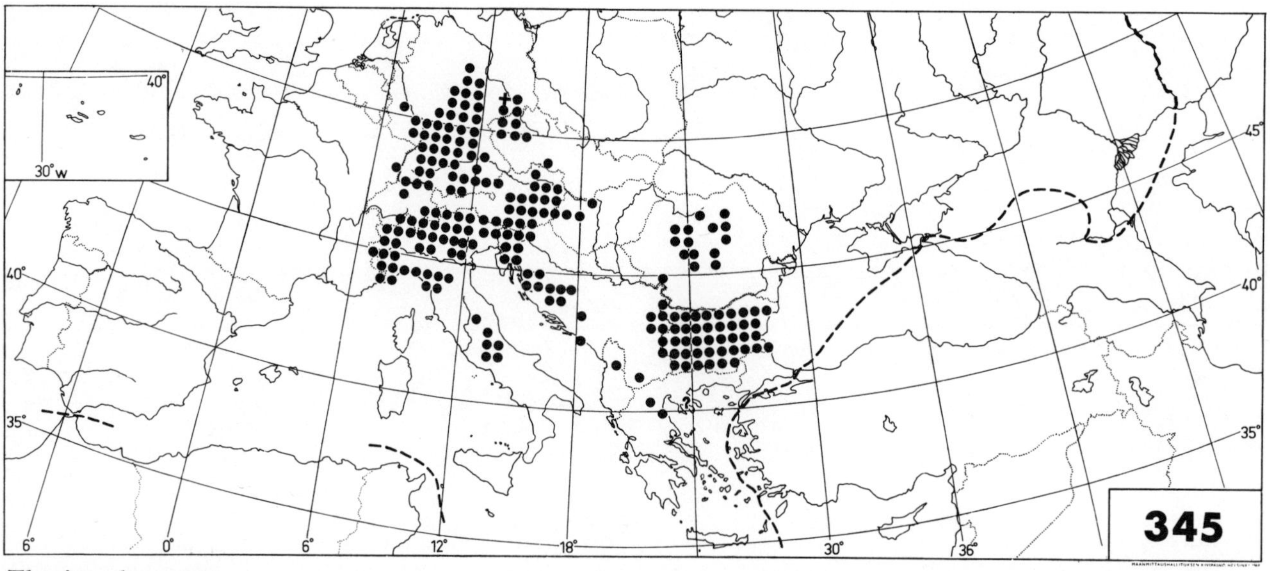

Thesium bavarum

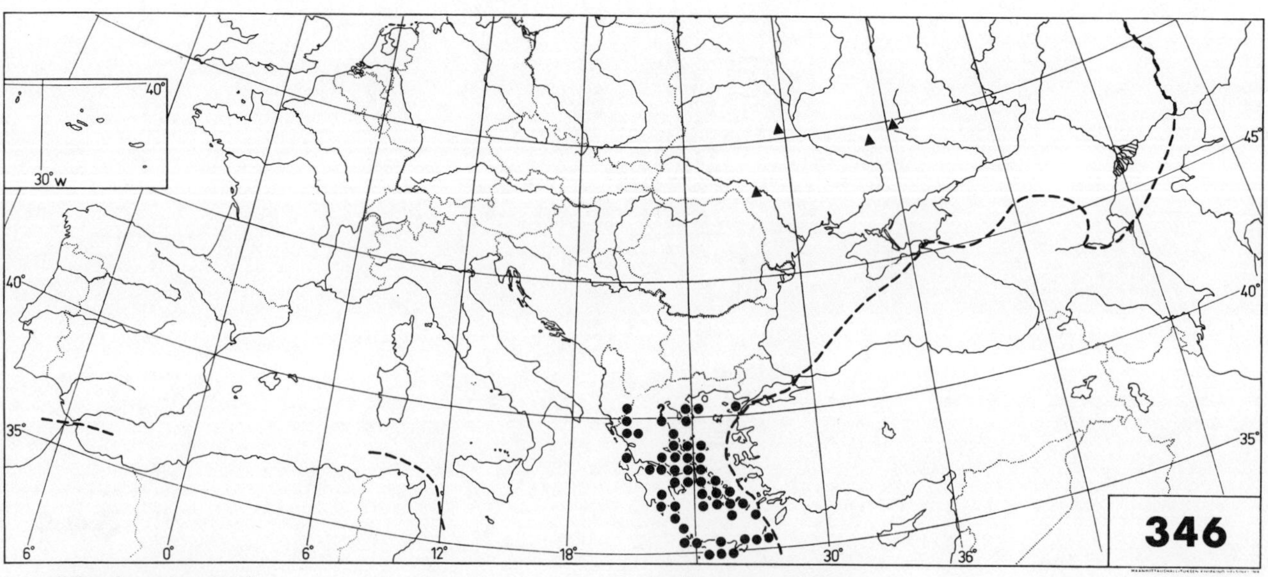

• = **Thesium bergeri** ▲ = **T. procumbens**

Thesium humifusum DC. — Map 347.

Thesium divaricatum auct. non Jan ex Mert. & Koch; T. divaricatum var. humifusum Duby

Notes. Extinct in Ho (given as present in Fl. Eur.).

Total range. Endemic to Europe.

Thesium divaricatum Jan ex Mert. & Koch — Map 348.

Thesium divaricatum subsp. vandasii Rohlena; T. linophyllon var. divaricatum (Jan ex Mert. & Koch) Fiori; T. humifusum subsp. divaricatum (Jan ex Mert. & Koch) Bonnier. Incl. T. nevadense Willk.

Notes. Bl added (not given in Fl. Eur.), Co, Cr and Rm omitted (given in Fl. Eur.).

Total range. Hendrych 1968: 41.

Thesium italicum A. DC. — Map 348.

Thesium humifusum subsp. italicum (A. DC.) Rouy; T. linophyllon var. italicum (A. DC.) Fiori

Taxonomy. R. Hendrych, Novit. Bot. (Praha) 1964: 19—21 (1964).

Notes. Co omitted (given in Fl. Eur.), because the records refer to *Thesium corsalpinum* Hendrych.

Total range. Endemic to Sardegna.

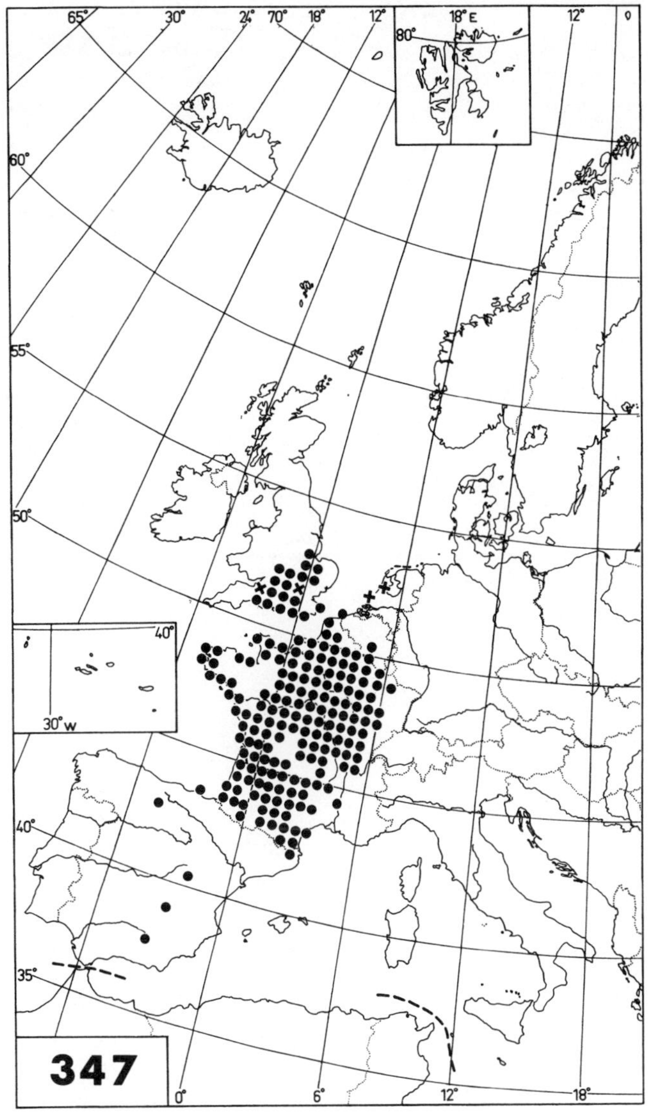

Thesium humifusum

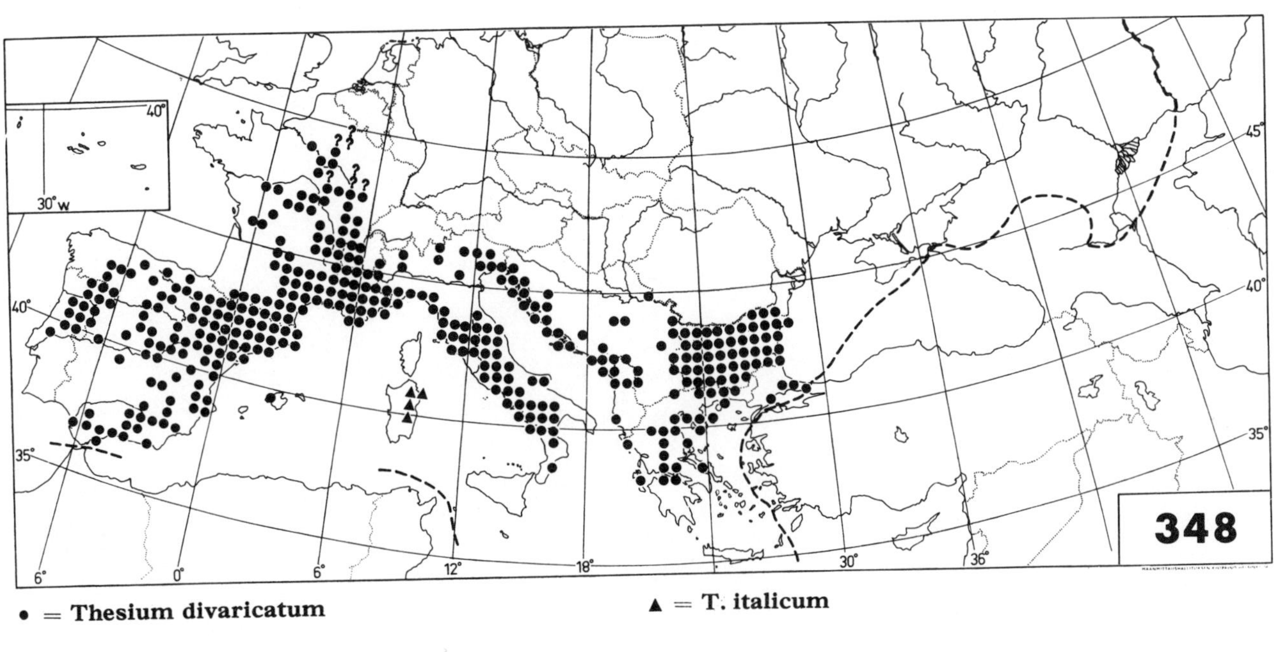

• = **Thesium divaricatum** ▲ = **T. italicum**

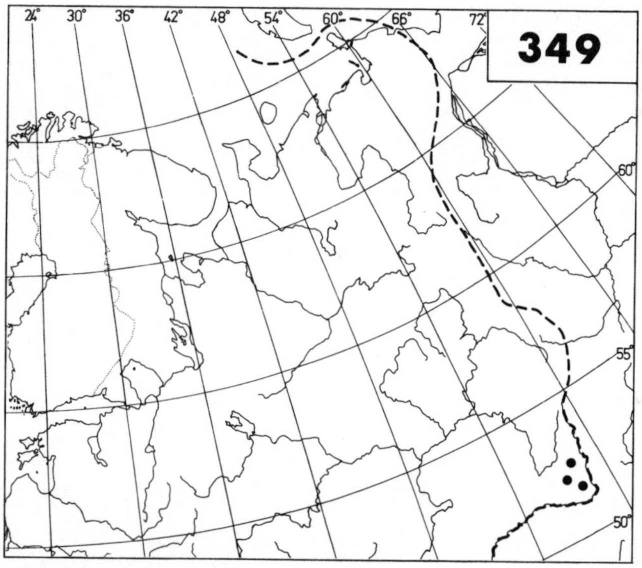

Thesium refractum

Thesium refractum C. A. Meyer — Map 349.

Notes. Rs(C) (the species not mentioned in Fl. Eur.).

Total range. Hendrych 1968: 49.

Thesium arvense Horvátovszky — Map 350.

Thesium ramosum Hayne; T. linophyllon var. ramosum (Hayne) Fiori, pro parte. Incl. T. bulgaricum Velen.

Taxonomy. R. Hendrych, Acta Univ. Carol. (Biol.) 1968: 243—262 (1969).

Notes. Rs(K) and Tu added (not given in Fl. Eur.).

Total range. R. Hendrych, Acta Univ. Carol. (Biol.) 1968: 249 (1969).

Thesium dollineri Murb.

Thesium humile auct. non Vahl

Taxonomy and nomenclature. R. Hendrych, Acta Univ. Carol. (Biol.) 1970: 359—382 (1972).

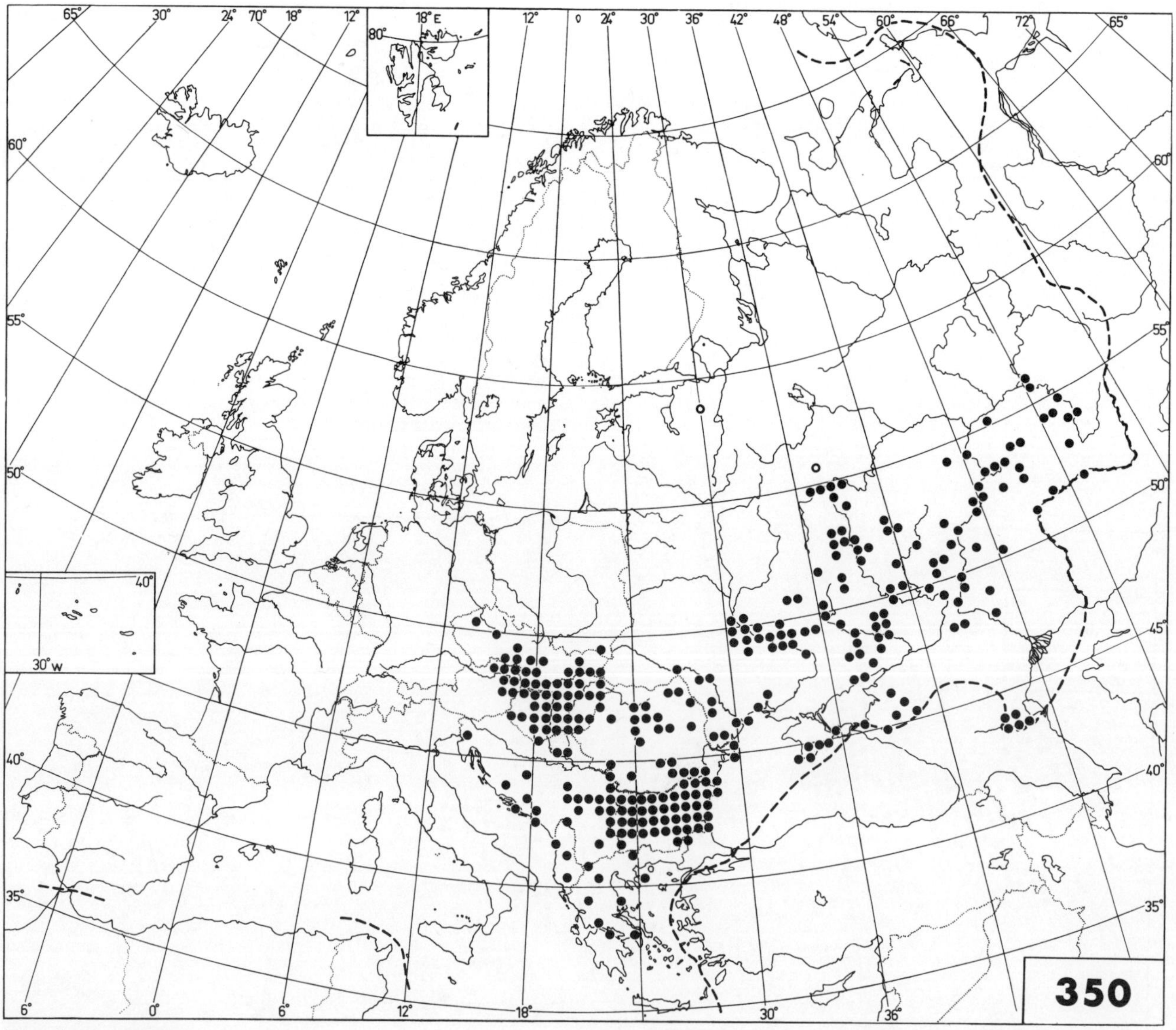

Thesium arvense

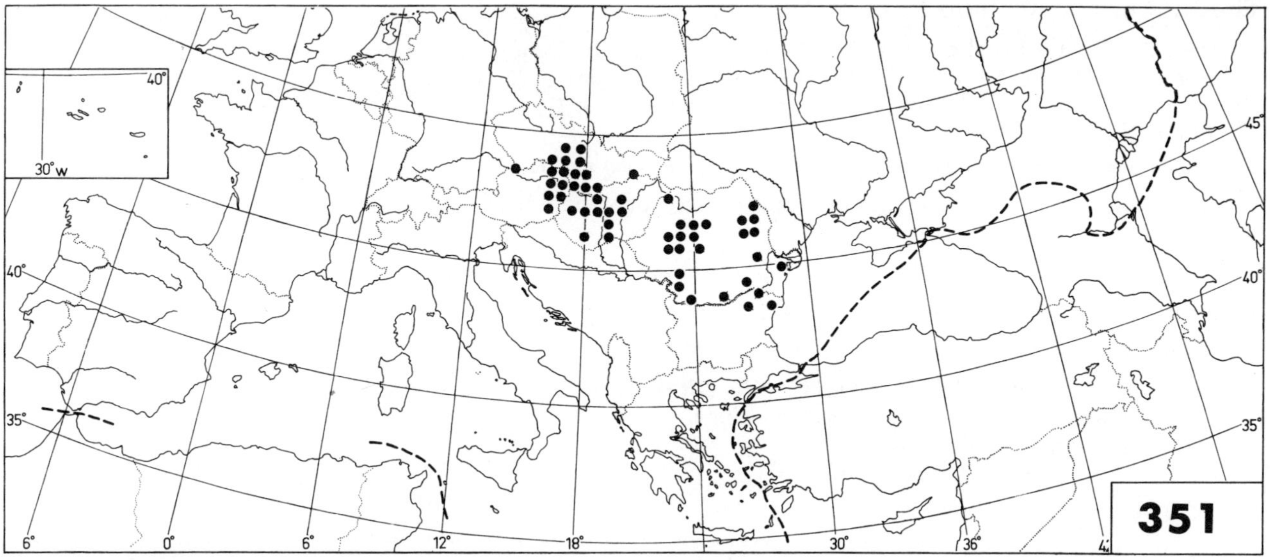

Thesium dollineri subsp. **dollineri**

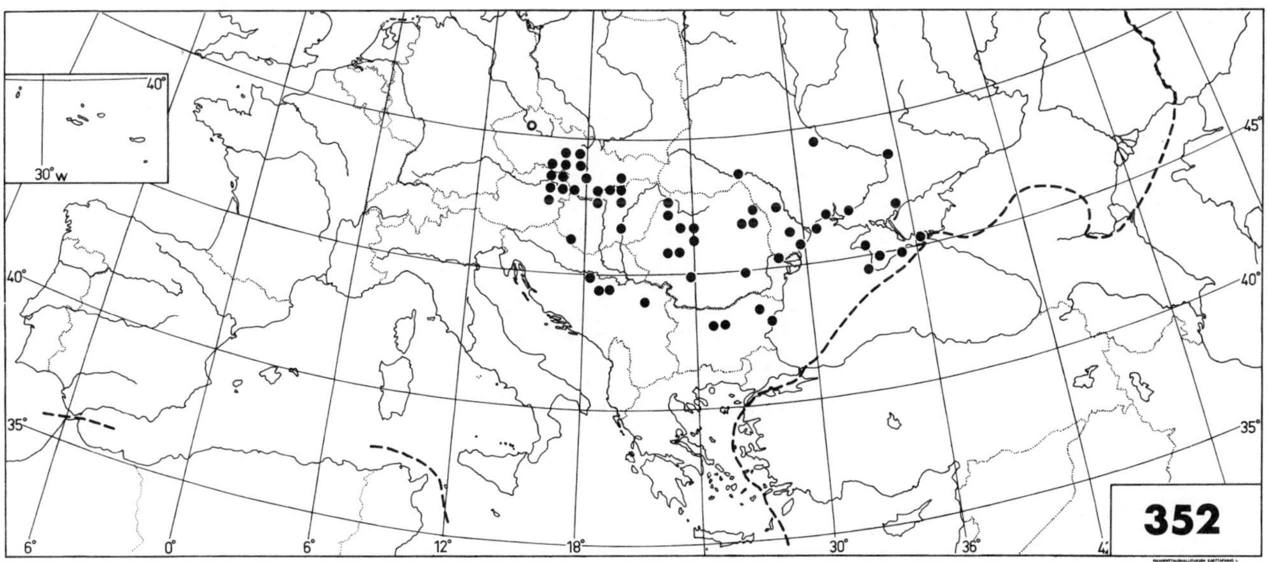

Thesium dollineri subsp. **simplex**

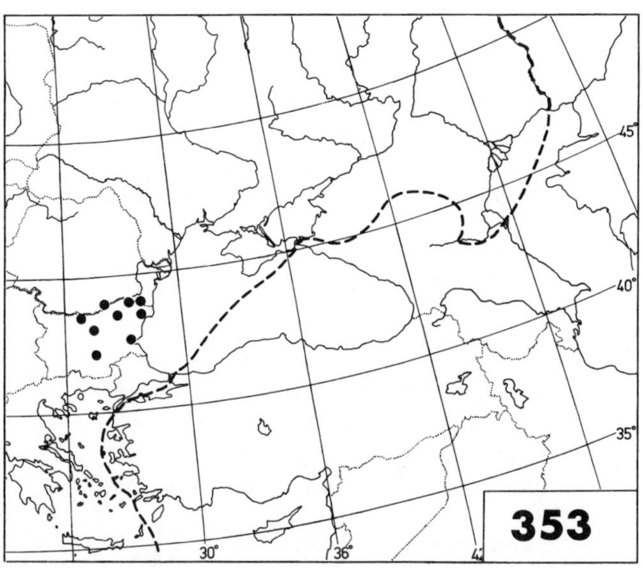

Thesium dollineri subsp. **moesiacum**

Notes. The subspecies have been mapped according to R. Hendrych, Acta Univ. Carol. (Biol.) 1970: 359—382 (1972).

Total range. Endemic to Europe.

T. dollineri subsp. **dollineri** — Map 351.

Notes. Au, Bu, Cz, Hu, Rm (Fl. Eur.: »Throughout the range of the species»).

T. dollineri subsp. **simplex** (Velen.) Stoj. & Stefanov — Map 352.

Thesium simplex Velen. Excl. T. moesiacum Velen.

Notes. Au, Bu, Cz, Hu, Ju, Rm, Rs(W, K).

T. dollineri subsp. **moesiacum** (Velen.) Stoj. & Stefanov — Map 353.

Thesium moesiacum Velen.; T. simplex Velen. subsp. moesiacum (Velen.) Kozuharov & Kuzmanov

Taxonomy. R. Hendrych, Acta Univ. Carol. (Biol.) 1970: 379—381 (1972).

Notes. Bu (included in subsp. *simplex* in Fl. Eur.).

Thesium brachyphyllum Boiss. — Map 354.

Notes. Rs(E) added (not given in Fl. Eur.). See also under *Thesium parnassi.*
Total range. R. Hendrych, Folia Geobot. Phytotax. (Praha) 1: 73 (1966).

Thesium parnassi A. DC. — Map 354.

Thesium linophyllon var. parnassi (A. DC.) Fiori; T. linophyllon var. ramosum (Hayne) Fiori, pro parte

Notes. Some of the N. Greek records mapped here apparently belong to *Thesium brachyphyllum* (B. Aldén, in litt.).
Total range. Endemic to Europe.

Thesium sommieri Hendrych — Map 355.

Thesium ramosum var. leve Sommier

Taxonomy. R. Hendrych, Preslia 36: 114—122 (1964).
Notes. It (the species not mentioned in Fl. Eur.).
Total range. Endemic to Europe.

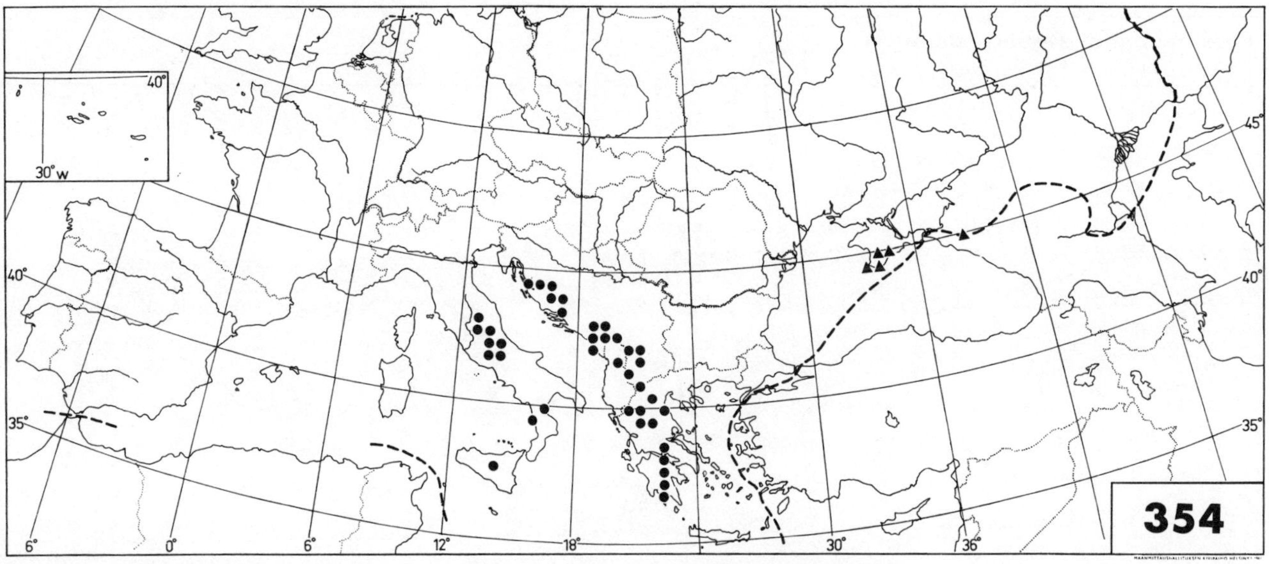

• = **Thesium parnassi** ▲ = **T. brachyphyllum**

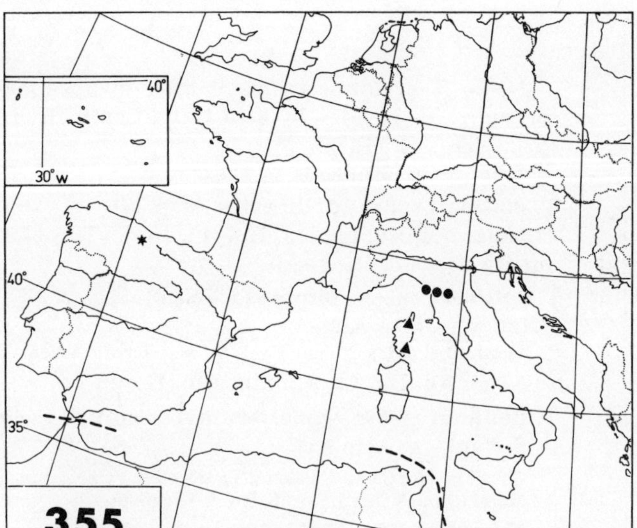

• = **Thesium sommieri**
▲ = **T. kyrnosum**
★ = **T. hispanicum**

Thesium kyrnosum Hendrych — Map 355.

Thesium alpinum auct. non L.; T. italicum auct. non A. DC.; T. ramosum auct. non Hayne

Taxonomy. R. Hendrych, Novit. Bot. Horti Bot. Univ. Carol. Prag. 1964: 19—21 (1964).
Notes. Co (given in Fl. Eur. as *Thesium alpinum*). For detailed list of localities, see J. Gamisans, Candollea 28: 56 (1973).
Total range. Endemic to Corse.

Thesium hispanicum Hendrych — Map 355.

Taxonomy. R. Hendrych, Folia Geobot. Phytotax. (Praha) 1: 70—77 (1966).
Notes. Hs (the species not mentioned in Fl. Eur.).
Total range. Endemic to Europe.

Thesium humile Vahl — Map 356.

Notes. No records from Ga (?Ga in Fl. Eur.).
Total range. Hendrych 1968: 47.

Thesium macedonicum Hendrych — Map 356.

Taxonomy. R. Hendrych, Biol. Glasnik 17: 11—15 (1964).
Notes. Ju (the species not mentioned in Fl. Eur.).
Total range. Endemic to Europe.

Thesium auriculatum Vandas — Map 357.

Taxonomy. R. Hendrych, Acta Horti Bot. Prag. 1963: 33—46 (1964).
Total range. Endemic to Europe.

Thesium kerneranum Simonkai — Map 357.

Total range. Endemic to Europe. For details, see R. Hendrych, Novit. Bot. Horti Bot. Univ. Carol. Prag. 1963: 7—9 (1963).

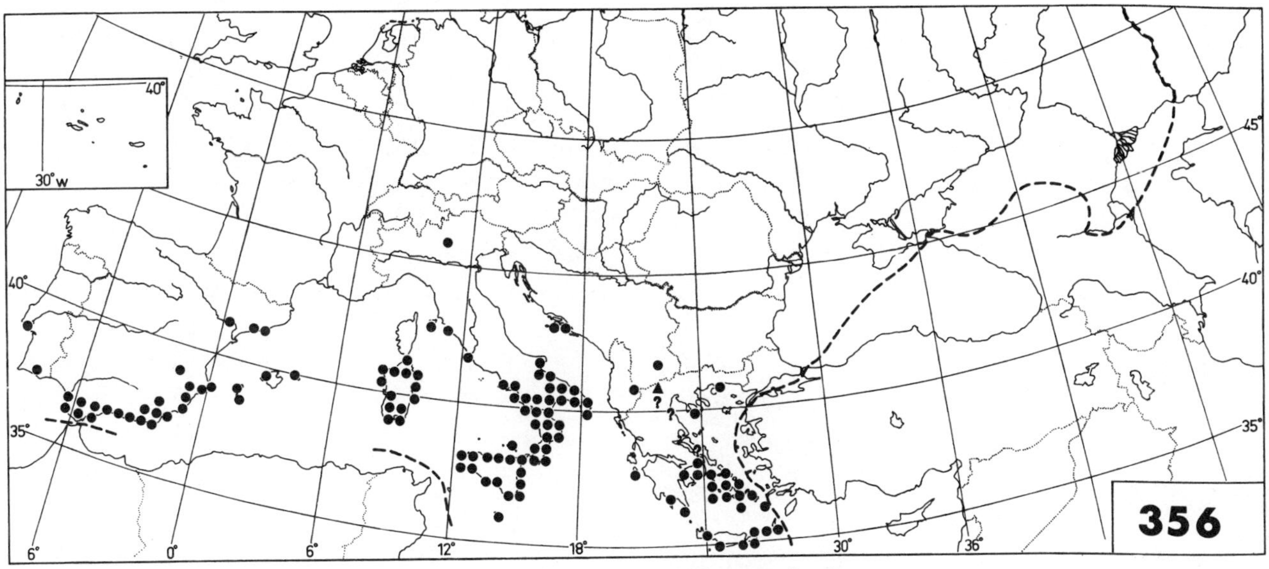

• = **Thesium humile** ▲ = **T. macedonicum**

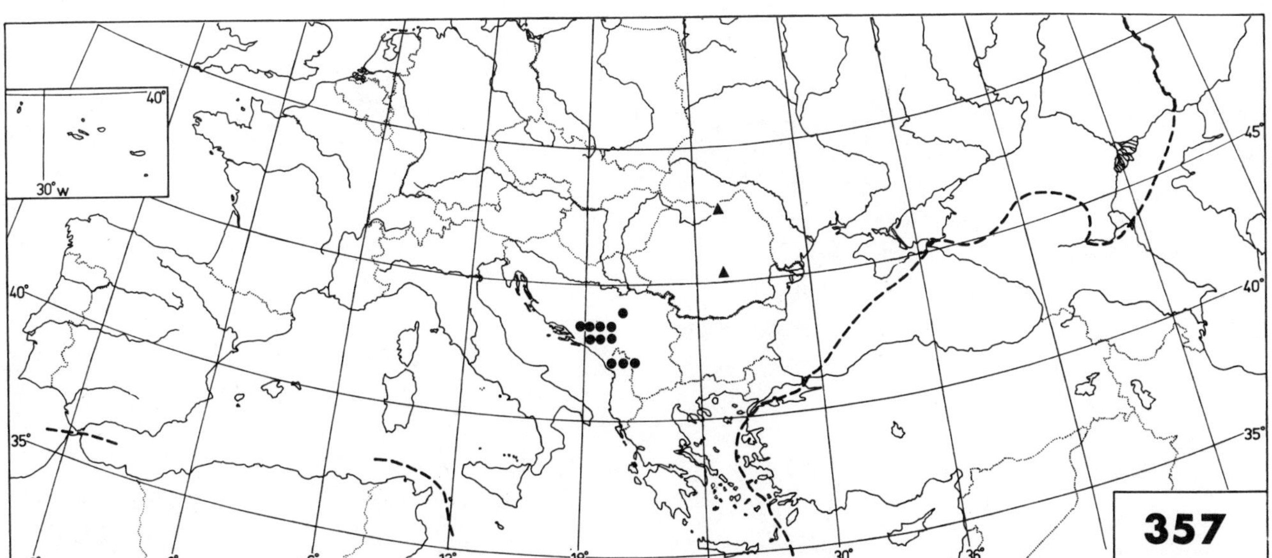

• = **Thesium auriculatum** ▲ = **T. kerneranum**

Thesium pyrenaicum Pourret — Map 358.

Thesium grandiflorum (A. DC.) Hand.-Mazz.; T. pratense Ehrh. ex Schrader; T. alpinum var. pyrenaicum (Pourret) Fiori

Taxonomy. R. Hendrych, Acta Univ. Carol. (Biol.) 1969: 93—117 (1970).
Total range. Endemic to Europe.

T. pyrenaicum subsp. **pyrenaicum** — Map 358.

Notes. Au, Be, Cz, Ga, Ge, He, †Ho, Hs, It, Lu, Po. »Throughout the range of the species» (Fl. Eur.).

T. pyrenaicum subsp. **grandiflorum** (A. DC.) Hendrych — Map 358.

Thesium grandiflorum (A. DC). Hand.-Mazz.; T. pyrenaicum subsp. alpestre O. Schwarz, nom. illeg.

Taxonomy and nomenclature. R. Hendrych, Novit. Bot. Horti Bot. Univ. Carol. Prag. 1964: 9 (1964) and Acta Univ. Carol. (Biol.) 1969: 93—117 (1970).

Notes. Au, It, Ju. Mapped mainly according to the information given by R. Hendrych, Acta Univ. Carol. (Biol.) 1969: 115—116 (1970). »E. Alps and mountains of N. Jugoslavia» (Fl. Eur.).

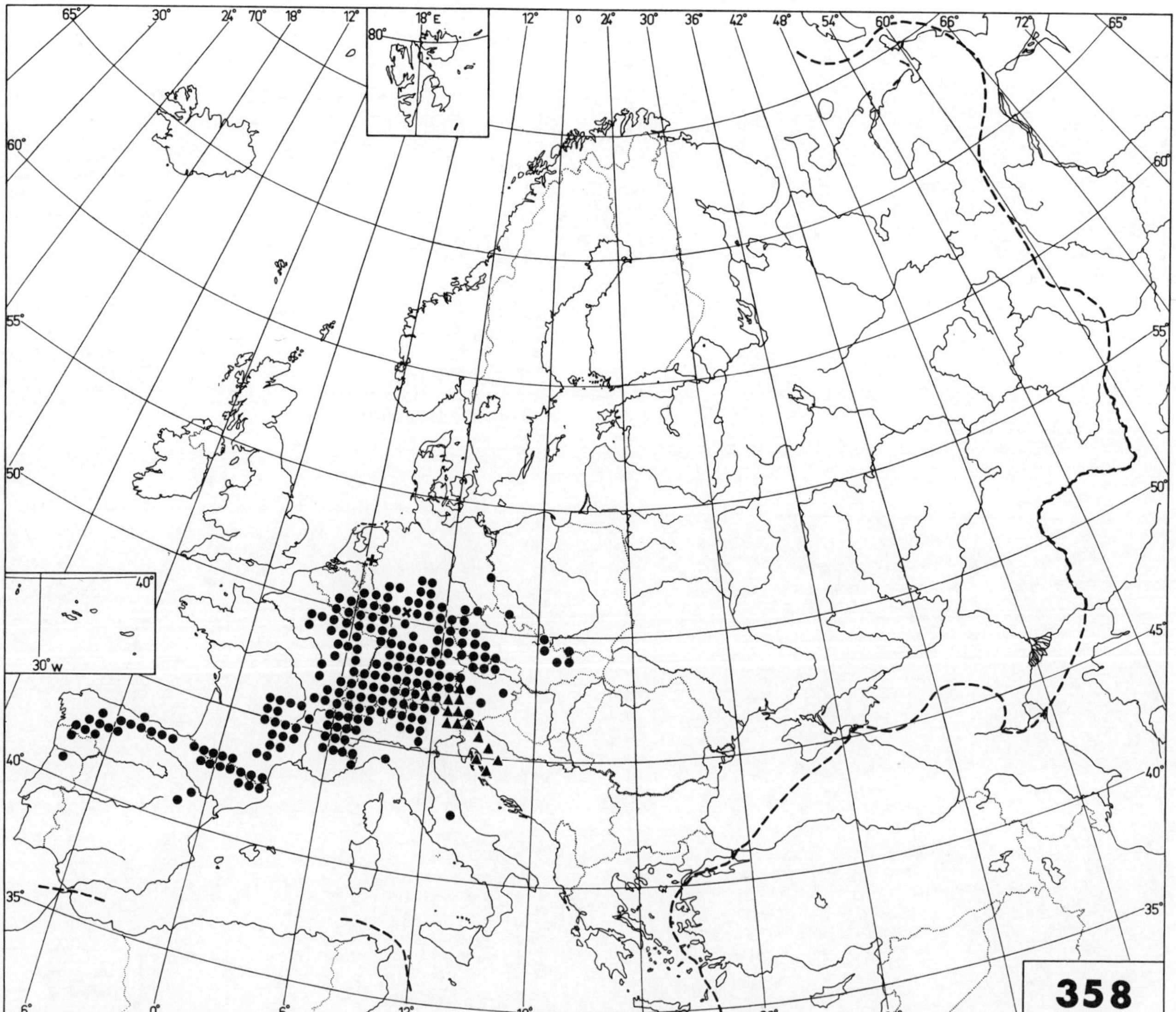

Thesium pyrenaicum

● = subsp. **pyrenaicum**
▲ = subsp. **grandiflorum** and (?) **pyrenaicum**

Thesium alpinum L. — Map 359.

Incl. Thesium tenuifolium Sauter ex Koch

Taxonomy. R. Hendrych, Acta Univ. Carol. (Biol.) 1966 (2): 107—138 (1967).

Notes. Co omitted (given in Fl. Eur.), because the material belongs to *Thesium kyrnosum*; extinct in Da (Da not given in Fl. Eur.); Rs(B) added (not given in Fl. Eur.).

Total range. MJW 1965: map 125c; R. Hendrych, Acta Univ. Carol. (Biol.) 1966 (2): 112 and 114 (1967); N. A. Miniaev, Vestn. Leningrad. Univ. (Biol.) 1969: 16 (1969).

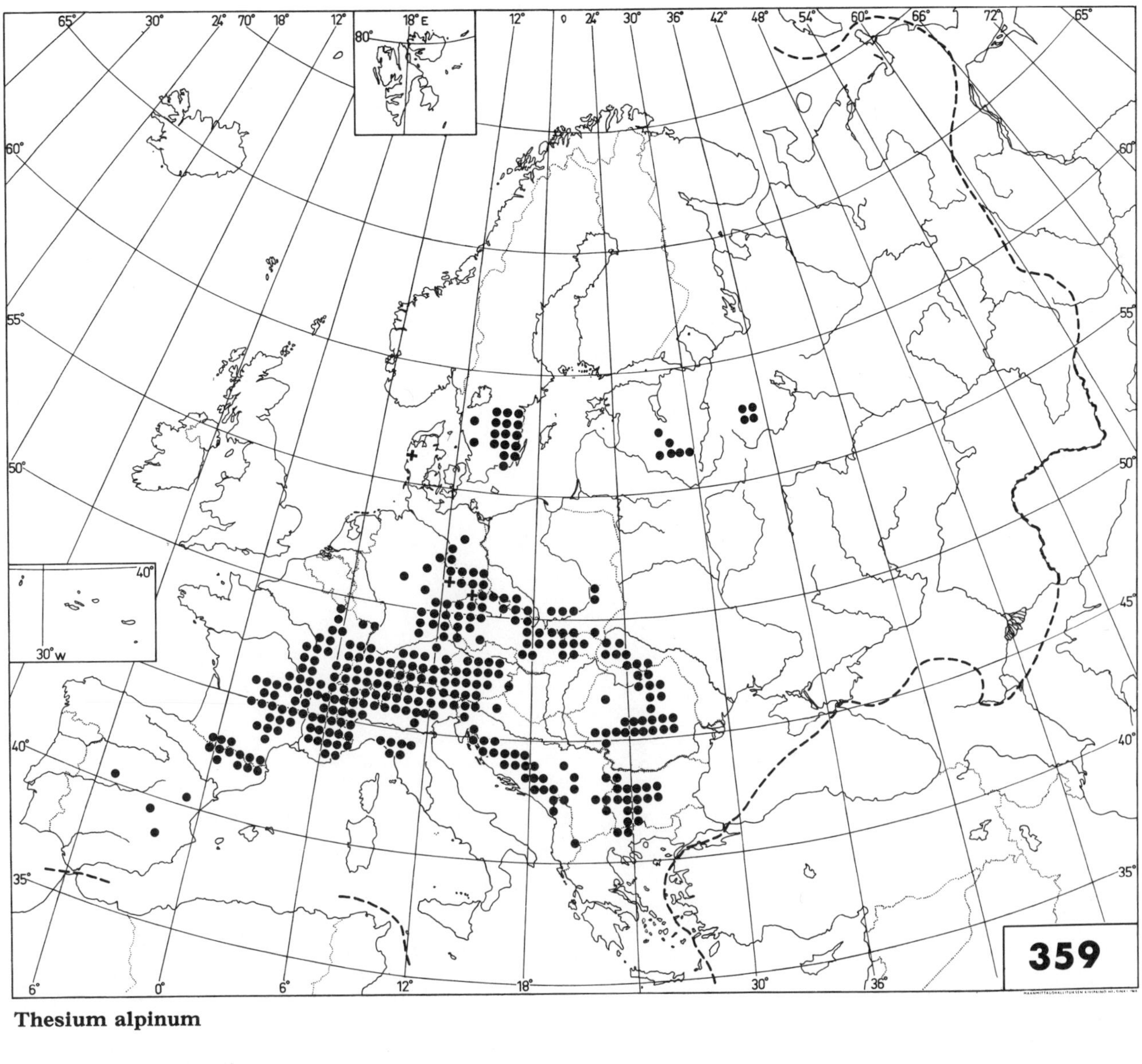

Thesium alpinum

Thesium corsalpinum Hendrych — Map 360.

Thesium ramosum var. leve auct. non Sommier; T. italicum auct. non A. DC.

Taxonomy. R. Hendrych, Novit. Bot. Horti Bot. Univ. Carol. Prag. 1964: 19—21 (1964).

Notes. Co (not distinguished from *Thesium italicum* in Fl. Eur.). For detailed list of localities, see J. Gamisans, Candollea 28: 56 (1973).

Total range. Endemic to Corse.

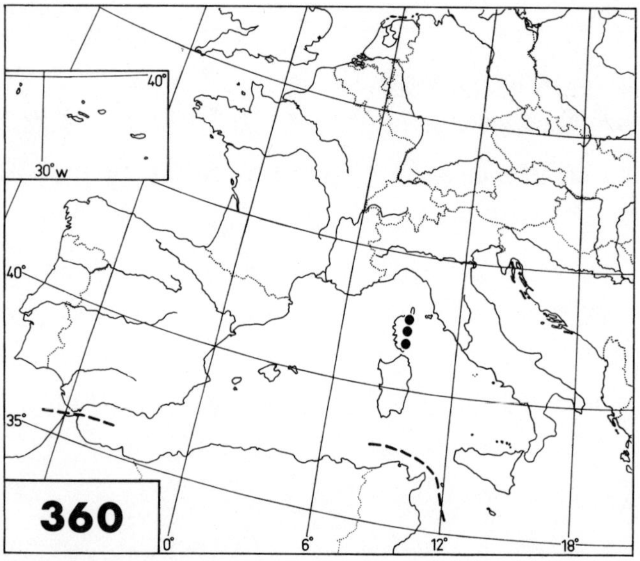

Thesium corsalpinum

Thesium rostratum Mert. & Koch — Map 361.

Notes. Ju added (not given in Fl. Eur.), Rm omitted (?Rm in Fl. Eur.).

Total range. Endemic to Europe.

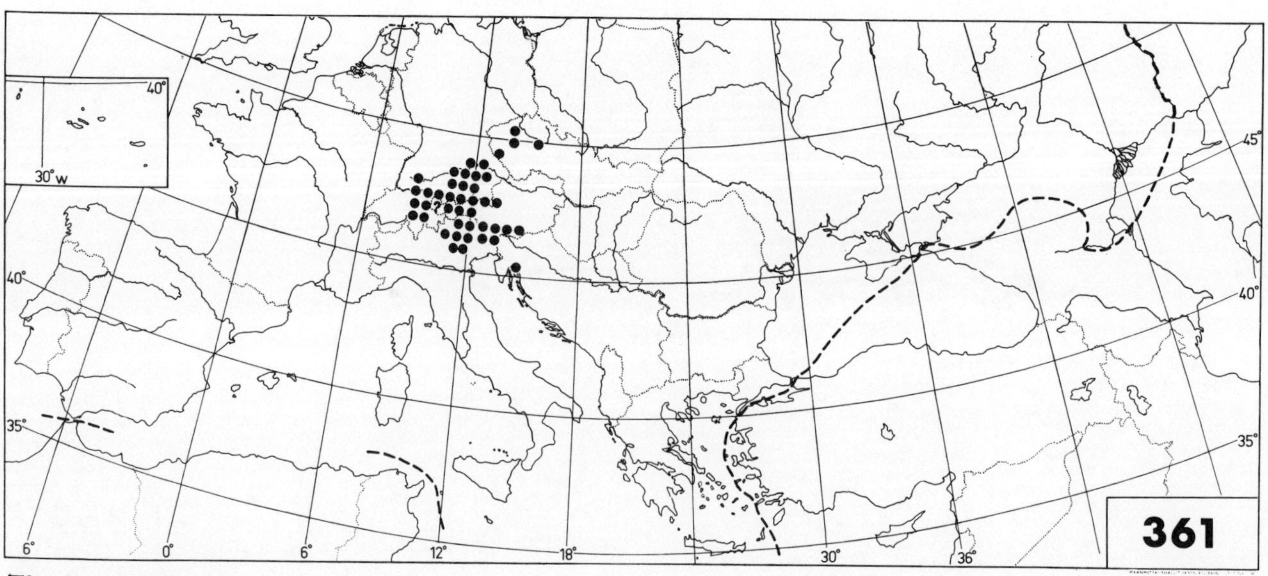

Thesium rostratum

LORANTHACEAE

Loranthus europaeus Jacq. — Map 362.

Total range. MJW 1965: map 126b; E. J. Jäger, Feddes Repert. 81: 86 (1970).

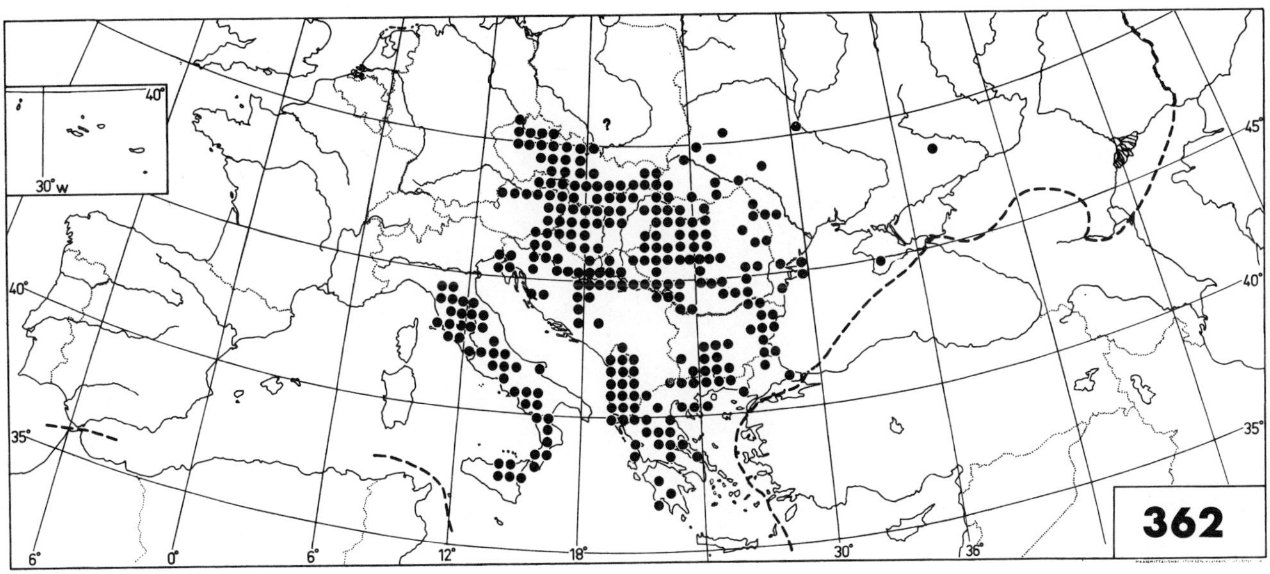

Loranthus europaeus

Viscum cruciatum Sieber ex Boiss. — Map 363.

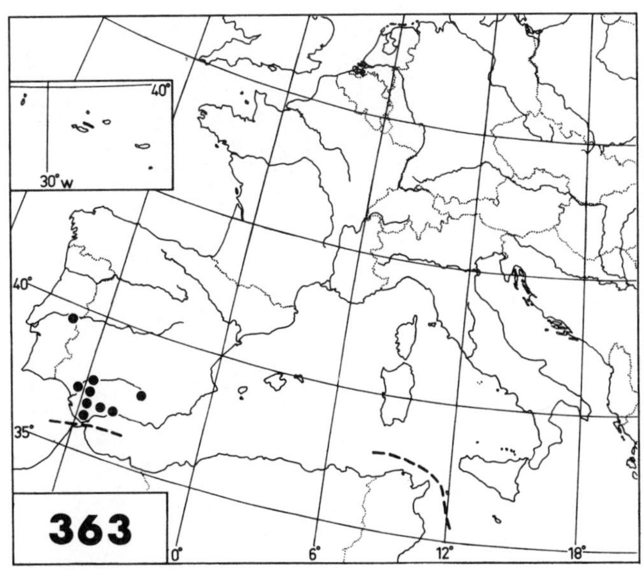

Viscum cruciatum

Viscum album L. — Map 364.

Incl. Viscum laxum Boiss. & Reuter

Notes. Besides the collective map for the species, only preliminary and incomplete maps could be prepared of the distribution of the subspecies. Of these, the two for the conifericolous subsp. *abietis* and subsp. *austriacum* are reproduced here. Bl omitted (given in Fl. Eur.), [Hb] and Rs(E) added (not given in Fl. Eur.).

Total range. MJW 1965: map 126c; G. Lang, Feddes Repert. 81: 329 (1970).

V. album subsp. **album**

Notes. »On dicotyledonous trees. Throughout the range of the species» (Fl. Eur.). Not mapped separately.

V. album subsp. **abietis** (Wiesb.) Abromeit — Map 365.

Viscum laxum var. abietis (Wiesb.) Hayek

Notes. Au, Bu, Co, Cz, Ga, Ge, Gr, He, Hs, Hu, It, Ju, Po, Rm. »On *Abies* spp. C. & S. Europe» (Fl. Eur.).

V. album subsp. **austriacum** (Wiesb.) Vollmann — Map 366.

Viscum laxum var. pini (Wiesb.) Hayek

Notes. Au, Bu, Cr, Cz, Ge, Gr, Hs, Hu, It, Ju, Po, Rm. »On *Pinus* and *Larix* spp. C. & S. Europe northwards to c. 54° N» (Fl. Eur.).

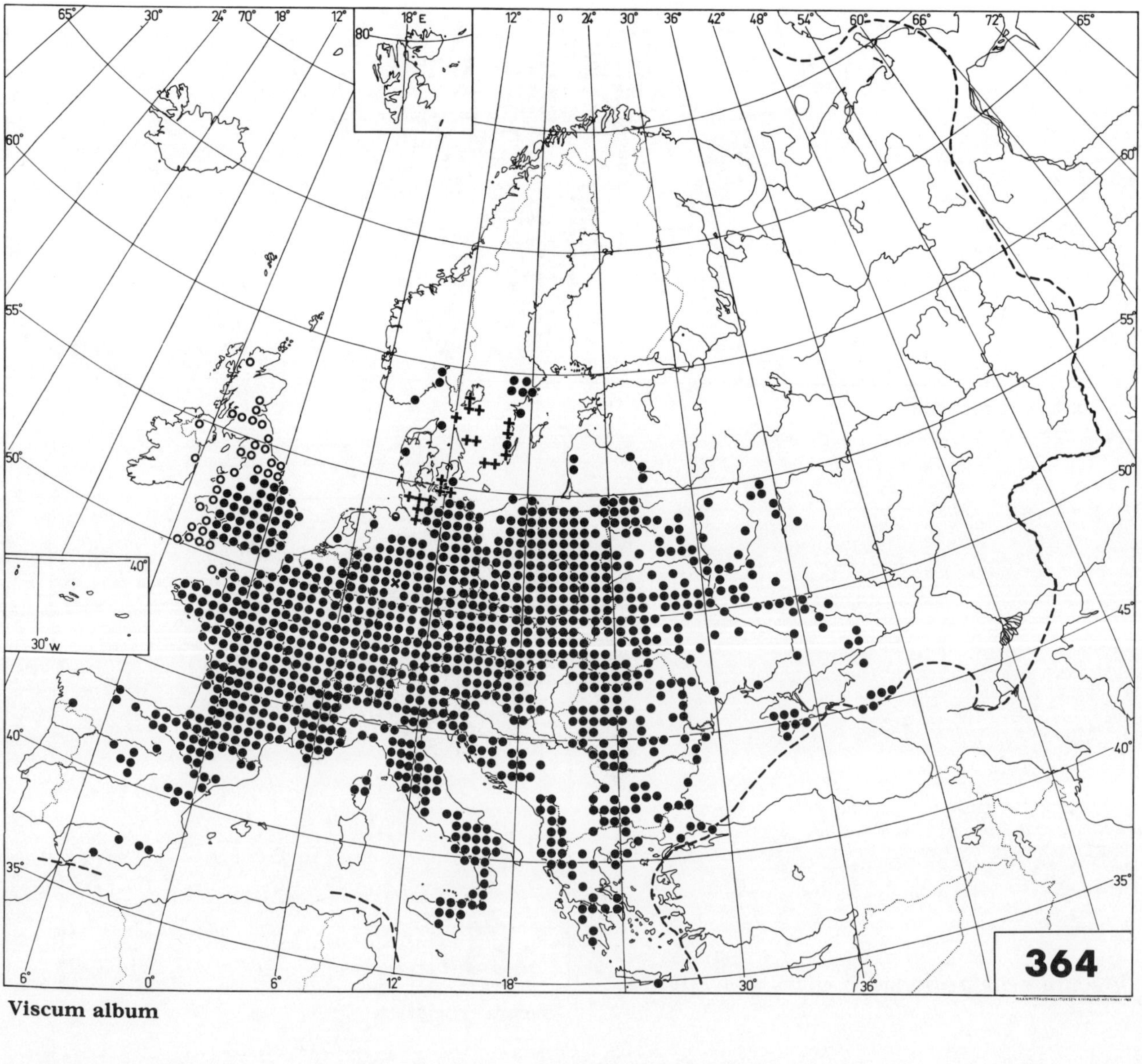

Viscum album

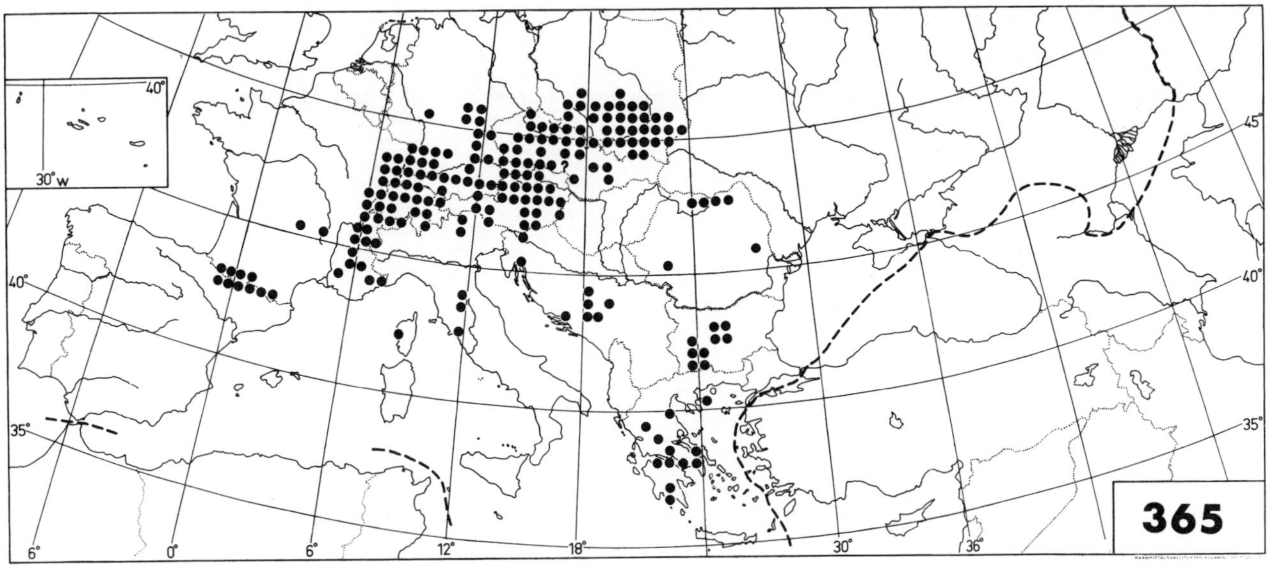

Viscum album subsp. **abietis**

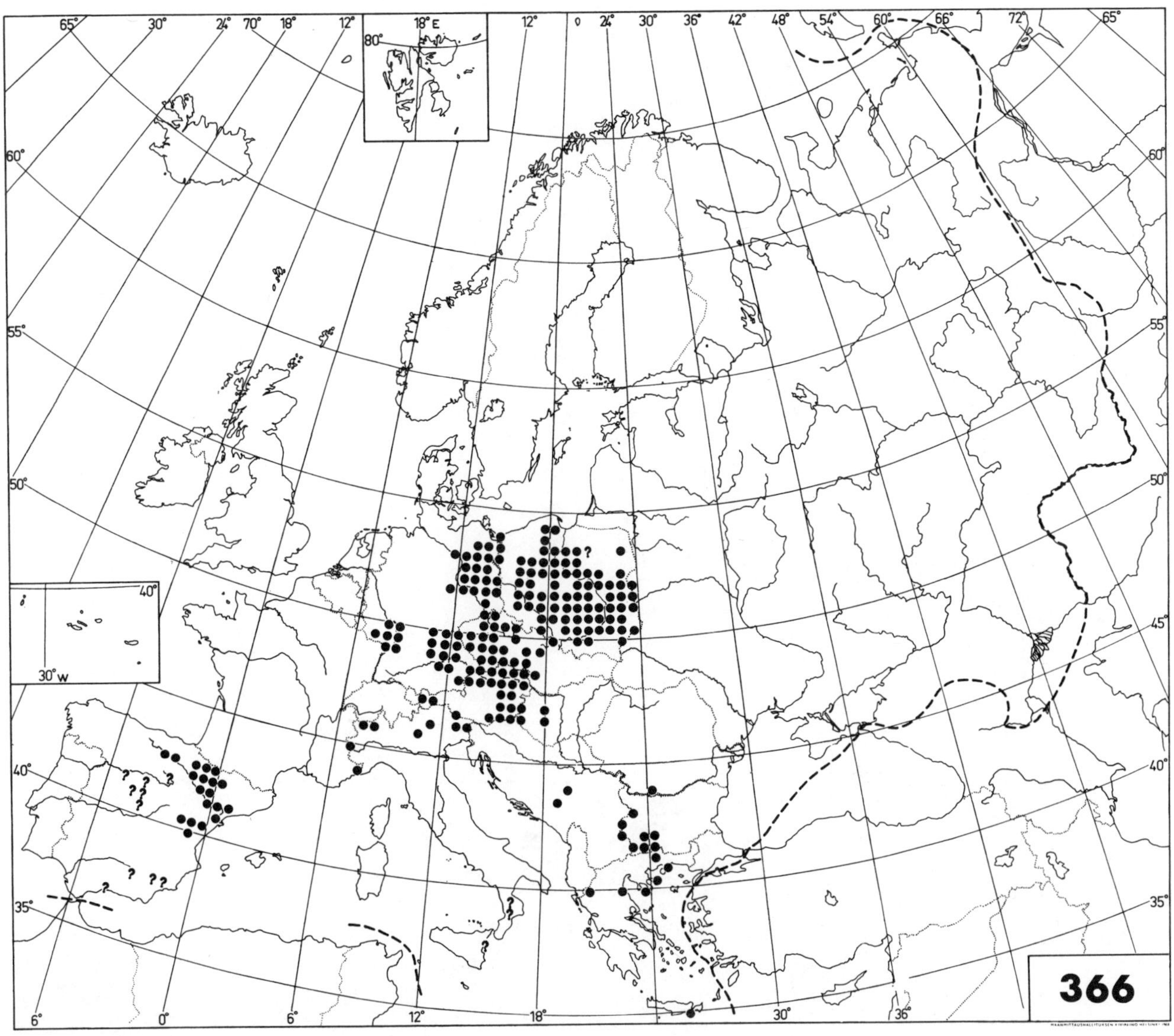

Viscum album subsp. **austriacum**

Arceuthobium oxycedri (DC.) Bieb. — Map 367.

Razoumofskya oxycedri (DC.) F. W. Schulz

Taxonomy. F. G. Hawksworth & D. Wiens, U. S. Dept. Agric. Forest Service Agric. Handb. 401: 167—169 (1972).
Notes. It and Lu omitted (?It, ?Lu in Fl. Eur.).
Total range. F. Markgraf, Ber. Deutsch. Bot. Ges. 52: 73 (1934).

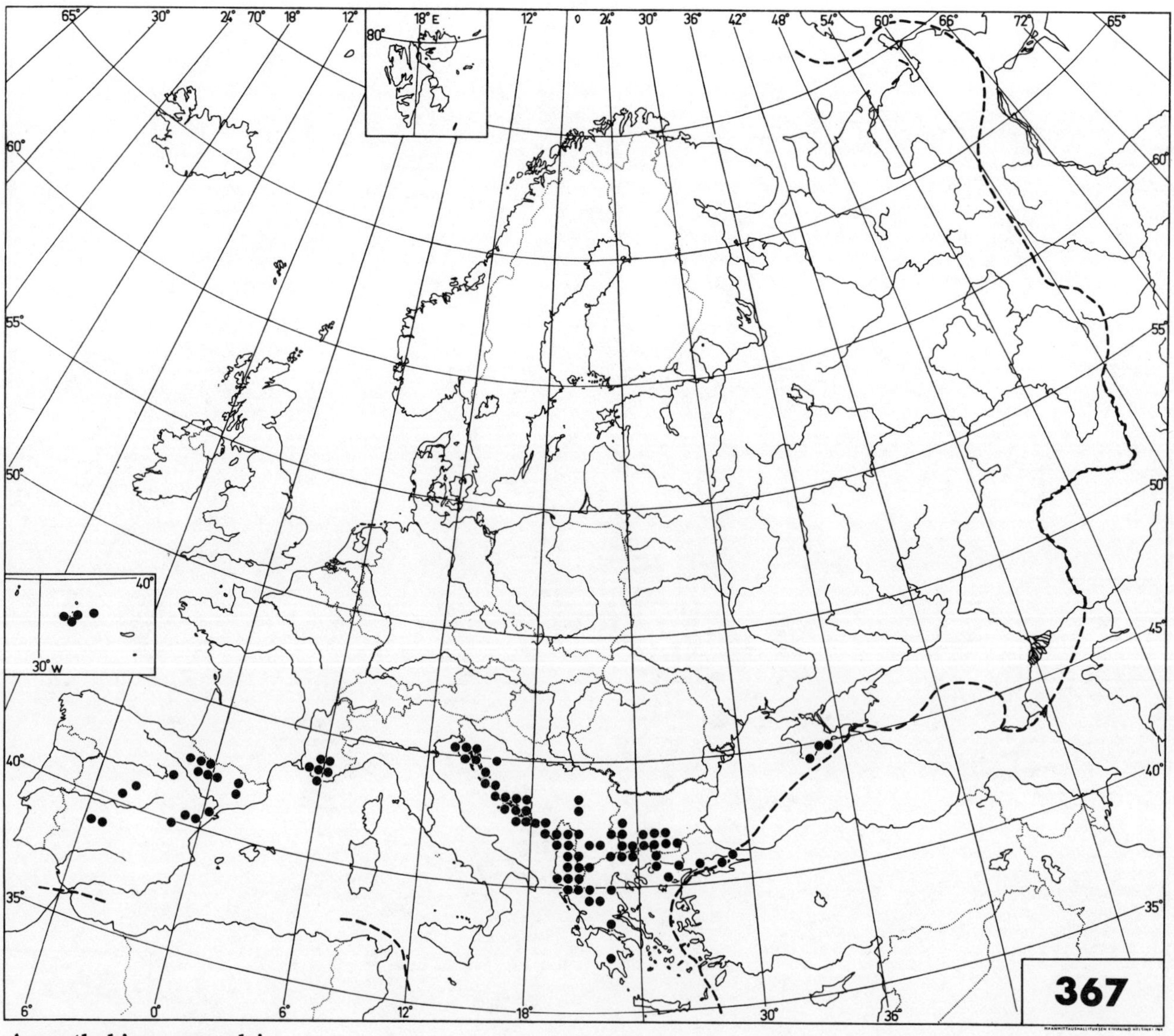

Arceuthobium oxycedri

ARISTOLOCHIACEAE

Asarum europaeum L. — Map 368.

Incl. Asarum ibericum Steven ex Woronow; A. europaeum subsp. caucasicum (Duchartre) Soó; A. europaeum var. caucasicum Duchartre; A. europaeum var. intermedium C. A. Meyer

Taxonomy. Asarum ibericum possibly deserves the rank of subspecies; see e.g. S. Seybold, Göttinger Flor. Rundbriefe 8: 50—53 (1974). For diverging views, e.g. S. Priszter, Ann. Biol. Univ. Debrecen 1: 201—207 (1950), and MJW 1965 Text: 466 may be consulted.

Notes. Bl and Lu omitted (given in Fl. Eur.), considered naturalized in Br, No and Su (*Br, No, Su in Fl. Eur.); Fe added (not given in Fl. Eur.): I. Kukkonen & P. Uotila, Ann. Bot. Fennici 9: 37—40 (1972); Ga added (given in the text in Fl. Eur.). S. Seybold, Göttinger Flor. Rundbriefe 8: 50—53 (1974), gives *Asarum ibericum* for Au, Cz, Ga, Ge, He, Hu, It, Ju and ?Po.

Total range. MJW 1965: map 126d; E. J. Jäger, Feddes Repert. 81: 76 (1970); A. Buhl *et al.*, Hercynia N.F. 11: 97 (1974).

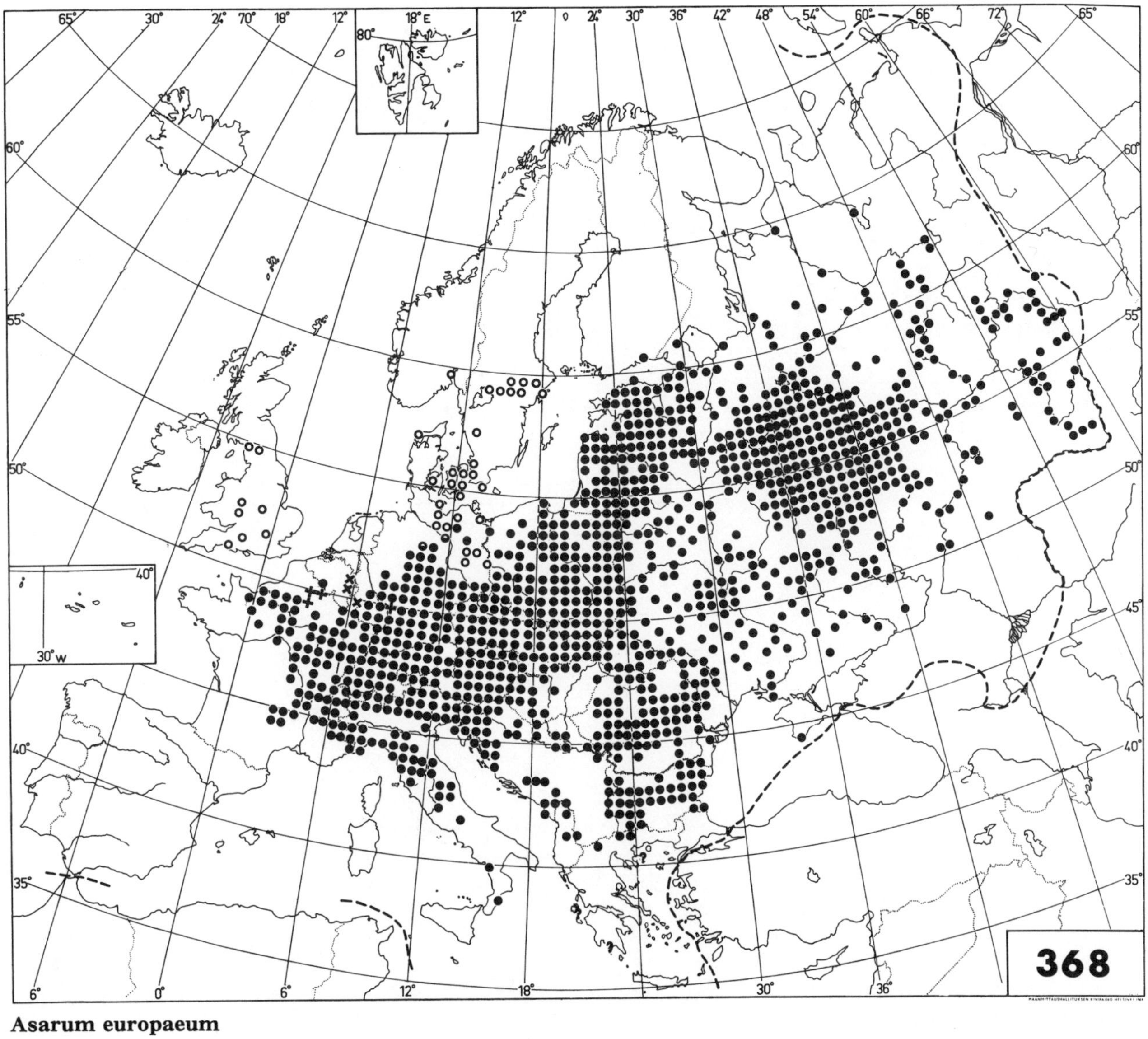

Asarum europaeum

Aristolochia sempervirens L. — Map 369.

Incl. Aristolochia altissima Desf.

Aristolochia baetica L. — Map 369.

Aristolochia cretica Lam. — Map 370.

Aristolochia pontica Lam.

?Aristolochia bodamae var. cordifolia Dingler

Notes. No European records confirmed (?Gr in Fl. Eur.).
Total range. P. H. Davis & M. S. Khan, Notes Roy. Bot. Gard. Edinb. 23: 526 (1961).

Aristolochia bodamae Dingler — Map 370.

Total range. P. H. Davis & M. S. Khan, Notes Roy. Bot. Gard. Edinb. 23: 526 (1961).

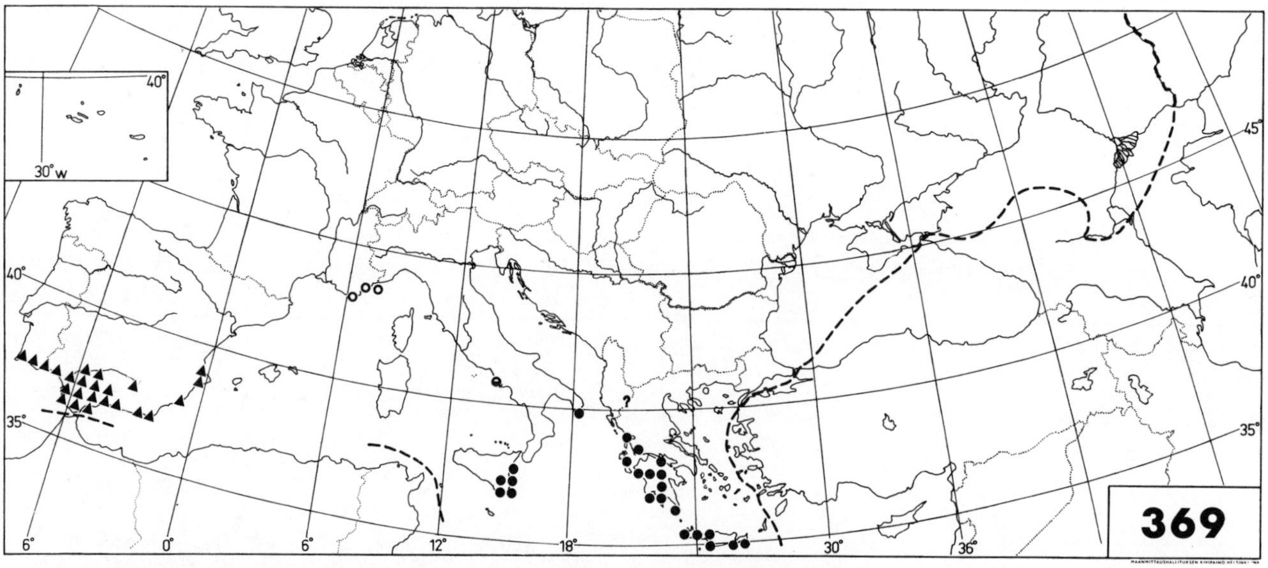

▲ = **Aristolochia baetica** • = **A. sempervirens**

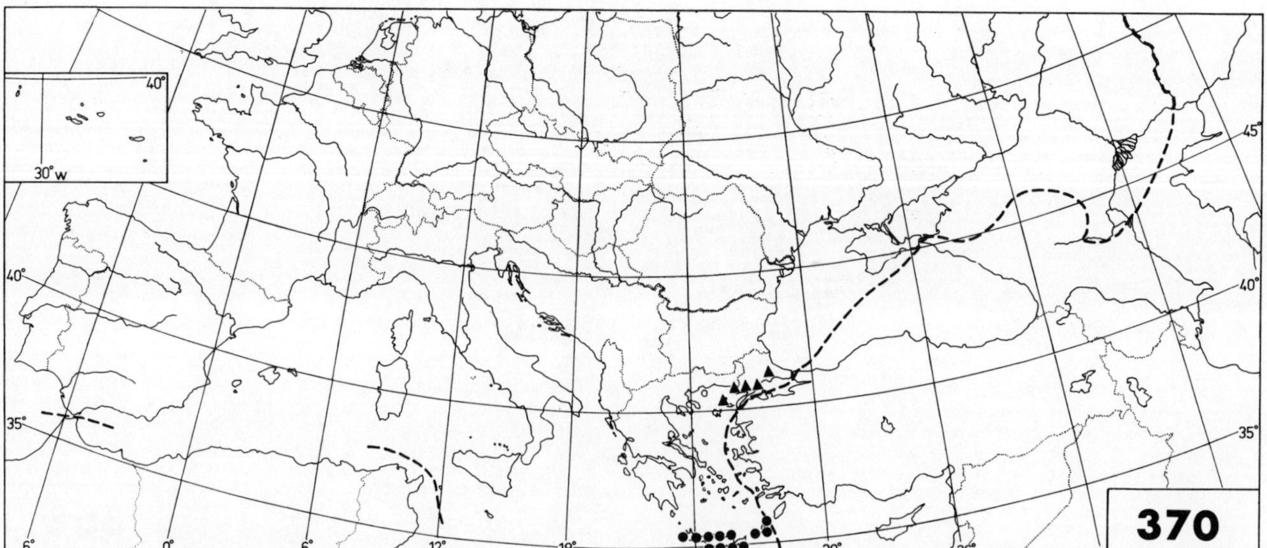

• = **Aristolochia cretica** ▲ = **A. bodamae**

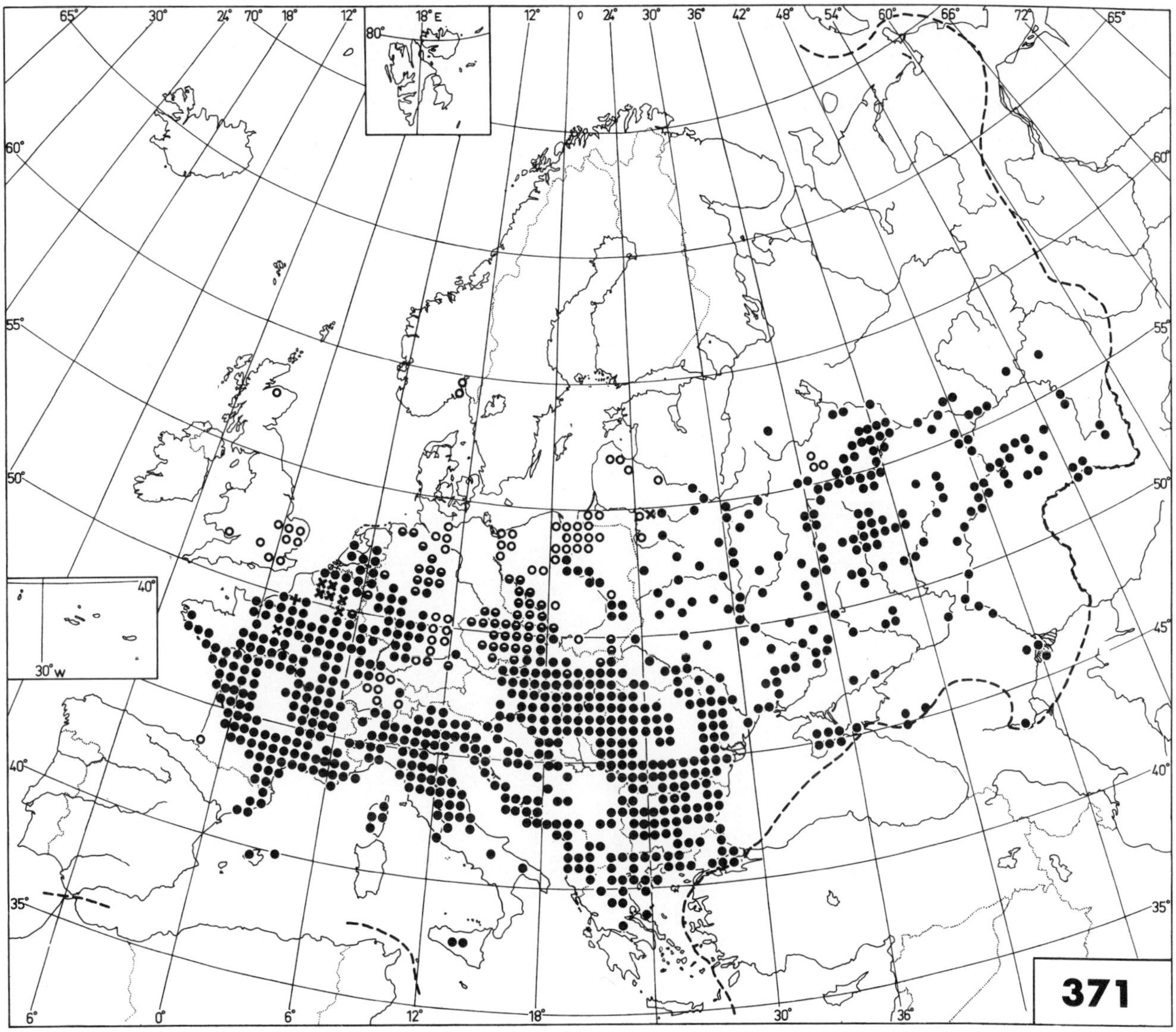

Aristolochia clematitis

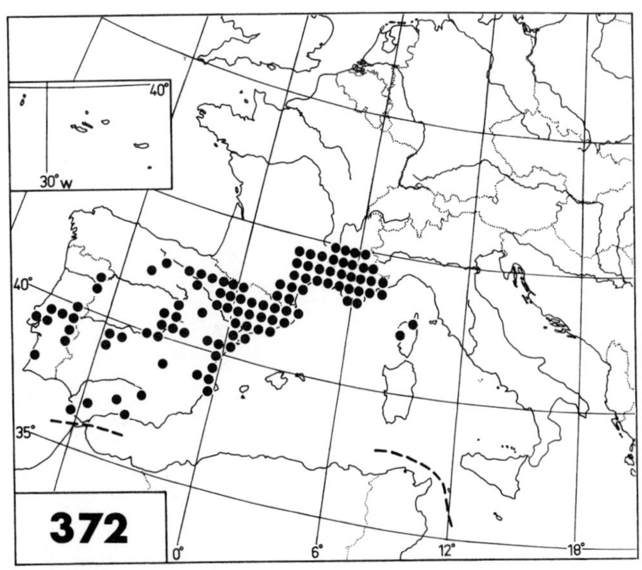

Aristolochia pistolochia

Aristolochia clematitis L. — Map 371.

Notes. Distinction between native and naturalized occurrences vague or not made in a number of countries. Bl confirmed (?Bl in Fl. Eur.), not native in Br (given as native in Fl. Eur.), Cr and Lu omitted (given in Fl. Eur.), [No] and Si added (not given in Fl. Eur.).

Aristolochia pistolochia L. — Map 372.

Notes. Sa omitted (given in Fl. Eur.).
Total range. Endemic to Europe.

Aristolochia rotunda L. — Map 373.

Notes. Cr, Lu and Sa omitted (given in Fl. Eur.).

Aristolochia pallida Willd. — Map 374.

Notes. Co and Cr omitted (given in Fl. Eur.).

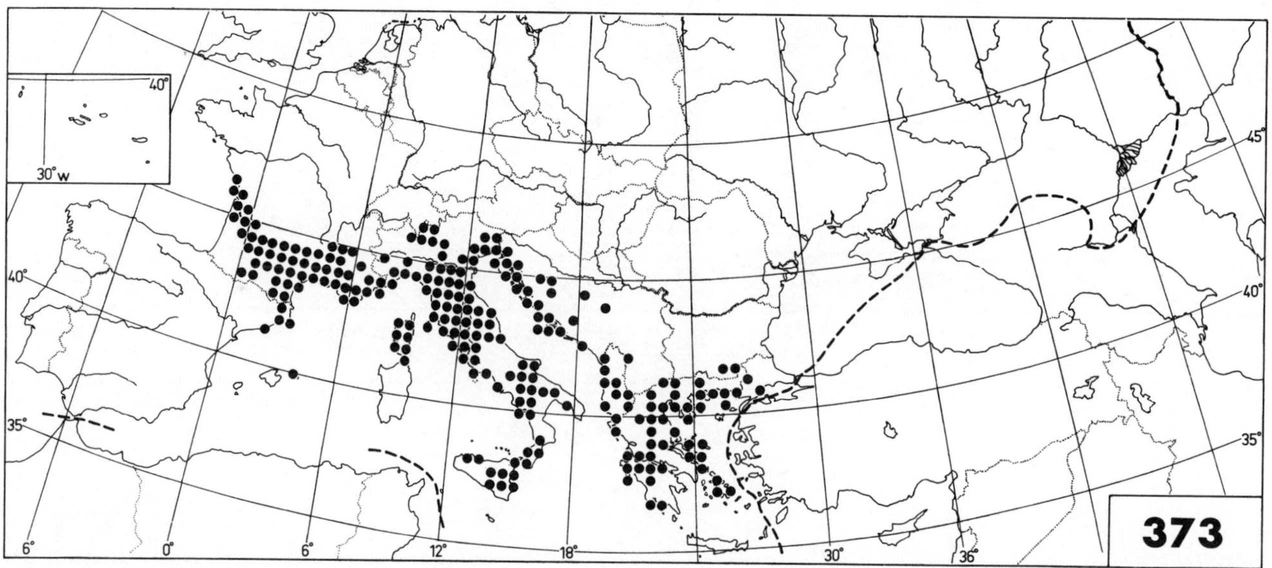

Aristolochia rotunda

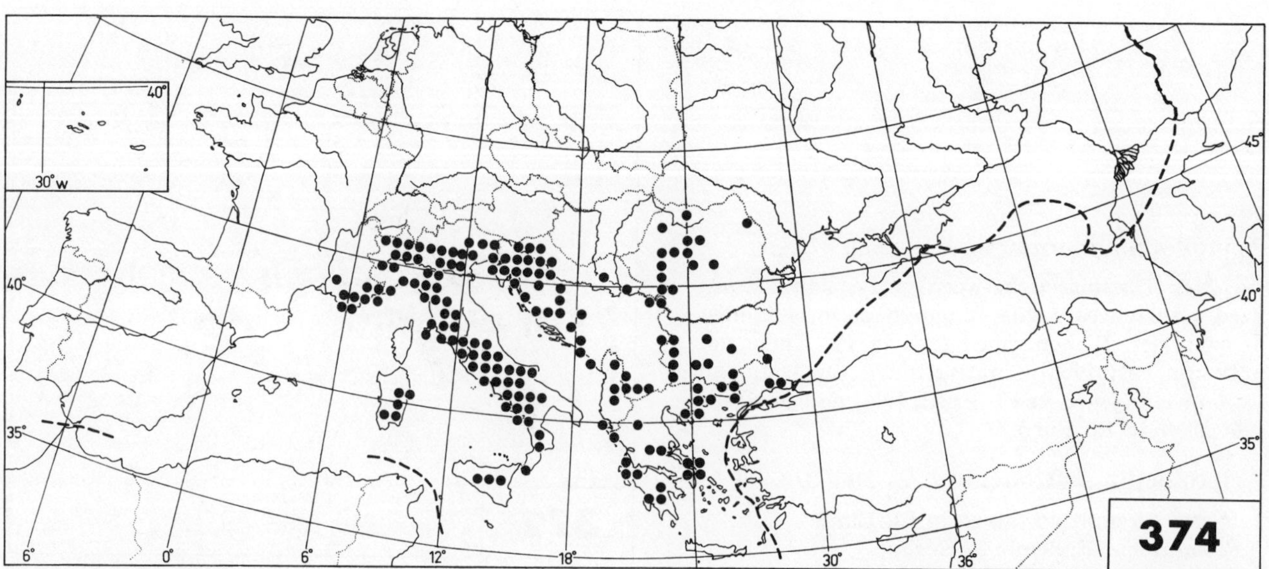

Aristolochia pallida

Aristolochia macedonica Bornm. — Map 375.

Aristolochia melanoglossa Bornm. non Speg.

Total range. Endemic to Europe.

Aristolochia croatica Horvatić — Map 375.

Total range. Endemic to Europe.

Aristolochia sicula Tineo — Map 375.

Total range. Endemic to Sicilia.

Aristolochia longa L. — Map 376.

Notes. Cr omitted (given in Fl. Eur.), possibly overmapped in Gr owing to confusion with *Aristolochia pallida.*

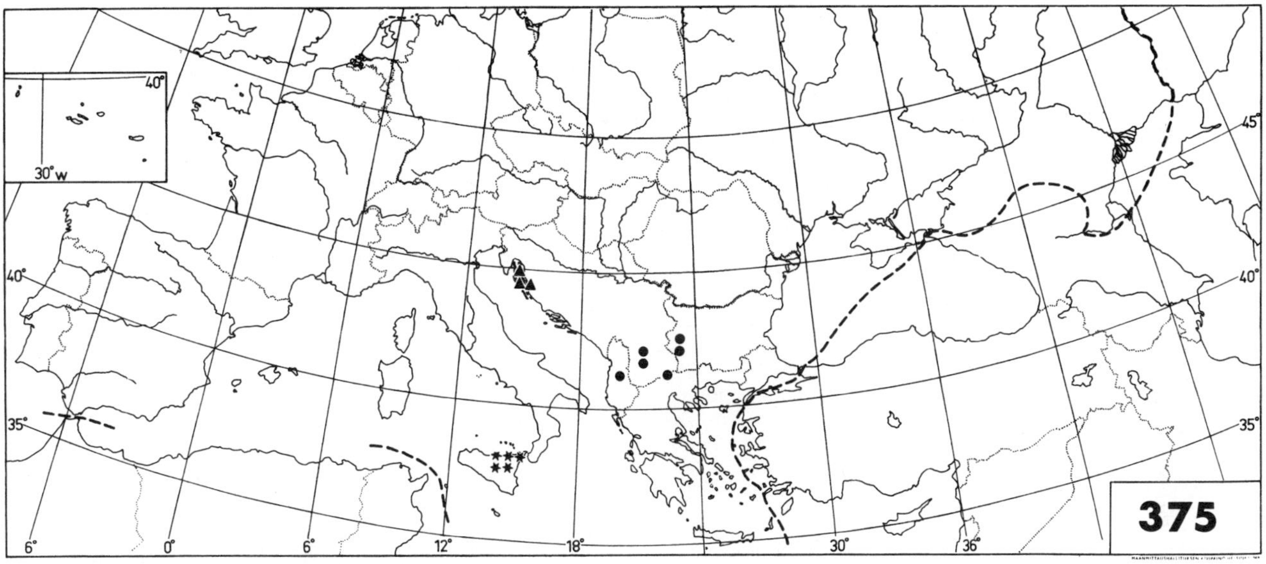

• = **Aristolochia macedonica** ★ = **A. sicula**
▲ = **A. croatica**

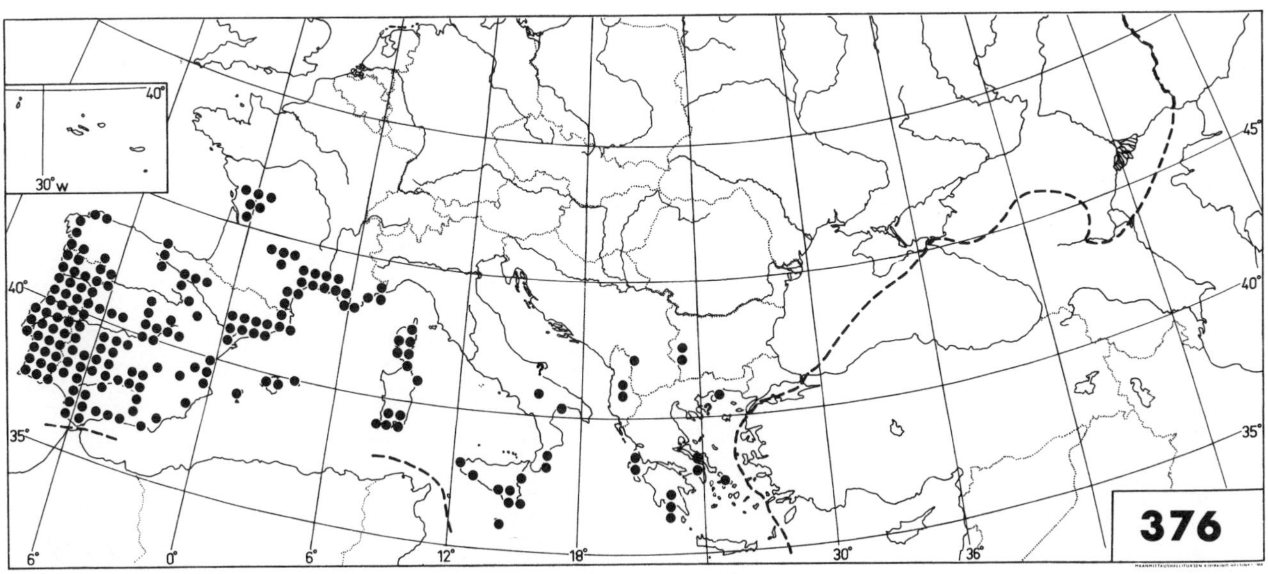

Aristolochia longa

Aristolochia bianorii Sennen & Pau — Map 377.

Total range. Endemic to Islas Baleares.

Aristolochia microstoma Boiss. & Spruner — Map 378.

Total range. Endemic to Europe. Not found on the Kikladhes, although given in Fl. Eur.

Aristolochia parvifolia Sm. — Map 379.

Aristolochia tournefortii Jaub. & Spach

Notes. Cr, Gr (the species not mentioned in Fl. Eur.).

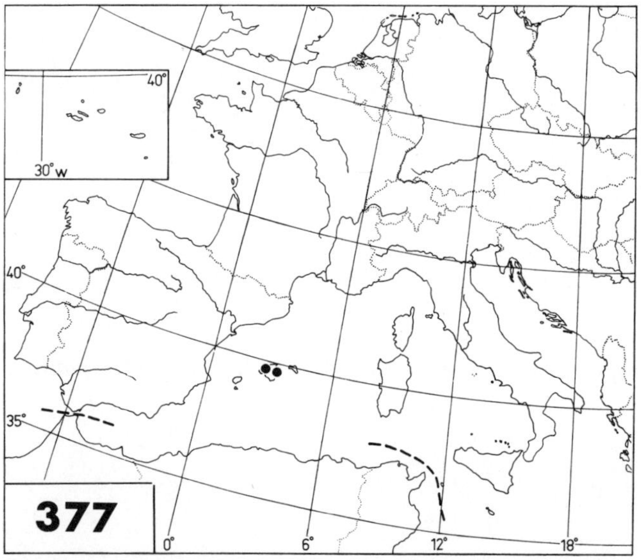

Aristolochia bianorii

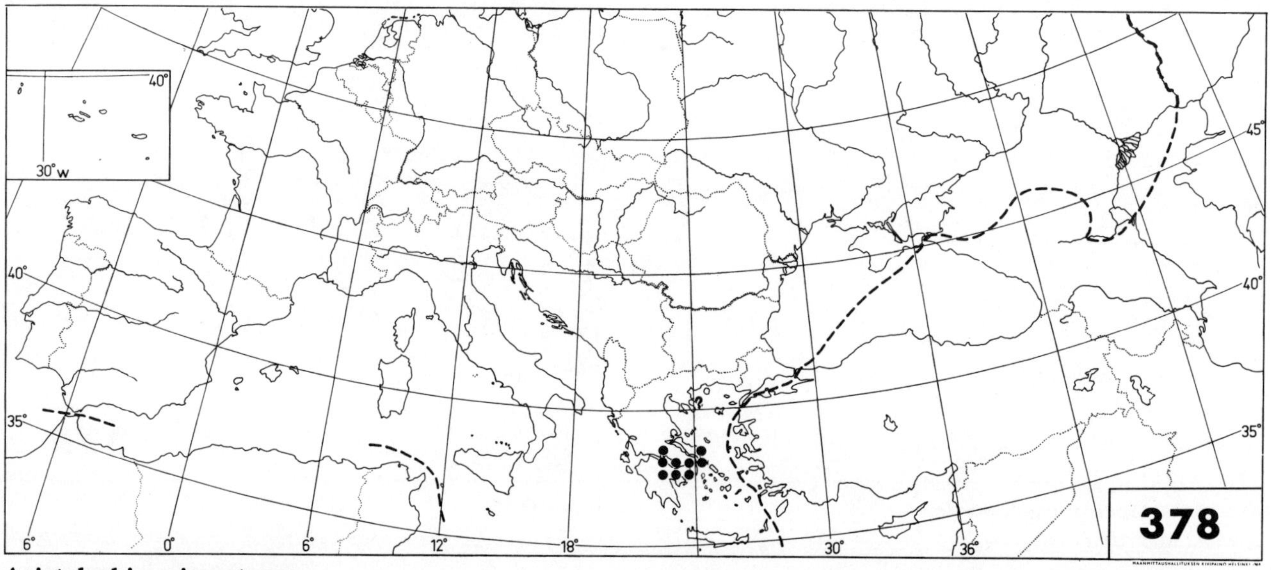

Aristolochia microstoma

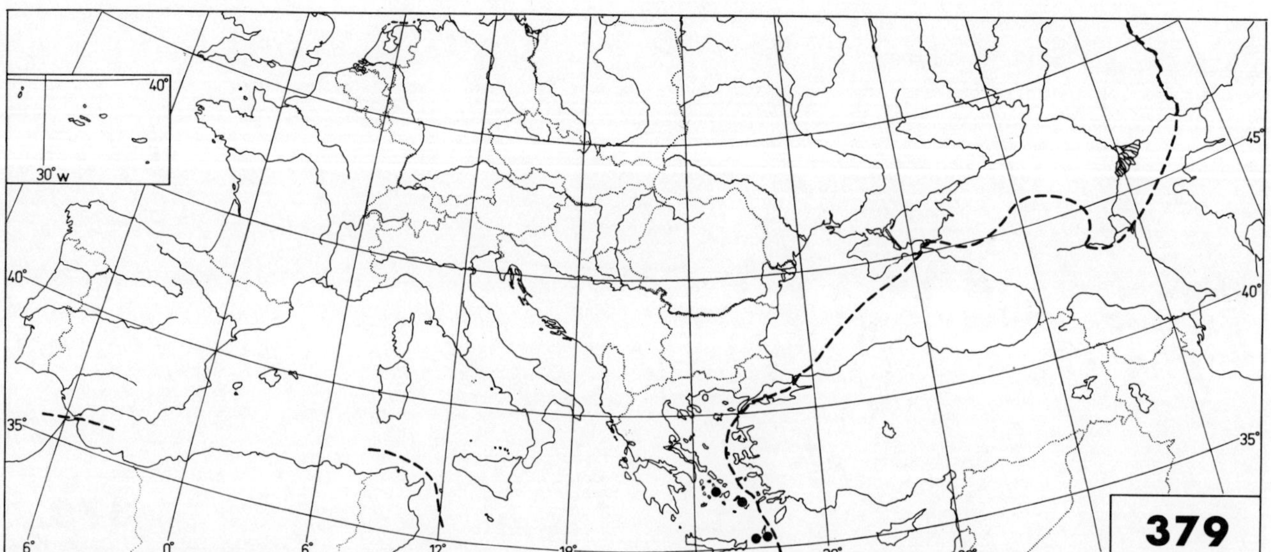

Aristolochia parvifolia

RAFFLESIACEAE (CYTINACEAE)

Cytinus L., coll. — Map 380.
Cytinus hypocistis + C. ruber

Cytinus hypocistis (L.) L. — Map 381.
Notes. Al added (not given in Fl. Eur.).

C. hypocistis subsp. **hypocistis** — Map 381.

Notes. Al, Bl, Co, Cr, Ga, Gr, Hs, It, Ju, Sa, Si, Tu. »On white-flowered species of *Cistus*; perhaps also on *Halimium* spp. From W. France to Jugoslavia and Sicilia» (Fl. Eur.). Map provisional and incomplete.

C. hypocistis subsp. **pityusensis** Finschow — Map 381.

Taxonomy. G. Finschow, Veröff. Überseemus. Bremen A 4 (12): 109–115 (1974).
Notes. Bl (not in Fl. Eur.). On *Cistus clusii* only.
Total range. So far known only from Ibiza, Islas Baleares.

C. hypocistis subsp. **macranthus** Wettst. — Map 381.
Notes. Lu. »On *Halimium* spp. Portugal; ? Spain» (Fl. Eur.).

C. hypocistis subsp. **orientalis** Wettst. — Map 381.
Notes. Cr, Gr. »On *Cistus parviflorus*. Kriti and Gavdhos» (Fl. Eur.).

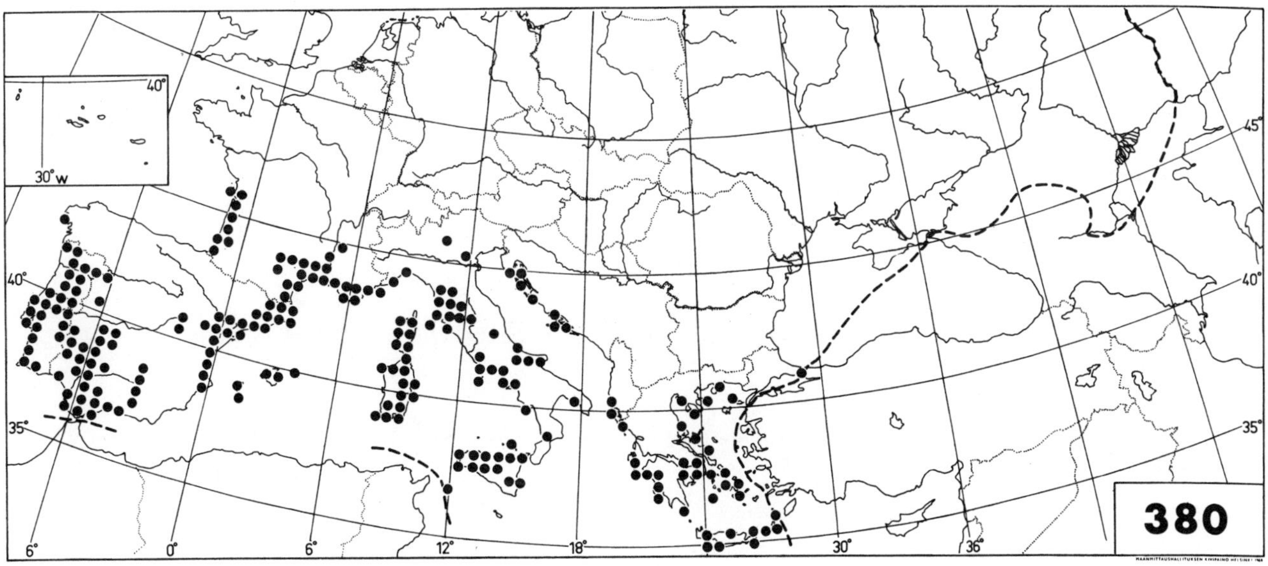

Cytinus (the genus)

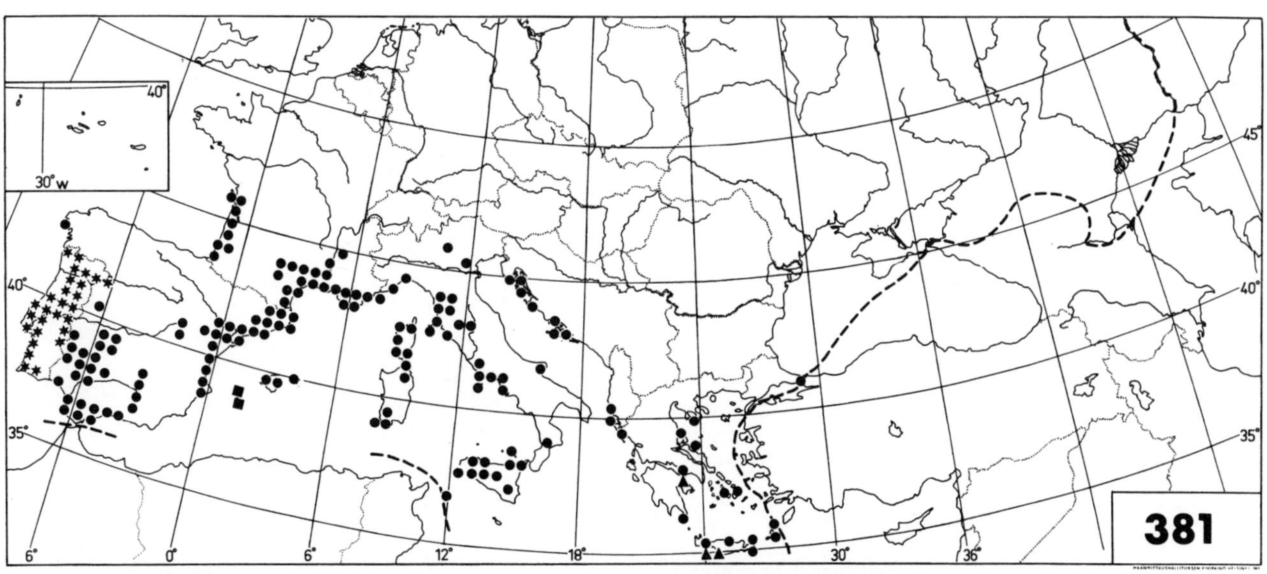

Cytinus hypocistis
• = subsp. **hypocistis**
■ = subsp. **pityusensis**
★ = subsp. **macranthus**
▲ = subsp. **orientalis**

Cytinus ruber (Fourr.) Komarov — Map 382.

Cytinus hypocistis subsp. kermesinus Guss.

Notes. Cr added (not given in Fl. Eur.), Ju omitted (given in Fl. Eur.). »On pink-flowered species of *Cistus*» (Fl. Eur.).

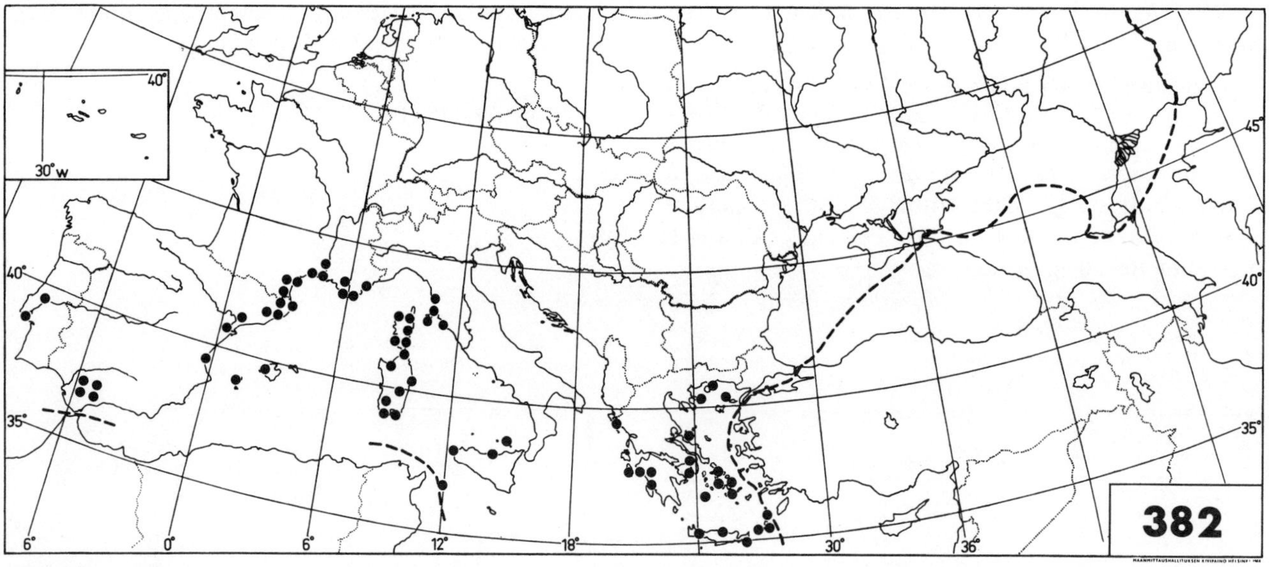

Cytinus ruber

BALANOPHORACEAE (CYNOMORIACEAE)

Cynomorium coccineum L. — Map 383.

Notes. Bl added (not given in Fl. Eur.), Cr omitted (?Cr in Fl. Eur.). See G. Eberle, Natur u. Museum 95 (1): 31–36 (Frankfurt a.M. 1965).

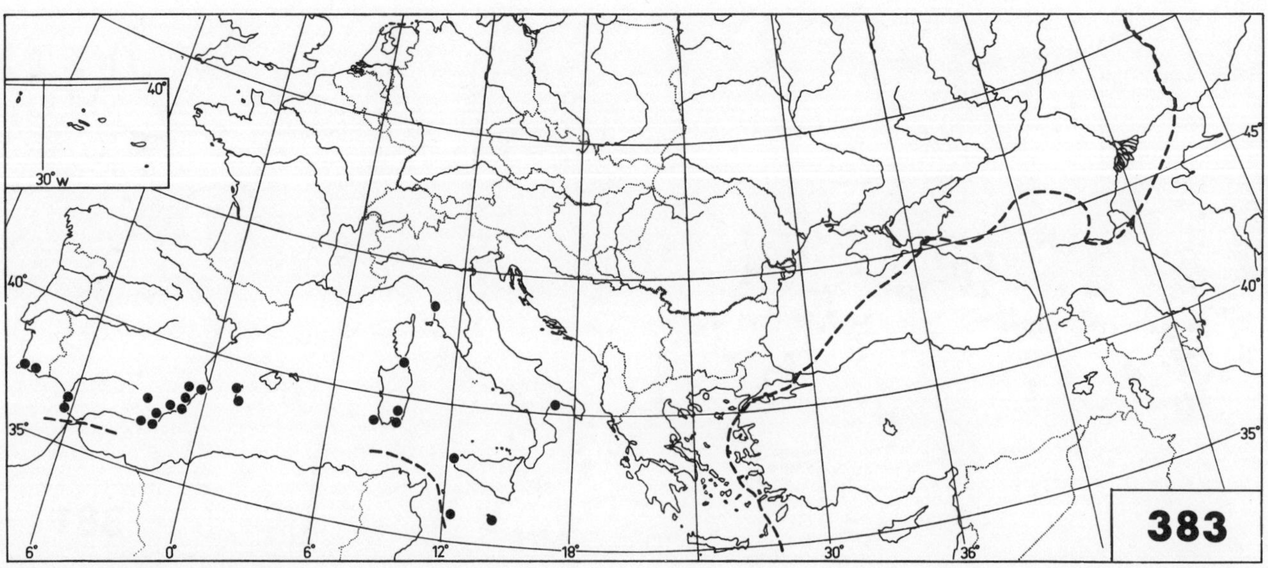

Cynomorium coccineum

4

POLYGONACEAE

CONTENTS

THE COMMITTEE FOR MAPPING THE FLORA OF EUROPE

SECRETARIAT, HELSINKI

J. Jalas (Chairman of the Committee)
J. Suominen (Secretary General)

ADVISERS

T. W. Böcher, København
A. R. Clapham, Arkholme
P. Critopoulos, Athínai
H. Ellenberg, Göttingen
E. Hultén, Stockholm
H. Meusel, Halle
G. Moggi, Firenze
A. Takhtajan, Leningrad
D. A. Webb, Dublin

COMMITTEE MEMBERS ACTING AS REGIONAL COLLABORATORS

Albania (Al)
I. Mitrushi, Tiranë

Austria (Au)
H. Niklfeld, Wien
Assisted by collaborators in Mapping the Flora of Central Europe

Belgium (Be, excl. Luxembourg)
E. van Rompaey (†), Antwerpen

British Isles (Br, Hb, Channel Islands)
F. H. Perring, Abbots Ripton
D. A. Webb, Dublin (Hb)

Bulgaria (Bu)
S. Kožuharov, Sofia
Assisted by:
A. V. Petrova, Sofia

Czechoslovakia (Cz)
J. Futák, Bratislava
J. Holub, Průhonice
Assisted by:
J. Houfek, Průhonice
F. Mladý, Průhonice
V. Skalický, Praha

Denmark (Da, Fa)
A. Hansen, København

Finland (Fe)
J. Suominen, Helsinki
Assisted by:
T. Ulvinen, Oulu
P. Uotila, Helsinki

France (Co, Ga)
P. Dupont, Nantes
Maps checked by several French botanists

German Democratic Republic (Ge: DDR)
H. MEUSEL, Halle
A. BUHL, Halle
E. WEINERT, Halle
Assisted by:
W. FISCHER, Potsdam
F. FUKAREK, Greifswald
W. HEMPEL, Dresden
H. D. KNAPP, Halle
S. RAUSCHERT, Halle

Federal Republic of Germany (Ge: BRD)
H. ELLENBERG, Göttingen
H. HAEUPLER, Göttingen
J. KRACH, München
K. LEWEJOHANN, Göttingen
P. SCHÖNFELDER, Regensburg
Assisted by ca. 600 collaborators in Mapping the Flora of Central Europe

Greece (Cr, Gr)
P. CRITOPOULOS, Athínai (—1977)
W. GREUTER, Berlin
D. PHITOS, Patras (1977—)
Assisted by:
B. ALDÉN, Göteborg
M. GUSTAFSSON, Lund
L.-Å. GUSTAVSSON, Lund
J. PERSSON, Göteborg
K.-H. RECHINGER, Wien

Hungary (Hu)
R. v. SOÓ, Budapest
A. TERPÓ, Budapest

Iceland (Is)
E. EINARSSON, Reykjavík
Assisted by:
H. BJÖRNSSON, Öraefi
I. DAVÍDSSON, Reykjavík
B. JÓHANNSSON, Reykjavík
H. KRISTINSSON, Akureyri
S. STEINDÓRSSON, Akureyri

Italy (It, Sa, Si)
E. NARDI, Firenze
M. RAFFAELLI, Firenze

Jugoslavia (Ju)
P. FUKAREK, Sarajevo
E. MAYER, Ljubljana
Assisted by:
D. TRPIN, Ljubljana

Luxembourg (Be, excl. Belgium)
L. REICHLING, Luxembourg

Netherlands (Ho)
F. ADEMA, Leiden
J. MENNEMA, Leiden

Norway (No, Sb)
R. Y. BERG, Oslo
K. FAEGRI, Bergen
Assisted by:
R. ELVEN, Tromsø
E. FREMSTAD, Bergen
O. I. RØNNING, Trondheim (Sb)
K.-E. SIBBLUND, Oslo

Poland (Po)
J. KORNAŚ, Kraków
Assisted by:
M. CIACIURA, Wrocław
D. FIJAŁKOWSKI, Lublin
A. FREY, Kraków
J. JASNOWSKA, Szczecin
K. KĘPCZYŃSKI, Toruń
M. KOPIJ, Warszawa
T. KRZACZEK, Lublin
J. MĄDALSKI, Wrocław
R. OLACZEK, Łódź
L. OLESIŃSKI, Olsztyn
A. PACYNA, Kraków
Z. SCHWARZ, Gdańsk
A. SOKOŁOWSKI, Białowieża
R. SOWA, Łódź
H. TRZCIŃSKA-TACIK, Kraków
A. ZAJĄC, Kraków
E. U. ZAJĄC, Kraków
W. ŻUKOWSKI, Poznań

Portugal (Az, Lu)
J. DO AMARAL FRANCO, Lisboa
Assisted by:
M. L. DA ROCHA-AFONSO, Lisboa

Romania (Rm)
N. Boşcaiu, Cluj

Spain (Bl, Hs)
E. F.-Galiano, Madrid
R. Folch i Guillèn, Barcelona (Bl, València and Catalonia)
P. Montserrat, Jaca
Assisted by:
B. Cabezudo, Sevilla
B. Valdés, Sevilla

Sweden (Su)
B. Jonsell, Uppsala
Assisted by:
M. Thulin, Uppsala

Switzerland (He)
O. Hegg, Bern
M. Welten, Bern
Assisted by:
R. Sutter, Bern

Turkey (European part; Tu)
H. Demiriz, Istanbul

U.S.S.R. (Rs(N,B,C,W,K,E))
A. I. Tolmachev, Leningrad
V. I. Czopik, Kiev
V. N. Tikhomirov, Moskva
Assisted by:
J. E. Alekseev, Moskva
A. Borodina, Leningrad
A. V. Cziczev, Moskva
I. J. Fatare, Riga
Ju. D. Gusev, Leningrad
G. B. Klavina, Riga
N. V. Kozlovskaya, Minsk
L. Laasimer, Tartu
A. N. Laschenkova, Syktyvkar
A. Lekavičius, Vilnius
N. B. Oktyabreva, Moskva
T. Plieva, Leningrad
N. K. Shvedchikova, Moskva
Z. G. Ulle, Syktyvkar
A. M. Verenko, Kiev
J. A. Vojtjuk, Kiev

FINNISH CONSULTATIVE COMMITTEE, HELSINKI

T. Ahti
P. Isoviita
J. Jalas (Chairman)
I. Kukkonen
H. Luther
R. Ruuhijärvi
J. Suominen (Secretary)

EXPLANATORY NOTES

MAPPING TECHNIQUES

The area mapped is that of Flora Europaea. The Azores and Spitsbergen appear as insets.

The mapping unit used is the 50-km square of the UTM (Universal Transverse Mercator) grid maps, covering Europe on a scale of 1 : 1 000 000. Europe comprises ca. 4 400 squares, which are used without respect to political boundaries. Along the compensation zones of every sixth meridian, the breadth (W-E) of the squares varies between 40 and 60 km.

A base map, 1 : 10 000 000, was used in collecting the primary data in the various countries, and in compiling and publishing the final maps, which have a scale of ca. 1 : 31 000 000. A sample specimen of this base map was distributed as an appendix to Vol. 1.

Deviations from the UTM grid system are listed in Vol. 1 (p. 10) and, with improvements and additions for Greece (map), in Vol. 3 (p. 7—9). The deviations are of three kinds. a) In coastal areas the squares with less than 10 % (250 km^2) of land are included in the neighbouring mainland square. However, islands far from the coast and long peninsulas have their own squares. b) Mapping symbols for isolated islands are put on the actual position of the island, irrespective of the grid. c) Individual mountain tops forming distinct biogeographical entities are not divided between two or more squares, despite the grid.

The mapping symbols are:

● ▲ native occurrence
○ △ introduction (established alien)
◒ status unknown or uncertain
+ extinct
× probably extinct, or, at least, not recorded since 1930
? record uncertain as regards identification or locality

For the second taxon in a map, and for special cases, the symbols ▲, ★, etc. are used. If not otherwise stated, the special symbols (+, ×, ?, etc.) belong to the taxon marked by ●.

Any reliable record was used when entering a symbol in a square, not only those based on herbarium specimens but also reliable published or unpublished sight records.

TAXA INCLUDED AND DEVIATIONS FROM FLORA EUROPAEA

The species and subspecies included are according to Flora Europaea (Vol. 1, 1964), with some deviations, mainly caused by new data published after 1964.

Included, although not given in Fl. Eur., is *Polygonum tenoreanum* Nardi & Raffaelli, which was first described in 1977. Included are also the native N.E. European species *P. humifusum* Pallas and *P. laxmannii* Lepechin, which are not mentioned in Fl. Eur. Of these, *P. humifusum,* at least, belongs to the group of taxa well known as subject to diverging taxonomical interpretations, as do *Rumex alpestris* subsp. *lapponicus* (Hiitonen) Jalas, *R. acetosa* subsp. *serpentinicola* (Rune) Nordh. (subspecific status given in 1966), *R. acetosa* subsp. *vinealis* (Timb.-Lagr. & Jeanb.) O. Bolòs & Vigo (subspecific status dating from 1974), and *R. aquaticus* subsp. *protractus* (Rech. fil.) Rech. fil. (not mapped separately). A special group of species, characteristic of the genera of the present volume, is that consisting of established aliens, including *Polygonum nepalense, P. capitatum, P. molle* and *Fagopyrum dibotrys* (not mapped), which are not mentioned in Fl. Eur. Here, as elsewhere, there may naturally be different opinions regarding the correct decision.

Omitted, although given in Fl. Eur.: *Polygonum tinctorium* (given as an established alien, but no records available).

Nomenclatural deviations from Fl. Eur. include the use of the generic name *Fallopia* Adanson instead of *Bilderdykia* Dumort., as well as *Rumex alpestris* Jacq. (instead of *R. arifolius* All.) and *R. nebroides* Campd. (instead of *R. gussonei* Arcangeli).

A good example of a *major change in the distributional pattern* given in Fl. Eur. is the record from Norway of *Rheum rhaponticum.*

THE TEXTUAL COMMENTS

The *synonyms* are those given in the text and index of Fl. Eur., supplemented with a selection of nomenclatural combinations formed in the most recent literature or otherwise representing and elucidating different views on the status and affinities of the taxon involved.

Under the headings *Taxonomy* and *Nomenclature* are given new recently published views and facts, differing from or complementing the treatment in Fl. Eur.

Notes give important new or omitted records (mainly indicated with the territorial abbreviations), especially when they complement or correct Fl. Eur. When the distributional information given by Fl. Eur. is not particularly detailed, as tends to be the case at the level of subspecies, all the pertinent territories are listed for the taxa mapped. When possible, the major additions and omissions at country level have been explained by reference to the relevant literature.

As in Fl. Eur., the following signs are used in connexion with the two-letter geographical abbreviations: [ ] = not native; * = status doubtful, possibly native; ? = occurrence doubtful; † = extinct.

Total range is treated in two different ways. First, in the case of European endemics, no references are given to distribution maps published earlier. Taxa endemic to areas within continental Europe are simply given as »endemic to Europe», and more detailed information is provided for those restricted to geographically well-separated island areas, e.g. »endemic to Sicilia». Second, when the total range exceeds the European boundaries, original total range maps or maps covering a major part of the total range (e.g. Eurasian part of a circumpolar range) have been cited, if any. However, the hopeless task of providing a complete list has not been attempted, especially not as concerns maps published previous to Hegi 1958, Hultén AA 1958 and MJW 1965 (for the abbreviations, see below).

Abbreviations used for frequently quoted literature:

Fl. Arct. URSS 1966 = A.I. Tolmatchev (ed.), Flora Arctica URSS. 5. Salicaceae — Portulacaceae. — 208 pp. Mosquae — Leninopoli 1966.

Fl. Bor.-or. Eur. URSS 1976 = A. I. Tolmatchev (ed.), Flora regionis boreali-orientalis territoriae Europaeae URSS. 2. Cyperaceae — Caryophyllaceae. — 316 pp. Leninopoli 1976.

Fl. Eur. = T. G. Tutin, V. H. Heywood et al. (eds.), Flora Europaea. 1. Lycopodiaceae to Platanaceae. — xxxiv + 464 pp. Cambridge 1964.

Fl. France 1973 = M. Guinochet & R. de Vilmorin, Flore de France. 1. — 366 pp. Paris 1973.

Fl. Turkey 1967 = P. H. Davis, Flora of Turkey. 2. — xii + 581 pp. Edinburgh 1967.

Haraldson 1978 = K. Haraldson, Anatomy and taxonomy in Polygonaceae subfam. Polygonoideae Meisn. emend. Jaretzky. — Symb. Bot. Upsal. 22(2): 1—95. 1978.

Hegi 1958 = K. H. Rechinger (ed.), G. Hegi, Illustrierte Flora von Mitteleuropa, 2. Aufl. 3(1) Polygonaceae: 352—436. München 1958. (maps by H. Meusel)

Hultén AA 1958 = E. Hultén, The Amphi-Atlantic Plants. — Kungl. Svenska Vet.-Akad. Handl., Ser. 4, 7(1): 1—340. 1958.

Hultén Alaska 1968 = E. Hultén, Flora of Alaska and neighboring territories. — 1008 pp. Stanford 1968.

Hultén CP 1971 = E. Hultén, The Circumpolar Plants. II. Dicotyledons. — Kungl. Svenska Vet.-Akad. Handl., Ser. 4, 13(1): 1—463. 1971.

Hylander 1966 = N. Hylander, Nordisk Kärlväxtflora. 2. — xi + 456 pp. Stockholm 1966.

MJW 1965 = H. Meusel, E. Jäger & E. Weinert, Vergleichende Chorologie der zentraleuropäischen Flora. — Text 583 pp., Karten 258 pp. Jena 1965.

Nova Fl. Port. 1971 = J. do Amaral Franco, Nova Flora de Portugal (Continente e Açores). 1. Lycopodiaceae — Umbelliferae. — xxiv + 648 pp. Lisboa 1971.

ACKNOWLEDGEMENTS

We are greatly obliged to the members of the Committee for Mapping the Flora of Europe, whose contributions have been a *sine qua non* for the completion of another volume of Atlas Florae Europaeae, and have been warmly appreciated, even though sometimes unavoidably delayed. Our special thanks are due to J. Holub (Průhonice) and H. Niklfeld (Wien) who, in addition to looking after and sorting out the material for their respective territories, have given the Secretariat invaluable assistance by providing critical comments on the manuscript and map drafts and supplementary material even from outside their own countries. Generous help with special plants or areas has also been received from a number of botanists not formally connected with the Mapping Committee. Here we would like to mention our special gratitude to T. Plieva (Leningrad), for helping us in so many ways with the material from Rs.

The expenditure of the Secretariat has been covered, as before, by annual allowances from the Finnish National Research Council for Sciences. The Finnish Biological Society Vanamo has continued to act as publisher for Atlas Florae Europaeae. The publication of the present volume was partly subsidized by the Finnish Cultural Foundation.

As earlier, working facilities for the Secretariat have been provided by the Department of Botany, University of Helsinki.

Technical assistance in preparing this volume has been given by Mrs. Liisa Mäkelä, M.A., Mr. Jari Oksanen, M.A., and Mrs. Paula Oksanen, M.A.

DICOTYLEDONES (cont.)

POLYGONACEAE

Koenigia islandica L. — Map 384.

Generic delimitation. Haraldson 1978: 69—70.

Taxonomy. Koenigia islandica var. *arctica* Hadač, Studia Bot. Čechica 5 : 3 (1942), from Sb, was recently given specific rank as *K. hadacii* Löve & Löve, Bot. Not. 128: 506 (1975). It is claimed to be diploid with 2n = 14 (instead of tetraploid; the exact wild origin of the plant(s?) studied is not given). Whether both ploidy levels are present in Sb is not known.

Total range. Hultén Alaska 1968: 375; W. J. Cody, Naturaliste Canadien 98: 146 (1971); Hultén CP 1971: map 55.

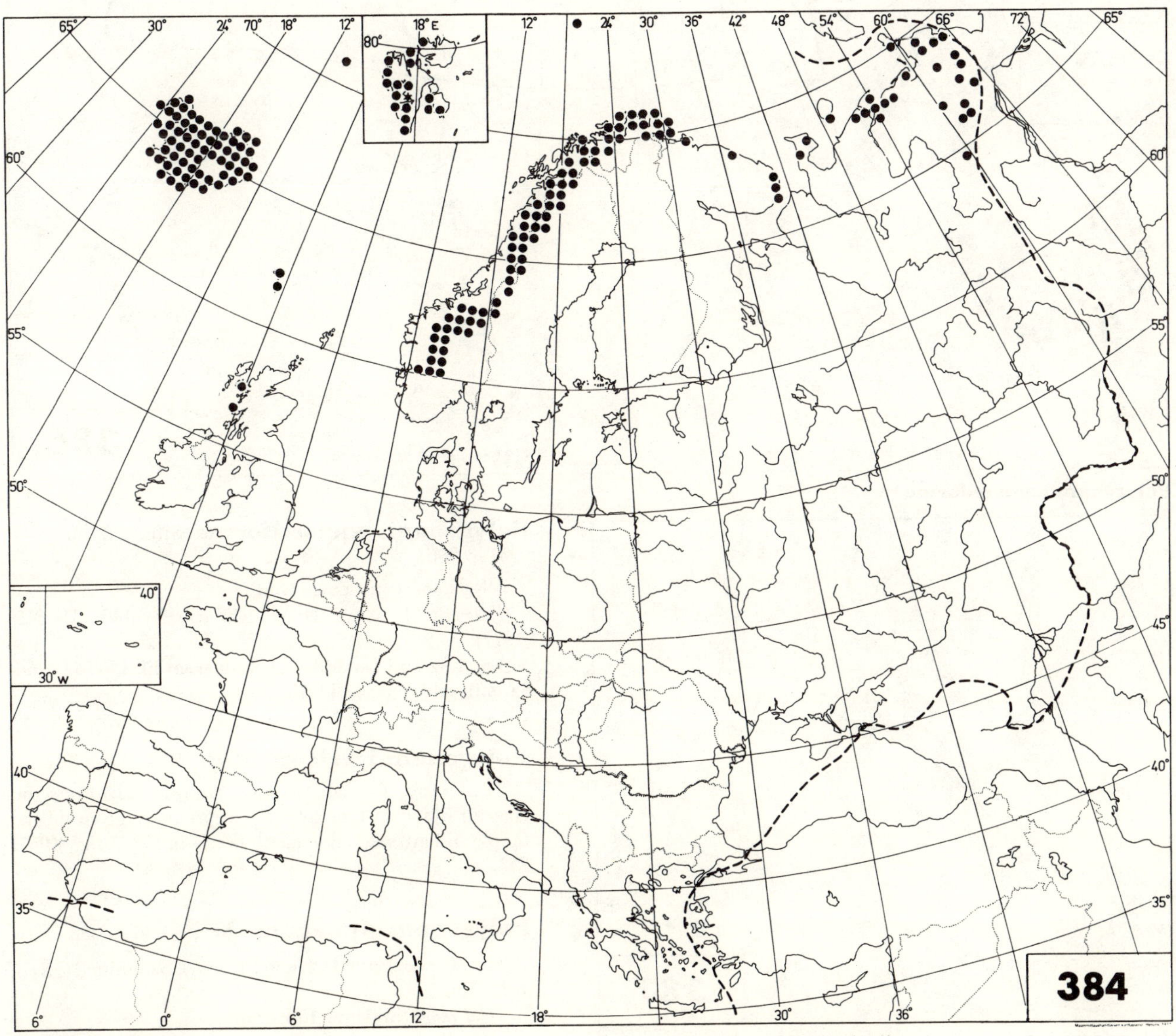

Koenigia islandica ★ = type locality of **K. hadacii**

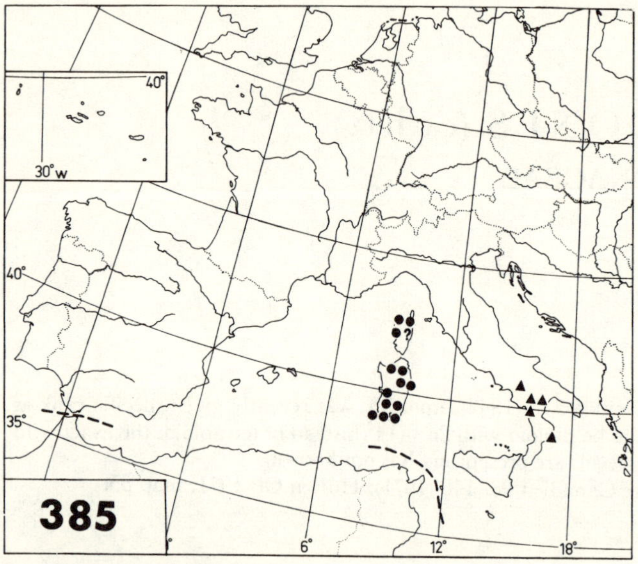

● = **Polygonum scoparium**
▲ = **P. tenoreanum**

Polygonum scoparium Req. ex Loisel. — Map 385.

Generic delimitation of *Polygonum* s. lato: Haraldson 1978: 1—95.

Taxonomy. E. Nardi & M. Raffaelli, Webbia 31: 515 (1977); M. Raffaelli, Boll. Soc. Sarda Sci. Nat. 17 (Cahiers Optima Leafl. 78): 289—294 (1978).

Total range. Endemic to Corse and Sardegna.

Polygonum tenoreanum Nardi & Raffaelli — Map 385.

Polygonum arenarium subsp. elegans Nyman; P. elegans Ten. non Aiton fil.; P. equisetiforme var. elegans (Nyman) Fiori. P. arenarium auct. non Waldst. & Kit.

Taxonomy. E. Nardi & M. Raffaelli, Webbia 31: 513—527 (1977). 2n = 20.

Notes. It (the species not mentioned in Fl. Eur.): E. Nardi & M. Raffaelli, Webbia 31: 518—523 (1977).

Total range. Endemic to Europe.

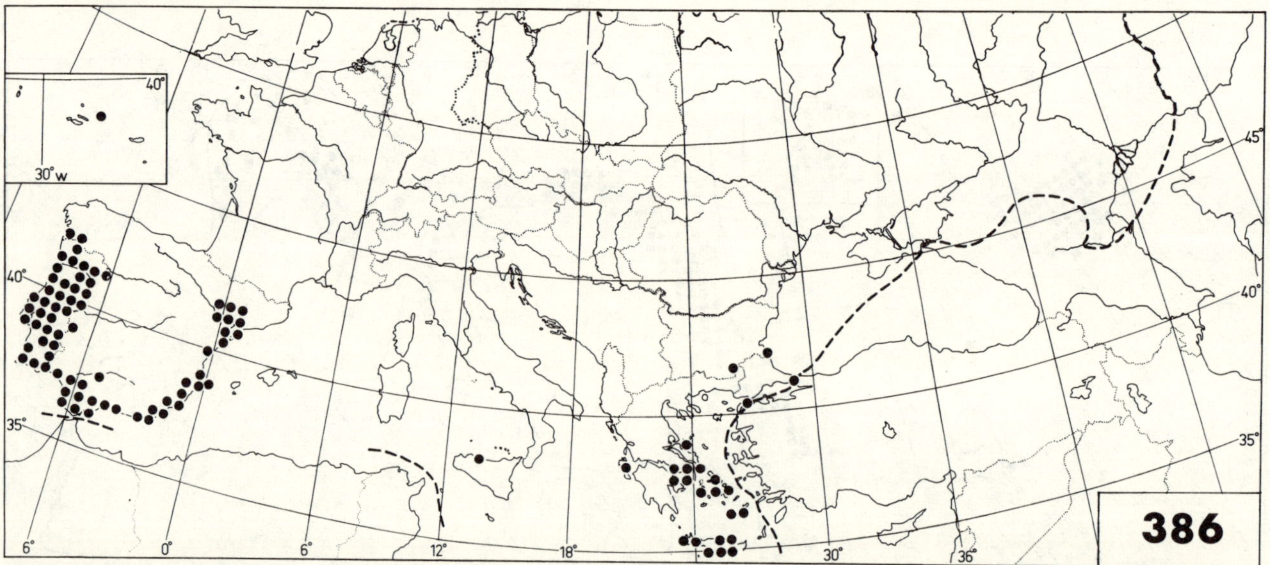

Polygonum equisetiforme

Polygonum equisetiforme Sibth. & Sm. — Map 386.

Polygonum controversum Guss.

Taxonomy. E. Nardi & M. Raffaelli, Webbia 31: 515 (1977).

Notes. Az and Lu added (not given in Fl. Eur.), It and Sa omitted (given in Fl. Eur.).

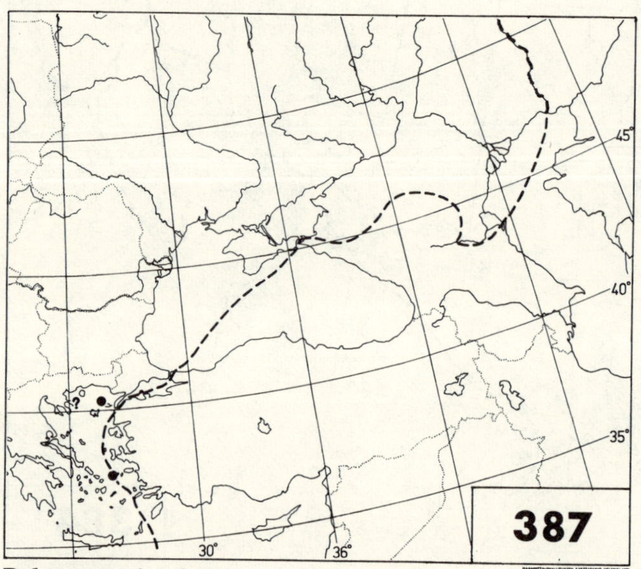

Polygonum icaricum

Polygonum icaricum Rech. fil. — Map 387.

Total range. Outside Europe known only from the Aegean island of Ikaria (and exceptionally given on the map), and given as endemic to Ikaria in Fl. Turkey 1967: 275.

Polygonum romanum Jacq. — Map 388.

Notes. Ga confirmed (?Ga in Fl. Eur.), Sa omitted (given in Fl. Eur.).

Total range. Endemic to Europe.

Polygonum longipes Halácsy & Charrel

Taxonomy. Placed in the synonymy of *Polygonum arenastrum.*

Polygonum gussonii Tod. ex Cesati, Passer. & Gibelli

? Polygonum herniarioides Delile

Taxonomy. According to Fl. Eur., plants from Si and ?It, somewhat resembling »small specimens of *Polygonum longipes*» but of dubious taxonomical status, may possibly, if perennial, be referred to *P. gussonii.* Otherwise, these too may be conspecific with *P. arenastrum.*

Notes. Not mapped.

Polygonum maritimum L. — Map 389.

Notes. Not extinct in Br (†Br in Fl. Eur.). Hb and Ho added (not given in Fl. Eur.): I.K. Ferguson & L. F. Ferguson, Irish Natur. Journal 18: 95 (1974). Rs(W) omitted (given in Fl. Eur.).

Total range. B. T. Styles, Watsonia 5: 210 (1962).

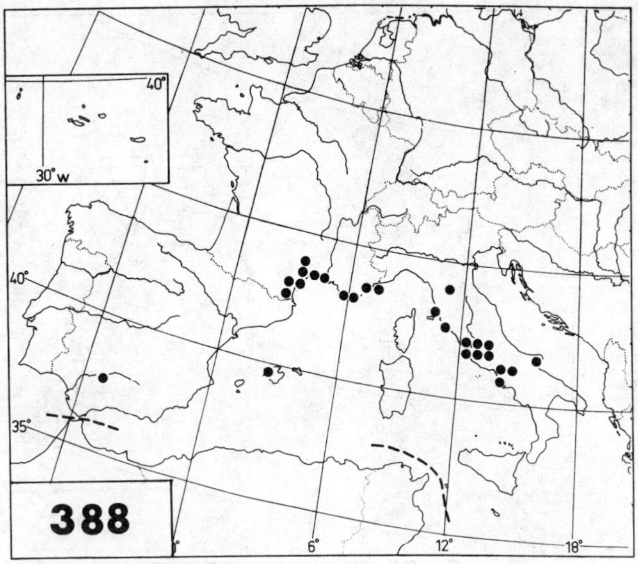

Polygonum romanum

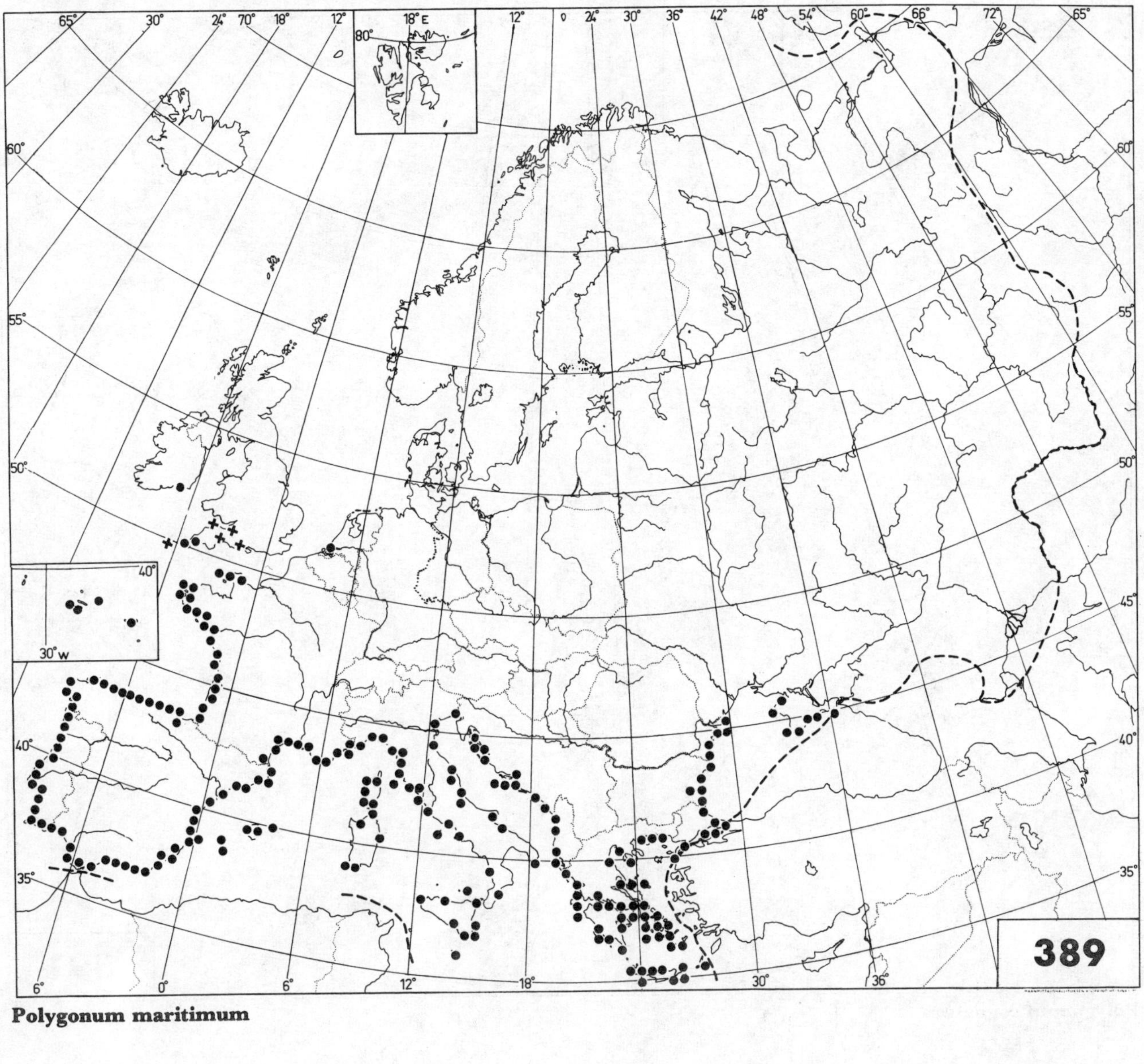

Polygonum maritimum

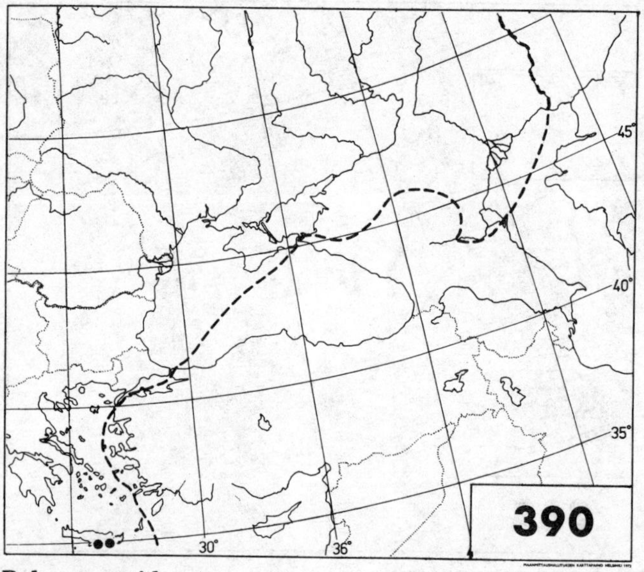

Polygonum idaeum

Polygonum idaeum Hayek — Map 390.

Notes. Occurrence in Gr not confirmed (? Gr in Fl. Eur.).
Total range. Endemic to Europe.

Polygonum cognatum Meissner — Map 391.

Polygonum alpestre C. A. Meyer

Native of S.W. Asia.

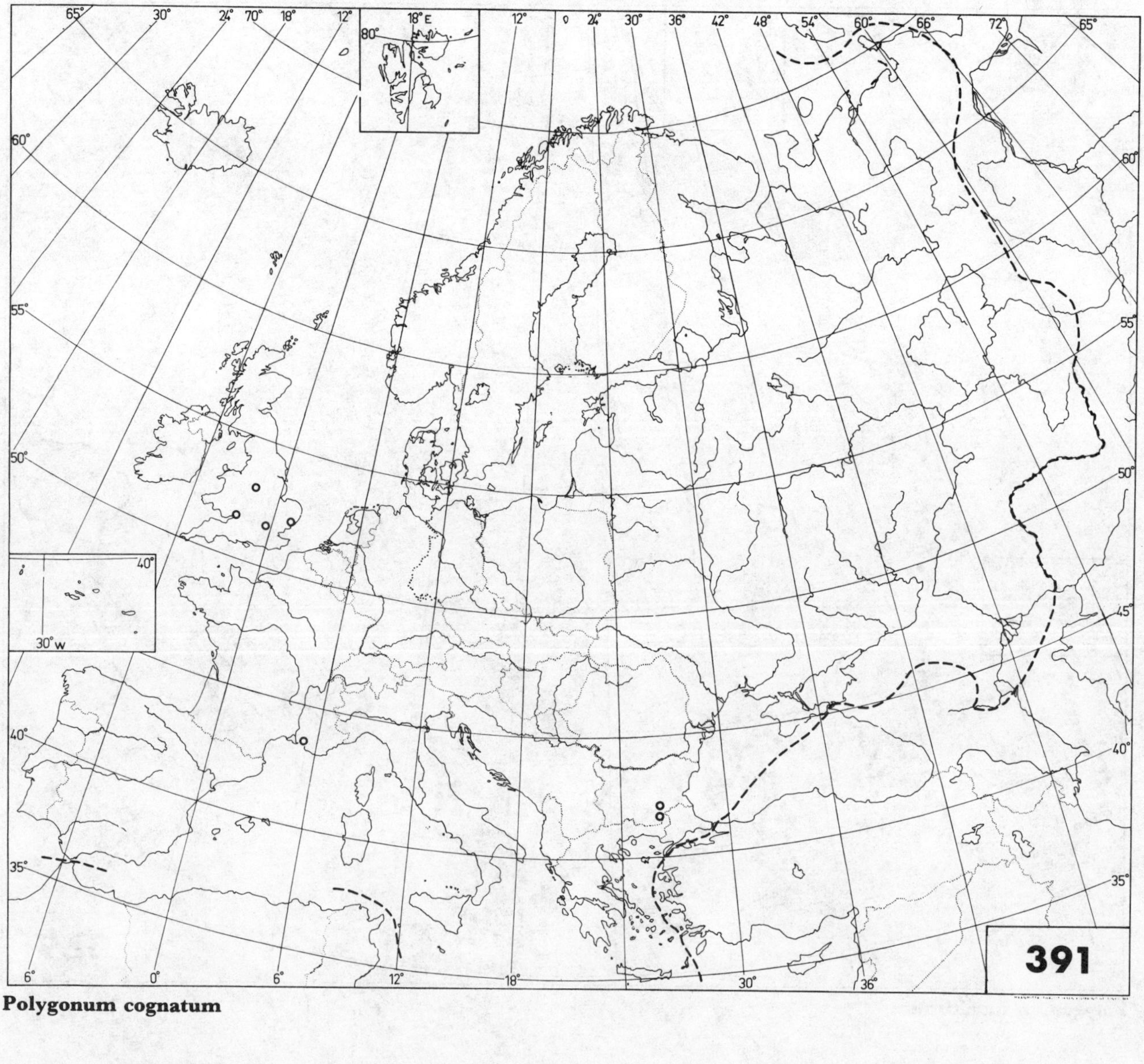

Polygonum cognatum

Polygonum floribundum Schlecht. ex Sprengel — Map 392.

Polygonum salsugineum Bieb. — Map 393.

Notes. Rs(C) added (not given in Fl. Eur.). No data available from Rs(K), although present there according to N. I. Rubtsev, Opred. vysshikh rast. Kryma, p. 118 (Leningrad 1972).

Polygonum samarense H. Gross — Map 394.

Total range. Endemic to Europe.

Polygonum aschersonianum H. Gross — Map 394.

Polygonum salsugineum auct. non Bieb.

Total range. Endemic to Europe.

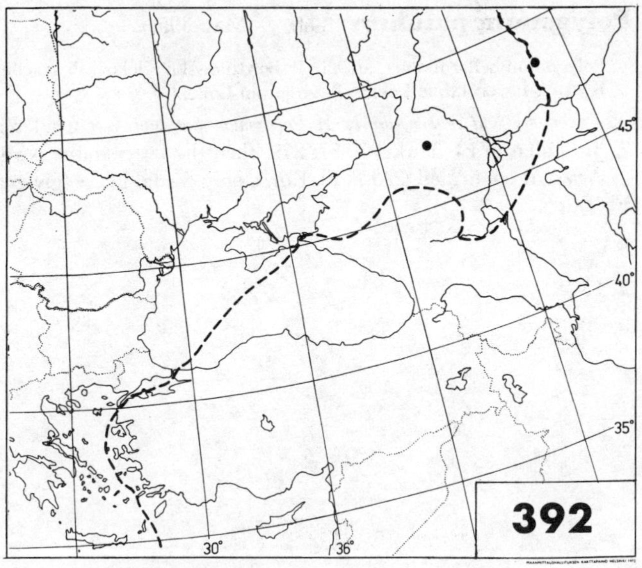

Polygonum floribundum

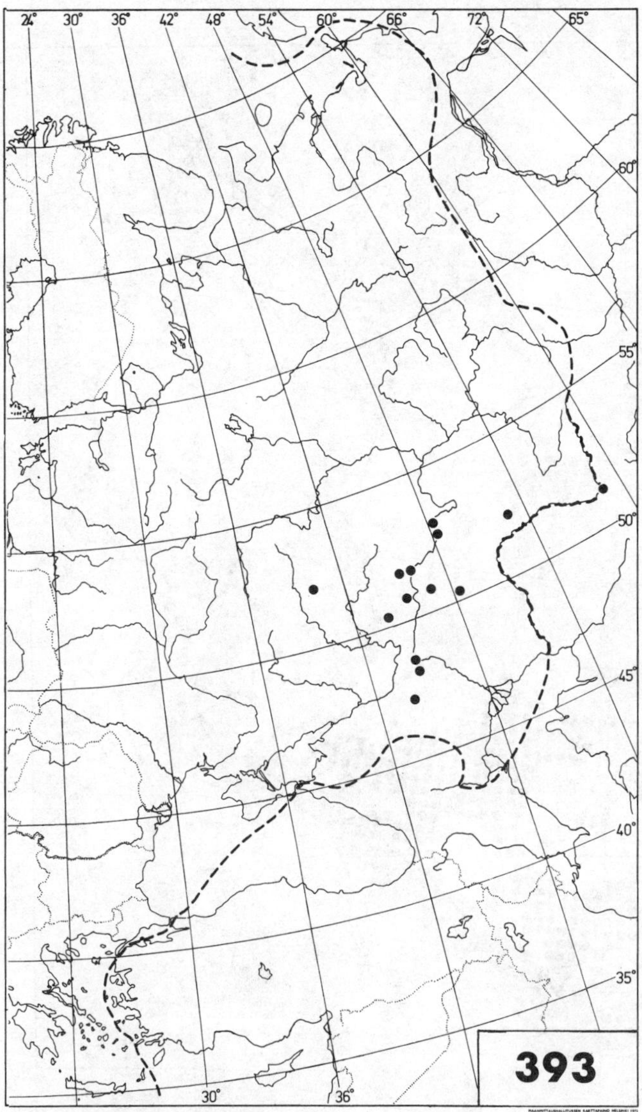

Polygonum salsugineum

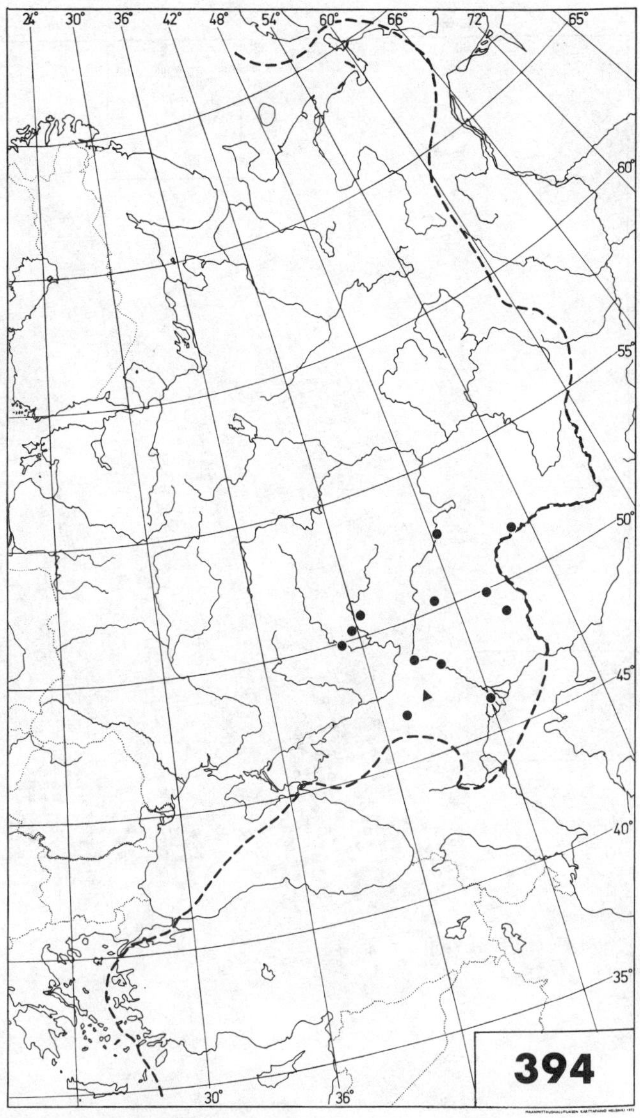

● = Polygonum samarense
▲ = P. aschersonianum

Polygonum patulum Bieb. — Map 395.

Polygonum bellardii auct. an All.; P. bordzilowskii Klokov; P. gracilius Klokov; P. kitaibelianum Sadler; P. kotovii Klokov; P. novoascanicum Klokov; P. spectabile Lehm.; P. virgatum Loisel.

Taxonomy and nomenclature. If *Polygonum patulum* is considered conspecific with *P. bellardii* All., as it is by M. J. E. Coode & J. Cullen in Fl. Turkey 1967: 279, then the latter name is valid for the taxon.

Notes. Al confirmed (?Al in Fl. Eur.), only casual in Ge (given as native in Fl. Eur.). Presence in Po doubtful (given as present in Fl. Eur.).

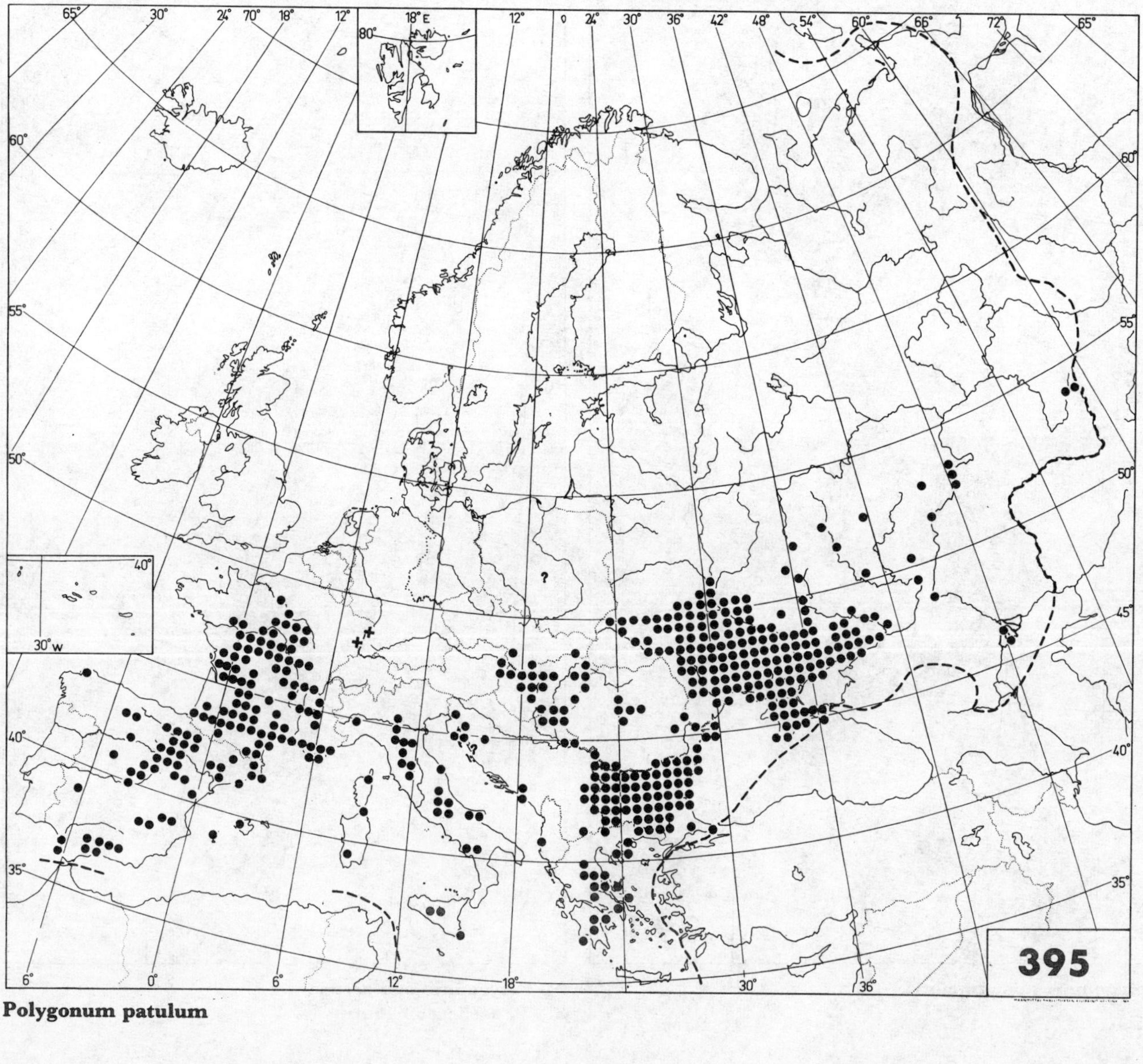

Polygonum patulum

Polygonum arenarium Waldst. & Kit. — Map 396.

Incl. Polygonum albanicum Jáv.

Taxonomy. D. A. Webb & A. O. Chater, Feddes Repert. 68 (Fl. Eur. Notulae Syst. 2): 188 (1963); N. Tzvelev, Nov. Syst. Pl. Vasc. (Leningrad) 15: 141—142 (1979).

Notes. Presence in Co doubtful (not given in Fl. Eur.; present according to Fl. France 1973: 228). Lu omitted (given in Fl. Eur.), Rs(C) added (not given in Fl. Eur.). No data available from Rs(K) although present there according to N. I. Rubtsev, Opred. vysshikh rast. Kryma, p. 120 (Leningrad 1972) (as *Polygonum janatae* and *P. pseudoarenarium*). The material available does not allow separate mapping of the two subspecies recognized in Fl. Eur.

P. arenarium subsp. arenarium

Incl. Polygonum janatae Klokov, P. pseudoarenarium Klokov and P. venantianum G. C. Clementi

Notes. »E. C. & S. E. Europe» (Fl. Eur.).

P. arenarium subsp. pulchellum (Loisel.) D. A. Webb & Chater

Polygonum pulchellum Loisel.

Notes. »Mediterranean region; Portugal» (Fl. Eur.).

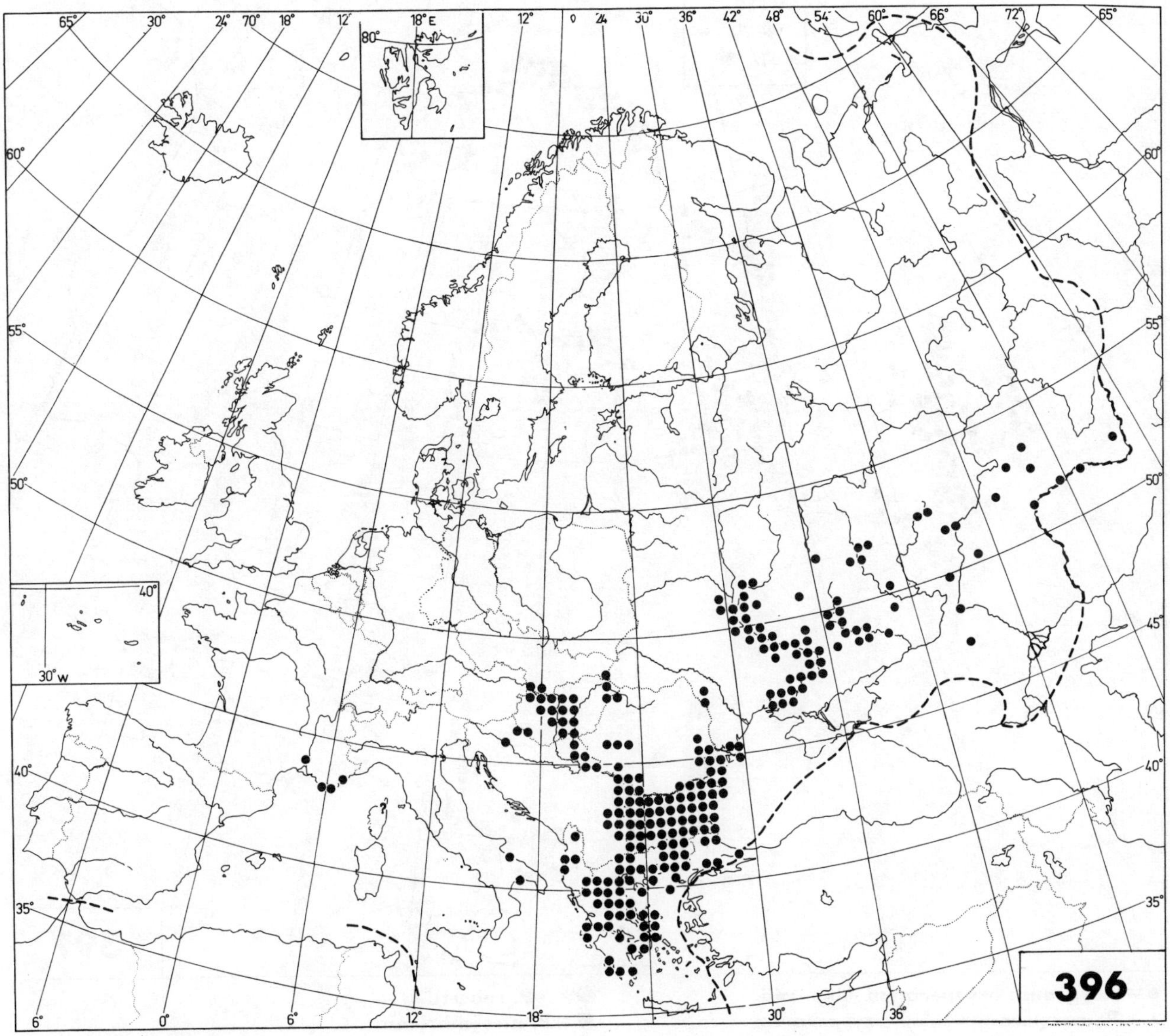

Polygonum arenarium

Polygonum oxyspermum Meyer & Bunge ex Ledeb. — Map 397.

Incl. Polygonum robertii Loisel.

Taxonomy. R. Nordhagen, Avh. Norske Vid.-Akad. Oslo Mat.-Nat. Kl. N. S. 5: 1—40 (1963); D. A. Webb & A. O. Chater, Feddes Repert. 68 (Fl. Eur. Notulae Syst. 2): 188—189 (1963); Hylander 1966: 364—368.

Notes. The Mediterranean plant called *Polygonum robertii* Loisel. has been tentatively mapped with *P. oxyspermum* although possibly not belonging here; cf. B. T. Styles, Watsonia 5: 210—211 (1962). Records from It and Si refer to *»P. oxyspermum* subsp. *raii»*. See also the text for *P. mesembricum.* Gr, Rs(C) and Si added (not given in Fl. Eur.), Rs(W) and Sa omitted (given in Fl. Eur.). No data available from Rs(K) although given as present in N. I. Rubtsev, Opred. vysshikh rast. Kryma, p. 117 (Leningrad 1972) (as *P. robertii).*

P. oxyspermum subsp. **oxyspermum** — Map 397.

Notes. Da, Fe, Ge, No, Po, Rs(B, C), Su. »Coasts of the Baltic; Kattegat; S. Norway» (Fl. Eur.).

Total range. Most probably endemic to Europe (although not marked as such in Fl. Eur.). Of the non-European occurrences given in Hultén AA 1958: map 270, those at the mouth of the River Yenisei have been referred to *Polygonum oxyspermum* subsp. *raii* (as *P. norvegicum);* those on Cape Breton Island, Nova Scotia, Canada, may originate from immigration from Europe.

P. oxyspermum subsp. **raii** (Bab.) D. A. Webb & Chater — Map 397.

Polygonum raii Bab.; P. robertii auct. non Loisel. Incl. P. norvegicum Sam. ex Lid; P. raii subsp. norvegicum Sam.

Notes. Br, Da, Ga, Ge, Hb, No, Rs(N). »Shores of N. W. Europe, from N. W. France to arctic Russia; also locally in W. Mediterranean region» (Fl. Eur.).

Total range. Á. Löve, Canad. Jour. Bot. 34: 507 (1956); Hultén AA 1958: map 269.

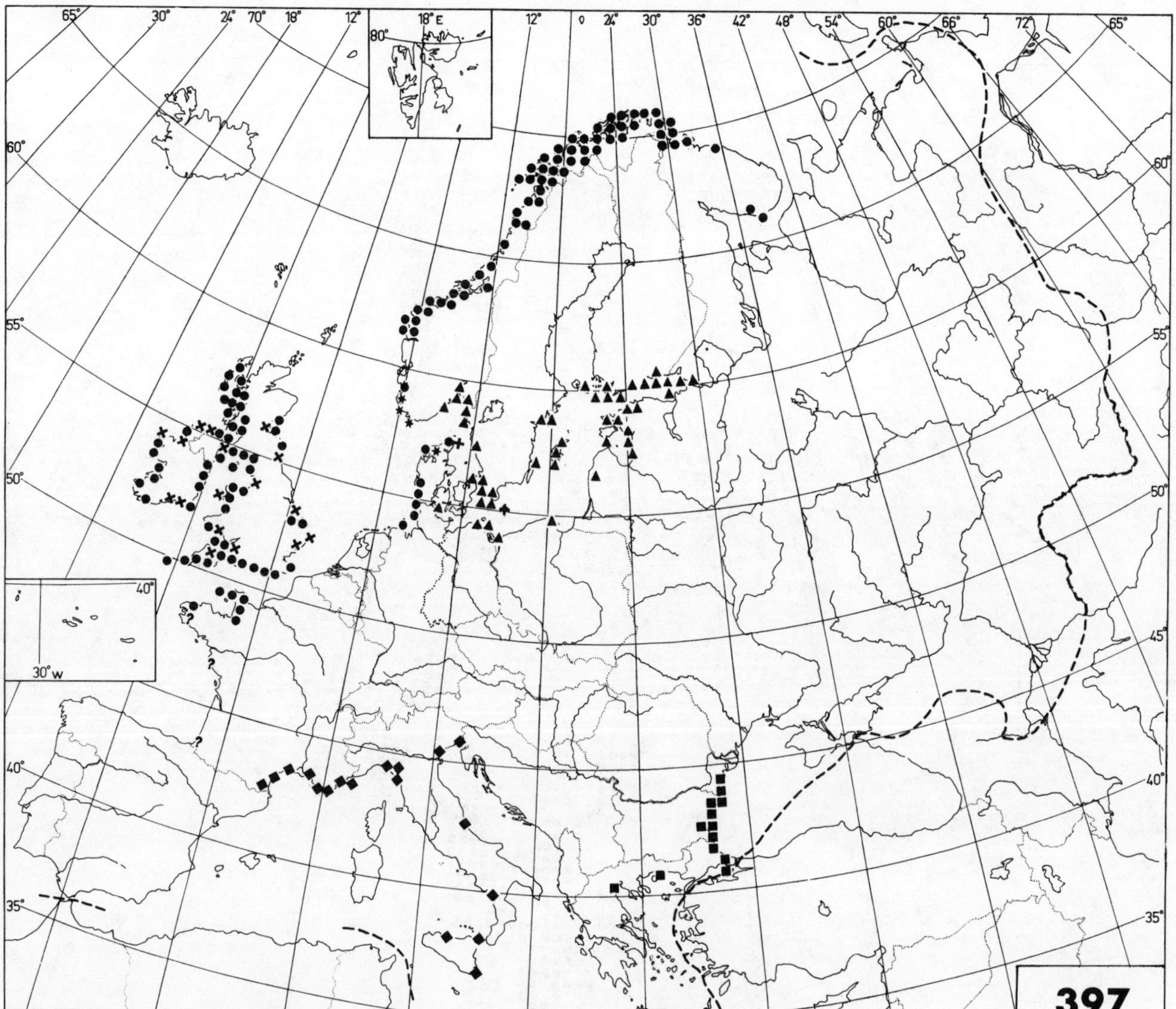

● = **Polygonum oxyspermum** subsp. **raii**
▲ = **P. oxyspermum** subsp. **oxyspermum**
★ != subsp. **raii** + subsp. **oxyspermum**
◆ = **»P. robertii»**
■ = **P. mesembricum**

Polygonum mesembricum Chrtek — Map 397.

Polygonum raii auct. non Bab.; P. robertii auct. non Loisel. Incl. P. euxinum Chrtek

Taxonomy. The plants from the Black Sea coast, called *Polygonum mesembricum* Chrtek, »may perhaps constitute a third subspecies [of *P. oxyspermum*]» (Fl. Eur.). J. Chrtek, Preslia 32: 366—368 (1960); I. Asenov in Fl. Reipubl. Pop. Bulg. 3: 223—224 (1966).

Notes. Bu, Gr, Rm (mapped according to I. Grinţescu in Fl. Rep. Pop. Rom. 1: 440—443 (1952), as »*Polygonum Rayi* Bab.»), Tu. Concerning Rs(K), see N. Tzvelev, Nov. Syst. Pl. Vasc. (Leningrad) 15: 133 (1979). »Black Sea coast (Turkey to Krym)» (Fl. Eur.).

Polygonum graminifolium Wierzb. ex Heuffel — Map 398.

Notes. [Au] omitted (given in Fl. Eur.). Not present in Bu (?Bu in Fl. Eur.). Cz added (not given in Fl. Eur.): J. Holub, Preslia 38: 79 (1966).

Total range. Endemic to Europe.

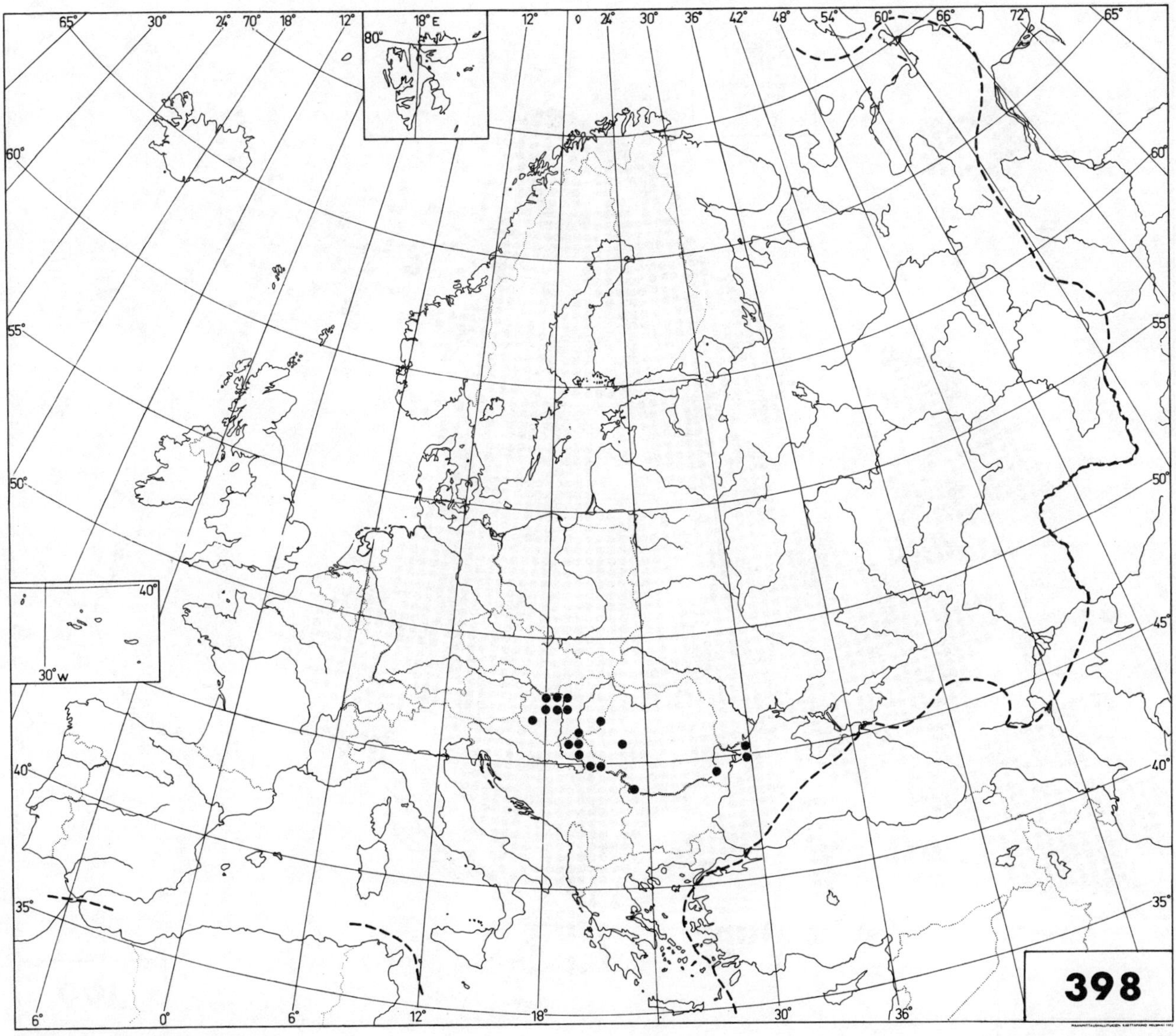

Polygonum graminifolium

Polygonum aviculare L., s. lato — Map 399.

Polygonum arenastrum + P. aviculare s. str. + P. boreale + P. longipes + P. neglectum + P. rurivagum

Taxonomy. Hylander 1966: 359—364; M. J. E. Coode & J. Cullen in Fl. Turkey 1967: 277; N. Tzelev, Nov. Syst. Pl. Vasc. (Leningrad) 15: 135—138 (1979).

Notes. Only casual in Sb *(Polygonum aviculare* given from »probably all territories» in Fl. Eur.). The map is incomplete since data are lacking from the E. and S.E. parts of Rs.

Total range. Hultén Alaska 1968: 390; Hultén CP 1971: map 245.

Polygonum aviculare L., s. str.

Polygonum heterophyllum Lindman; P. littorale auct. p.p.

Notes. »Probably throughout Europe, but only as an alien in the extreme north. ?All territories.» (Fl. Eur.).

Polygonum boreale (Lange) Small

Polygonum arenastrum subsp. boreale (Lange) Á. Löve; P. aviculare var. boreale Lange; P. heterophyllum subsp. boreale (Lange) Löve & Löve

Notes. »Northern part of N. W. Europe from Zetland northwards. Br Fa Fe Is No Su.» (Fl. Eur.).

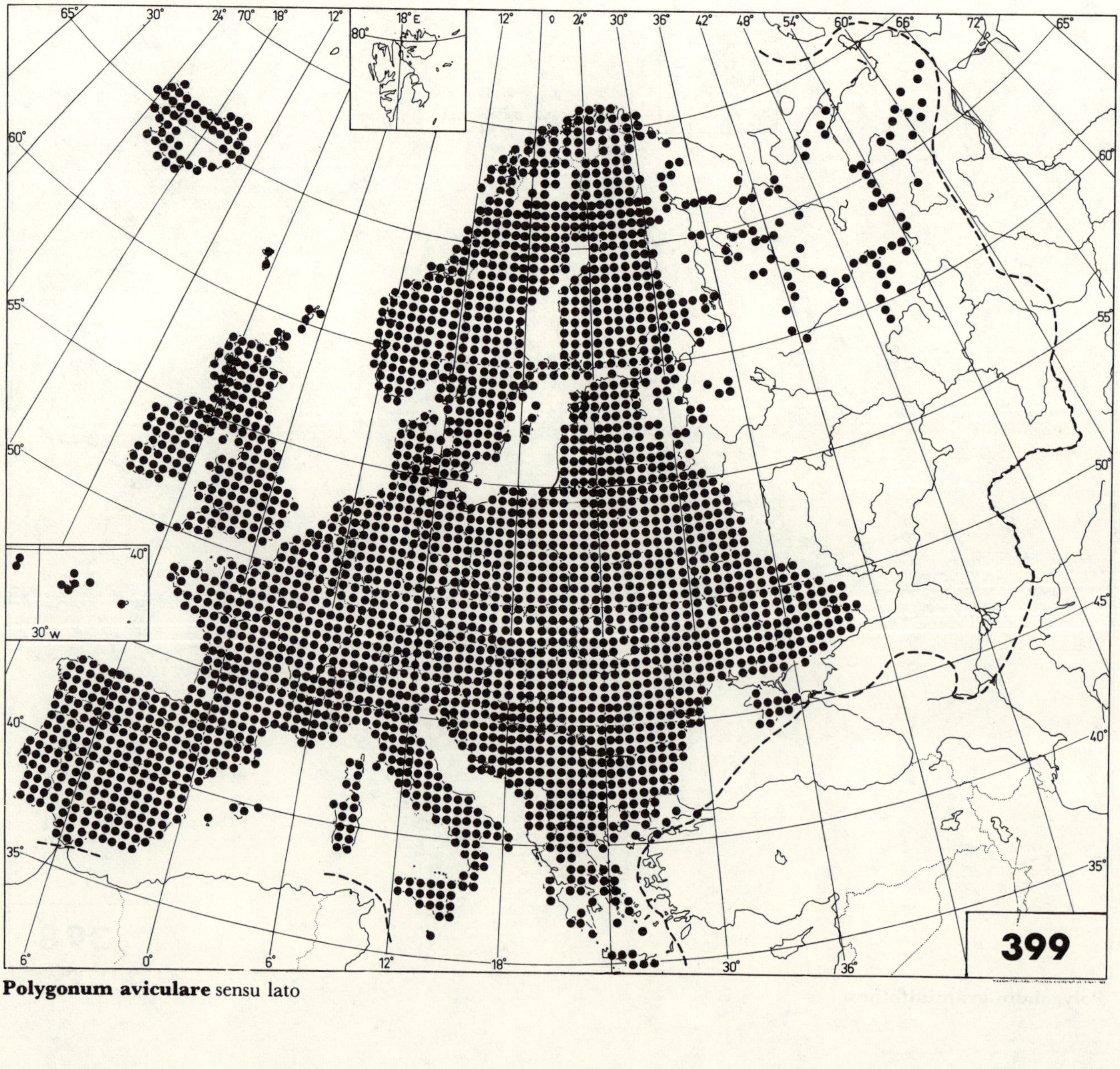

Polygonum aviculare sensu lato

Polygonum rurivagum Jordan ex Boreau

Polygonum aviculare subsp. rurivagum (Jordan ex Boreau) O. Bolòs & Vigo; P. bellardii auct. non All.; P. heterophyllum subsp. rurivagum (Jordan ex Boreau) Lindman

Taxonomy. D. A. Webb & A. O. Chater, Feddes Repert. 68 (Fl. Eur. Notulae Syst. 2): 188 (1963); O. Bolòs & J. Vigo, Buttl. Inst. Catalana Hist. Nat. 38 (Sec. Bot. 1): 86 (1974).
Notes. »Distribution uncertain; mainly in W. & S. C. Europe, but extending north-eastwards to Sweden» (Fl. Eur.).

Polygonum neglectum Besser

Taxonomy. Polygonum neglectum Besser is put into the synonymy of *P. bellardii* All. by N. Tzvelev, Nov. Syst. Pl. Vasc. (Leningrad) 15: 135 (1979).
Notes. »Plants from U.S.S.R. and from C. Europe have been referred here, but the status of the taxon is not clear» (Fl. Eur.).

Polygonum arenastrum Boreau

Polygonum aequale Lindman; P. aviculare auct. non L.; P. aviculare subsp. aequale (Lindman) Asch. & Graebner; P. littorale auct. p.p. Incl. P. acetosellum Klokov, P. calcatum Lindman, P. longipes Halácsy & Charrel and P. propinquum Ledeb.

Notes. »Probably throughout most of Europe, except the extreme north; distribution imperfectly known.» (Fl. Eur.).

Polygonum humifusum Pallas — Map 400.

Polygonum aviculare var. propinquum Trautv. (an Ledeb.?)

Taxonomy. Fl. Arct. URSS 1966: 168—169. Diploid with $2n = 20$.
Notes. Rs(N) (the species not mentioned in Fl. Eur.): Fl. Bor.-or. Eur. URSS 1976: 179, map 255.
Total range. Fl. Arct. URSS 1966: 169; Hultén Alaska 1968: 392.

Polygonum acetosum Bieb. — Map 400.

Notes. Not present in Bu (?Bu in Fl. Eur.). In Rs(E), reported only from the middle Ural River (»near Astrakhan» in Fl. Eur.).

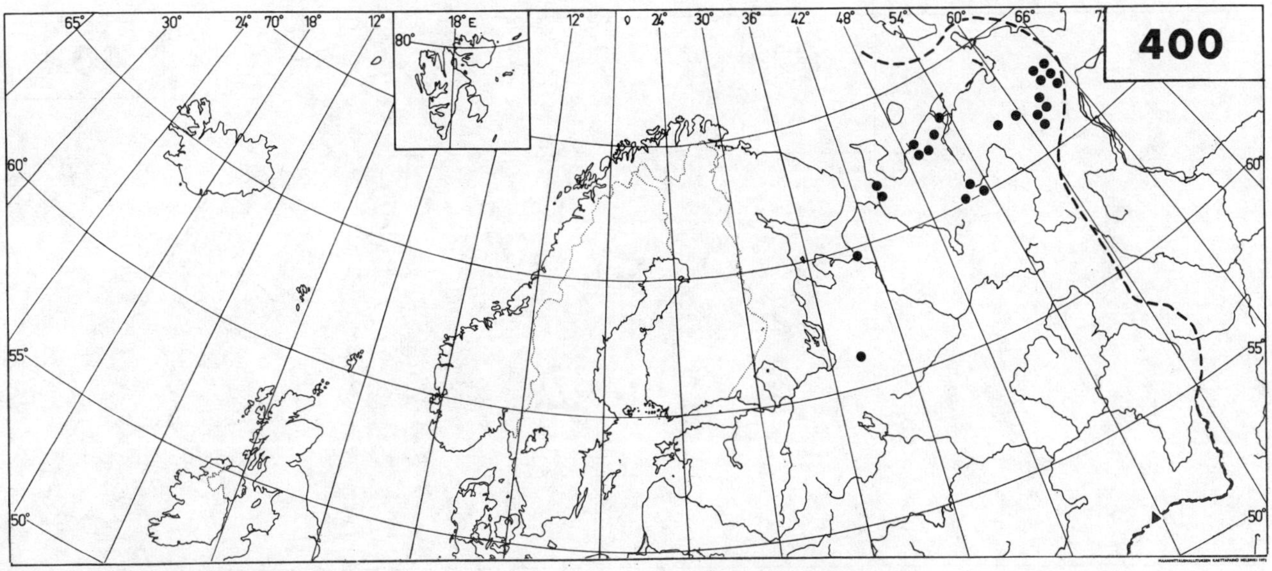

● = **Polygonum humifusum** ▲ = **P. acetosum**

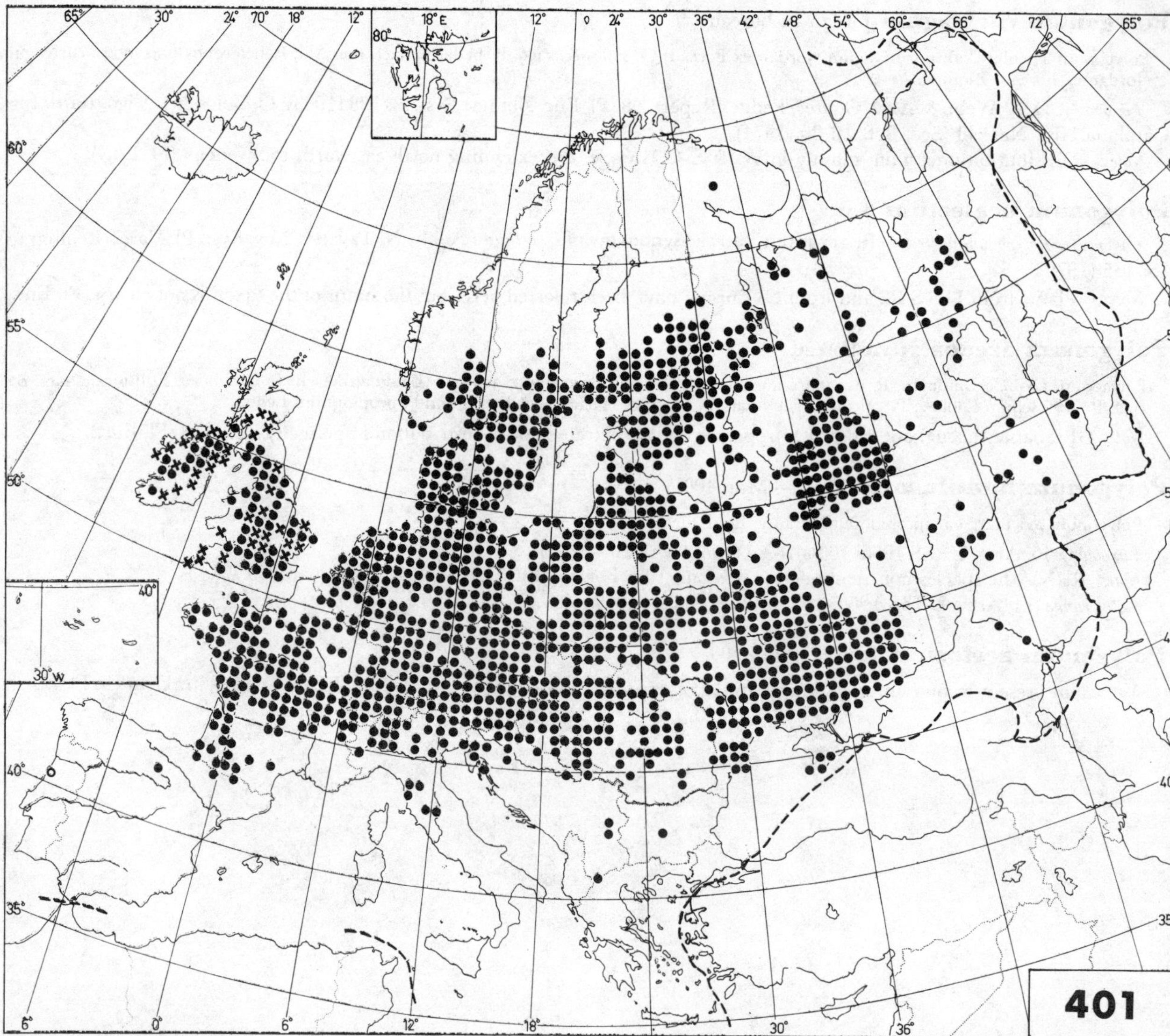

Polygonum minus

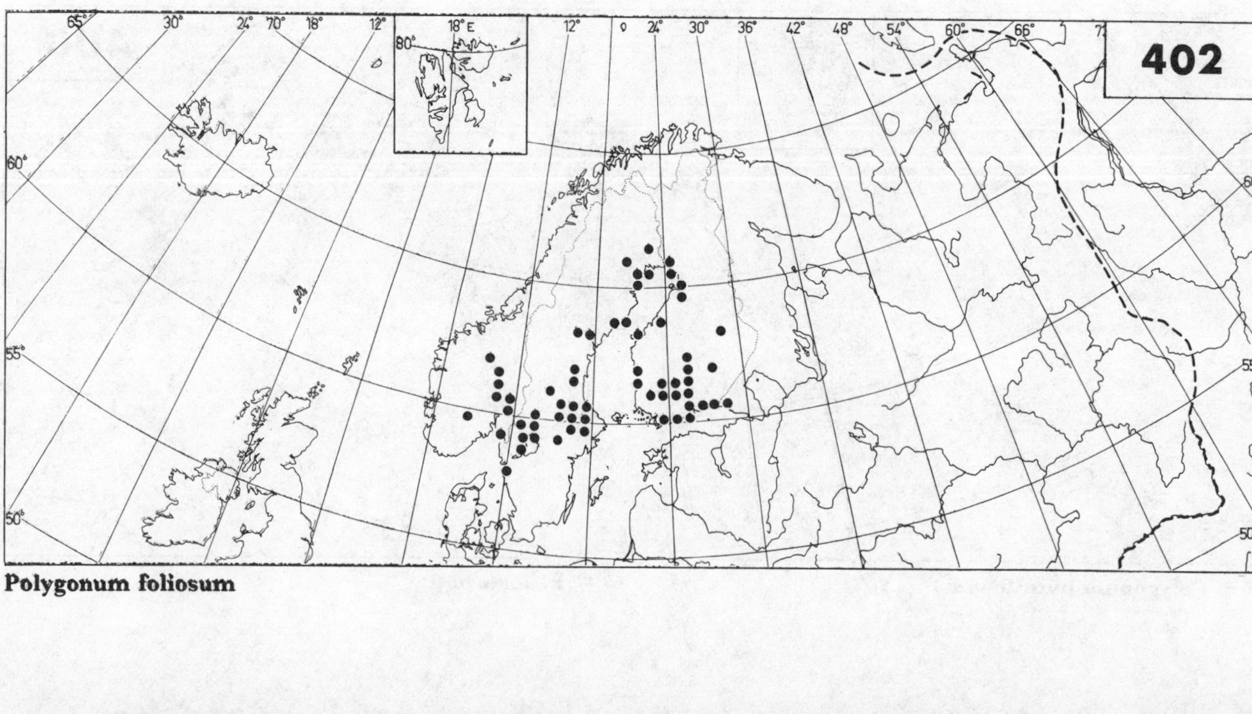

Polygonum foliosum

Polygonum minus Hudson — Map 401.

Persicaria minor (Hudson) Opiz

Generic delimitation of *Persicaria* Miller: Haraldson 1978: 70—74.
Notes. Gr added (not given in Fl. Eur.). Not native in Lu (given in Fl. Eur.).

Polygonum foliosum H. Lindb. — Map 402.

Persicaria foliosa (H. Lindb.) Kitagawa

Notes. Rs(N, B) omitted (given in Fl. Eur.), Rs(C) added (not given in Fl. Eur.). Also reported from the middle Don, Rs(E), outside the map area, but the identification needs checking.

Polygonum mite Schrank — Map 403.

Persicaria laxiflora (Weihe) Opiz

Notes. Rs(B) added (not given in Fl. Eur.), Rs(E), Sa and Si omitted (given in Fl. Eur.), Tu not confirmed (?Tu in Fl. Eur.). The two dots from Rs(N), and perhaps some of the other northernmost records from Rs are doubtful; see Fl. Bor.-or. Eur. URSS 1976: 181—182, map 263.

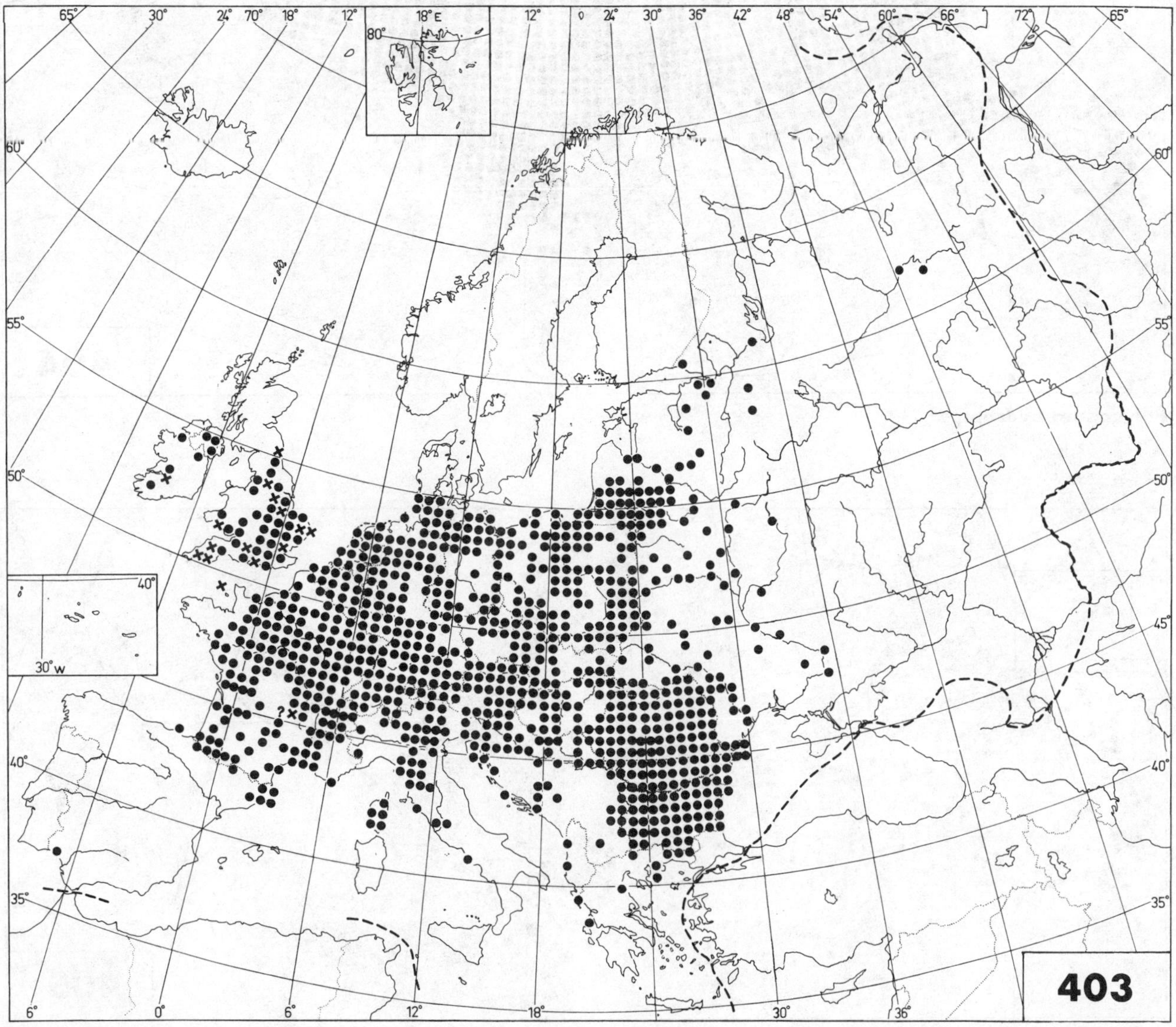

Polygonum mite

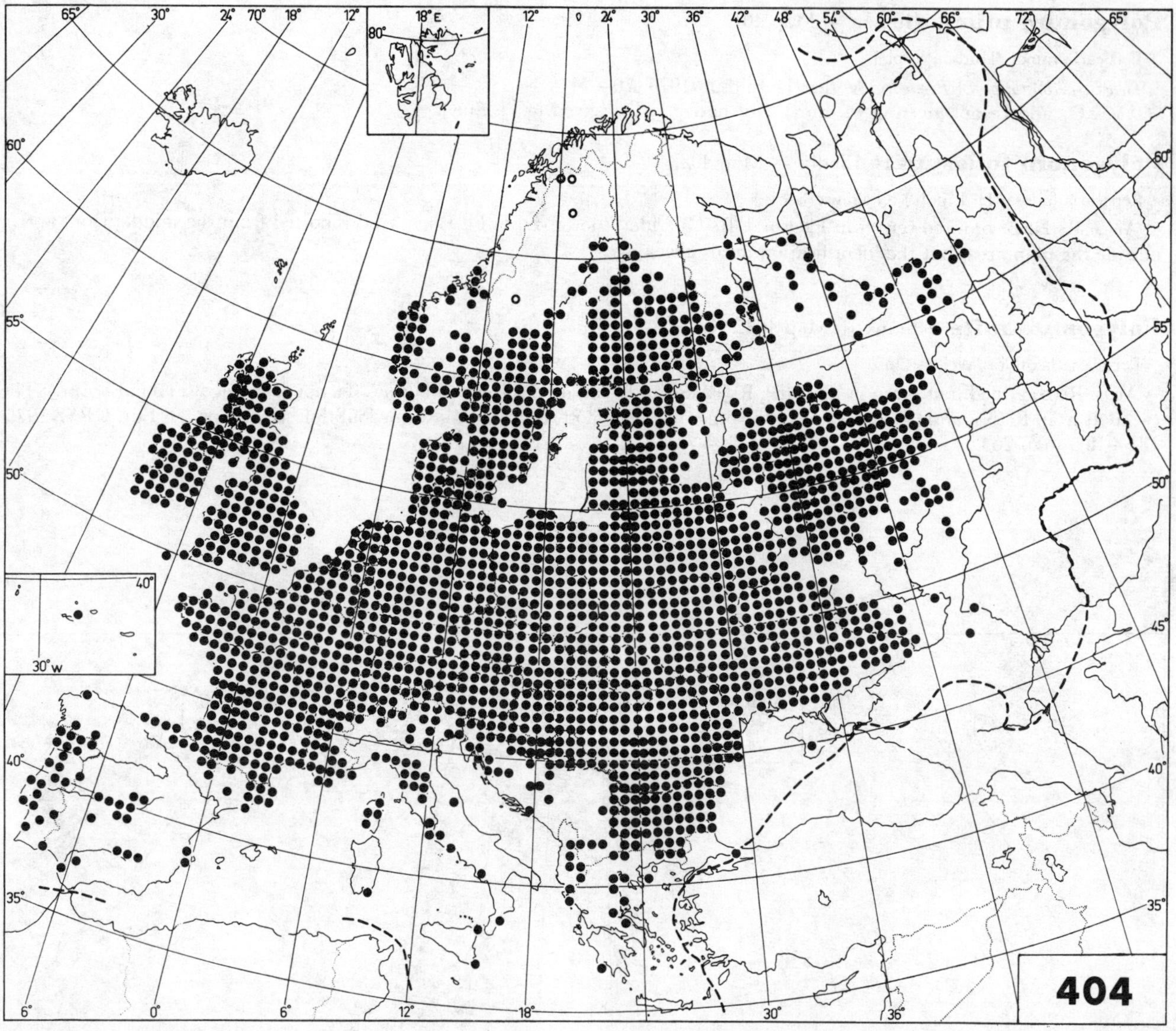

Polygonum hydropiper

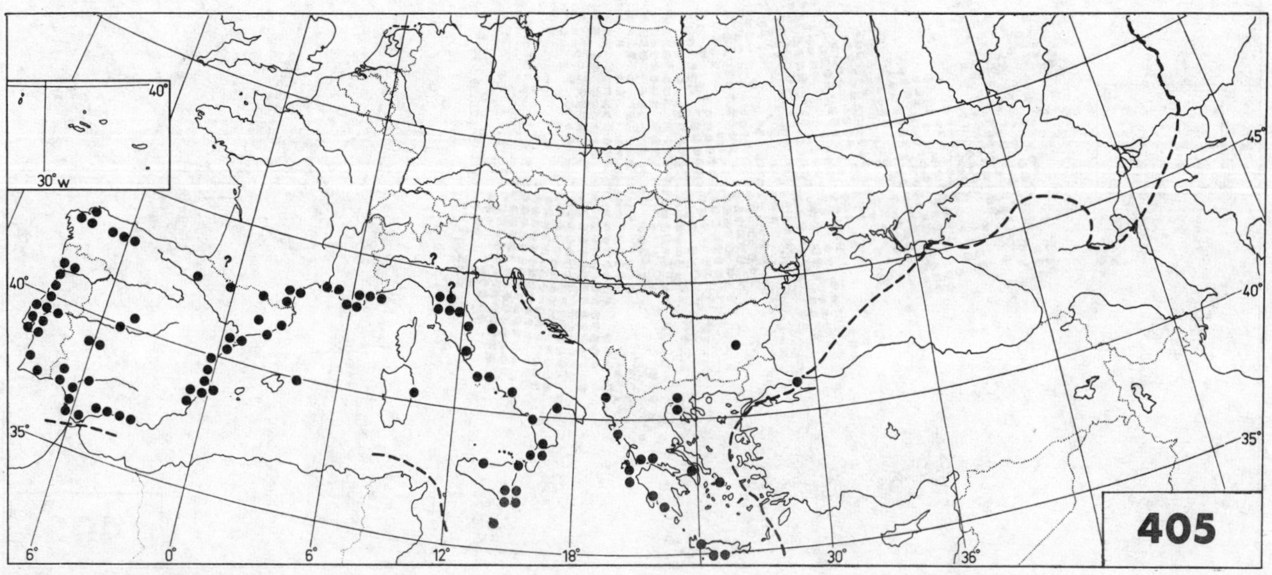

Polygonum salicifolium

Polygonum hydropiper L. — Map 404.

Persicaria hydropiper (L.) Opiz

Notes. Az and Sa added (not given in Fl. Eur.).

Total range. MJW 1965: map 131b (Eurasian part); Hultén Alaska 1968: 389; Hultén CP 1971: map 209 (holarctic part).

Polygonum salicifolium Brouss. ex Willd. — Map 405.

Persicaria salicifolia (Brouss. ex Willd.) Assenov; P. serrulata (Lag.) Webb & Moq.; Polygonum scabrum Poir. non Moench; P. serrulatum Lag. Incl. P. serrulatoides H. Lindb.

Notes. Az omitted (given in Fl. Eur.), the records being due to misidentification of *Polygonum hydropiperoides*. Tu added (not given in Fl. Eur.).

Polygonum persicaria L. — Map 406.

Persicaria maculata (Rafin.) Löve & Löve; P. maculosa S. F. Gray; P. vulgaris Samp.; Polygonum maculatum Rafin.

Nomenclature. Haraldson 1978: 74.

Notes. Az added (not given in Fl. Eur.): Nova Fl. Port. 1971: 79; A. Hansen, Anuário Soc. Brot. 37: 108 (1971). Sa omitted (given in Fl. Eur.).

Total range. Hultén Alaska 1968: 388; Hultén CP 1971: map 254 (holarctic part).

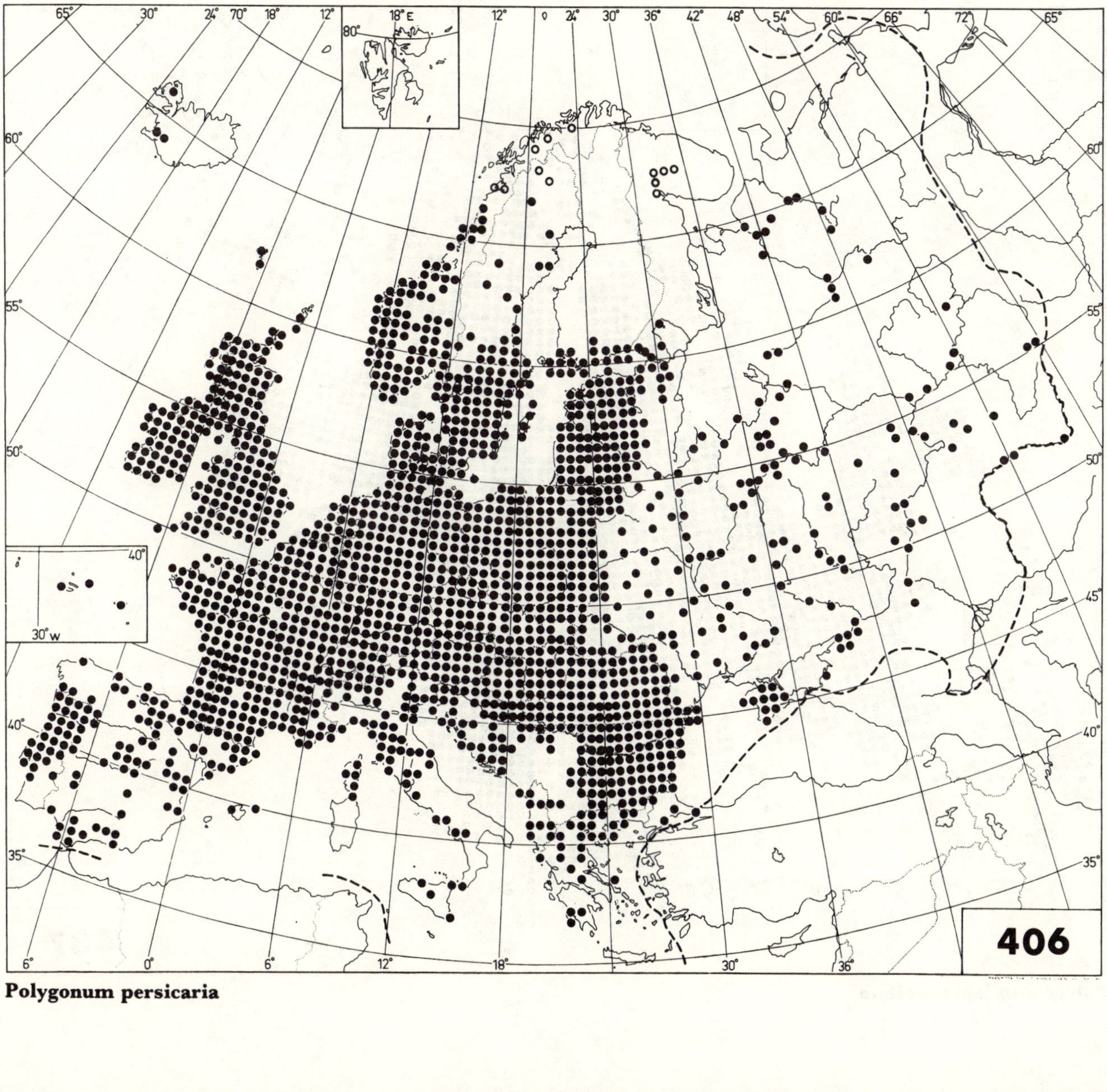
406

Polygonum persicaria

Polygonum lapathifolium L. — Map 407.

Persicaria lapathifolia (L.) S. F. Gray; Polygonum andrzejowskianum Klokov; P. brittingeri Opiz; P. danubiale A. Kerner; P. incanum F. W. Schmidt; P. linicola Sutulov; P. nodosum Pers.; P. pallidum With.; P. paniculatum Andrz.; P. scabrum Moench; P. tomentosum Schrank; P. zaporoviense Klokov

Taxonomy. J. Timson, Watsonia 5: 386—395 (1963).

Notes. Az added (not given in Fl. Eur.): Nova Fl. Port. 1971: 79; A. Hansen, Anuário Soc. Brot. 37: 98 (1971). Is omitted (given in Fl. Eur.).

Total range. MJW 1965: map 131a; Hultén Alaska 1968: 389; Hultén CP 1971: map 236.

Polygonum lapathifolium

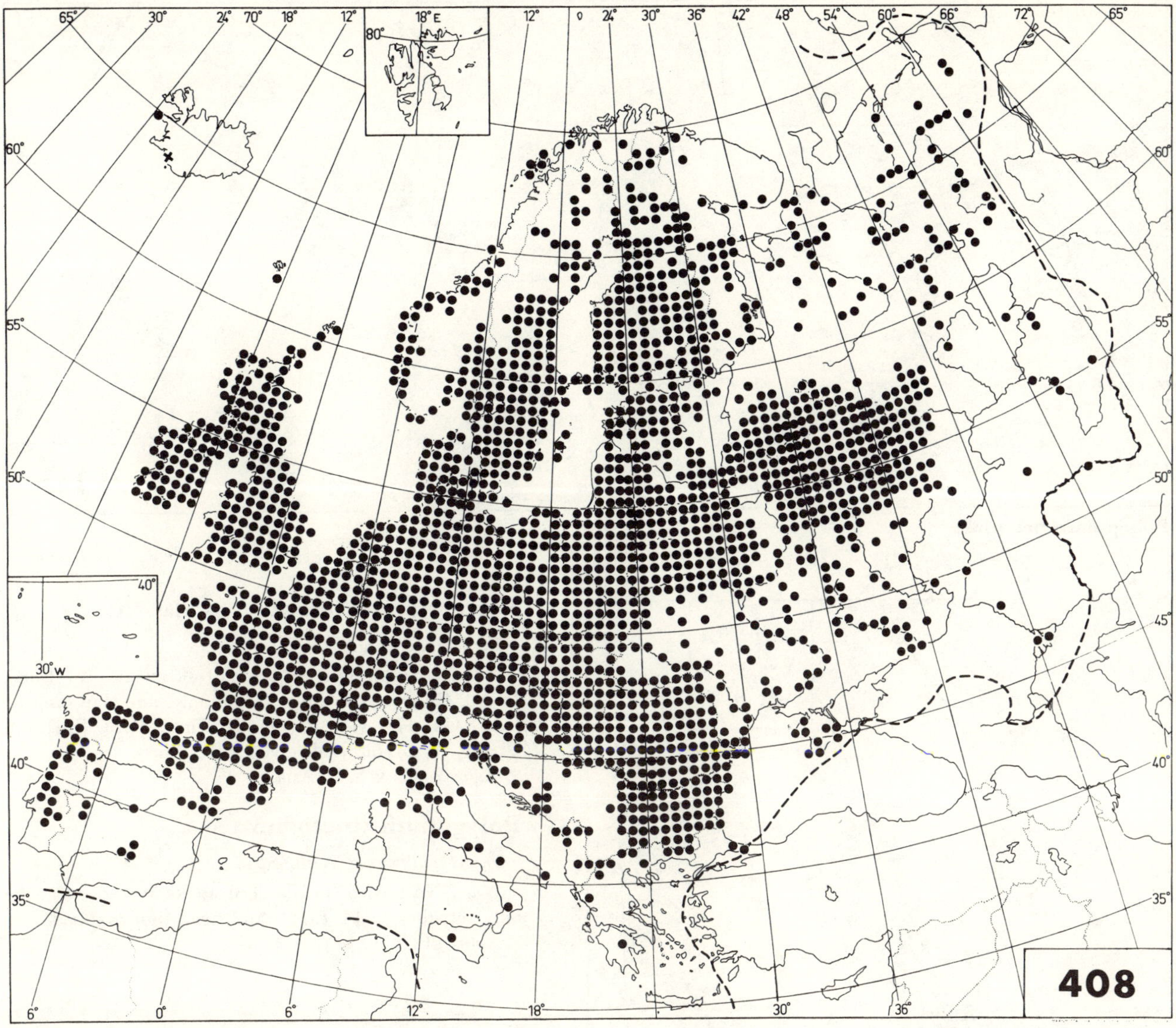

Polygonum amphibium

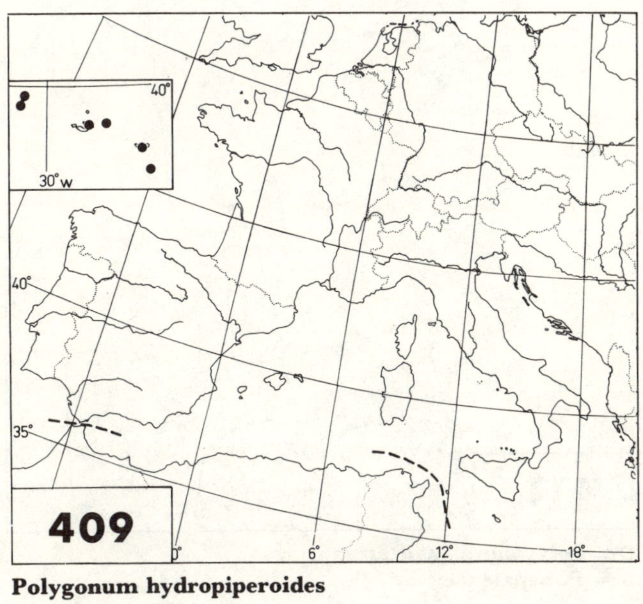

Polygonum hydropiperoides

Polygonum amphibium L. — Map 408.

Persicaria amphibia (L.) S.F. Gray

Notes. Sa omitted (given in Fl. Eur.).

Total range. MJW 1965: map 130d; Hultén Alaska 1968: 387; Hultén CP 1971: map 166.

Polygonum hydropiperoides Michx — Map 409.

Total range. MJW 1965: map 131b (not given from the Açores); Hultén Alaska 1968: 388.

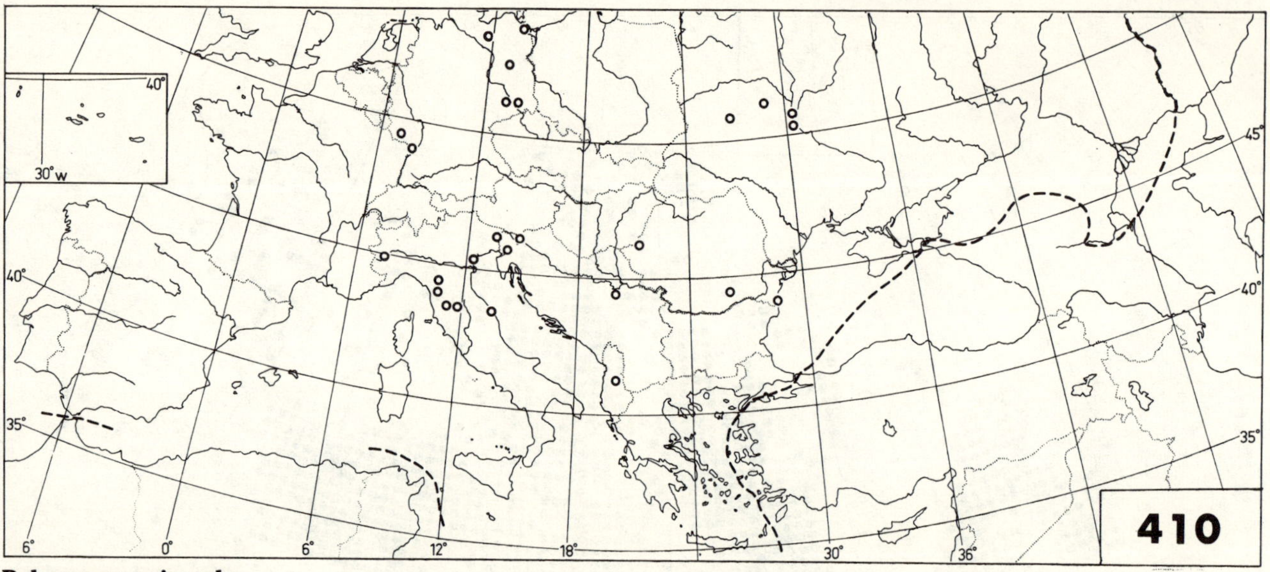

Polygonum orientale

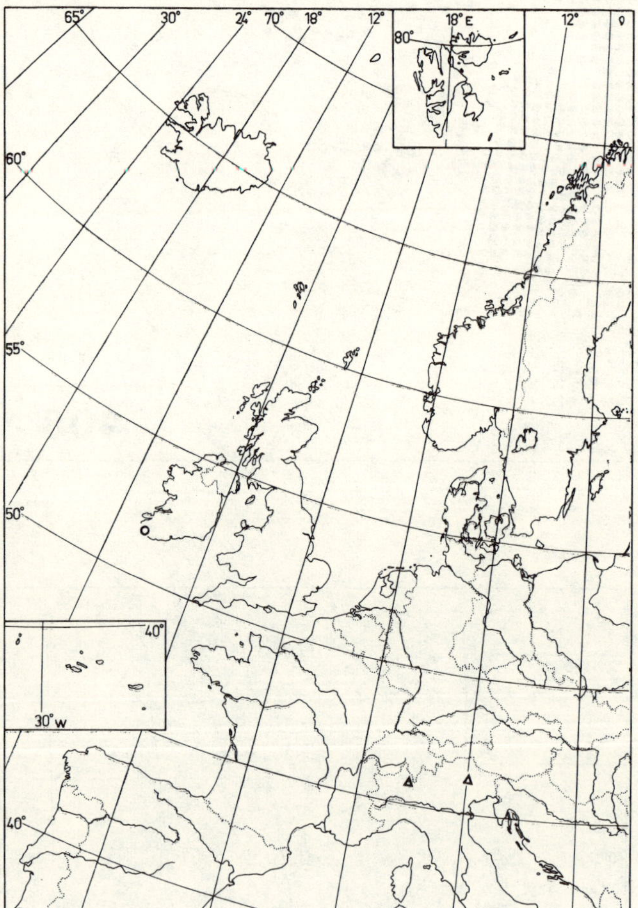

O = **Polygonum sagittatum**
Δ = **P. nepalense**

Polygonum orientale L. — Map 410.

Persicaria orientalis (L.) Vilmorin

Notes. [Al] and [Rs(C)] added (not given in Fl. Eur.). In Au, Ga, He, Ho, Po a casual only, and thus omitted ([Au, Ga, He, Ho, Po] in Fl. Eur.), [Cz], [Lu] and [Rs(W, K, E)] omitted (given in Fl. Eur.).

Native of E. & S. E. Asia.

Polygonum tinctorium Aiton

Persicaria tinctoria (Aiton) Spach

Notes. No records available from Rs(W) (given as the only territory in Fl. Eur.), and hence not mapped.

Native of China.

Polygonum sagittatum L. — Map 411.

Persicaria sagittata (L.) H. Gross; Truellum sagittatum (L.) Soják

Generic delimitation and nomenclature. J. Soják, Preslia (Praha) 46: 139—156 (1974); Haraldson 1978: 72—73.

Notes. On the verge of extinction: Fl. Donaldson, F. Donaldson & N. F. McMillan, Irish Natur. Journal 18: 331—332 (1976).

Native of E. Asia and North America.

Polygonum nepalense Meissner — Map 411.

Persicaria nepalensis (Meissner) H. Gross

Notes. [It] (the species not mentioned in Fl. Eur.).

Native of S. and E. Asia.

Polygonum capitatum Buch.-Hamilton ex D. Don — Map 412.

Persicaria capitata (Buch.-Hamilton ex D. Don) H. Gross

Notes. [Az, Lu] (the species not mentioned in Fl. Eur.).

Native of the Himalayas and E. Asia.

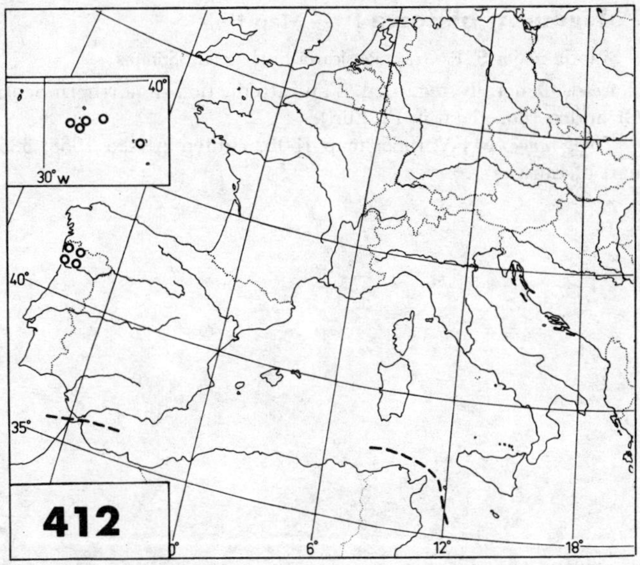

Polygonum capitatum

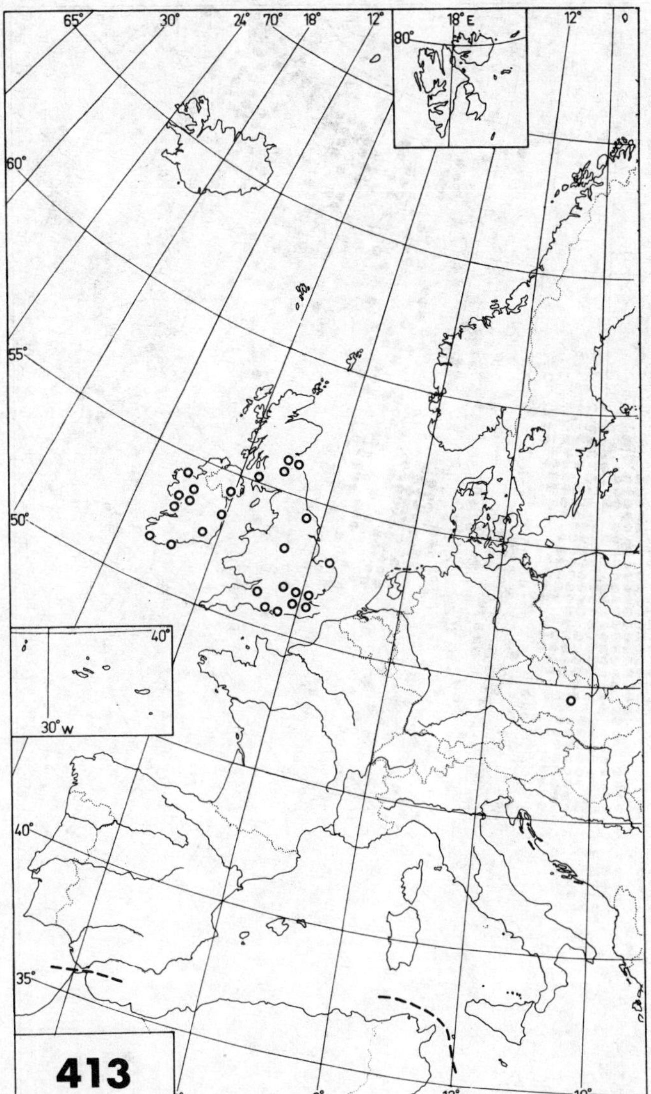

Polygonum amplexicaule

Polygonum amplexicaule D. Don — Map 413.

Bistorta amplexicaulis (D. Don) Greene

Generic delimitation of *Bistorta* Miller: Haraldson 1978: 75.

Notes. [Br] added (not given in Fl. Eur.). Also reported from one locality in Cz: M. Smejkal, Zprávy Cesk. Bot. Spolecn. (Praha) 3: 1—6 (1968).

Native of the Himalayas.

Polygonum bistorta L. — Map 414.

Bistorta major S. F. Gray; Persicaria bistorta (L.) Samp.

Notes. Generally regarded as an introduction in northern Scotland but no details available. *Fe confirmed (?Fe in Fl. Eur.), Gr added (not given in Fl. Eur.).

Total range. MJW 1965: map 130b; Hultén Alaska 1968: 385; Atlas Lekarstv. Rast. SSSR: 44—45 (Moskva 1976) (Soviet part of Eurasia).

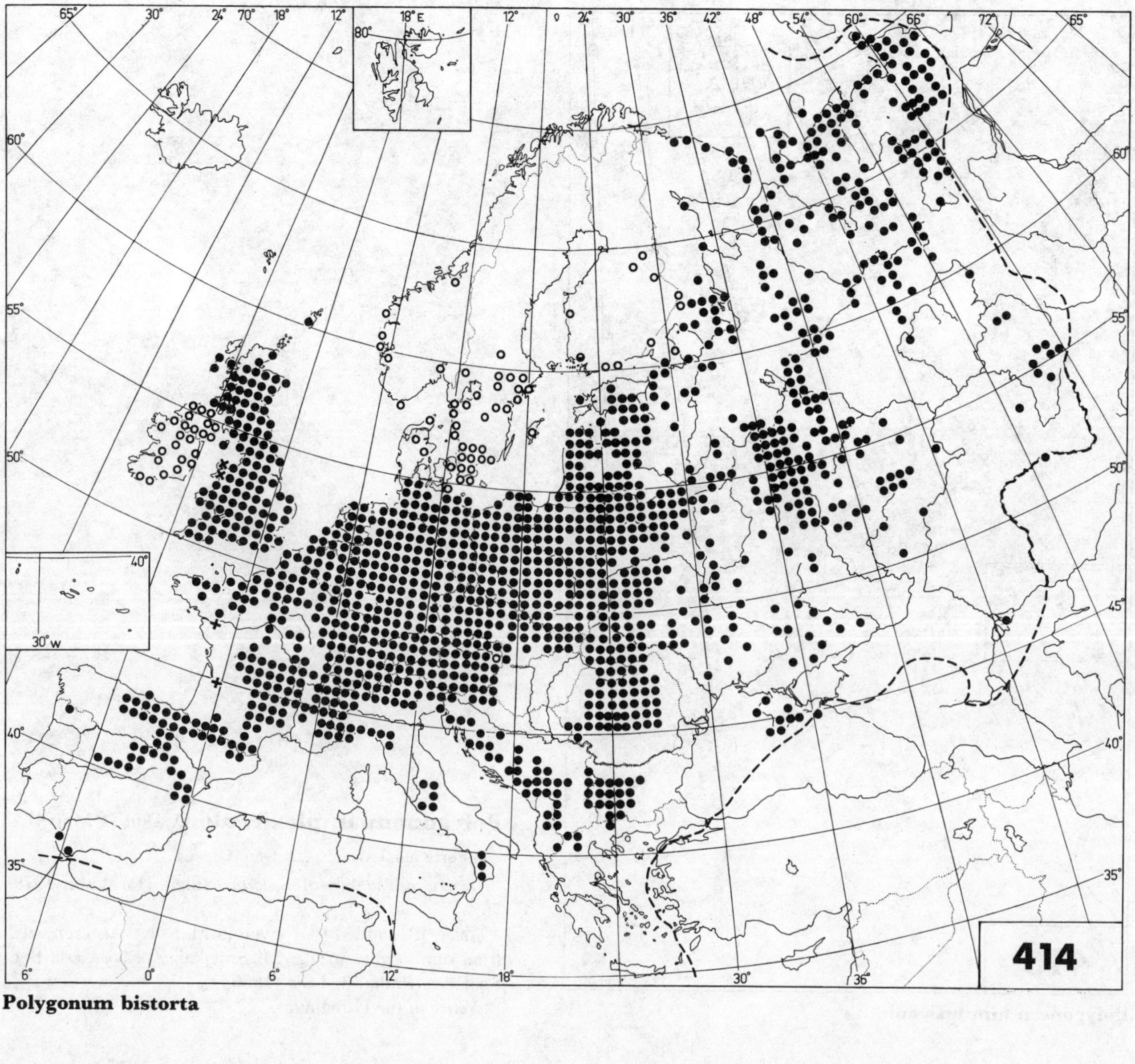

Polygonum bistorta

Polygonum viviparum L. — Map 415.

Bistorta vivipara (L.) S. F. Gray

Taxonomy. For the cryptic polyploidy (2n = 66, 88, 99) in Fa, see K. Engell, Bot. Tidsskr. 72: 113—118 (1978).
Notes. Al added (not given in Fl. Eur.), Rs(E) omitted (given in Fl. Eur.). Probably extinct in Da (given as extinct in Fl. Eur.).
Total range. Hegi 1958: 412; MJW 1965: map 130c; Hultén Alaska 1968: 385; Hultén CP 1971: map 67.

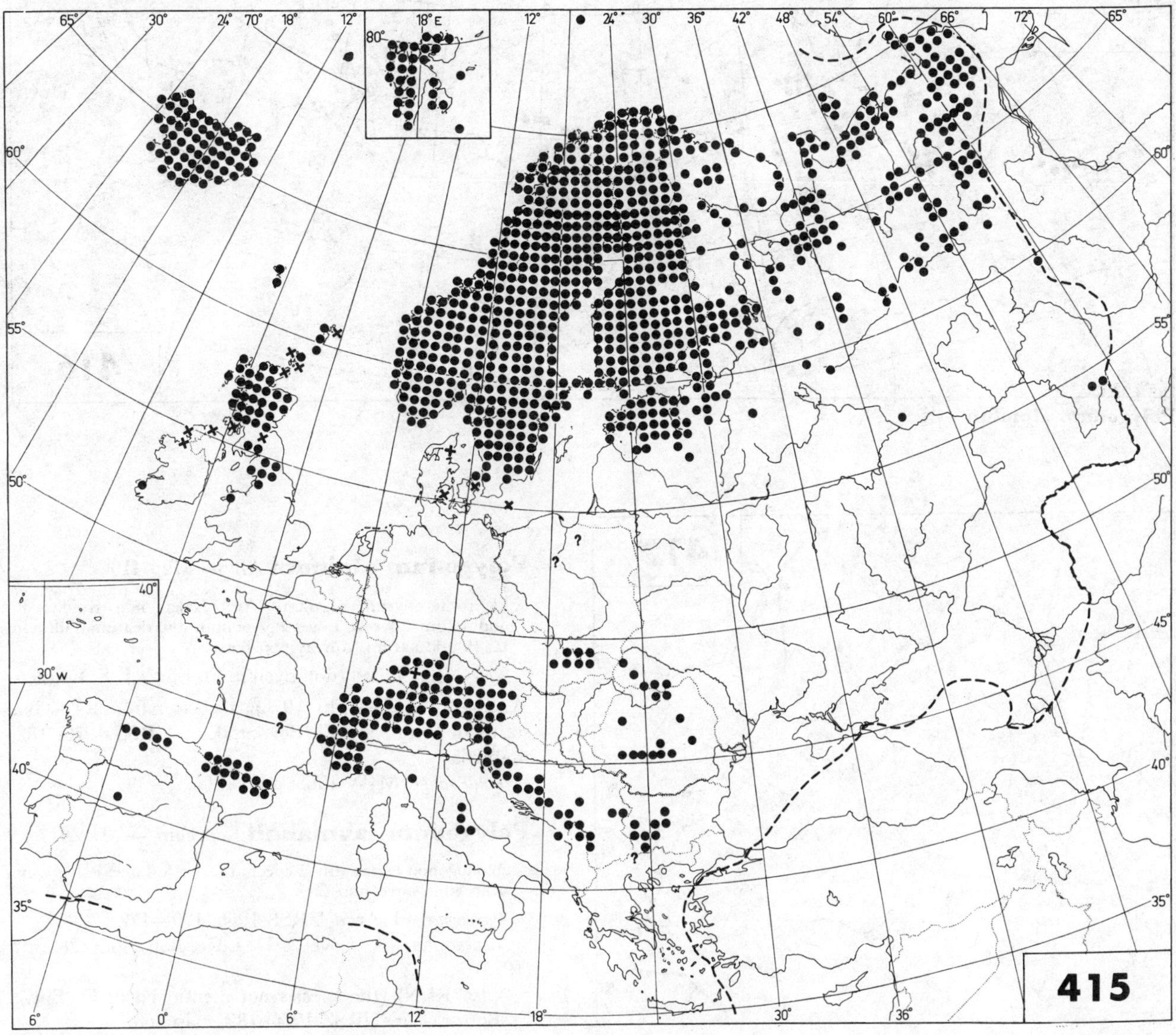

Polygonum viviparum

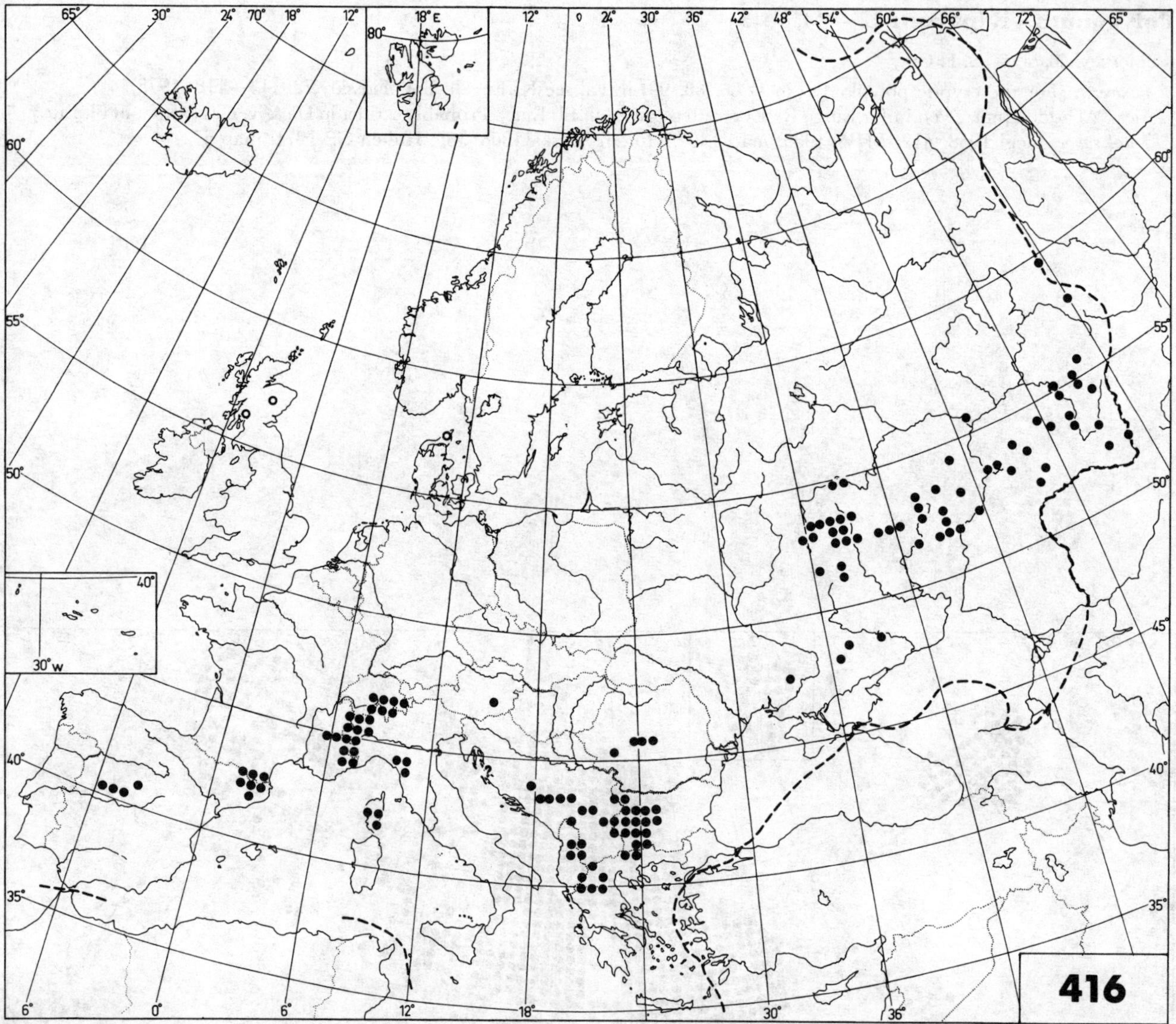

Polygonum alpinum

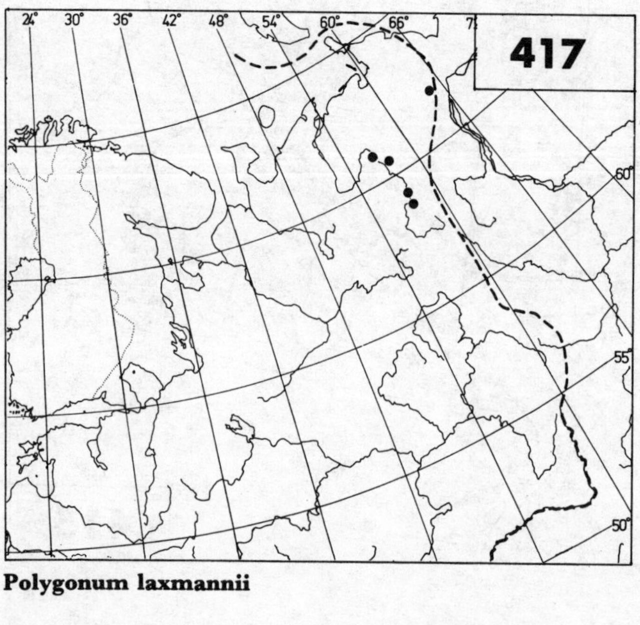

Polygonum laxmannii

Polygonum alpinum All. — Map 416.

Pleuropteropyrum alpinum (All.) Nakai non Koidz.; P. undulatum Löve & Löve; Polygonum divaricatum Vill. non L.; P. undulatum Murray non Berg.

Notes. [Br] added (not given in Fl. Eur.): J. E. Lousley, Watsonia 11: 144—146 (1976). Also given from Rs(N) but status doubtful; see Fl. Bor.-or. Eur. URSS 1976: 182, map 265.

Total range. MJW 1965: map 130a.

Polygonum laxmannii Lepechin — Map 417.

Aconogonon laxmannii (Lepechin) Löve & Löve; Polygonum ochreatum auct. non L.

Taxonomy. Fl. Arct. URSS 1966: 170—172.

Nomenclature. Á. Löve & D. Löve, Bot. Not. 128: 507 (1975).

Notes. Rs(N) (the species not mentioned in Fl. Eur.): Fl. Bor-or. Eur. URSS 1976: 182, map 266.

Total range. Fl. Arct. URSS 1966: 172.

Polygonum polystachyum Wall. ex Meissner — Map 418.

Aconogonon polystachyum (Wall. ex Meissner) Kral

Generic delimitation and nomenclature. M. Kral, Preslia (Praha) 41: 258—260 (1969); Haraldson 1978: 69.

Notes. [Be] omitted (given in Fl. Eur.): the two localities given in Atlas Fl. Belge & Luxemb., text p. 51 (1978), are in Ga. [Cz, Hb, No, Su] added (not given in Fl. Eur.). Concerning [Su], see Hylander 1966: 380, and Svensk Bot. Tidskr. 64 (Suppl.): 96 (1970).

Native of the Himalayas.

Polygonum molle D. Don — Map 419.

Aconogonon molle (D. Don) Hara; A. paniculatum (Blume) Haraldson; Polygonum frondosum Meissner; P. paniculatum Blume; P. rude Meissner

Nomenclature. Haraldson 1978: 69.

Notes. [Br] (the species not mentioned in Fl. Eur.): J. E. Lousley, Watsonia 11: 145 (1976).

Native of the Himalayas.

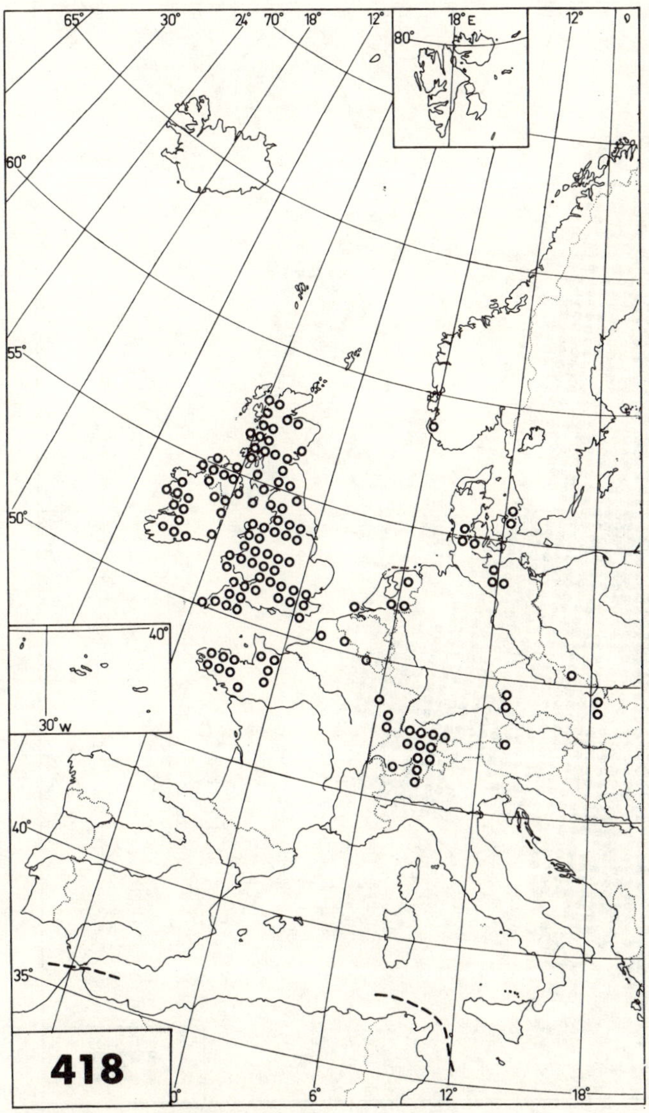

Polygonum polystachyum

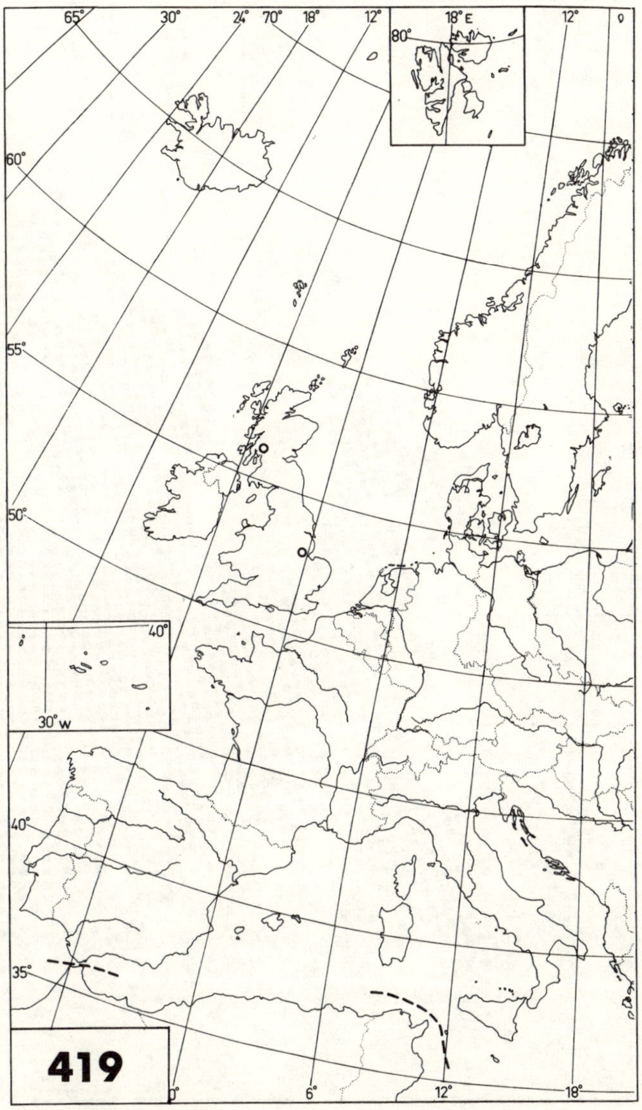

Polygonum molle

Fallopia convolvulus (L.) Á. Löve — Map 420.

Bilderdykia convolvulus (L.) Dumort.; Polygonum convolvulus L.; Tiniaria convolvulus (L.) Webb & Moq.

Generic delimitation and nomenclature. D. A. Webb & A. O. Chater, Feddes Repert. 68 (Fl. Eur. Notulae Syst. 2): 187—188 (1963); J. Dandy, Regnum Veget. 51: 101 (1967); J. Holub, Folia Geobot. Phytotax. (Praha) 6: 171—177 (1970); Haraldson 1978: 76—78.

Notes. Is omitted (given in Fl. Eur.). Some of the Azorean records belong to *Fallopia aubertii.*

Total range. Hultén Alaska 1968: 384; Hultén CP 1971: map 237 (holarctic part).

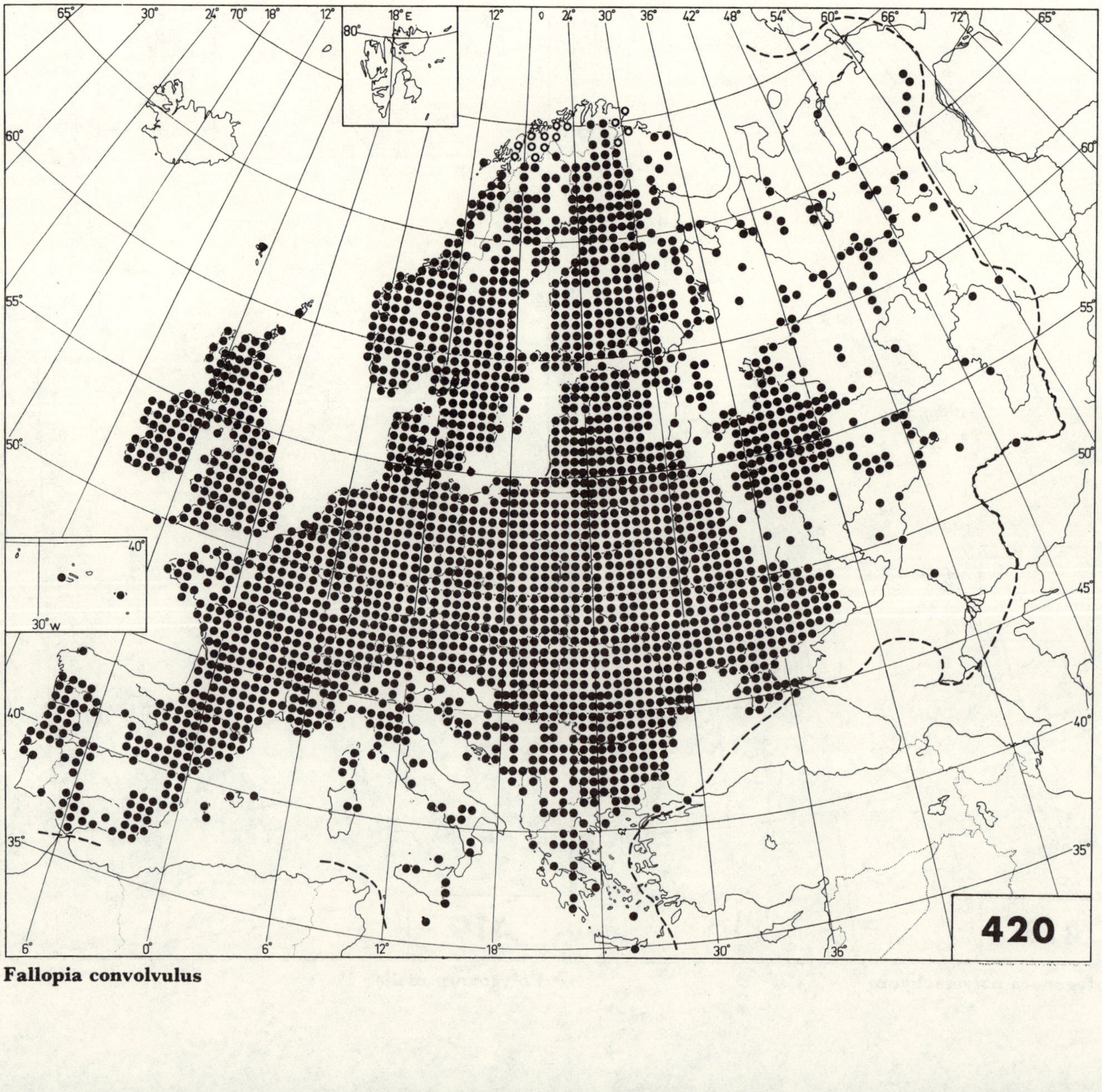

Fallopia convolvulus

Fallopia dumetorum (L.) J. Holub — Map 421.

Bilderdykia dumetorum (L.) Dumort.; Polygonum dumetorum L.; Tiniaria dumetorum (L.) Opiz

Total range. MJW 1965: map 131c.

Fallopia aubertii (Louis Henry) J. Holub

Bilderdykia aubertii (Louis Henry) Moldenke; Polygonum aubertii Louis Henry; P. baldschuanicum auct. non Regel

Taxonomy. D. A. Webb & A. O. Chater, Feddes Repert. 68 (Fl. Eur. Notulae Syst. 2): 187—188 (1963); Hylander 1966: 383—384.
Notes. Obviously not truly naturalized; data incomplete and uneven. Not mapped.
Native of W. China and Tibet.

Fallopia baldschuanica (Regel) J. Holub

Bilderdykia baldschuanica (Regel) D. A. Webb; Polygonum baldschuanicum Regel; Reynoutria baldschuanica (Regel) Moldenke

Taxonomy. See under *Fallopia aubertii.*
Notes. It remains uncertain whether this species is truly naturalized in Europe. Not mapped.
Native of Tadzhikistan.

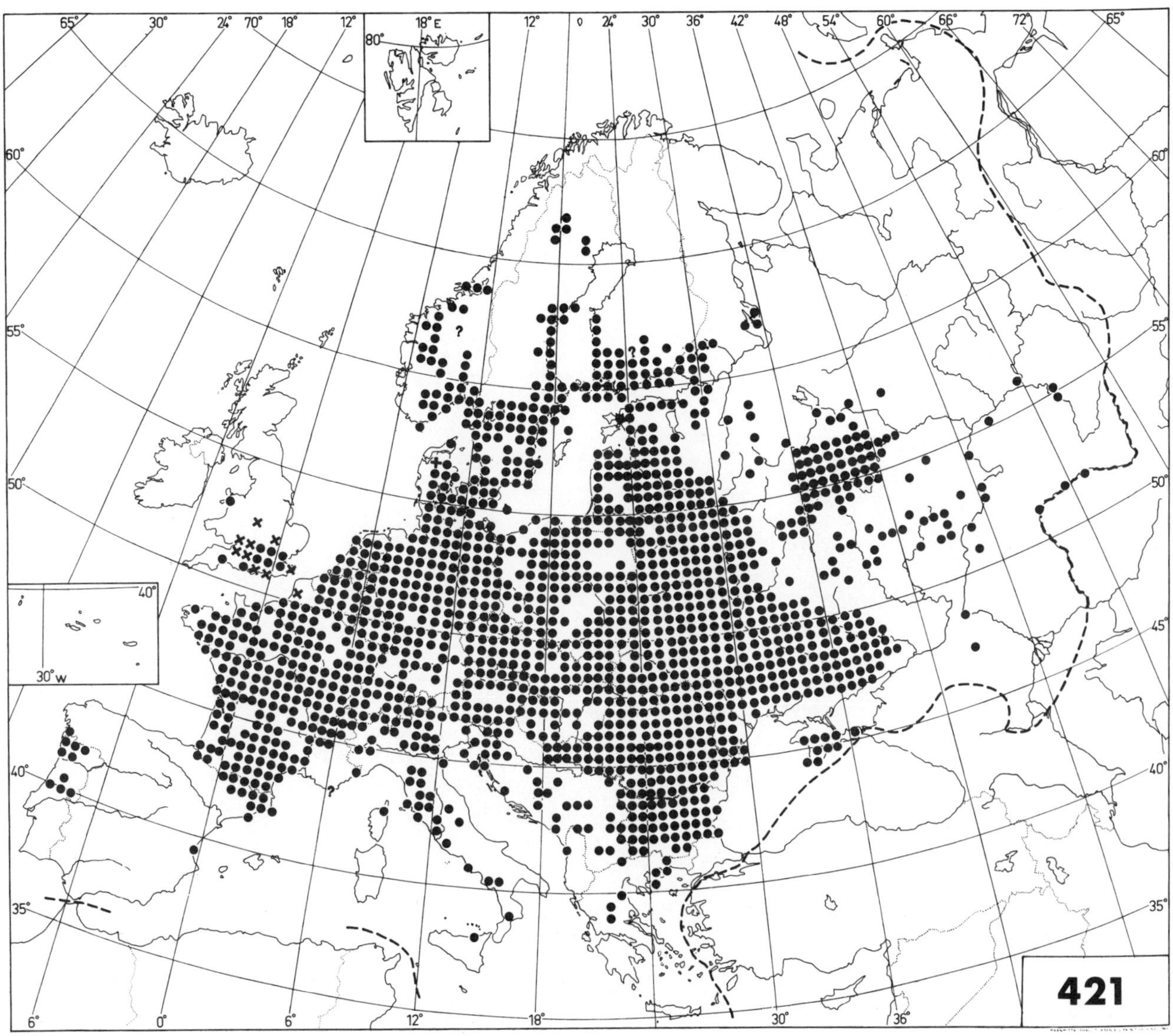

Fallopia dumetorum

Reynoutria japonica Houtt. — Map 422.

Pleuropterus cuspidatus (Siebold & Zucc.) H. Gross; Polygonum cuspidatum Siebold & Zucc.; Tiniaria japonica (Houtt.) Hedberg

Generic delimitation of *Reynoutria* Houtt.: Haraldson 1978: 75—76.

Notes. [Bu, Da, Hs, It, Ju, Rs(B, C), Su] added (not given in Fl. Eur.); concerning Hs, see J. Izco, Bol. Real Soc. Esp. Hist. Nat. Ser. Biol. 72: 25—28 (1976). Rs(N) omitted (given in Fl. Eur.).

Native of Japan.

Reynoutria sachalinensis (Friedrich Schmidt Petrop.) Nakai — Map 423.

Pleuropterus sachalinensis (Friedrich Schmidt Petrop.) H. Gross; Polygonum sachalinense Friedrich Schmidt Petrop.; Tiniaria sachalinensis (Friedrich Schmidt Petrop.) Janchen

Notes. [Au, Bu, Da, Ge, Rs(B, W), Su] added (not given in Fl. Eur.), [Hu] and [Rs(N)] omitted (given in Fl. Eur.).

Native of E. Asia.

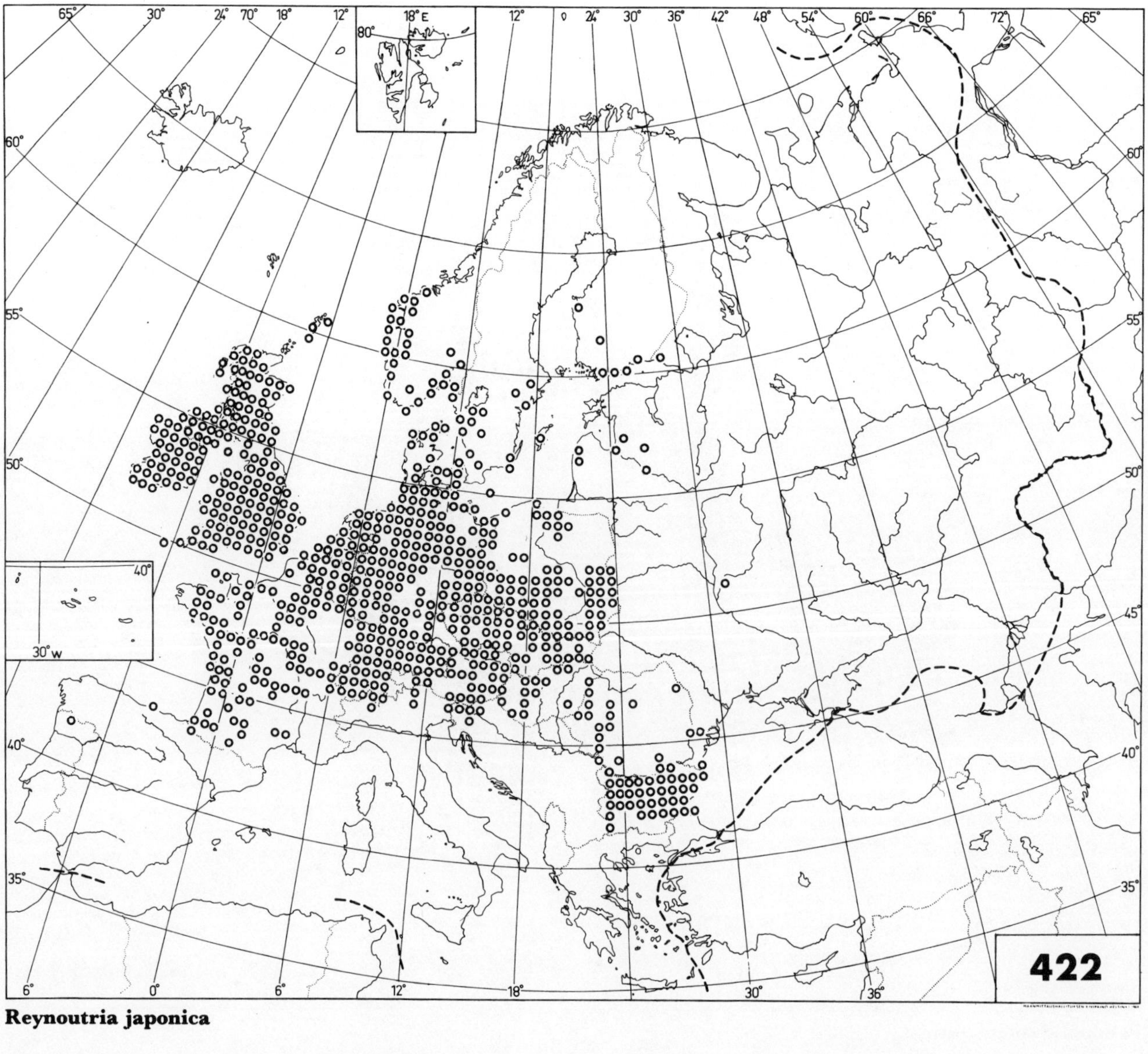

Reynoutria japonica

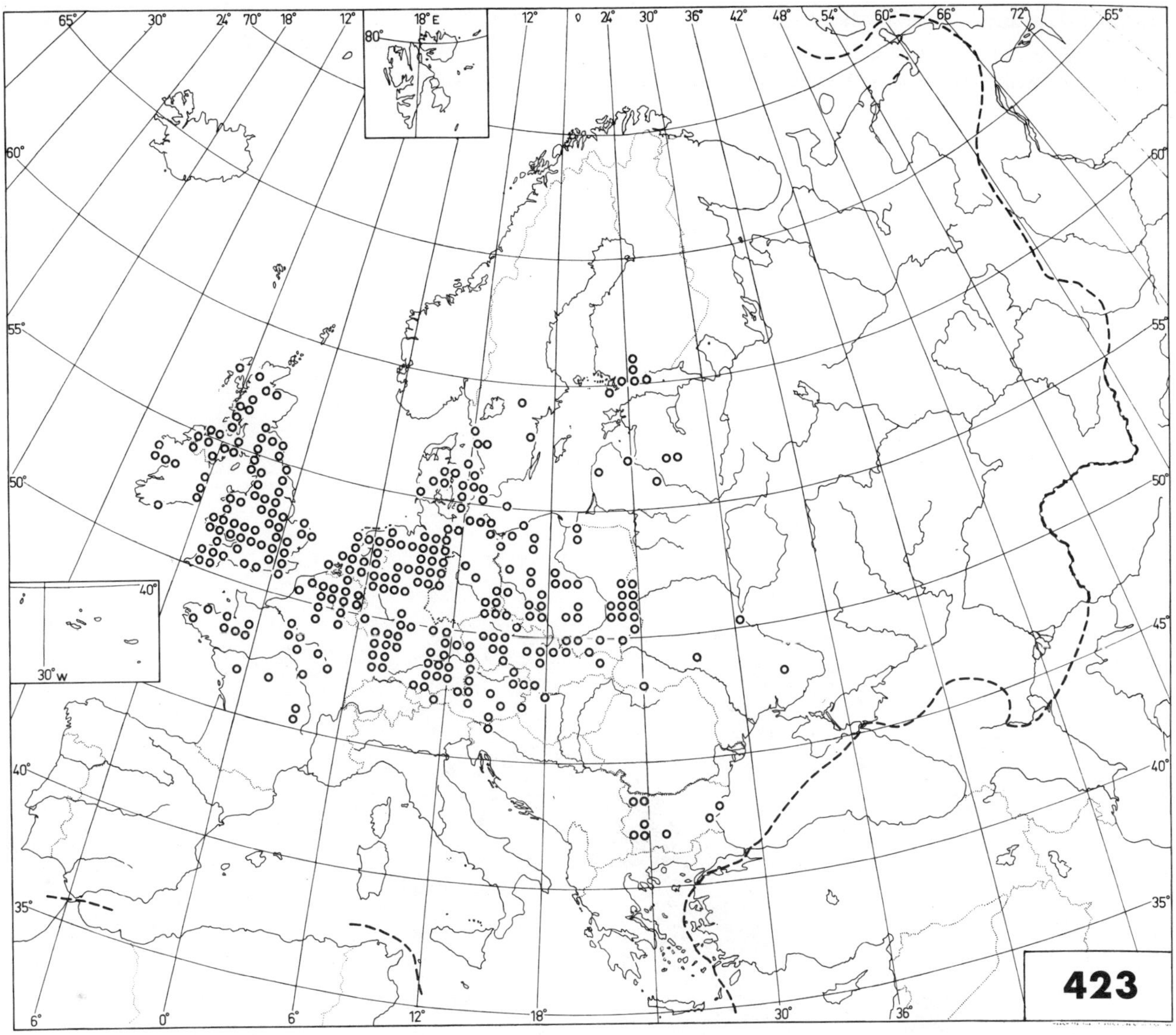

Reynoutria sachalinensis

Fagopyrum esculentum Moench

Fagopyrum sagittatum Gilib., nomen inval.; F. vulgare T. Nees; Polygonum fagopyrum L.

Generic delimitation of *Fagopyrum* Miller: Haraldson 1978: 81—83.

Notes. [Au, Bu, Da, Ge, Rs(B, W), Su] added (not given in Fl. Eur.), [Hu] and [Rs(N)] omitted (given in Fl. Eur.).

Native of E. Asia.

Fagopyrum tataricum (L.) Gaertner

Polygonum tataricum L.

Notes. Occurrence highly dependent on the constant cultivation of *Fagopyrum esculentum;* with the drastic decrease of this crop in Europe, *F. tataricum* has also disappeared over large areas. »Cultivated in much of Europe, though less often than *F. esculentum,* and widely distributed as a ruderal, and as a weed in cultures of *F. esculentum.»* (Fl. Eur.). Not mapped.

Native of Asia (Siberia to Himalaya).

Fagopyrum dibotrys (D. Don) Hara

Fagopyrum cymosum (Trev.) Meissner; Polygonum cymosum Trev.; P. dibotrys D. Don

Notes. [Br] (the species not mentioned in Fl. Eur.): J. E. Lousley, Watsonia 11: 145—146 (1976). Also reported for [Az] but there are some doubts as to whether the species is truly naturalized there: A. Hansen, Anuário Soc. Brot. 40: 18 (1974). Not mapped.

Native of Asia (China and Himalaya).

Oxyria digyna (L.) Hill — Map 424.

Notes. Al, Fe and Gr added (not given in Fl. Eur.): L.-Å. Gustavsson, Bot. Not. 131: 16 (1978).
Total range. MJW 1965: map 129d; Hultén Alaska 1968: 383; Hultén CP 1971: map 65.

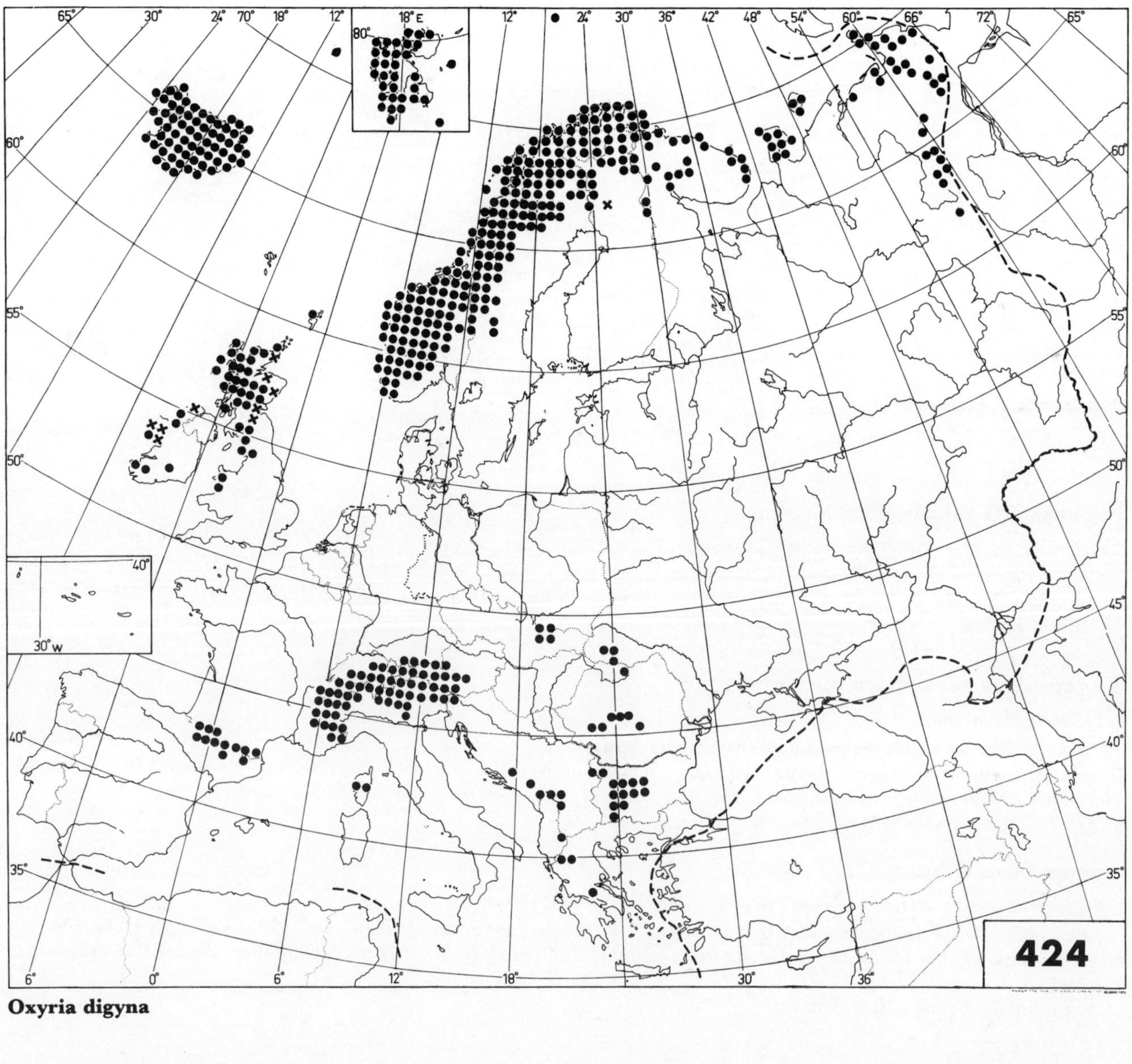

Oxyria digyna

Rheum tataricum L. fil. — Map 425.

Notes. [Rs(E)] omitted (given in Fl. Eur.), [Rs(W)] added (not given in Fl. Eur.).

Rheum rhaponticum L. — Map 426.

Taxonomy. R. Nordhagen, Skr. Vid.-Akad. Oslo, N. S. 31: 1—90 (1973).

Notes. Although given in Fl. Eur., the native status in Bu has been doubted; see B. A. Kuzmanov & S. I. Kozuharov, Boissiera 19: 321—322 (1971). No added (not given in Fl. Eur.): R. Nordhagen, Skr. Vid.-Akad. Oslo, N. S. 31: 1—90 (1973). See also TPC Red Data Sheets Europe (preprint Kew 1977).

Total range. Endemic to Europe. Related taxa in Central Asia.

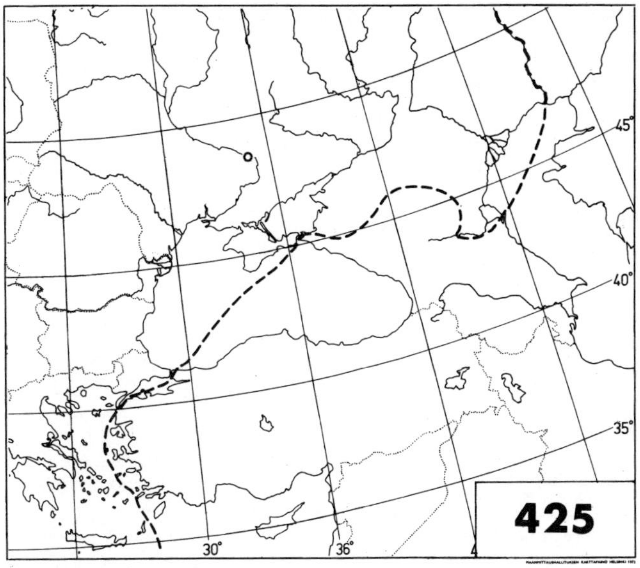

Rheum tataricum

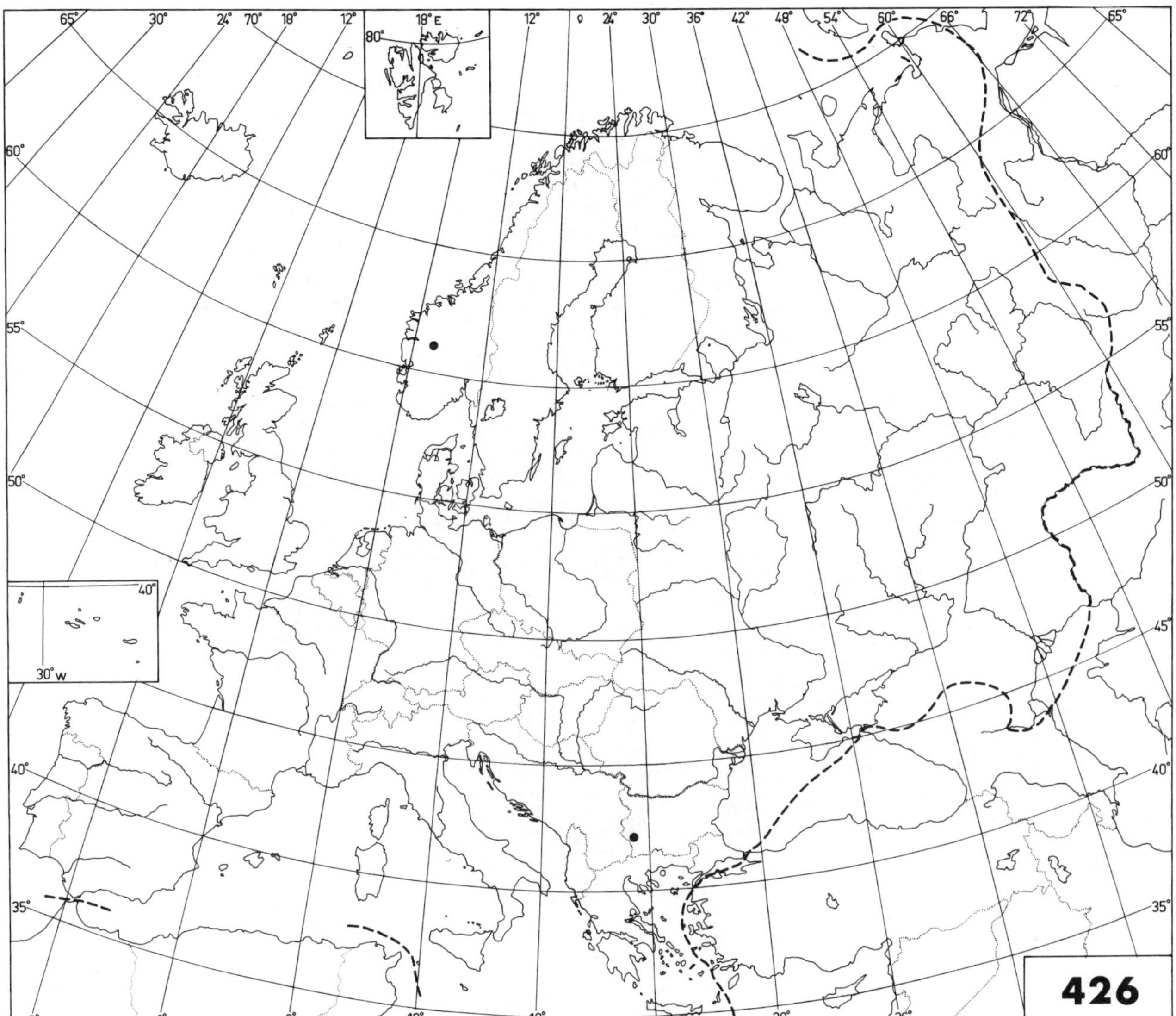

Rheum rhaponticum

Rumex acetosella L., s. lato — Map 427.

Rumex acetosella s. str. + R. angiocarpus + R. fascilobus Klokov + R. graminifolius + R. tenuifolius

Taxonomy. Hylander 1966: 353—354; W. Harris, New Zealand Jour. Bot. 7: 125—141 (1969), 11: 115—144 (1973); Hultén CP 1971: 306, 401—402; J. C. M. den Nijs, Acta Bot. Neerl. 23: 655—675 (1974), 25: 417—447 (1976); C. Gardou & G. Bigot, Candollea 31: 257—271 (1976); A. Borodina, Nov. Syst. Pl. Vasc. (Leningrad) 15: 110—112 (1979).

Notes. Not recorded from Bl and Rs(K), only casual in Sb (given as present »throughout Europe» in Fl. Eur.).

Total range. MJW 1965: map 127a; Hultén Alaska 1968: 376; Hultén CP 1971: map 297.

Rumex angiocarpus Murb.

Rumex acetosella subsp. angiocarpus (Murb.) Murb.; R. salicifolius auct.

Taxonomy. According to a recent study, the alleged correlation between angiocarpy and diploidy does not hold true; see J.C.M. den Nijs, Acta Bot. Neerl. 23: 655—675 (1974), 27: 148 (1978).

Notes. »C. & W. Europe; W. Mediterranean region» (Fl. Eur.). In addition, diploids with 2n = 14 are found in Bu (gymnocarpous) and Gr (angiocarpous): J.C.M. den Nijs, Acta Bot. Neerl. 27: 148 (1978).

Total range. Á. Löve, Hereditas 30: fig. 1 (1944); Hultén Alaska 1968: 376; Hultén CP 1971: map 297.

Rumex tenuifolius (Wallr.) Á. Löve

Rumex acetosella β tenuifolius Wallr.; R. acetosella subsp. tenuifolius (Wallr.) O. Schwarz

Notes. »N. & E. Europe» (Fl. Eur.). Also present in C. Europe: Hegi 1958: 359; J. Holub, Preslia 38: 79 (1966). In addition, the tetraploid chromosome number 2n = 28, considered characteristic of *Rumex tenuifolius*, has been recorded from Be, Bu, Ga, Gr, Ho and Hs: J.C.M. den Nijs, Acta Bot. Neerl. 23: 655—675 (1974), 27: 148 (1978).

Total range. Á. Löve, Hereditas 30: figs. 2 and 5 (1944); Hultén CP 1971: map 297.

Rumex acetosella L., s. str.

Notes. »Throughout Europe» (Fl. Eur.).

Total range. Á. Löve, Hereditas 30: fig. 3 (1944); Hultén CP 1971: map 297.

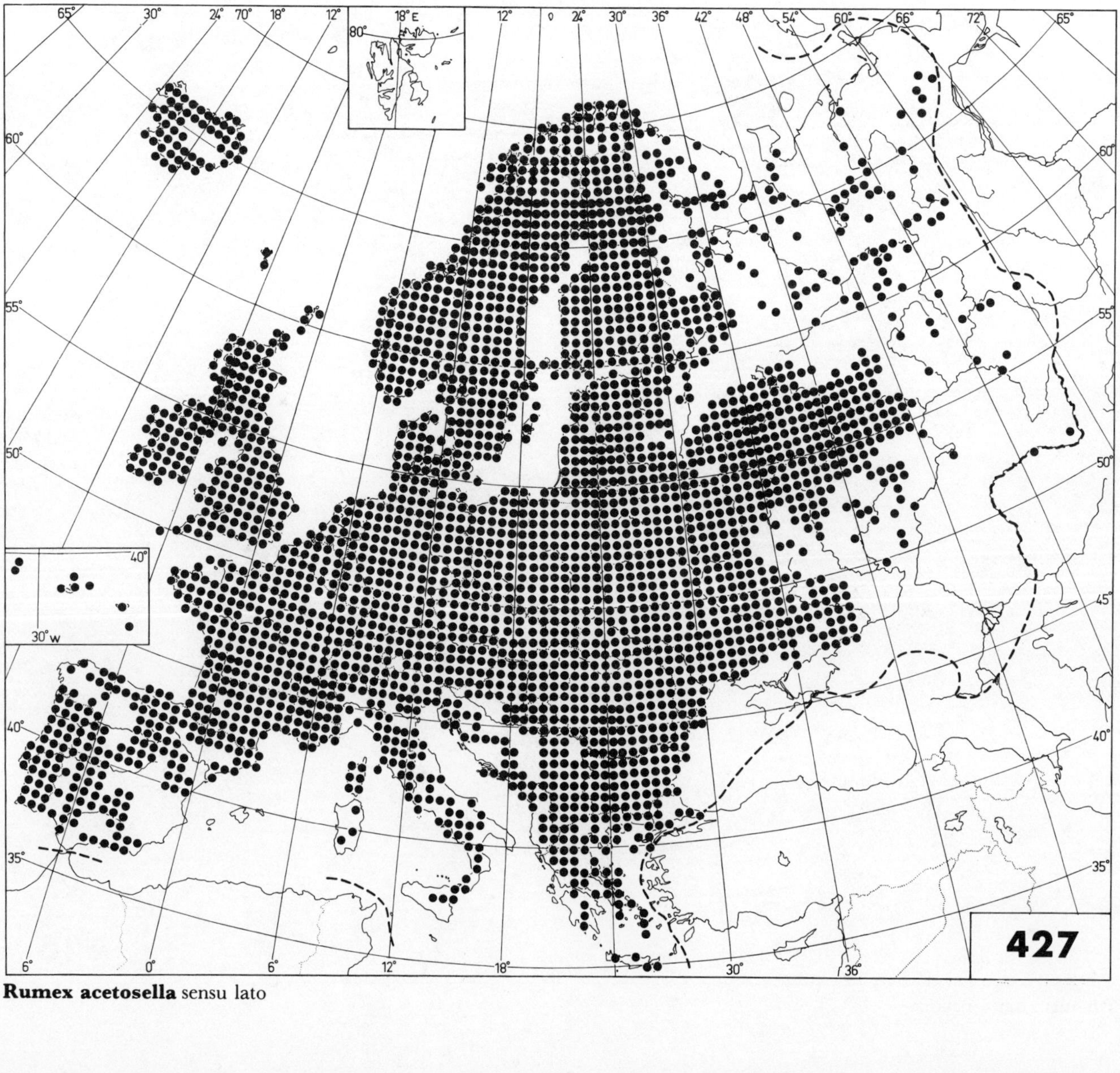

Rumex acetosella sensu lato

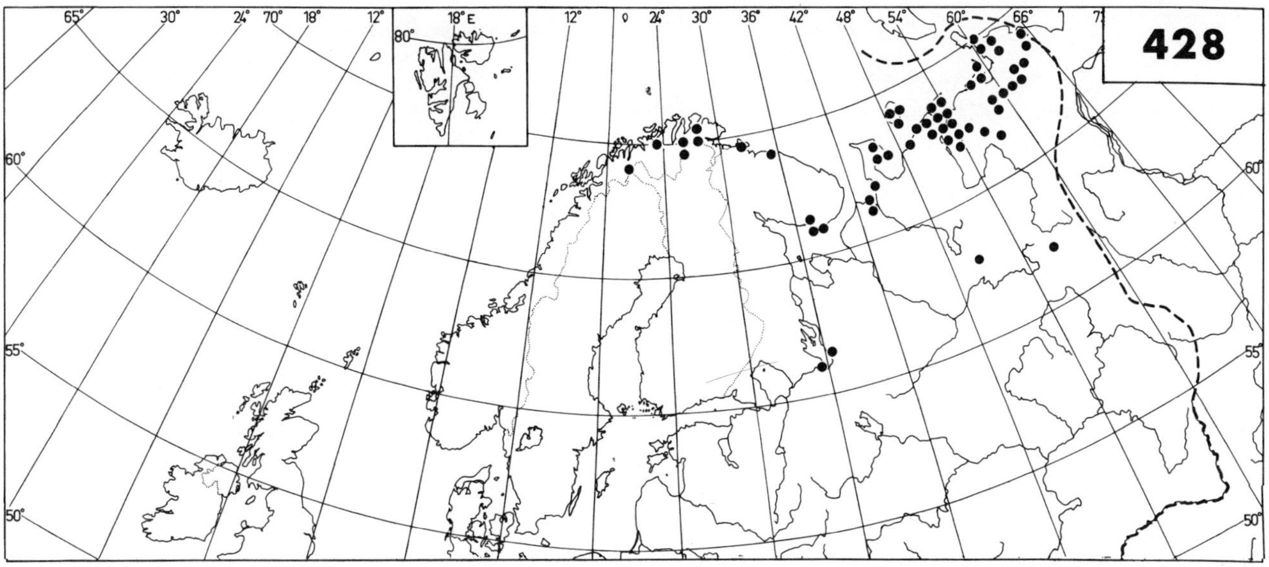

Rumex graminifolius

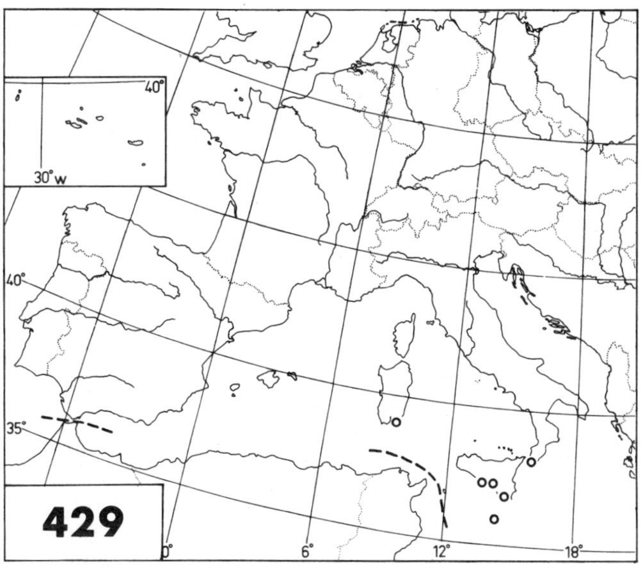

Rumex lunaria

Rumex graminifolius Rudolph ex Lamb. — Map 428.

Rumex acetosella subsp. graminifolius (Rudolph ex Lamb.) Cajander; R. angustissimus Ledeb.

Notes. No added (not given in Fl. Eur.): R. Elven, Blyttia 35: 99—104 (1977). Presence in Fe could not be confirmed (given in Fl. Eur.), although some of the Norwegian records are from sites along the Tana River opposite Finnish territory. No chromosome counts available for Norwegian or Finnish material. Rs(C) added (not given in Fl. Eur.).

Total range. Á. Löve, Hereditas 30: fig. 4 (1944); Fl. Arct. URSS 1966: 148 (Eurasian part); Hultén Alaska 1968: 376.

Rumex lunaria L. — Map 429.

Notes. [It] added (not given in Fl. Eur.).

Native of Islas Canarias. MJW 1965: 127b.

Rumex tingitanus L. — Map 430.

Notes. It omitted (given in Fl. Eur.).

Total range. MJW 1965: map 127b (Bl included and It excluded, contrary to treatment in Fl. Eur.).

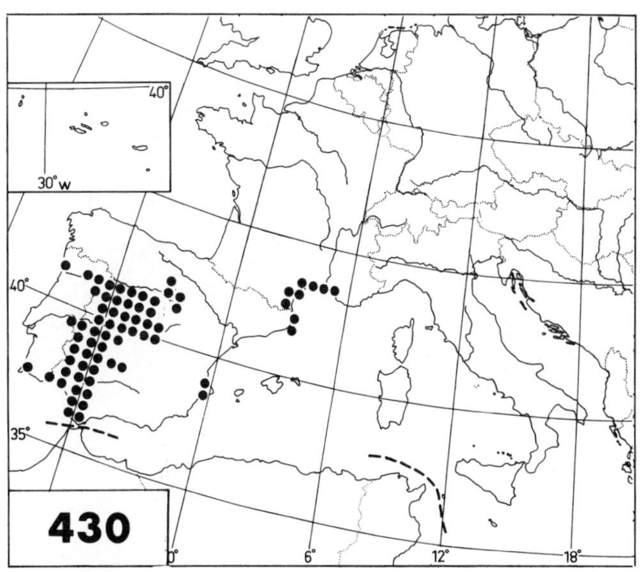

Rumex tingitanus

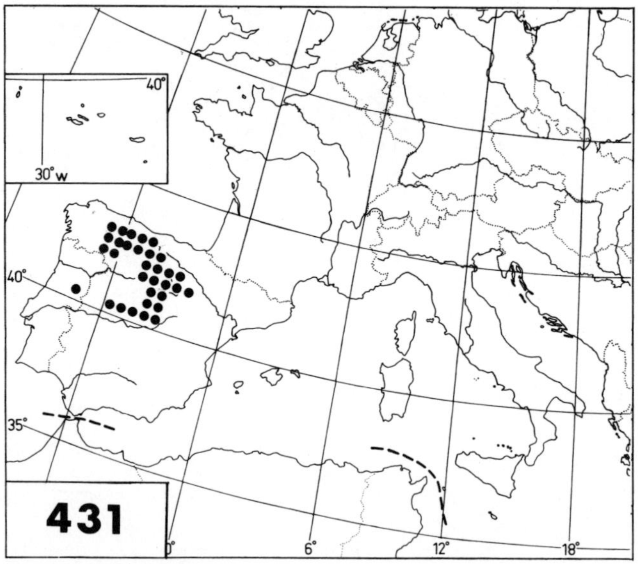

Rumex suffruticosus

Rumex suffruticosus Meissner — Map 431.
Acetosa suffruticosa (Gay ex Meissner) Á. Löve & Kapoor

Nomenclature. Á. Löve & B. M. Kapoor, Taxon 16: 521 (1967).

Notes. Lu confirmed (?Lu in Fl. Eur.): Nova Fl. Port. 1971: 84.

Rumex scutatus L. — Map 432.

Taxonomy. The Crimean plant was identified with the Caucasian *Rumex hastifolius* Bieb. by A. Borodina, Nov. Syst. Pl. Vasc. (Leningrad) 15: 104 (1979), and given subspecific rank, as *R. scutatus* subsp. *hastifolius* (Bieb.) Borod.

Notes. Considered native in Ho ([Ho] in Fl. Eur.), Hu and Lu omitted (given in Fl. Eur.). Rs(W) and Si added (not given in Fl. Eur.): A. Borodina, Nov. Syst. Pl. Vasc. (Leningrad) 15: 103 (1979). Most records of non-native occurrence in Cz are of early date.

Rumex induratus Boiss. & Reuter — Map 433.
Acetosa scutata (L.) Miller subsp. indurata (Boiss. & Reuter) Á. Löve & Kapoor; Rumex scutatus subsp. induratus (Boiss. & Reuter) Maire & Weiller

Nomenclature. Á. Löve & B. M. Kapoor, Taxon 16: 521 (1967).

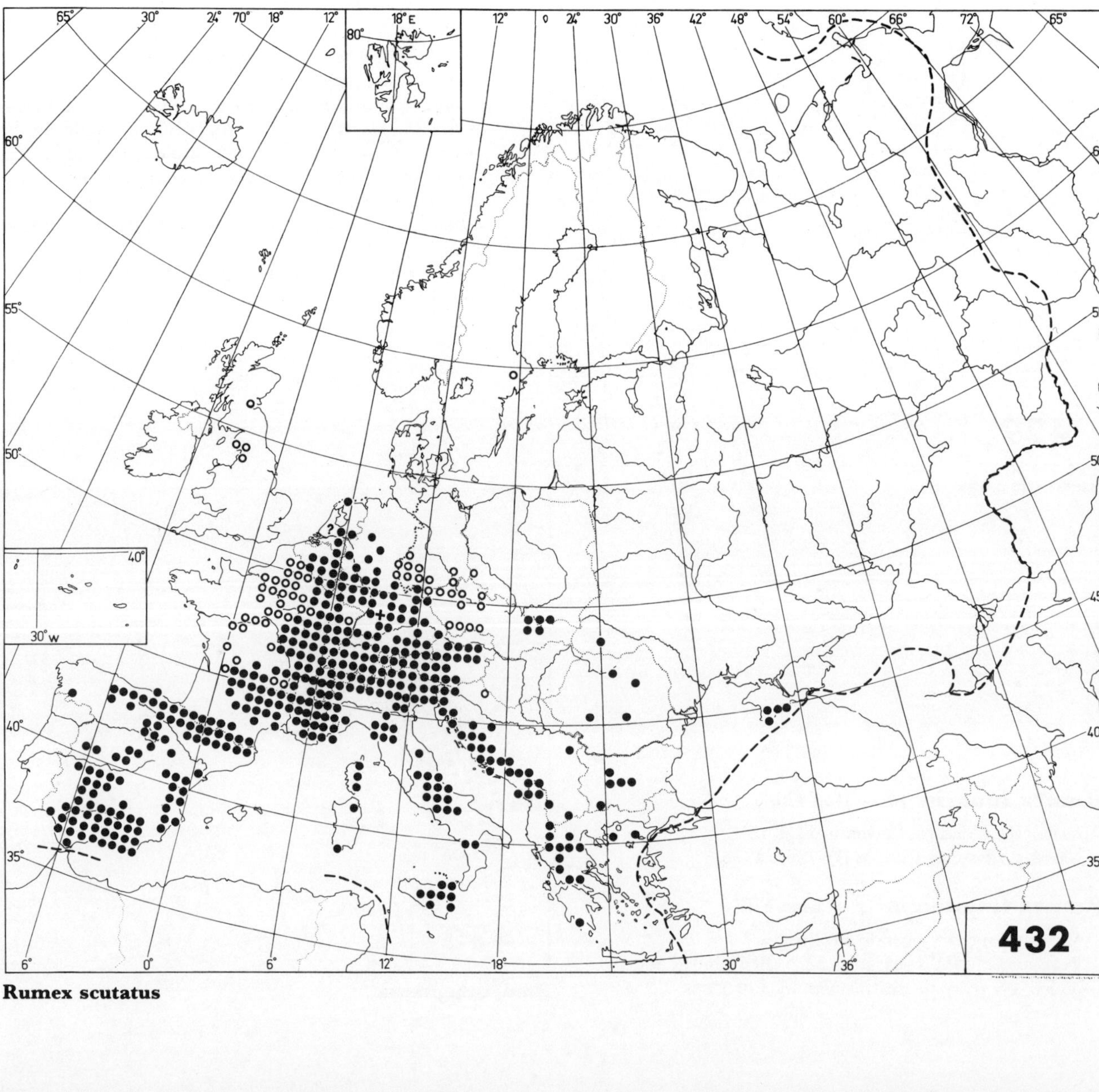

Rumex scutatus

Rumex tuberosus L. — Map 434.

Acetosa tuberosa (L.) Chaz.; Rumex euxinus Klokov

Taxonomy. A. Borodina, Nov. Syst. Pl. Vasc. (Leningrad) 15: 108—110 (1979).

Notes. Co and Sa omitted (given in Fl. Eur.), Cr, Rs(K) and Si added (not given in Fl. Eur.), extinct in Ga (given as present in Fl. Eur.).

Total range. MJW 1965: map 127d.

R. tuberosus subsp. **tuberosus** — Map 434.

Incl. Rumex tuberosus subsp. euxinus (Klokov) Borod.

Notes. Bu, †Ga, Gr, It, Ju, Rm, Rs(E), Si, Tu. Also present in Rs(K) although exact data not available. »Throughout most of the range of the species» (Fl. Eur.).

R. tuberosus subsp. **creticus** Rech. fil. — Map 434.

Acetosa tuberosa subsp. cretica (Rech. fil.) Á. Löve & Kapoor; Rumex creticus Boiss., nomen illeg.

Notes. Cr, Gr, Tu. »S. E. part of Mediterranean region» (Fl. Eur.).

Rumex nivalis Hegetschw. — Map 435.

Acetosa pratensis Miller subsp. nivalis (Hegetschw.) Á. Löve

Notes. Ge added (not given in Fl. Eur.).

Total range. Obviously endemic to Europe (not given as such in Fl. Eur.). MJW 1965: map 128b.

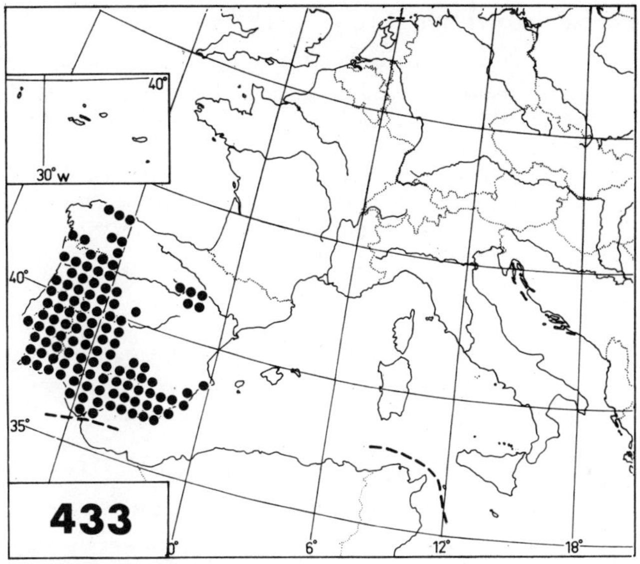

Rumex induratus

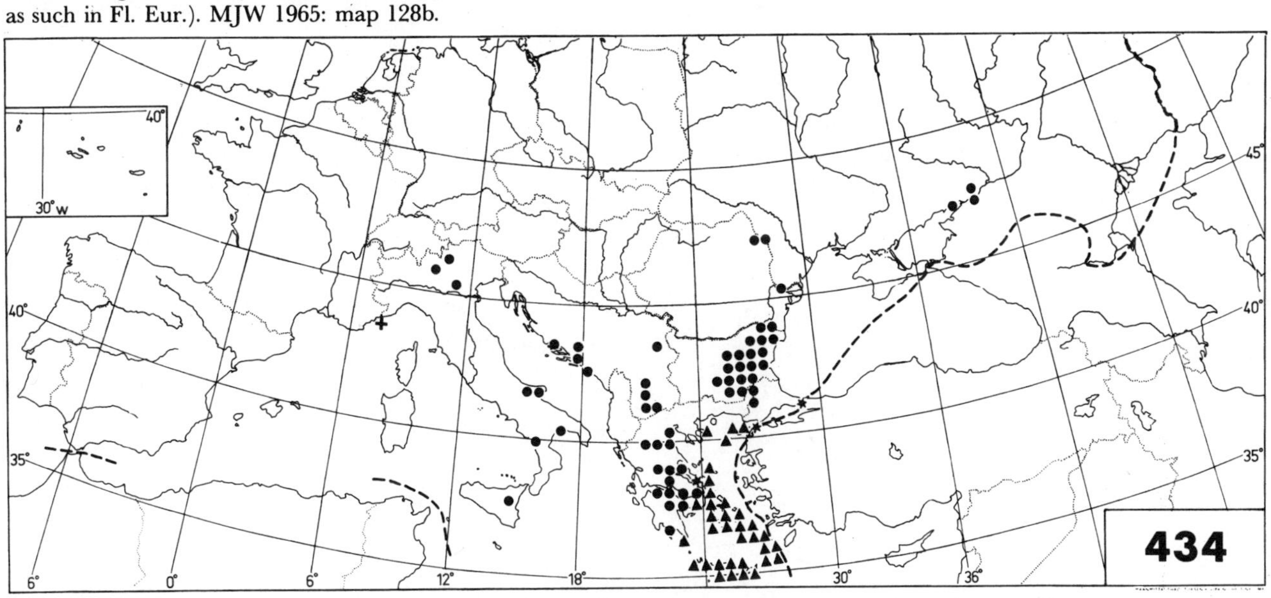

Rumex tuberosus ● =subsp. **tuberosus** ▲ =subsp. **creticus**

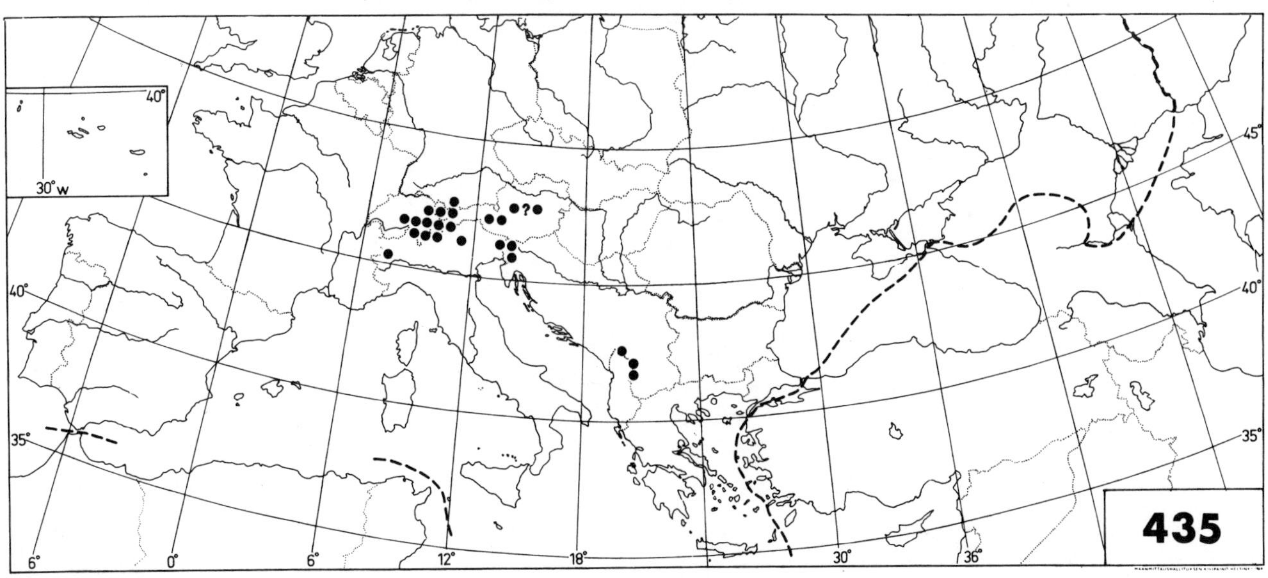

Rumex nivalis

Rumex alpestris Jacq. — Map 436.

Acetosa arifolia (All.) Schur; Rumex arifolius All.; R. carpaticus Rech. fil; R. montanus Desf. Incl. R. lapponicus (Hiitonen) Czernov

Taxonomy and nomenclature. E. Janchen, Phyton (Austria) 10: 272—274 (1964); M. Breistroffer, Bull. Soc. Bot. France 110 (sess. extraord.): 110 (1966); Hylander 1966: 348—351; Hultén CP 1971: 298; A. Borodina, Nov. Syst. Pl. Vasc. (Leningrad) 15: 105—107 (1979).

Notes. Fa and Is omitted (given in Fl. Eur.): Hylander 1966: 349; cf., however, Á. Löve, Islenzk Ferdaflóra, p. 202 (1970). Fe and Rs(W) added (not given in Fl. Eur.): A. Borodina, Nov. Syst. Pl. Vasc. (Leningrad) 15: 107 (1979). Lu and Sa not confirmed (?Lu, ?Sa in Fl. Eur.). No, Rs(N) and Su confirmed (?No, ?Rs(N) ?Su in Fl. Eur.).

Total range. MJW 1965: map 128a; Hultén Alaska 1968: 377; Hultén CP 1971: map 290.

R. alpestris subsp. **alpestris** — Map 436.

Acetosa pratensis subsp. alpestris (Jacq.) Á. Löve; Rumex acetosa subsp. arifolius (All.) O. Dahl

Notes. Au, Bu, Co, Cz, Ga, Ge, Gr, He, It, Ju, Po, Rm. Also present in Rs(W) although exact data not available.

Total range. MJW 1965: map 128a.

R. alpestris subsp. **lapponicus** (Hiitonen) Jalas — Map 436.

Acetosa alpestris subsp. lapponica (Hiitonen) Á. Löve; Rumex acetosa subsp. lapponicus Hiitonen; R. lapponicus (Hiitonen) Czernov

Nomenclature. J. Jalas, Ann. Bot. Fennici 14: 191 (1977).

Notes. Fe, No, Rs(N), Su. »Appears to . . . replace it [= *Rumex arifolius* All.] in Fennoscandia» (Fl. Eur.).

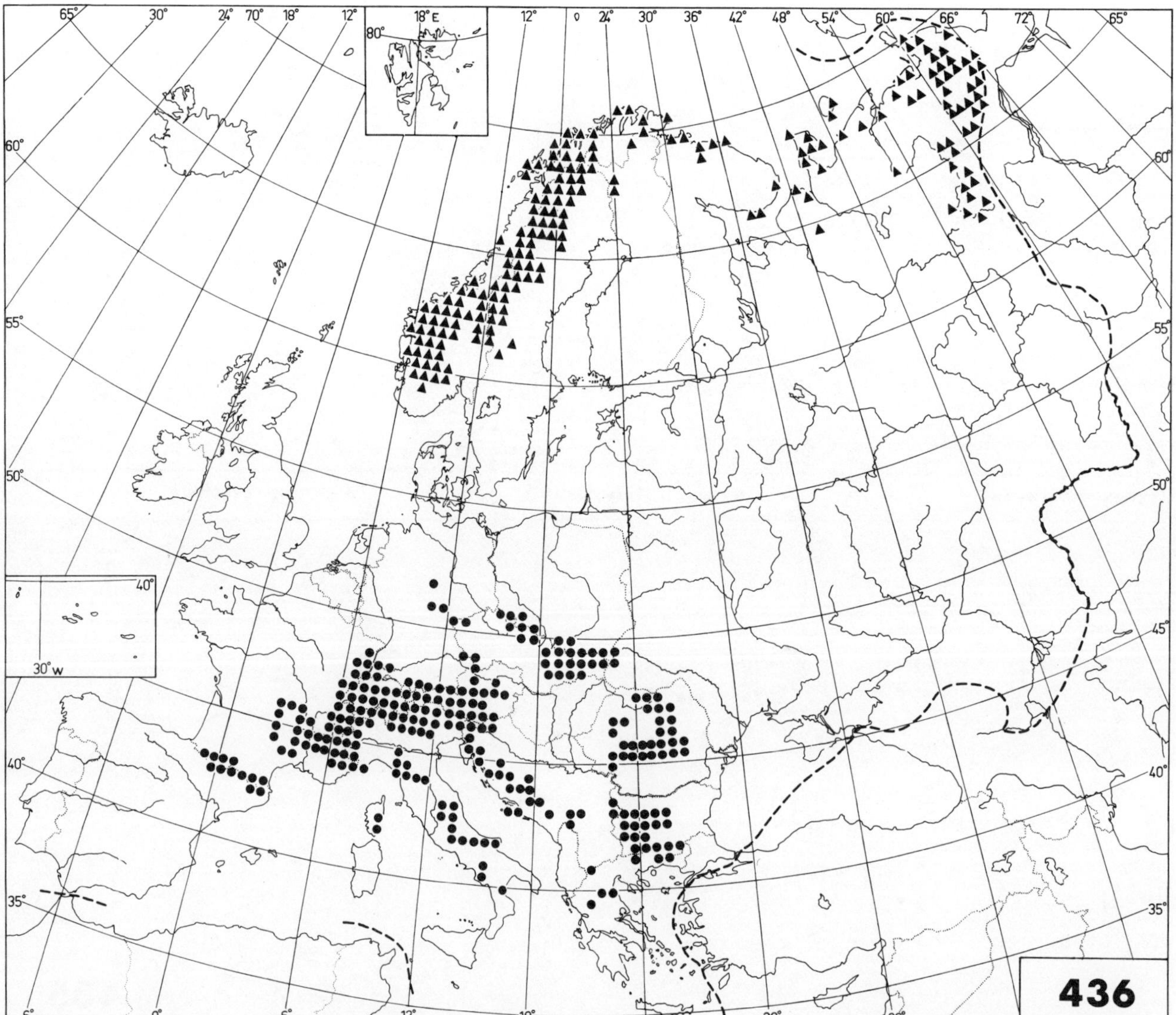

Rumex alpestris ● =subsp. **alpestris** ▲ =subsp. **lapponicus**

Rumex amplexicaulis Lapeyr. — Map 437.

Acetosa pratensis subsp. amplexicaulis (Lapeyr.) Á. Löve & Kapoor; Rumex acetosa subsp. amplexicaulis (Lapeyr.) O. Bolòs & Vigo; R. arifolius subsp. amplexicaulis (Lapeyr.) Nyman

Nomenclature. M. Breistroffer, Bull. Soc. Bot. France 110 (sess. extraord.): 110—111 (1966); Á. Löve & B. M. Kapoor, Taxon 16: 520 (1967); O. Bolòs & J. Vigo, Butll. Inst. Catalana Hist. Nat. 38 (Sec. Bot. 1): 85 (1974).

Rumex nebroides Campd. — Map 438.

Rumex gussonei Arcangeli; R. triangularis sensu Guss., non DC.; ? R. acetosa var. alpinus Boiss.

Taxonomy. K. H. Rechinger, Candollea 26: 173—181 (1971).

Notes. Ga confirmed (»somewhat similar plants occur rarely in the S. W. Alps and E. Pyrenees», according to Fl. Eur.). Si confirmed (?Si in Fl. Eur.).

Total range. The only non-European record is from N. E. Turkey, according to K. H. Rechinger, Candollea 26: 175 (1971).

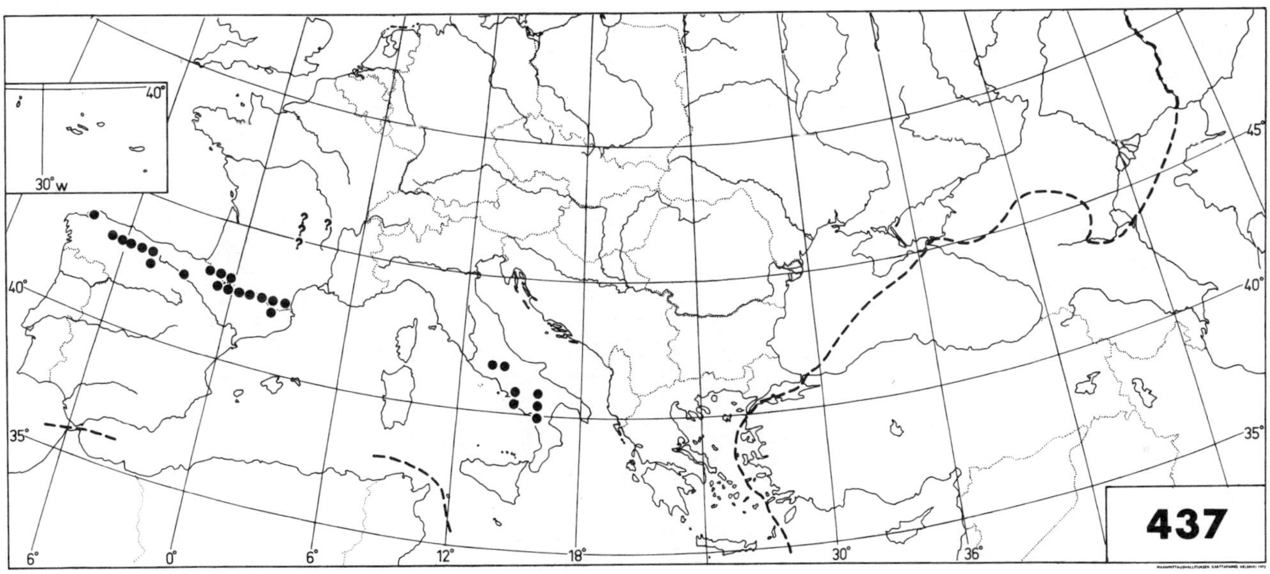

Rumex amplexicaulis

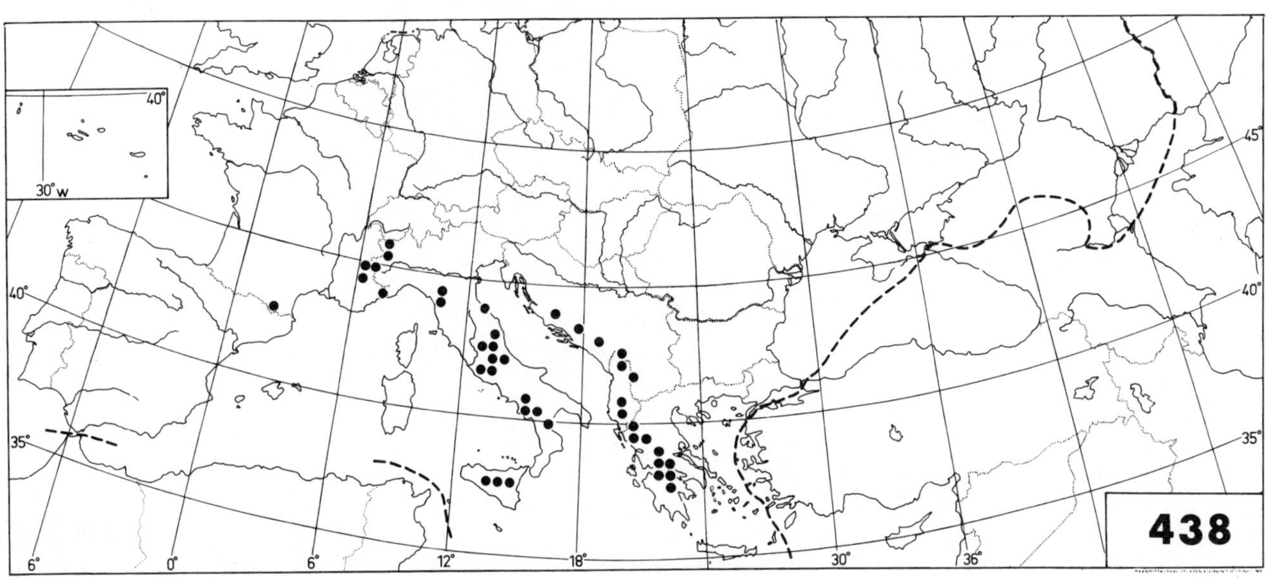

Rumex nebroides

Rumex acetosa L. — Map 439.

Acetosa pratensis Miller. Incl. Rumex biformis Lange, R. fontanopaludosus Kalela, R. hibernicus Rech. fil. and R. vinealis Timb.-Lagr. & Jeanb.

Taxonomy. Hylander 1966: 348—351: Hultén CP 1971: 298; E. Valdés-Bermejo & S. Castroviejo, Anal. Inst. Bot. Cavanilles 34: 326—327 (1977); A. Borodina, Nov. Syst. Pl. Vasc. (Leningrad) 15: 105 (1979).

Notes. Considered native in Is ([Is] in Fl. Eur., the native taxon referred to *Rumex arifolius)*: Hylander 1966: 349—351. Lu and Sa confirmed (?Lu, ?Sa in Fl. Eur.). Si and Tu added (not given in Fl. Eur. or in Fl. Turkey).

Total range. MJW 1965: map 127c (incl. *Rumex alpestris);* Hultén Alaska 1968: 377; Hultén CP 1971: map 289 (holarctic part).

R. acetosa subsp. **acetosa** — Map 439.

Incl. Acetosa alpestris subsp. islandica Löve & Löve, Rumex acetosa subsp. biformis (Lange) Castroviejo & Valdés-Bermejo and R. acetosa subsp. fontanopaludosus (Kalela) Hyl.

Notes. Al, Au, Be, Br, Bu, Co, Cz, Da, Fa, Fe, Ga, Ge, Hb, He, Ho, Hs, Hu, Is, It, Ju, Lu, No, Po, Rm, Rs(N, B, C, W, E), Sa, Si, Su, Tu.

R. acetosa subsp. **vinealis** (Timb.-Lagr. & Jeanb.) O. Bolòs & Vigo — Map 439.

Rumex vinealis Timb.-Lagr.

Nomenclature. O. Bolòs & J. Vigo, Butll. Inst. Catalana Hist. Nat. 38 (Sec. Bot. 1): 85 (1974).

Notes. Ga.

Total range. Endemic to Europe.

R. acetosa subsp. **serpentinicola** (Rune) Nordh. — Map 440.

Acetosa pratensis subsp. serpentinicola (Rune) Á. Löve & Kapoor; Rumex acetosa var. serpentinicola Rune

Taxonomy and nomenclature. O. Rune, Svensk Bot. Tidskr. 51: 51—53 (1957); Hylander 1966: 349—351; R. Nordhagen, Blyttia 24: 286—294 (1966); Á. Löve & B. M. Kapoor, Taxon 16: 520 (1967).

Notes. No, Su (the subspecies not mentioned in Fl. Eur.).

Total range. Endemic to Europe.

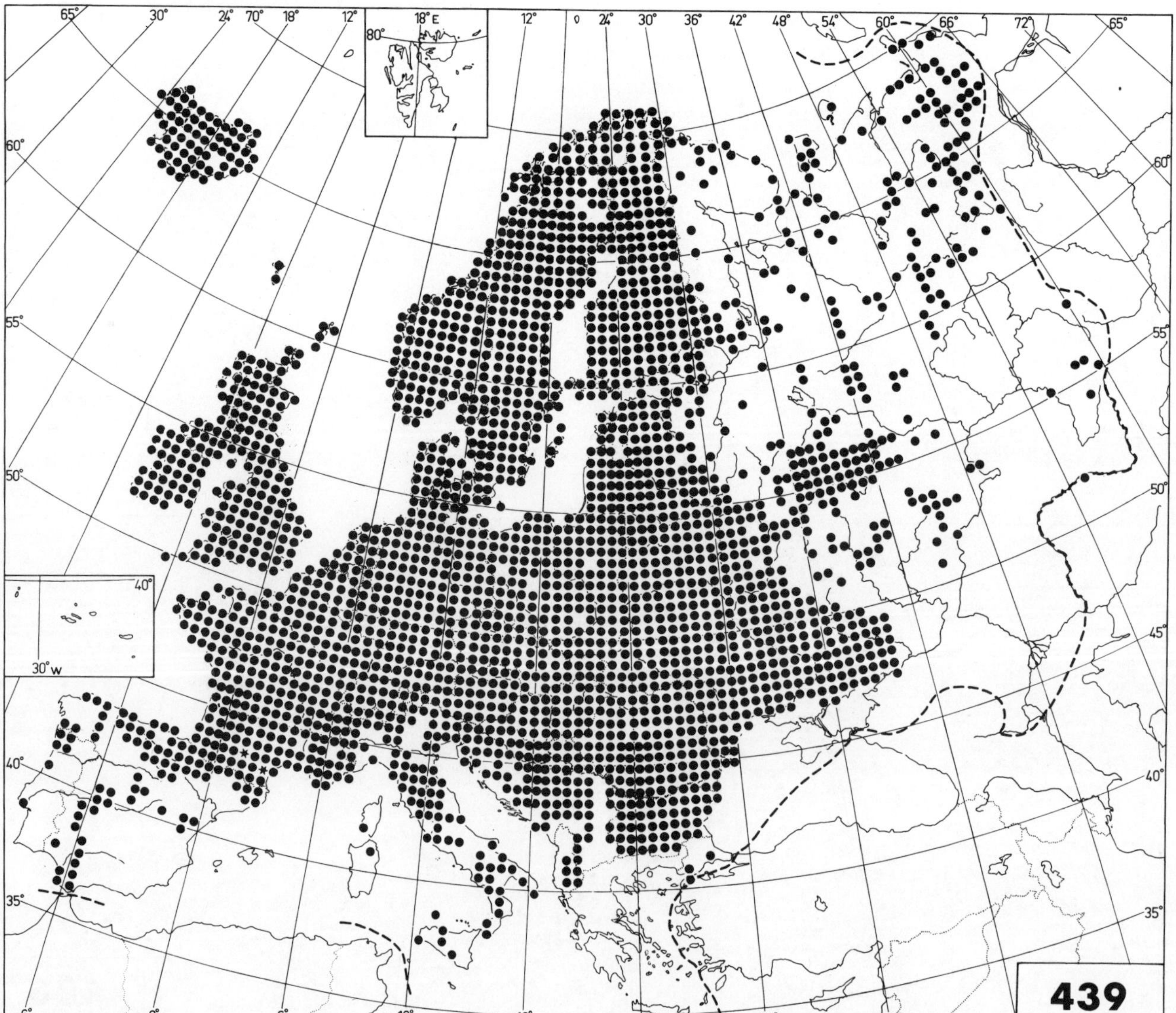

Rumex acetosa ★ =subsp. **acetosa** + subsp. **vinealis**

Rumex thyrsiflorus Fingerh. — Map 441.

Acetosa thyrsiflora (Fingerh.) Á. Löve; Rumex acetosa subsp. auriculatus (Wallr.) O. Dahl; R. acetosa subsp. thyrsiflorus (Fingerh.) Hayek; R. auriculatus (Wallr.) Murb; R. haematinus Kihlman; R. haplorhizus Czern.; R. nemorivagus Timb.-Lagr.

Taxonomy. Two subspecies are recognized by A. Borodina, Nov. Syst. Pl. Vasc. (Leningrad) 15: 107—108 (1979), viz. the widespread subsp. *thyrsiflorus*, and subsp. *haematinus* (Kihlman) Borod., which is given as endemic to Rs(N).

Notes. Be and Bu added (not given in Fl. Eur.), considered native in Ga and No ([Ga, No] in Fl. Eur.), [It] confirmed ([?It] in Fl. Eur.), Rs(K) omitted (given in Fl. Eur.).

Total range. MJW 1965: map 127d; Hultén CP 1971: map 289.

Rumex rugosus Campd.

Acetosa ambigua (Gren.) Á. Löve; Rumex acetosa subsp. ambigua (Gren.) Á. Löve; R. acetosa var. hortensis Dierbach; R. ambiguus Gren.

Notes. »Cultivated as a vegetable and established here and there as an escape» (Fl. Eur.). Not mapped.

Origin unknown; known only from cultivation.

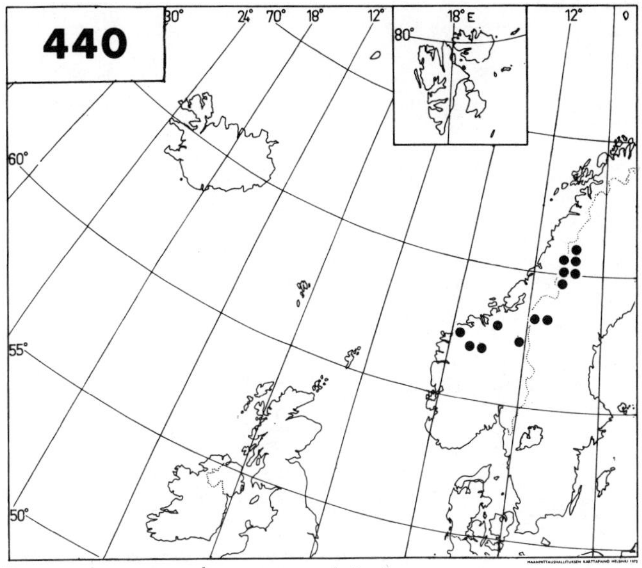

Rumex acetosa subsp. **serpentinicola**

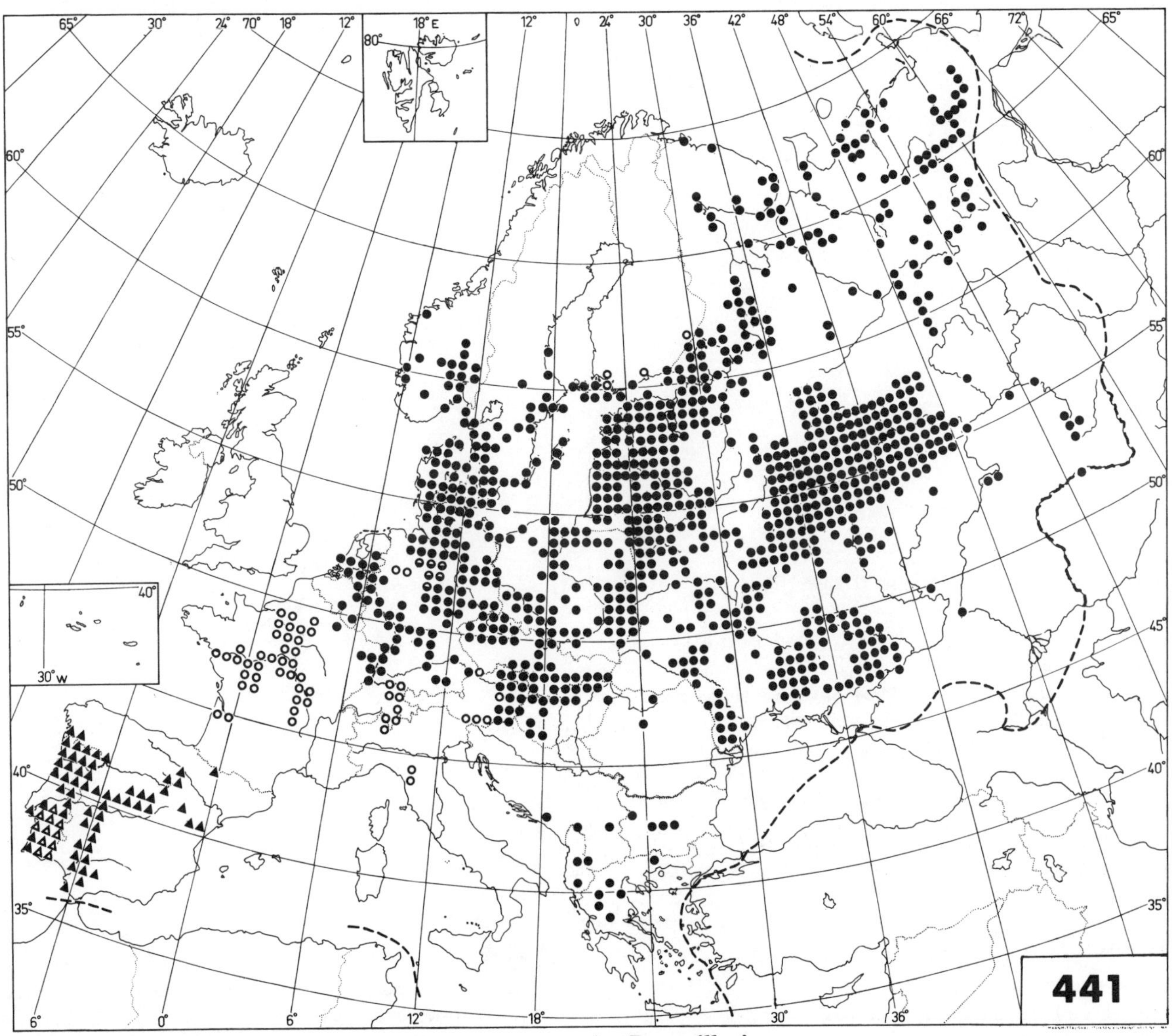

● = **Rumex thyrsiflorus**

▲ = **R. papillaris**

Rumex papillaris Boiss. & Reuter — Map 441.

Acetosa pratensis subsp. papillaris (Boiss. & Reuter) Á. Löve & Kapoor; Rumex acetosa subsp. papillaris (Boiss. & Reuter) Franco

Nomenclature. Á. Löve & B. M. Kapoor, Taxon 16: 520 (1967); Nova Fl. Port. 1971: 84, 549.

Rumex intermedius DC. — Map 442.

Rumex thyrsoides subsp. intermedius (DC.) Maire & Weiller

Taxonomy. The Portuguese plant was described as *Rumex intermedius* subsp. *lusitanicus* Franco in Nova Fl. Port. 1971: 549—550.

Notes. It confirmed (?It in Fl. Eur.). Lu added (not given in Fl. Eur.): Mem. Soc. Brot. 21: 138 (1970—71); Nova Fl. Port. 1971: 85.

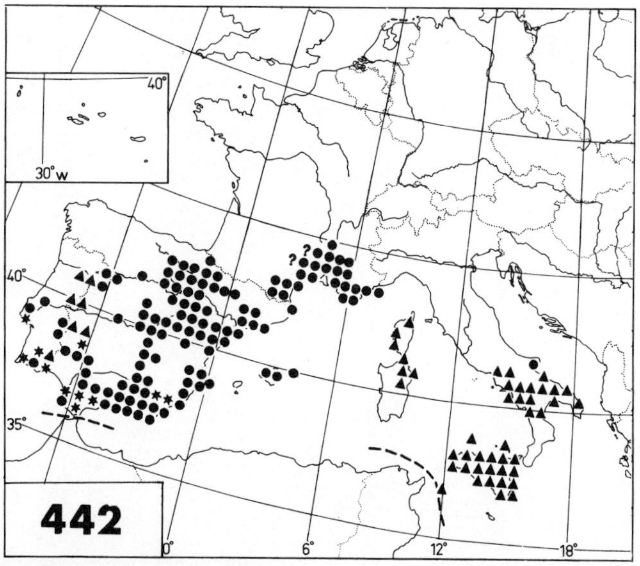

● = **Rumex intermedius**
▲ = **R. thyrsoides**
★ = **R. intermedius + R. thyrsoides**

Rumex thyrsoides Desf. — Map 442.

Acetosa thyrsoides (Desf.) Á. Löve & Kapoor

Nomenclature. Á. Löve & B. M. Kapoor, Taxon 16: 520 (1967).

Notes. Co and Lu added (not given in Fl. Eur.), Ga omitted (given in Fl. Eur.).

Rumex triangulivalvis (Danser) Rech. fil. — Map 444.

Rumex mexicanus auct. non Meissner; R. salicifolius Weinm. subsp. triangulivalvis Danser

Taxonomy. Hylander 1966: 332; H. Sukopp & H. Scholz, Ber. Deutschen Bot. Ges. 78: 457—458 (1966).

Notes. [Au, Be, Br, Da, Ho, Is and Su] omitted because occurring only as a casual (given in Fl. Eur.). Persisting since the 1920s in Fe (Helsinki), outside the area mapped (not given in Fl. Eur.).

Native of North America. N. Sarkar, Canad. Jour. Bot. 36: 960 (1958); Hultén Alaska 1968: 381 (under *Rumex mexicanus*).

Rumex frutescens Thouars — Map 444.

Rumex cuneifolius Campd.

Notes. Only casual in Da ([Da] in Fl. Eur.).

Native of South America.

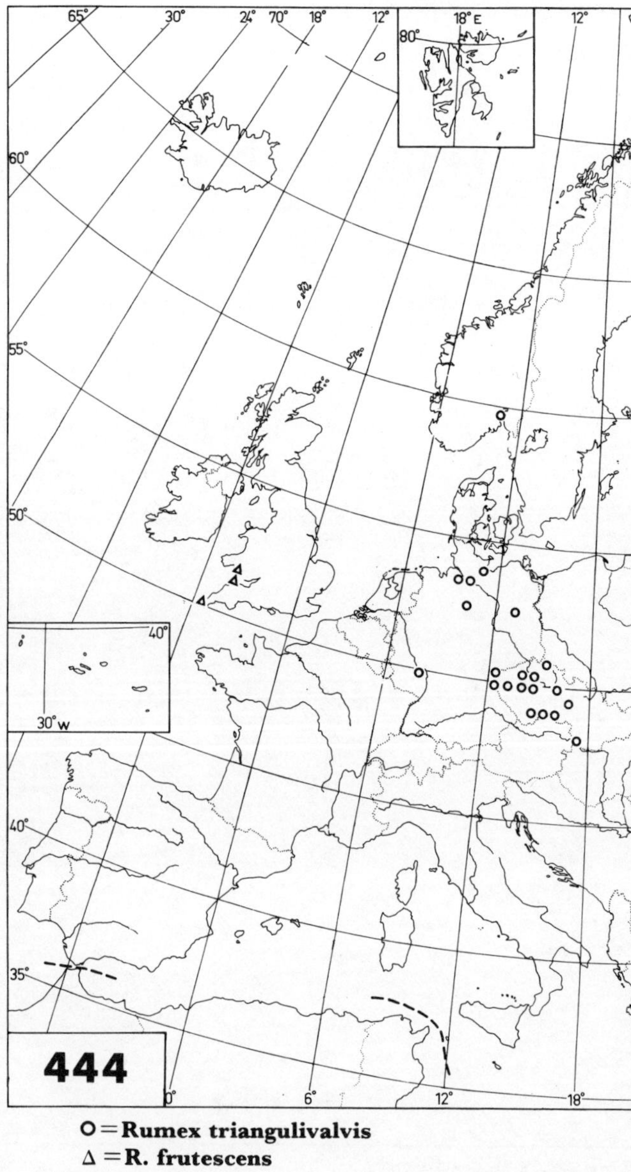

O = **Rumex triangulivalvis**
Δ = **R. frutescens**

Rumex vesicarius L. — Map 443.

Acetosa vesicaria (L.) Á. Löve

Notes. Not collected recently in Gr (?Gr in Fl. Eur.).

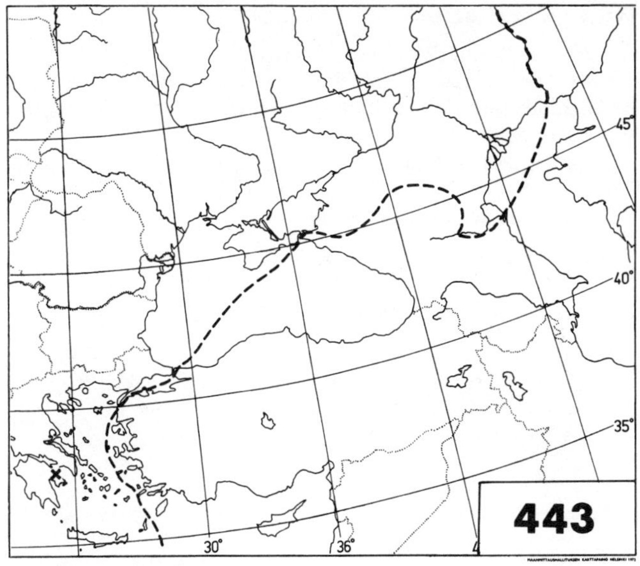

Rumex vesicarius

Rumex alpinus L. — Map 445.

Total range. MJW 1965: map 128d.

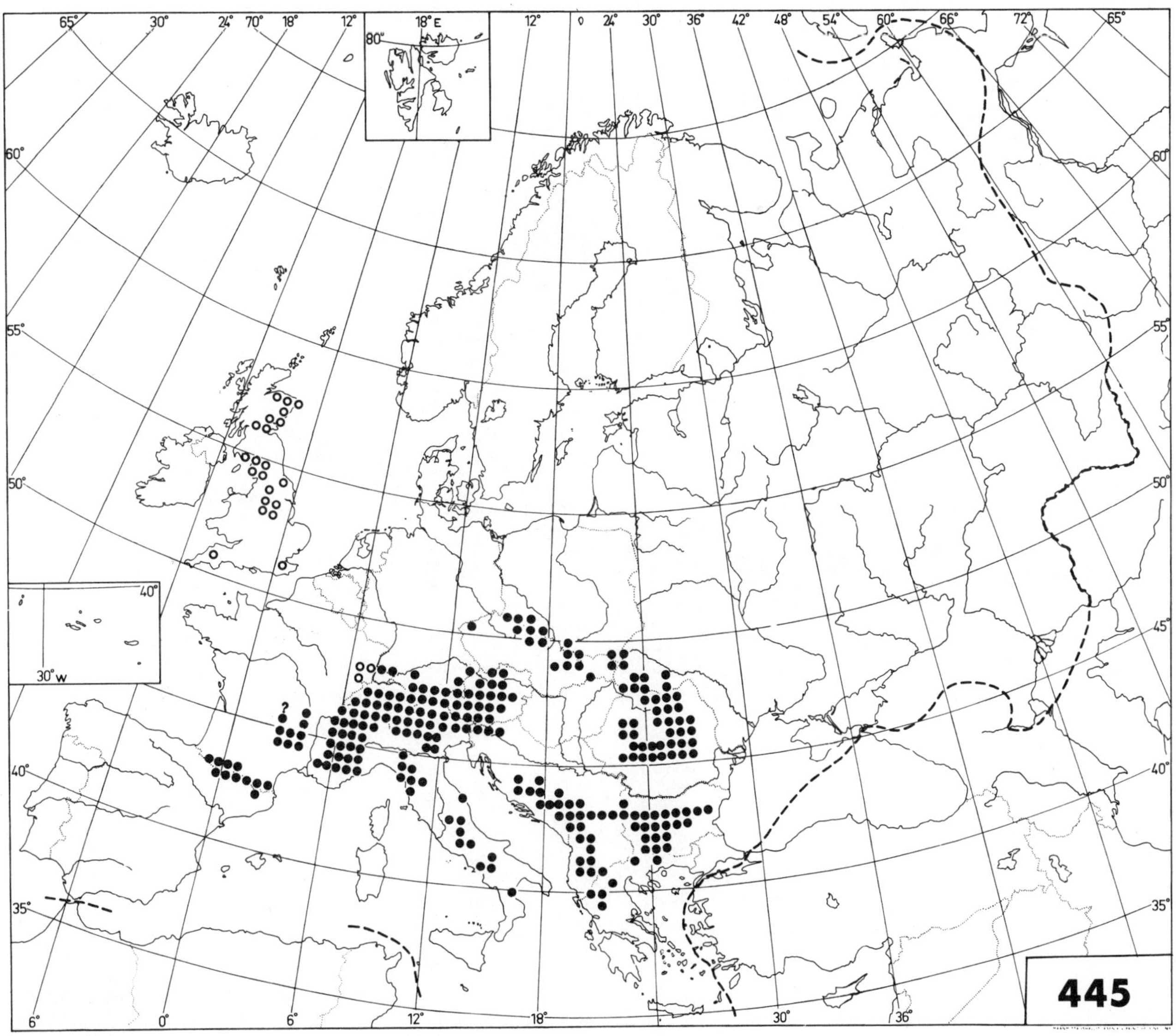

Rumex alpinus

Rumex aquaticus L. — Map 446.

Taxonomy. See under *Rumex arcticus.*

Notes. †Be, at least not seen for c. 100 years (given as present in Fl. Eur.): Atlas Fl. Belge Luxemb., text, p. 50 (1978). Bu, It and Rs(C) added (not given in Fl. Eur.), extinct in Ho (given as present in Fl. Eur.). The data available do not allow separate mapping of the subspecies.

Total range. Hegi 1958: 372 (Eurasian part); MJW 1965: map 128c; Hultén CP 1971: map 123.

R. aquaticus subsp. **aquaticus**

Notes. Territories as for the species.

R. aquaticus subsp. **protractus** (Rech. fil.) Rech. fil.

Rumex protractus Rech. fil.

Taxonomy. K. H. Rechinger, Candollea 12: 56—57 (1949); Fl. Arct. URSS 1966: 157; Fl. Bor.-or. Eur. URSS 1976: 174.

Notes. Rs(N) (the subspecies not mentioned in Fl. Eur.).

Total range. Hultén CP 1971: map 123 (Asiatic part).

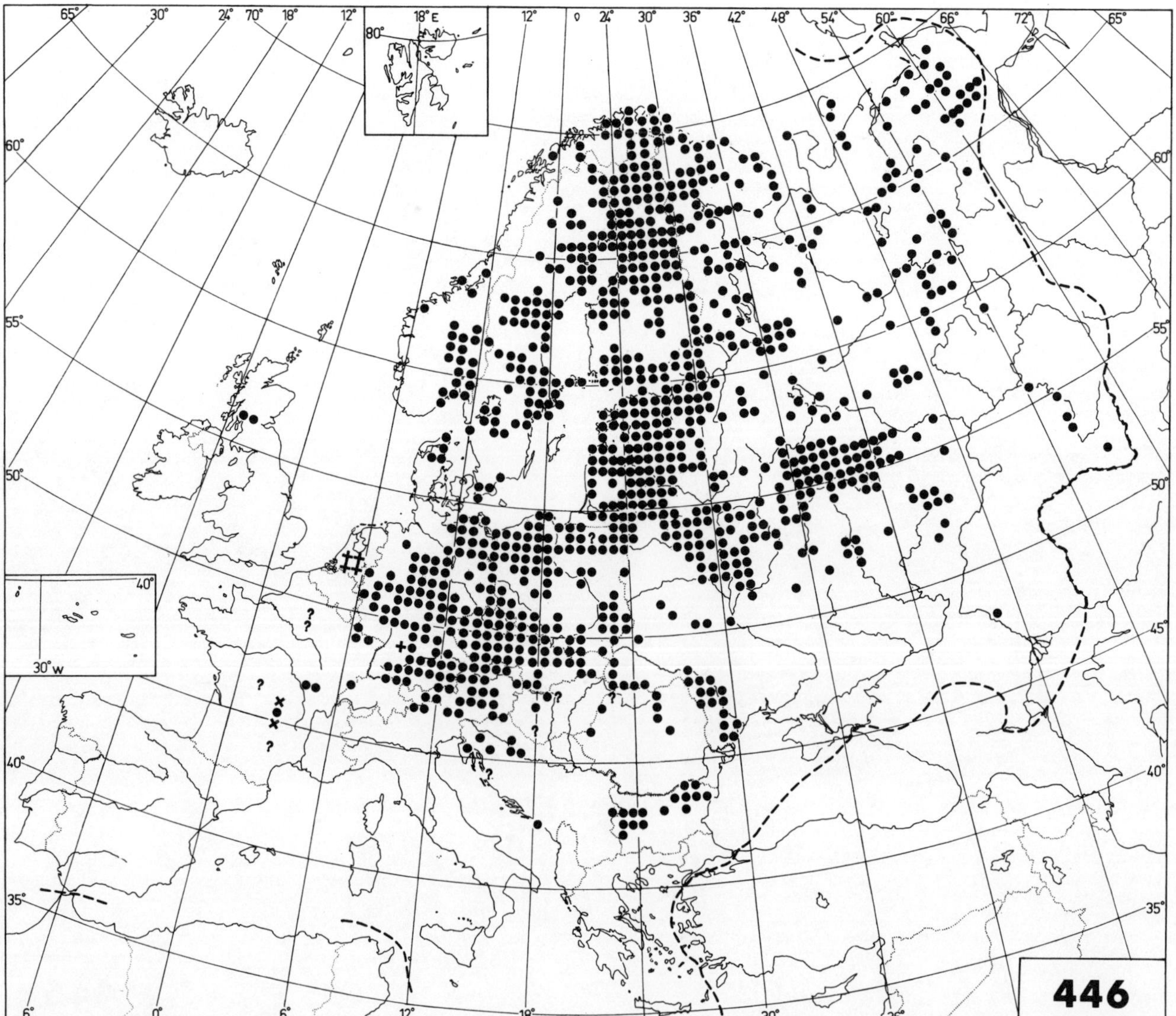

Rumex aquaticus

Rumex azoricus Rech. fil. — Map 447.

Total range. Endemic to the Açores.

Rumex caldeirarum Rech. fil.

Taxonomy. Insufficiently known. »Possibly a hybrid between *R. azoricus* and *R. obtusifolius* subsp. *obtusifolius*» (Fl. Eur.).

Notes. Not mapped.

Rumex arcticus Trautv. — Map 448.

Rumex aquaticus subsp. arcticus (Trautv.) Hiitonen; R. aquaticus var. kolaënsis Murb.; R. domesticus var. nanus Hooker

Taxonomy. K. H. Rechinger, Candollea 12: 57—60 (1949); Hylander 1966: 337; Fl. Arct. URSS 1966: 154—157; Hultén CP 1971: 16. The occurrence of obvious intermediates between *Rumex arcticus* and *R. aquaticus*, at least in the European part of the Russian Arctic, supports the view that *R. arcticus* should be treated as a subspecies under *R. aquaticus*. Intermediate material from the island of Kolguev, outside the known ranges of both species, was taken as a putative hybrid between them, and described as *R. aquaticus* subsp. *insularis* Tolm. in Fl. Arct. URSS 1966: 158. The Kolguev plant is here referred to *R. aquaticus*; cf. Fl. Bor.-or. Eur. URSS 1976: 174, maps 247—248.

Total range. Fl. Arct. URSS 1966: 156 (Eurasian part); Hultén Alaska 1968: 379; Hultén CP 1971: map 7.

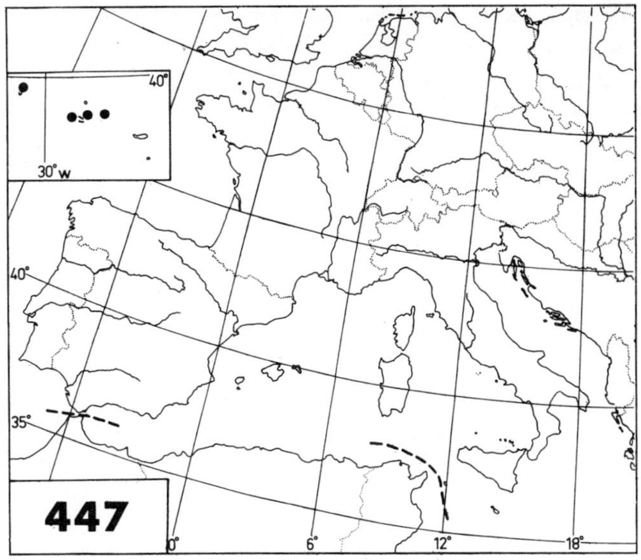

Rumex azoricus

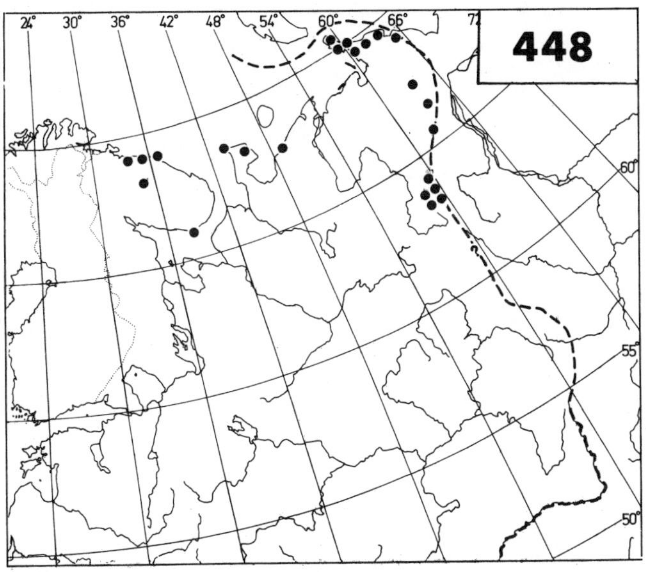

Rumex arcticus

Rumex balcanicus Rech. fil. — Map 449.

Notes. Al added (not given in Fl. Eur.).

Total range. Endemic to Europe.

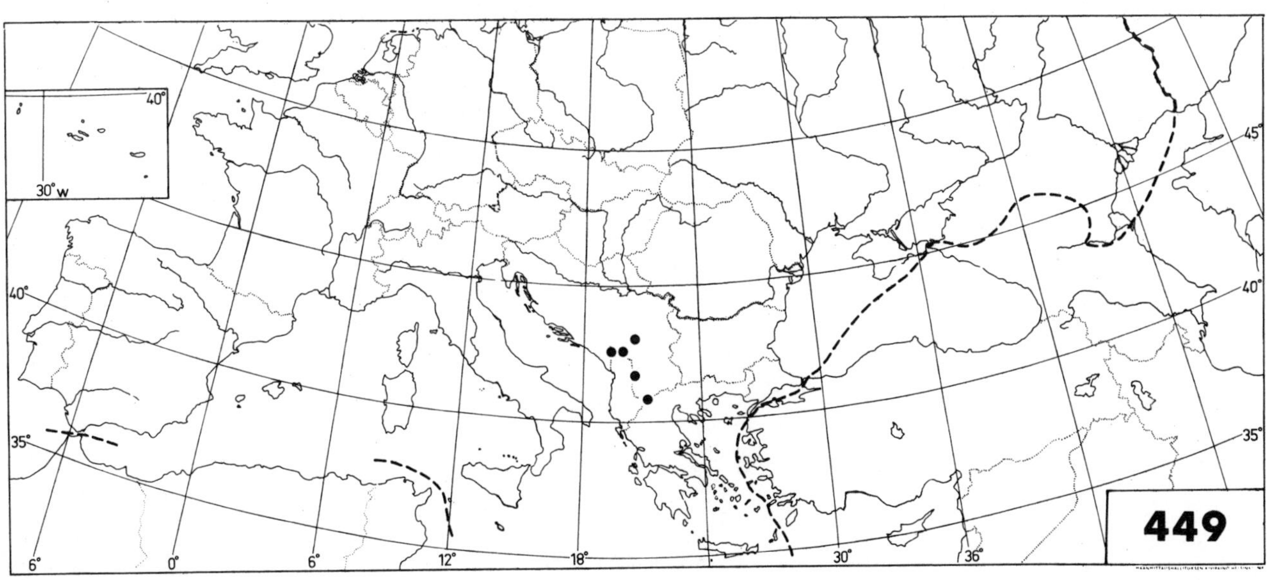

Rumex balcanicus

Rumex pseudonatronatus Borbás — Map 450.

Rumex fennicus (Murb.) Murb.; R. pseudonatronatus subsp. fennicus Murb.

Notes. Probably also in westernmost Rm (not given in Fl. Eur.): Fl. Rep. Pop. Rom. 1: 395—396 (1952).

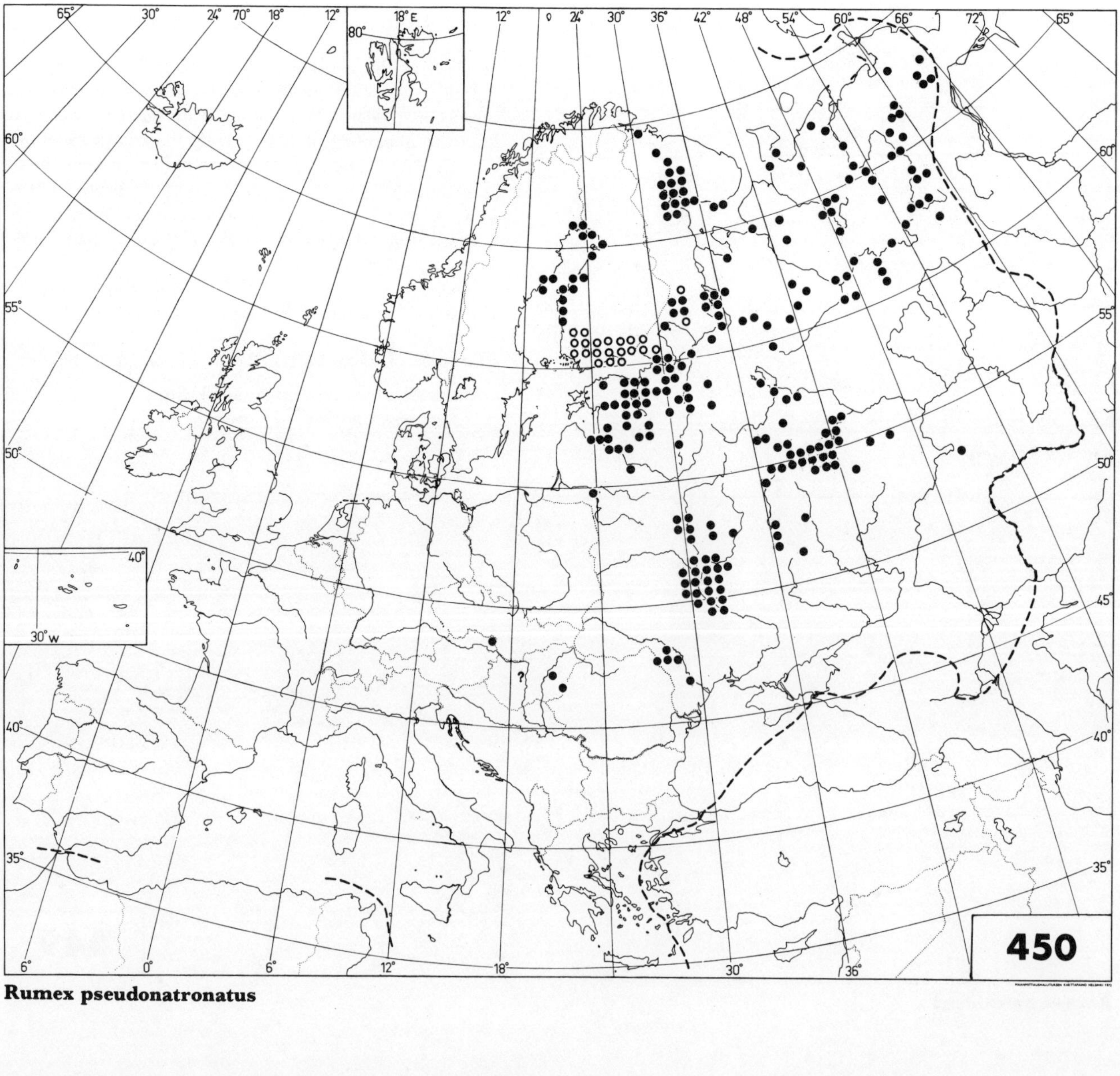

Rumex pseudonatronatus

Rumex cantabricus Rech. fil. — Map 451.

Notes. Ga added (not given in Fl. Eur.): J. Vivant, Bull. Soc. Bot. France 119: 335—338 (1972).

Total range. Endemic to Europe.

Rumex longifolius DC. — Map 452.

Rumex domesticus Hartman

Taxonomy. J. Jalas & R. K. Lindholm, An. Inst. Bot. Cavanilles 32: 197—202 (1975).

Notes. Considered native in Is (both Is and [Is] in Fl. Eur.). [Po] and Rs(E) added (not given in Fl. Eur.). The records of non-native occurrence in Cz are based only on early information in the literature. Recently recorded as an alien from Au by A. Neumann & A. Polatschek, Verh. Zool.-Bot. Ges. Wien 114: 53 (1974).

Total range. Hultén Alaska 1968: 379.

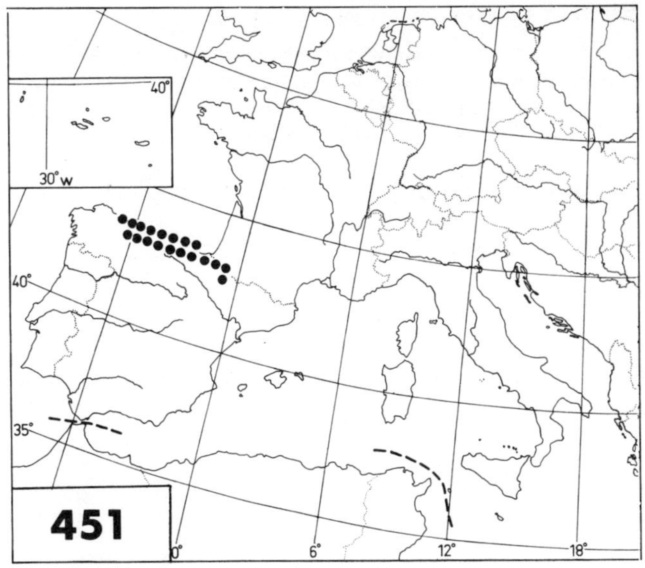

Rumex cantabricus

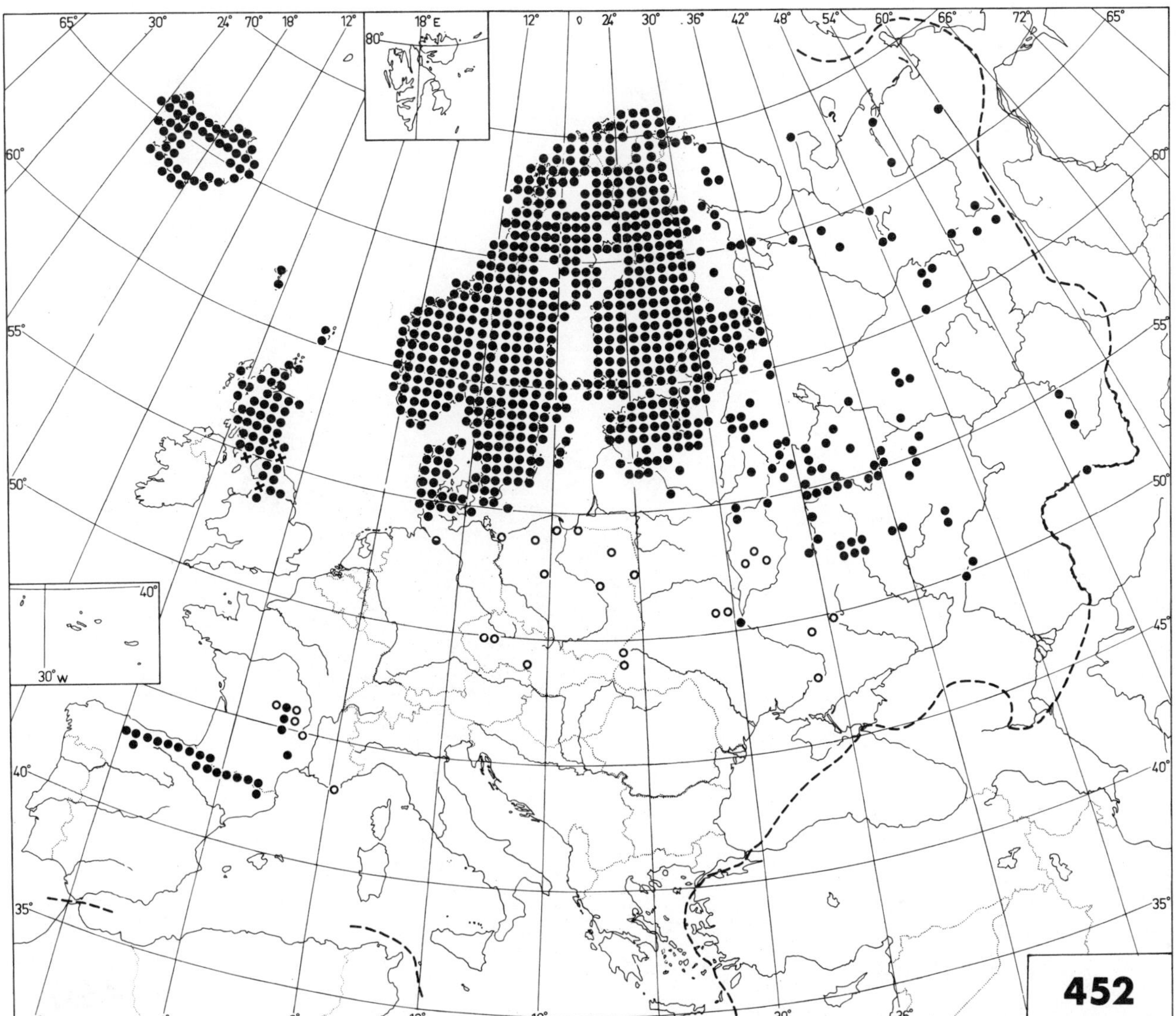

Rumex longifolius

Rumex confertus Willd. — Map 453.

Notes. [Au] and [Br] omitted (given in Fl. Eur.). ?It, Rs(N) and [Su] added (not given in Fl. Eur.): special attention should be paid to the remote occurrence in It; it may deserve closer study, especially its native status. Doubtful native in easternmost Po (given as native in Fl. Eur.).

Total range. H. Trczińska-Tacik, Fragm. Fl. Geobot. 9: 74, 76 (1963).

Rumex hydrolapathum Hudson — Map 454.

Notes. Al, Rs(N) and Sa added (not given in Fl. Eur.), Rs(K) omitted (given in Fl. Eur.).

Total range. MJW 1965: map 129b (outside Europe known only from one Caucasian locality).

Rumex cristatus DC. — Map 455.

Rumex graecus Boiss. & Heldr.

Notes. [Au] omitted (given in Fl. Eur.), obviously erroneously recorded for Hs (not given in Fl. Eur.), It not confirmed (?It in Fl. Eur.), Ju added (not given in Fl. Eur.), Tu added (neither in Fl. Eur. nor in Fl. Turkey 1967: 288).

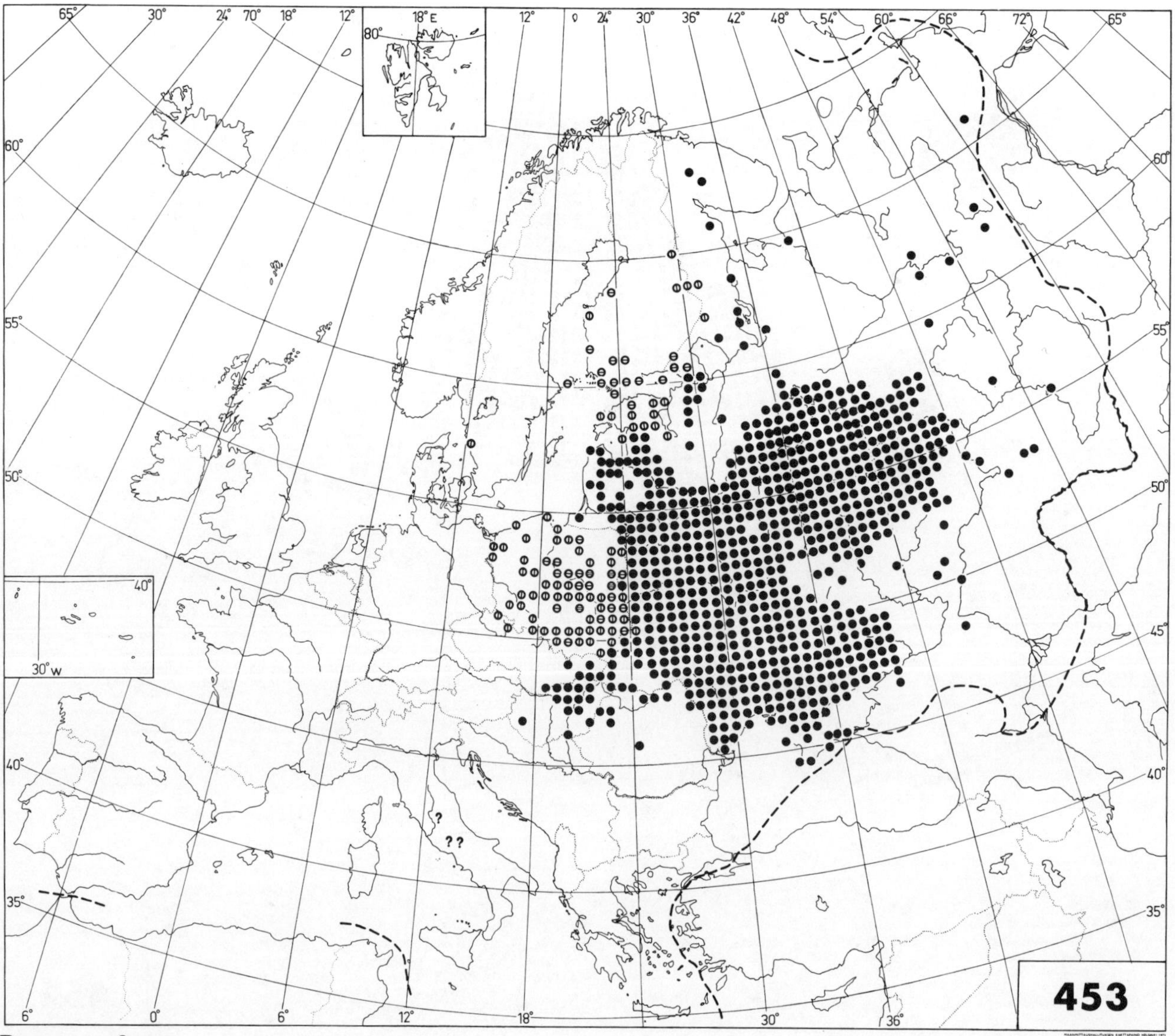

Rumex confertus

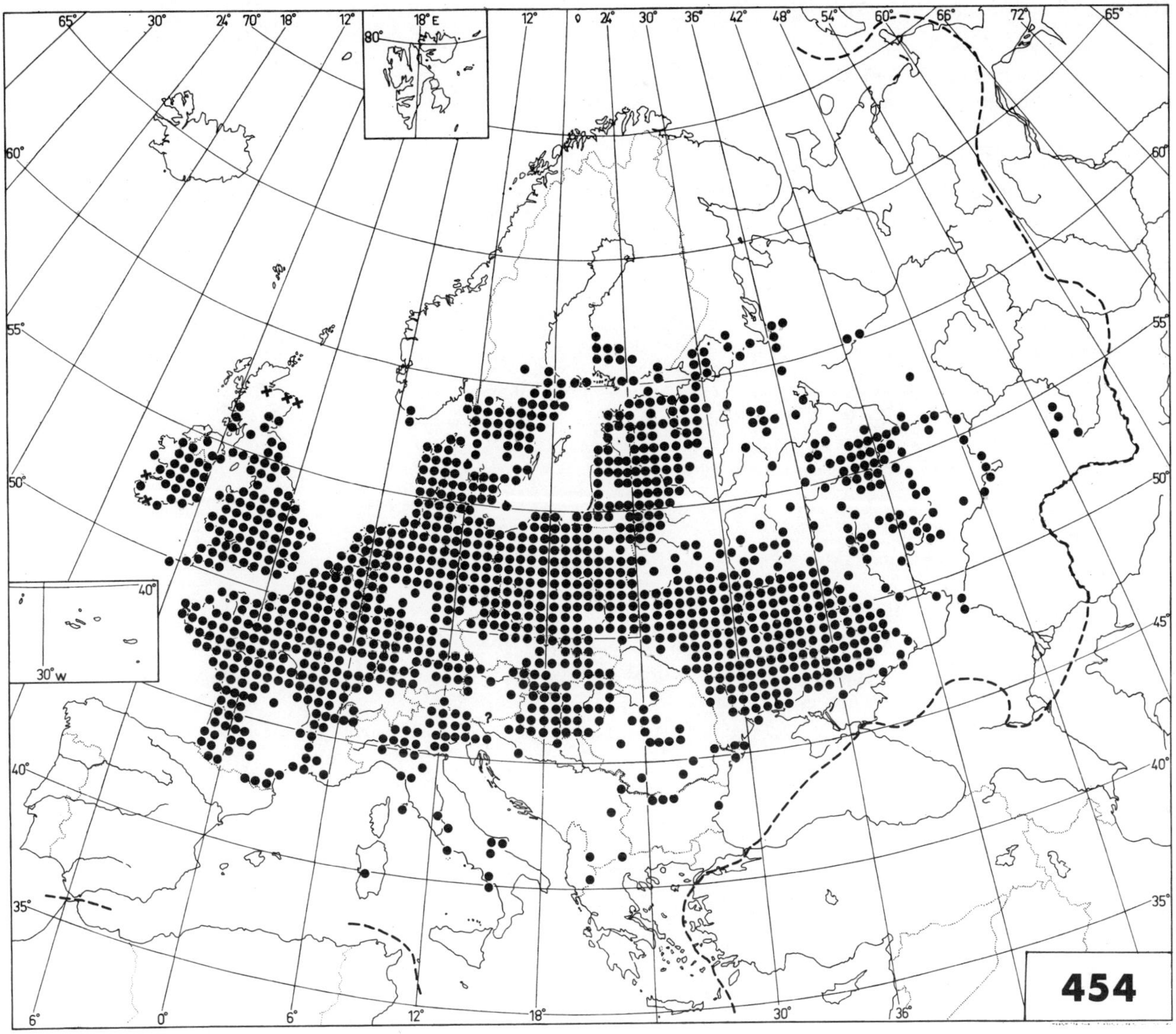

Rumex hydrolapathum

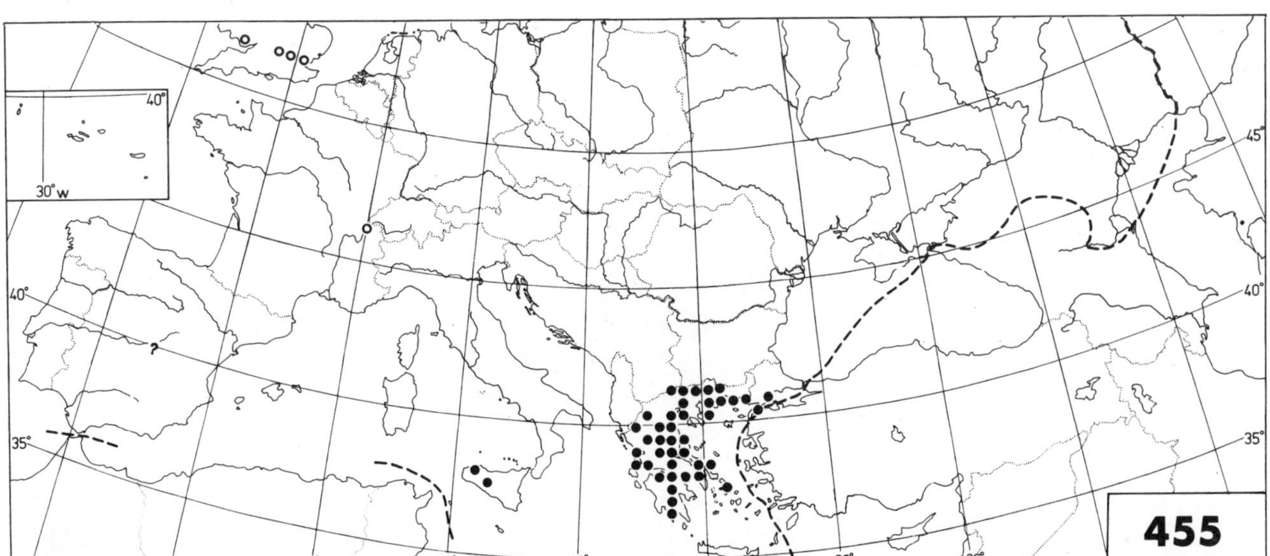

Rumex cristatus

Rumex kerneri Borbás — Map 456.

Rumex confertoides Bihari

Notes. [Au] omitted (given in Fl. Eur.).
Total range. Endemic to Europe.

Rumex patientia L. — Map 457.

Notes. [Ge, He, It, Po, Rs(C), Si] and *Tu added (not given in Fl. Eur.). The three subspecies recognized in Fl. Eur., and more or less sympatric in large areas of S. E. Europe, are not mapped separately. See the map in Hegi 1958: 376.

R. patientia subsp. **patientia**

Notes. »Au, Bu, Cz, Gr, Hu, Rm» (Fl. Eur.).

R. patientia subsp. **orientalis** (Bernh.) Danser

Rumex lonoczewskii Klokov; R. orientalis Bernh.

Notes. »Bu, Cz, Hu, Ju, Rm, Rs(W, K, E)» (Fl. Eur.).

R. patientia subsp. **recurvatus** (Rech.) Rech. fil.

Rumex recurvatus Rech.

Notes. »Bu, Cz, Rm» (Fl. Eur.).
Total range. Endemic to Europe (although not mentioned as such in Fl. Eur.): K. H. Rechinger, Candollea 12: 73 (1949).

Rumex crispus L. — Map 458.

Rumex odontocarpus Sándor

Total range. P. B. Cavers & J. L. Harper, Jour. Ecol. 52 (Biol. Fl. Brit. Isles 98): 757 (1964) (northern hemisphere); Hultén Alaska 1968: 378; Hultén CP 1971: map 260 (holarctic part).

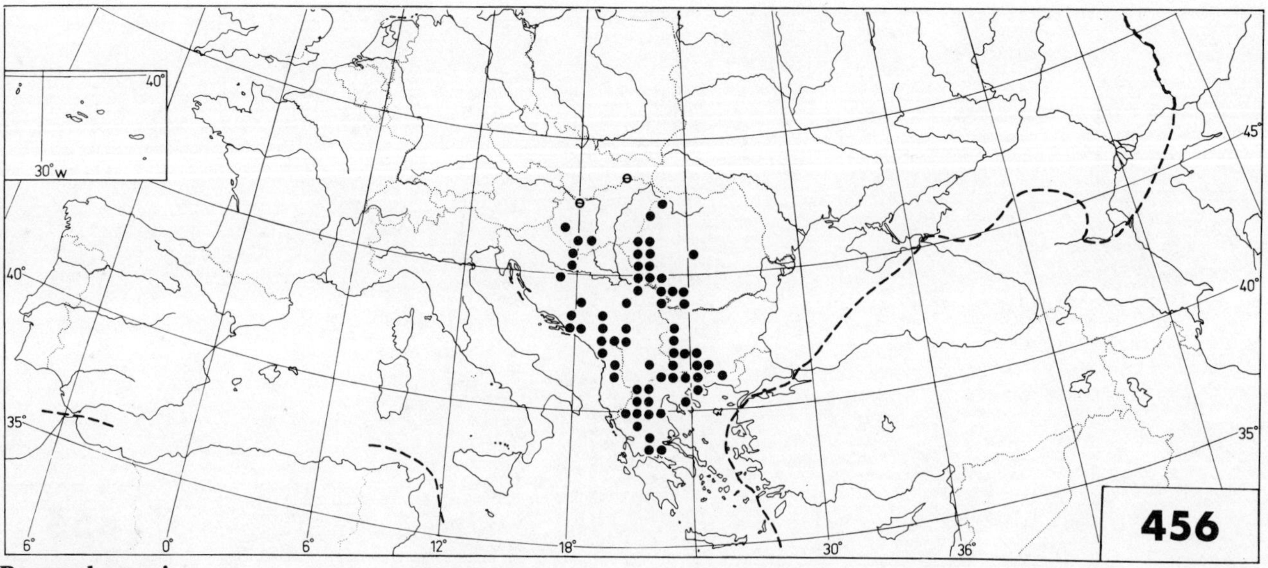

Rumex kerneri

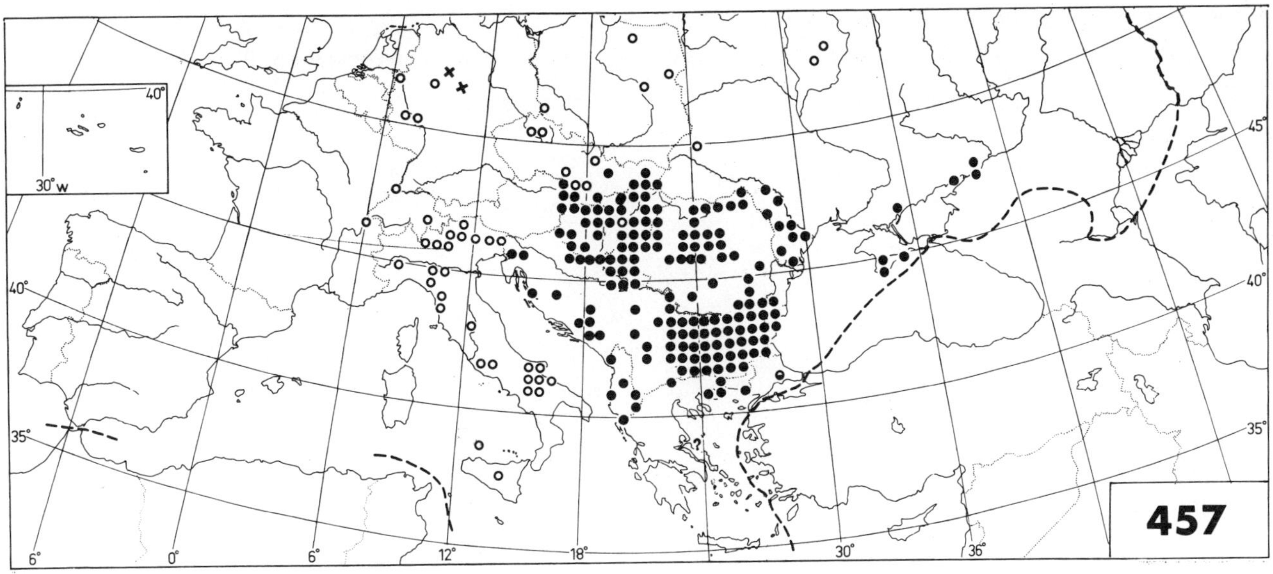

Rumex patientia

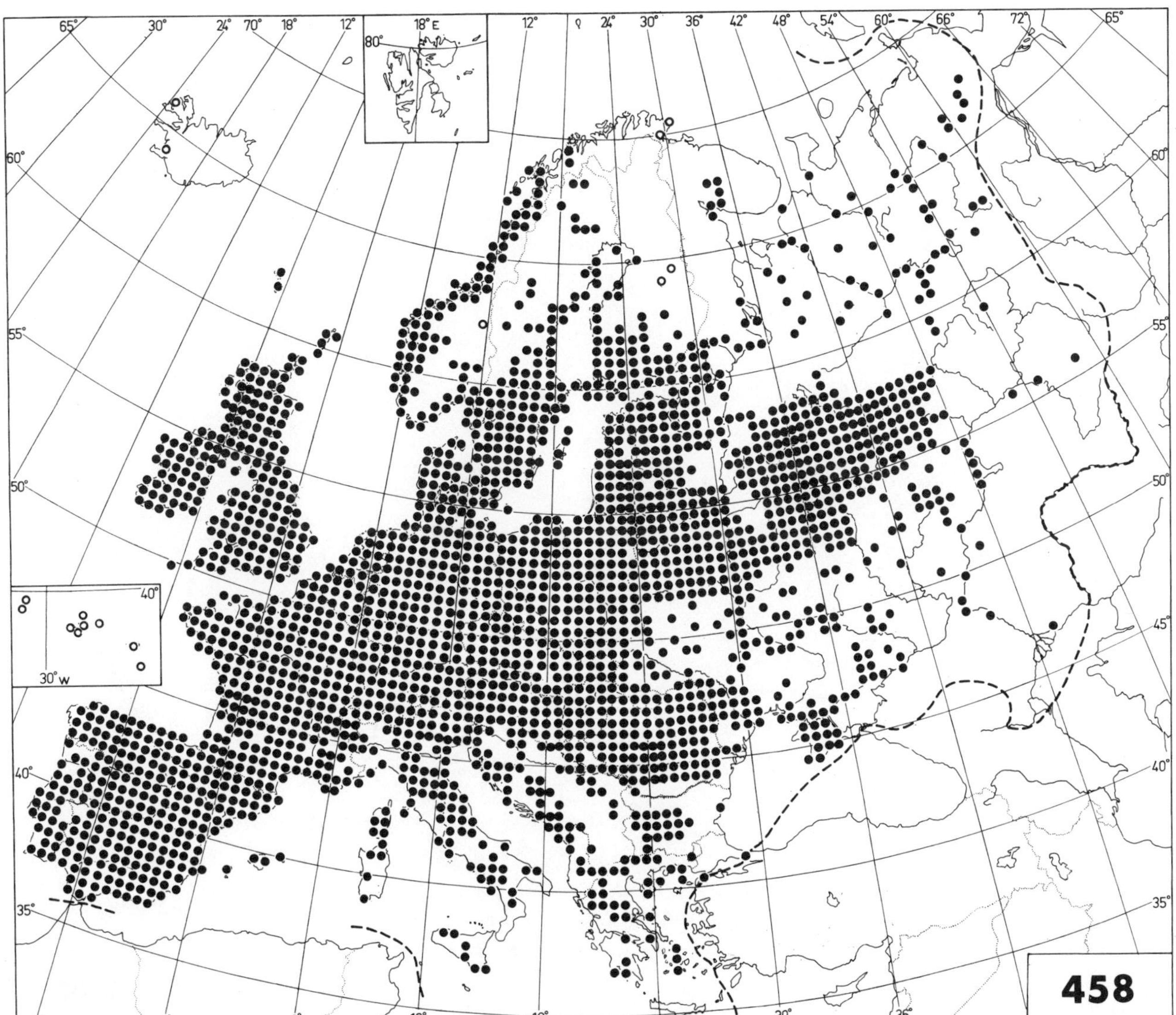

Rumex crispus

Rumex stenophyllus Ledeb. — Map 459.

Notes. Only casual in Da and Su ([Da, Su] in Fl. Eur.), and almost certainly so also in No and Rs(N, B). Rs(C, K) added (not given in Fl. Eur.). For Ge, see F. Hermann, Hercynia 3: 179—180 (1940).

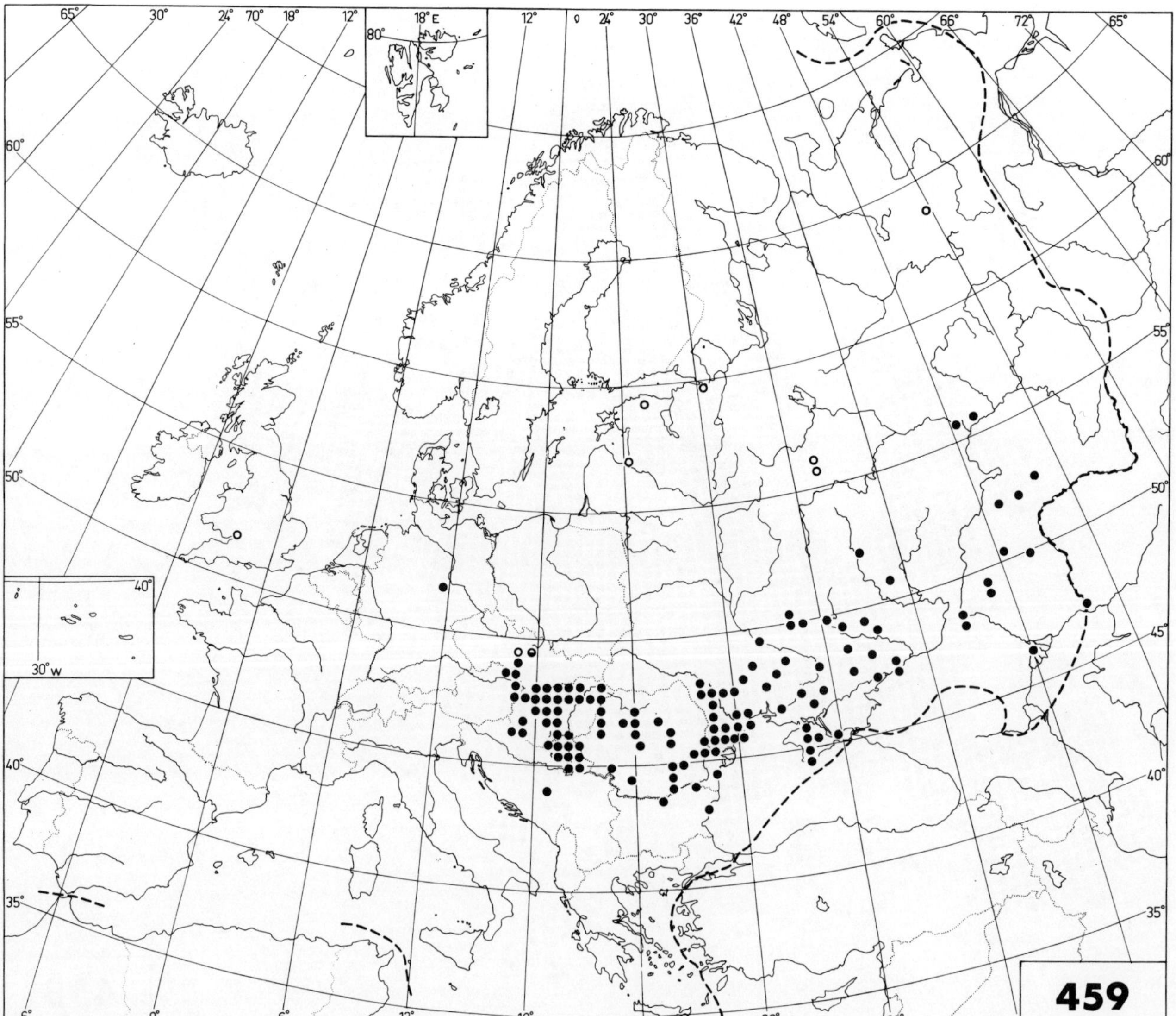

Rumex stenophyllus

Rumex conglomeratus Murray — Map 460.

Notes. Not established in Fe and No (»only as an alien in Az, Fe, No», according to Fl. Eur.). Rs(B) added (not given in Fl. Eur.).
Total range. Hegi 1958: 381; MJW 1965: map 129a.

R. conglomeratus subsp. **conglomeratus** — Map 460.

Notes. The range as for the species.

R. conglomeratus subsp. **nevadensis** H. Lindb. — Map 460.

Notes. Hs. »S. Spain (Sierra Nevada)» (Fl. Eur.).
Total range. Endemic to Europe.

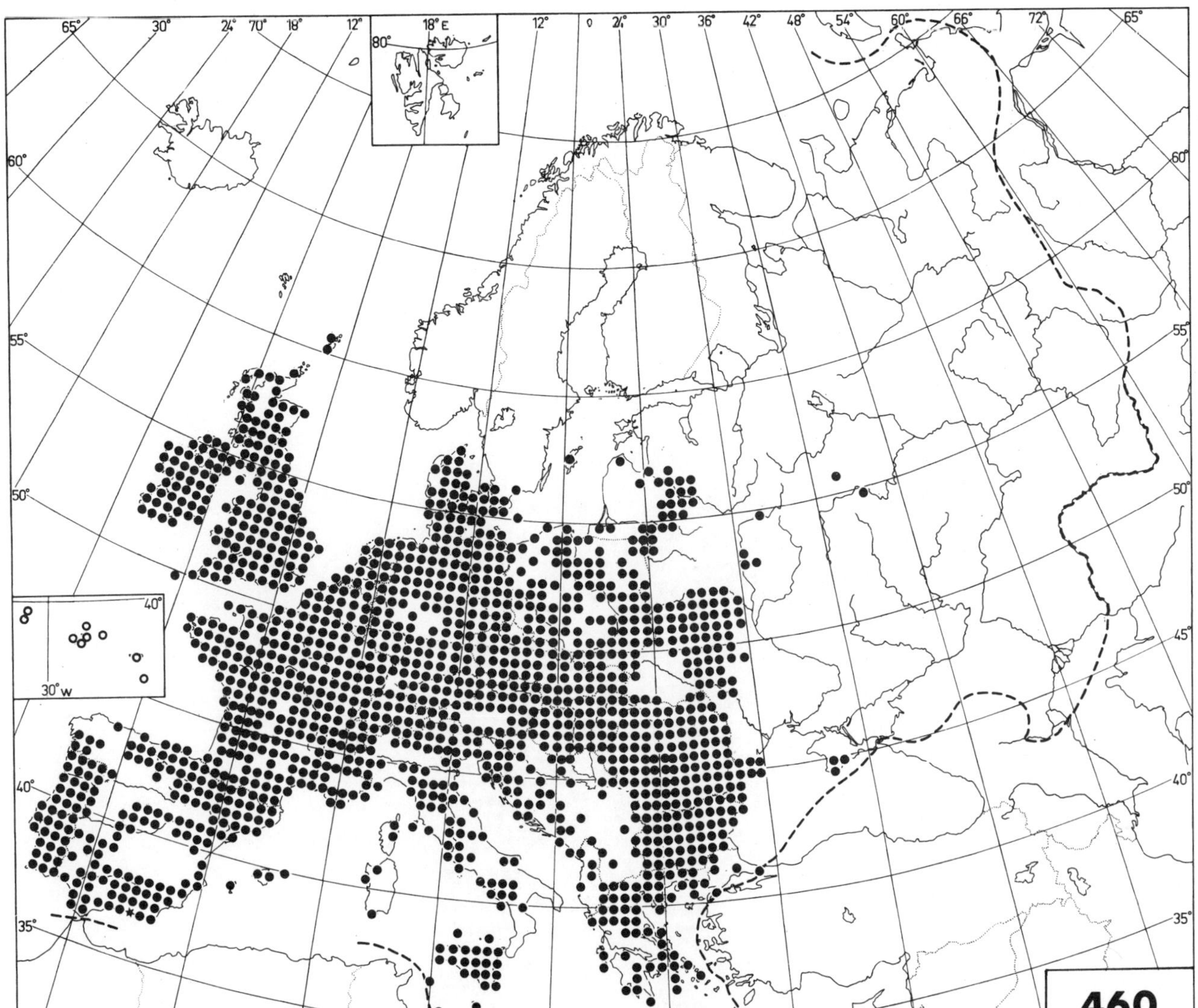

Rumex conglomeratus ★=subsp. **conglomeratus** + subsp. **nevadensis**

Rumex sanguineus L. — Map 461.

Rumex nemorosus Schrad.

Notes. Al and Rs(K) omitted (given in Fl. Eur.), Ga, Gr, Rs(E) and Tu added (not given in Fl. Eur.). The record for Az in A. Hansen, Anuário Soc. Brot. 40: 15 (1974), is obviously referable to *R. conglomeratus* (A. Hansen *in litt.).*

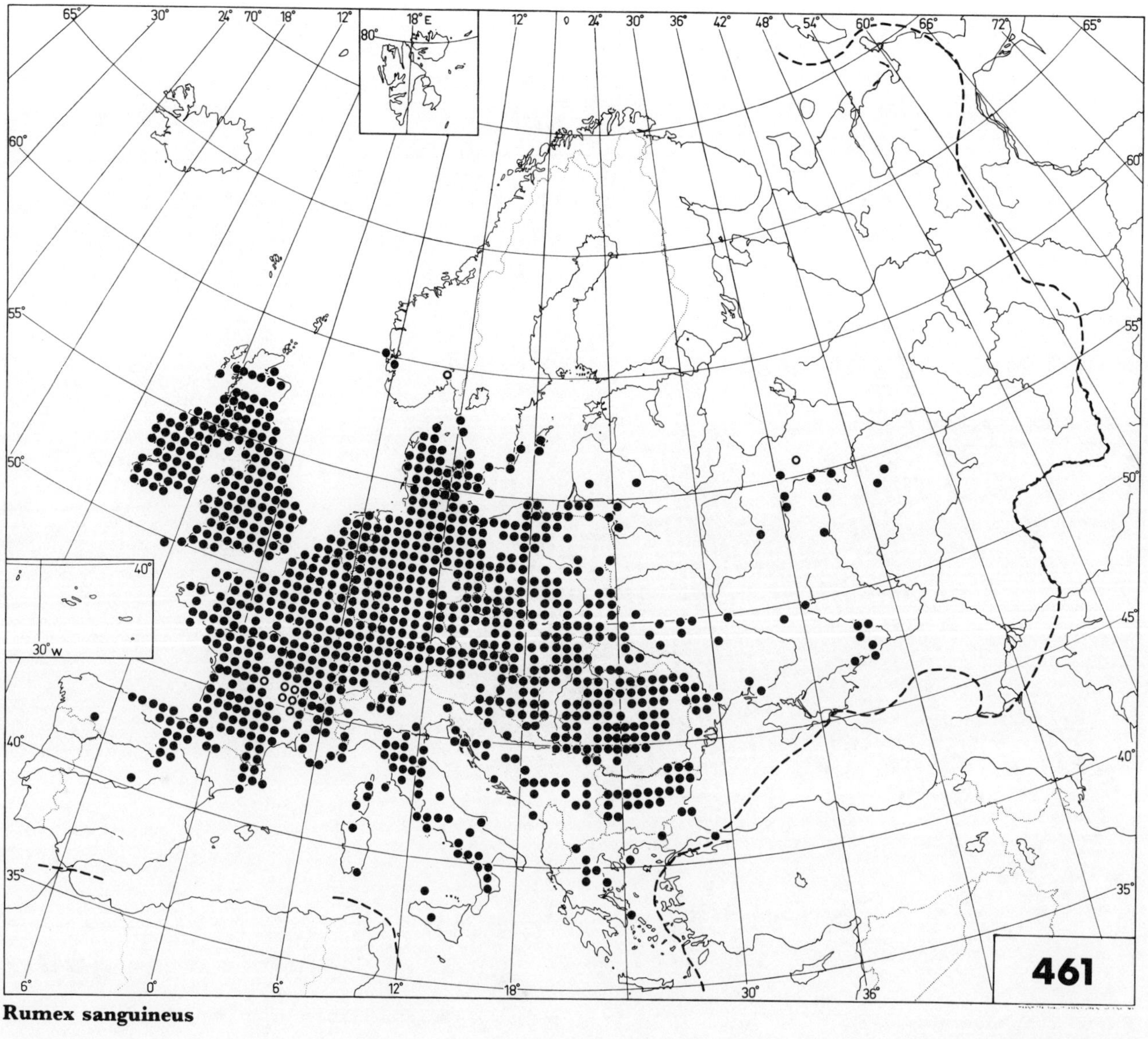

Rumex sanguineus

Rumex rupestris Le Gall — Map 462.

Rumex nepalensis Sprengel — Map 463.

Rumex peregrinus Boiss.

Notes. Ju confirmed (?Ju in Fl. Eur.).

Total range. K. H. Rechinger, Beih. Bot. Centr. 49(2): 32 (1932).

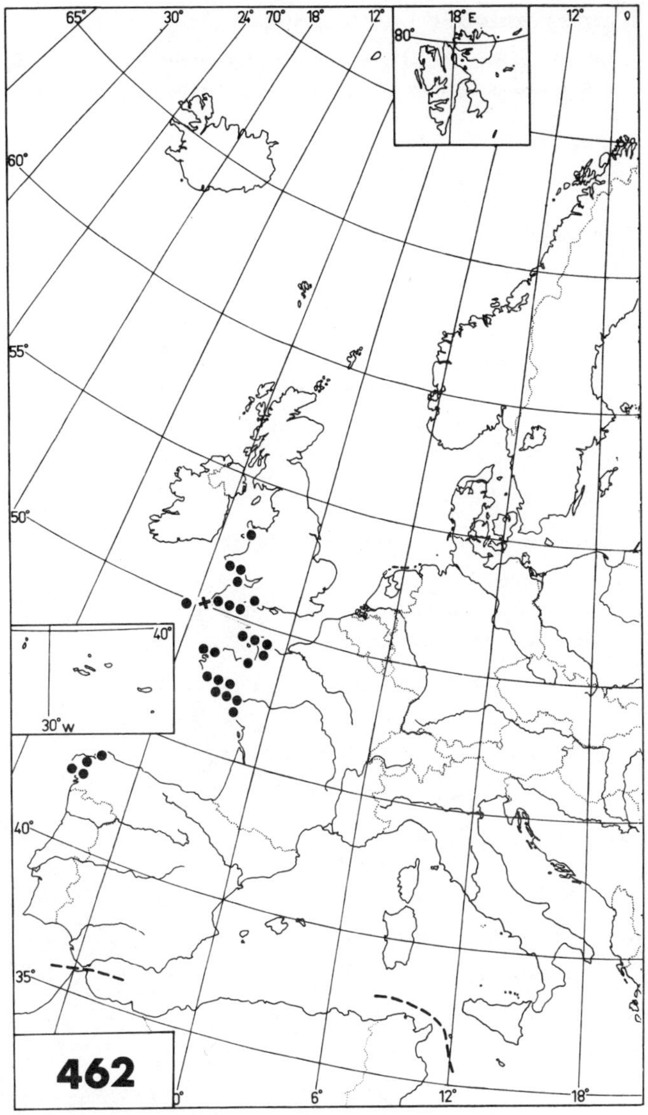

Rumex rupestris

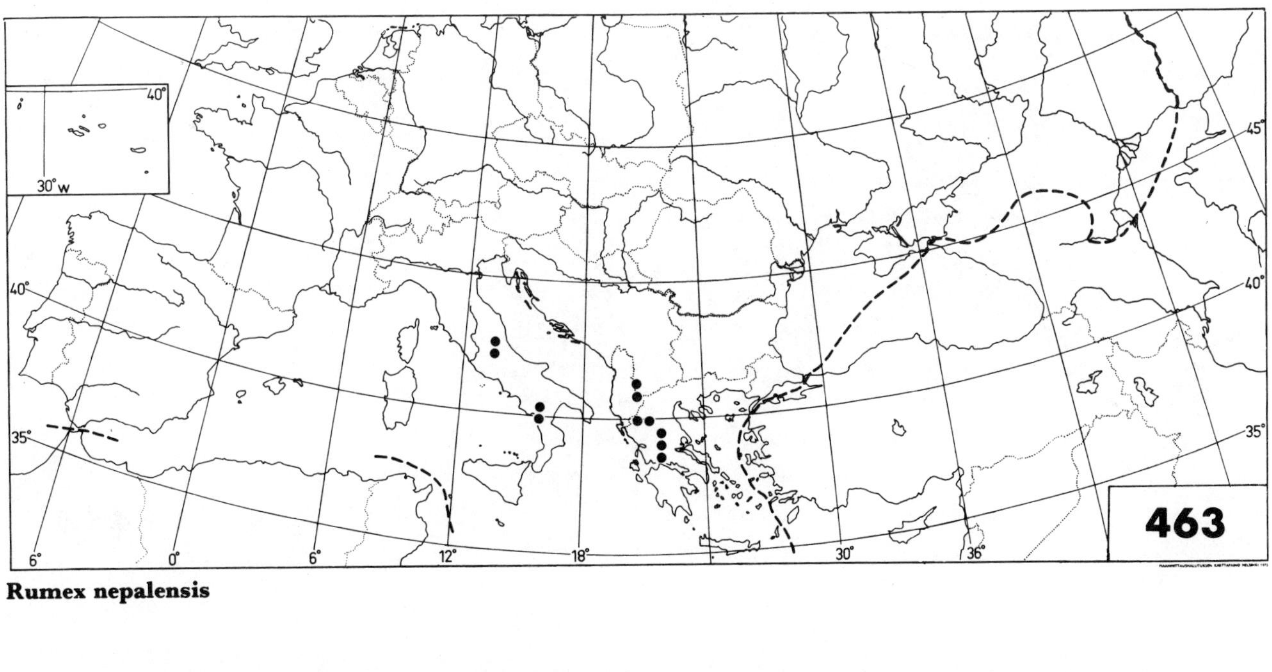

Rumex nepalensis

Rumex pulcher L. — Map 464.

Notes. Cz [Au, Da, Hb, Su] omitted (given in Fl. Eur.), Ge added (not given in Fl. Eur.). The individual subspecies have not been mapped separately.

R. pulcher subsp. **pulcher**

Notes. »Throughout the range of the species» (Fl. Eur.).

R. pulcher subsp. **divaricatus** (L.) Murb.

Rumex divaricatus L.

Notes. »S. Europe» (Fl. Eur.).
Total range. Hegi 1958: 389.

R. pulcher subsp. **anodontus** (Hausskn.) Rech. fil.

Rumex pulcher var. anodontus ('anodonta') Hausskn.

Notes. »Kriti» (Fl. Eur.).
Total range. Hegi 1958: 389.

R. pulcher subsp. **raulinii** (Boiss.) Rech. fil.

Rumex raulinii Boiss.

Notes. »Greece and Aegean region» (Fl. Eur.).
Total range. Hegi 1958: 389.

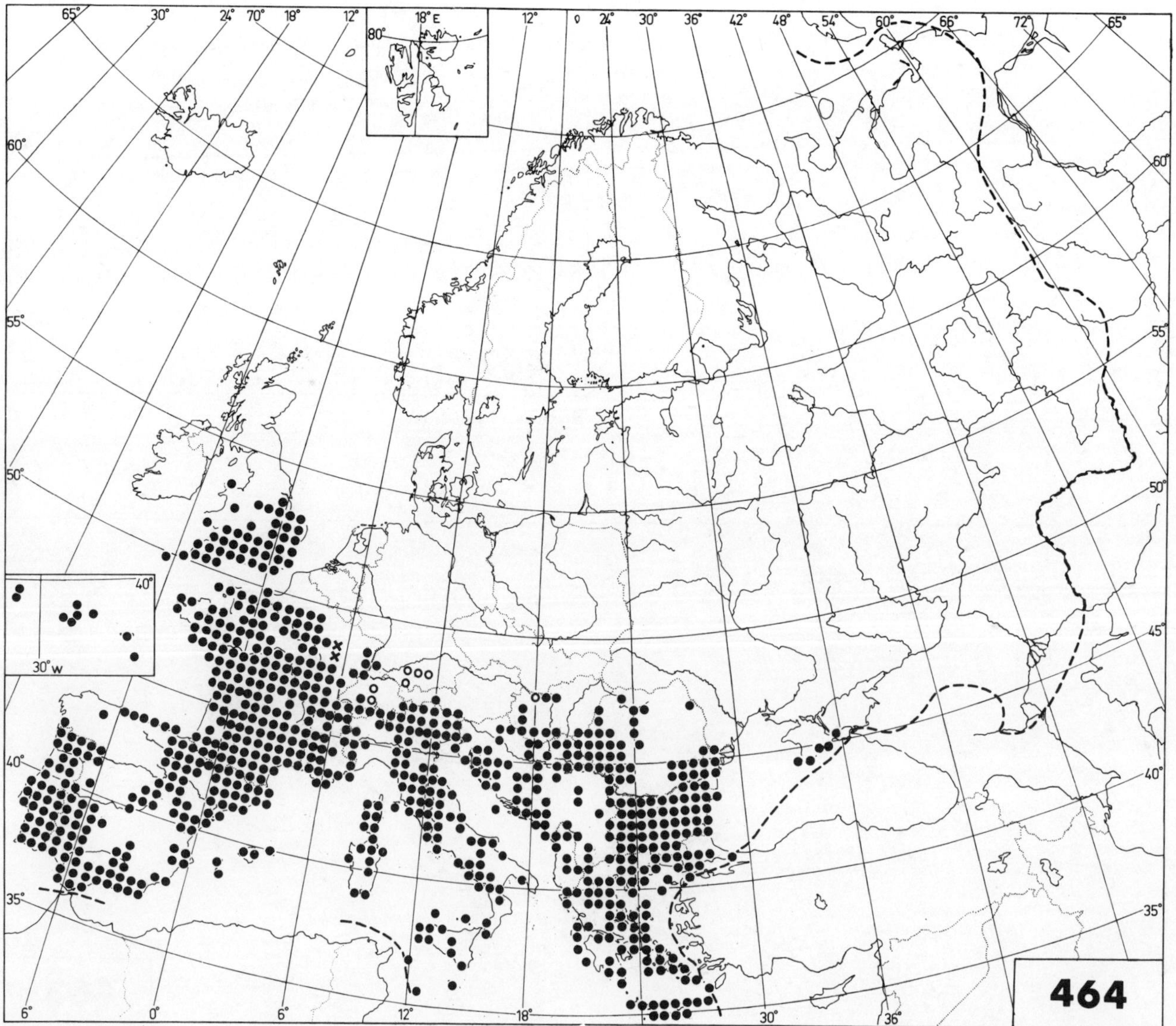

Rumex pulcher

Rumex obtusifolius L. — Map 465.

Notes. Not native in Fe and probably not native in Rs(N); only casual in Is; Si omitted. »All [the territories] except Cr, Sb, but only as an alien in Az, Is and perhaps elsewhere» (Fl. Eur.). Of the four subspecies, only subsp. *subalpinus* has been mapped separately.

Total range. Hegi 1958: 386 (excl. the non-European secondary ranges); P. B. Cavers & J. L. Harper, Jour. Ecol. 52 (Biol. Fl. Brit. Isles 98): 757 (1964); MJW 1965: map 129c; Hultén Alaska 1968: 378; Hultén CP 1971: map 248 (holarctic part).

R. obtusifolius subsp. **obtusifolius**

Rumex obtusifolius subsp. agrestis (Fries) Čelak.; R. obtusifolius subsp. friesii Rech.

Notes. »Native in W. Europe, but widely naturalized elsewhere» (Fl. Eur.).

R. obtusifolius subsp. **subalpinus** (Schur) Rech. fil. — Map 465.

Rumex sylvestris subsp. subalpinus (Schur) Simonkai

Nomenclature. J. Holub, Preslia 38: 79 (1966).

Notes. Al, Bu, Hu, Po, Rm, Rs(W). »Carpathians and mountains of Balkan peninsula» (Fl. Eur.).

Total range. Hegi 1958: 386; P. B. Cavers & J. L. Harper, Jour. Ecol. 52 (Biol. Fl. Brit. Isles 98): 739 (1964).

R. obtusifolius subsp. **transiens** (Simonkai) Rech. fil.

Rumex sylvestris var. transiens Simonkai

Notes. Established alien in Br: P. B. Cavers & J. L. Harper, Jour. Ecol. 52 (Biol. Fl. Brit. Isles 98): 738 (1964). The most common subsp. in [Fe]: J. Jalas, Suuri kasvikirja 2: 138—139 (Keuruu 1965). »C. Europe, Balkan peninsula, S. Scandinavia» (Fl. Eur.).

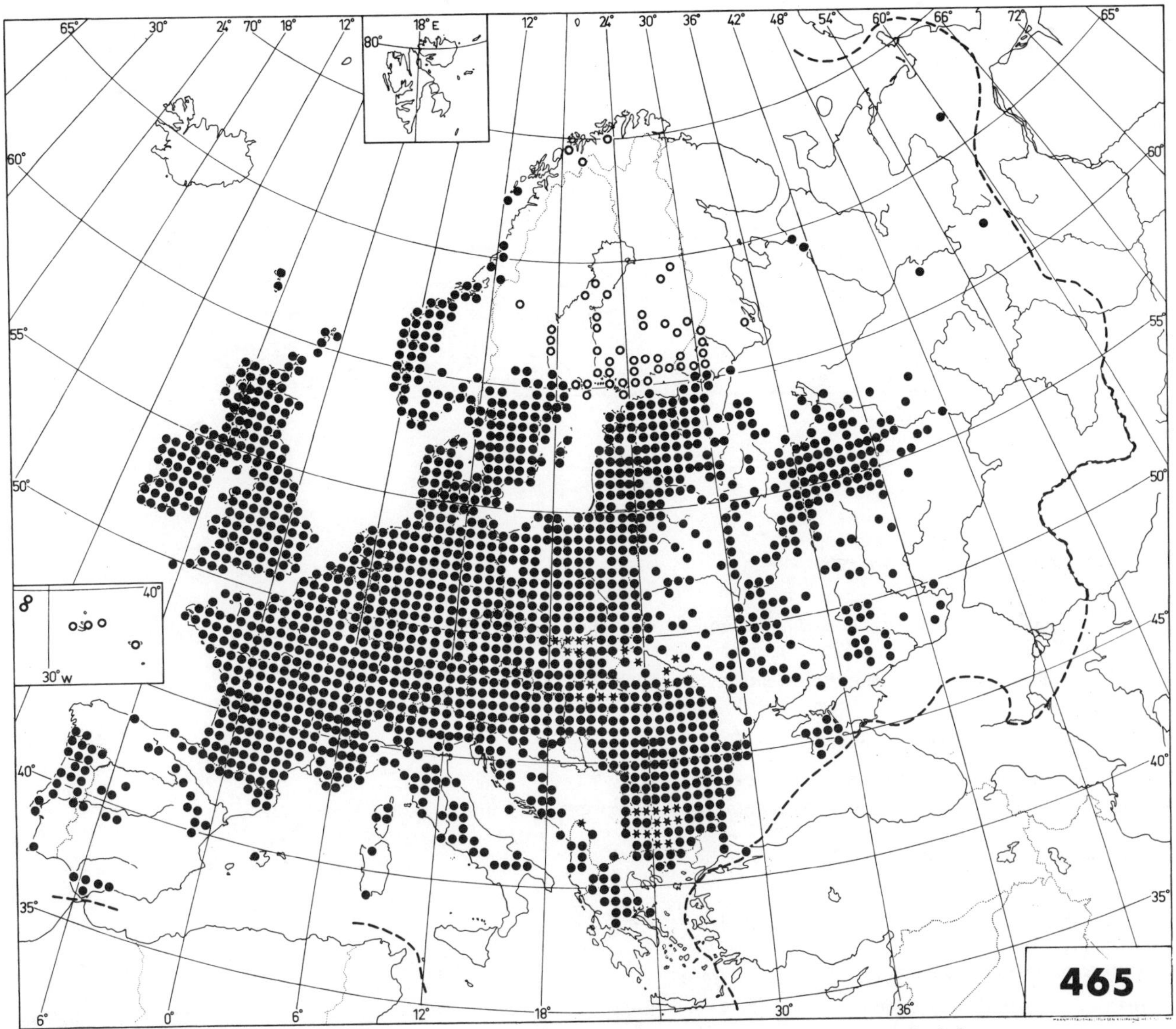

Rumex obtusifolius ★=subspp. indet. + subsp. **subalpinus**

R. obtusifolius subsp. **sylvestris** (Wallr.) Čelak.

Rumex obtusifolius var. microcarpus Dierb.; R. sylvestris Wallr.

Nomenclature. J. Holub, Preslia 38: 79 (1966).

Notes. Newly naturalized in Sa: M. Chiappini, Giorn. Bot. Ital. 70: 526—531 (1964). »Europe from Italy and Sweden eastwards» (Fl. Eur.).

Total range. Hegi 1958: 386; P. B. Cavers & J. L. Harper, Jour. Ecol. 52 (Biol. Fl. Brit. Isles 98): 739 (1964); MJW 1965: map 129c (W. limit).

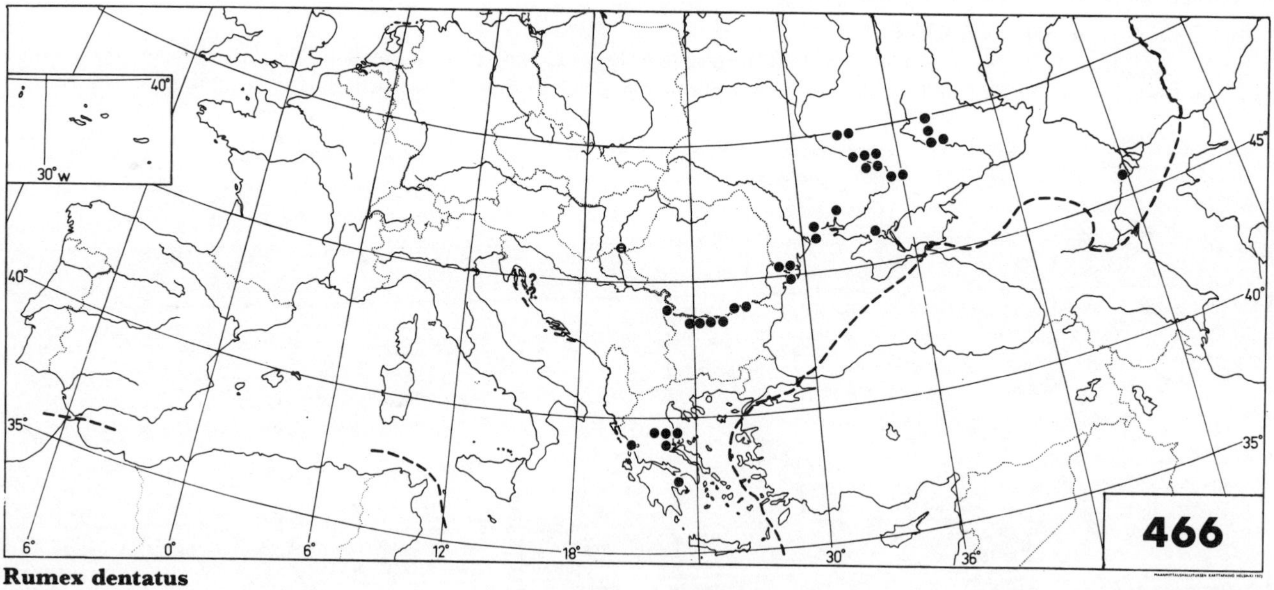

Rumex dentatus

Rumex dentatus L. — Map 466.

Rumex limosus Thuill.

Notes. Al omitted (given in Fl. Eur.), Bu added (not given in Fl. Eur.). Since the distinction between the subspecies is not clear, mapped as one entity.

Total range. K. H. Rechinger, Beih. Bot. Centr. 49(2): 129 (1932).

R. dentatus subsp. **halacsyi** (Rech.) Rech. fil.

Rumex halacsyi (limosus x pulcher) Rech.

Notes. »Romania to Greece; occasionally naturalized further west» (Fl. Eur.).

R. dentatus subsp. **reticulatus** (Besser) Rech. fil.

Rumex reticulatus Besser

Notes. »Ukraine (near Odessa); S. E. Russia (Volga delta)» (Fl. Eur.).

Rumex obovatus Danser

Notes. ». . . recorded as introduced in several places in N. Europe and is perhaps becoming naturalized» (Fl. Eur.). Not mapped.
Native of South America.

Rumex palustris Sm. — Map 467.

Rumex limosus auct. non Thuill.

Notes. Al and Cz added (not given in Fl. Eur.): J. Holub, Preslia 38: 79 (1966). He, Hs and No omitted (given in Fl. Eur.), Tu added (not given in Fl. Eur. or in Fl. Turkey 1967).

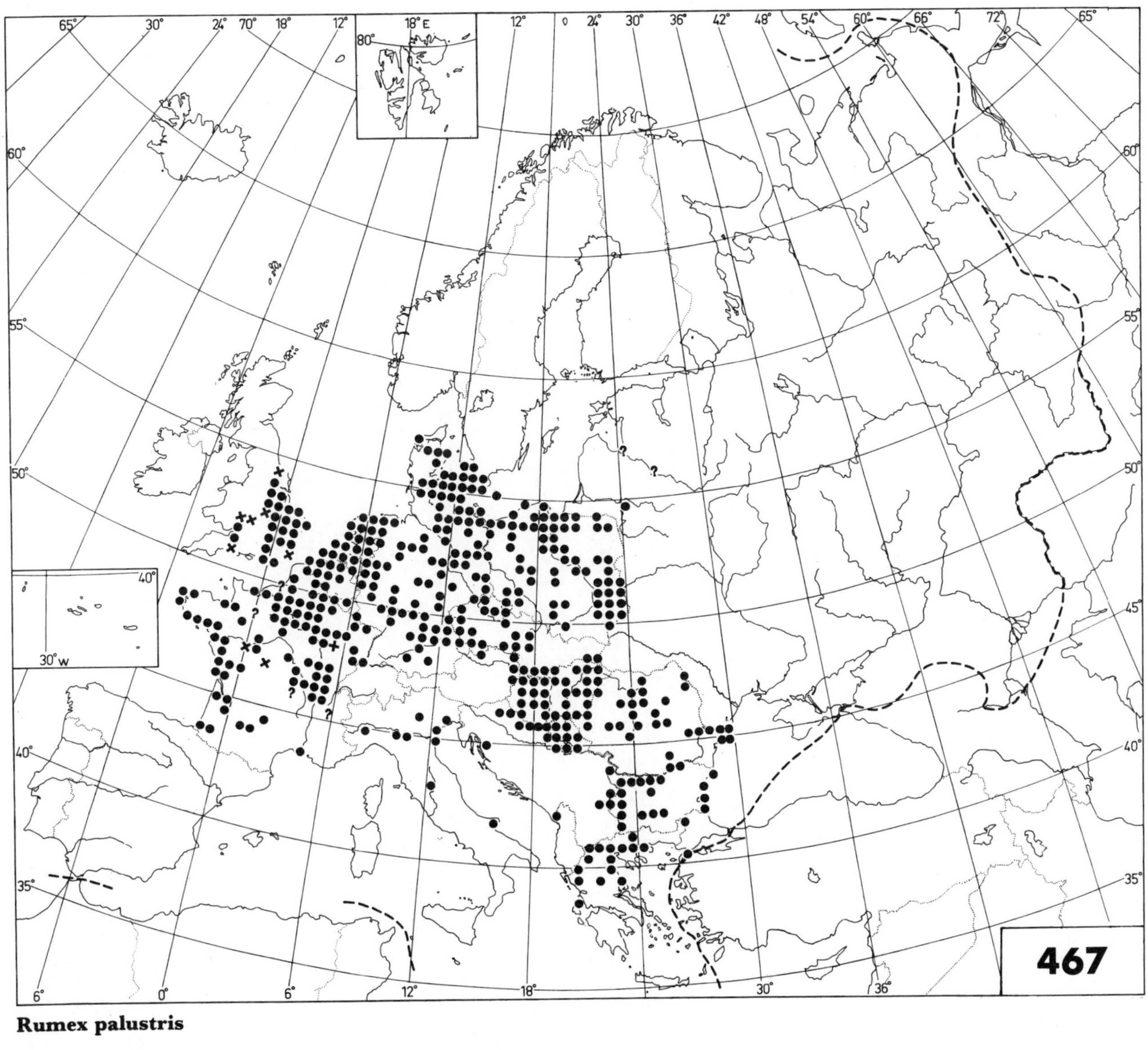

Rumex palustris

Rumex maritimus L. — Map 468.

Notes. Hs omitted (?Hs in Fl. Eur.), Rs(N) added (not given in Fl. Eur.).

Total range. Hultén Alaska 1968: 383 (erroneously under *Rumex maritimus* subsp. *fueginus*); Hultén CP 1971: map 165.

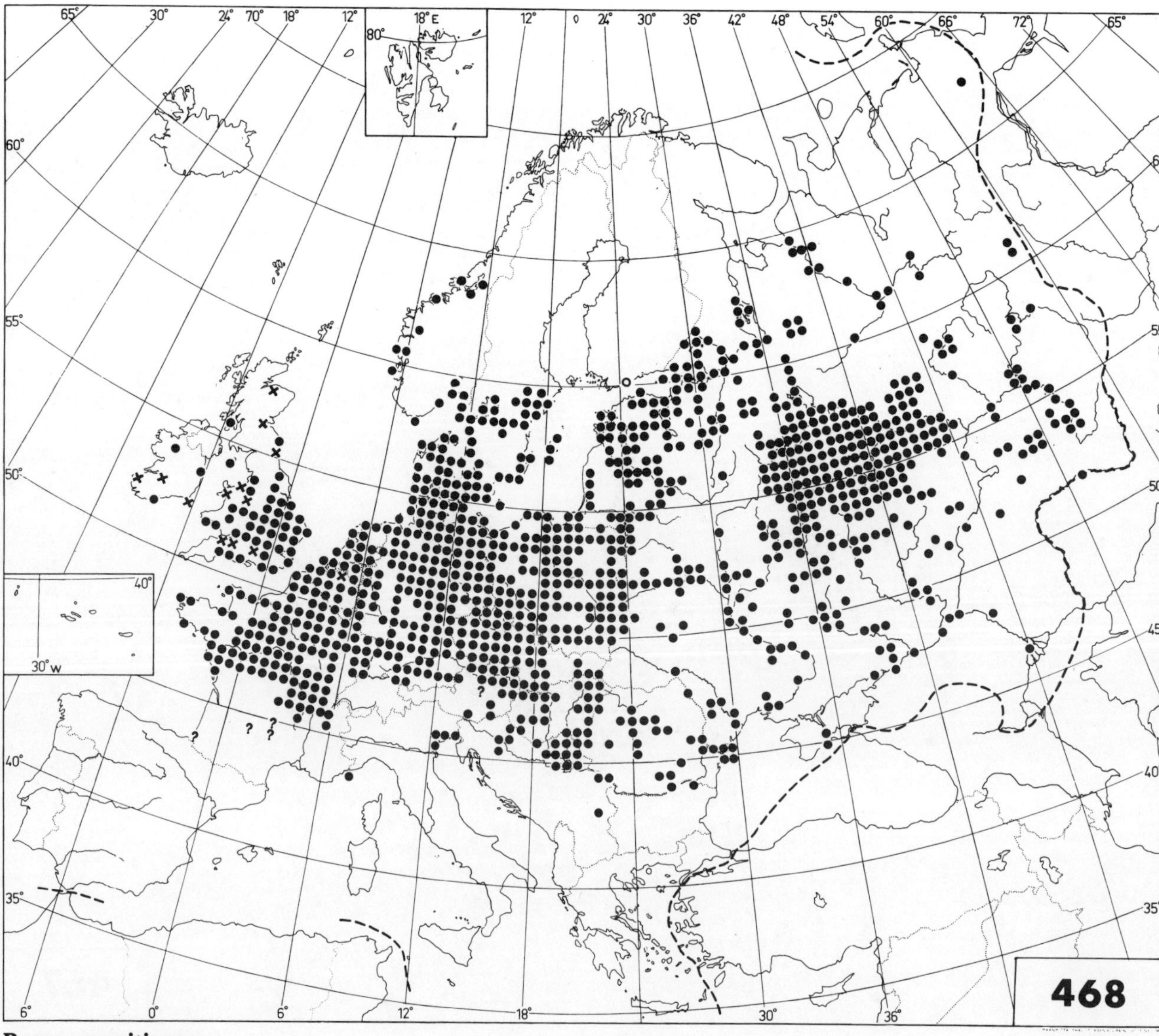

Rumex maritimus

Rumex rossicus Murb. — Map 469.

Rumex maritimus subsp. rossicus (Murb.) Krylov

Rumex marschallianus Reichenb. — Map 470.

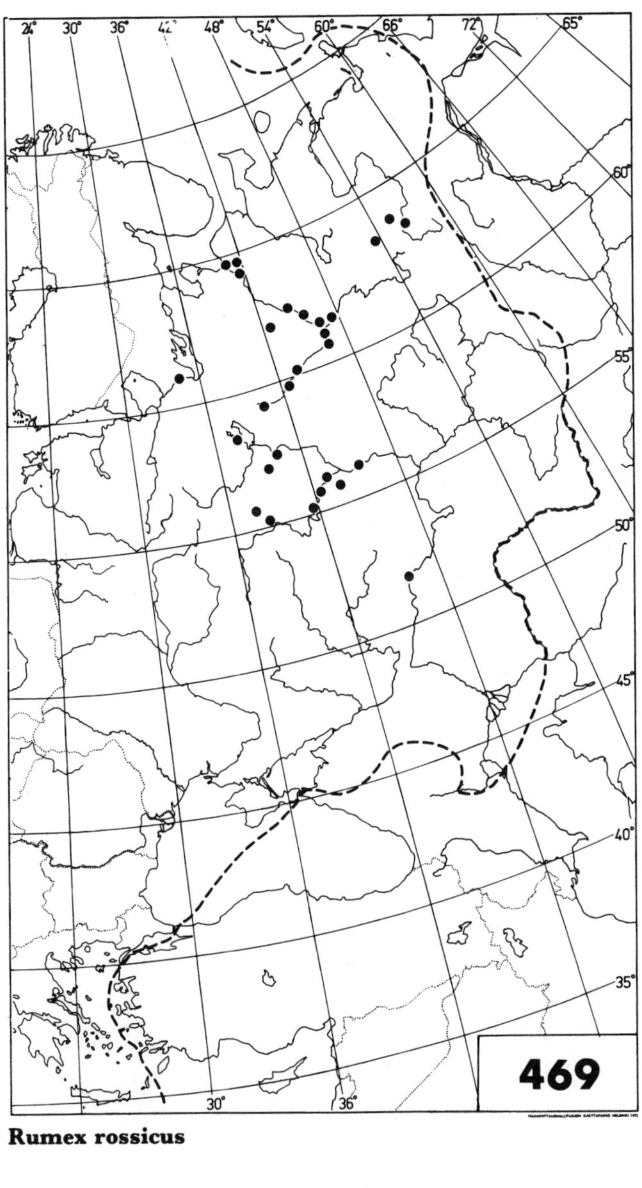

Rumex rossicus

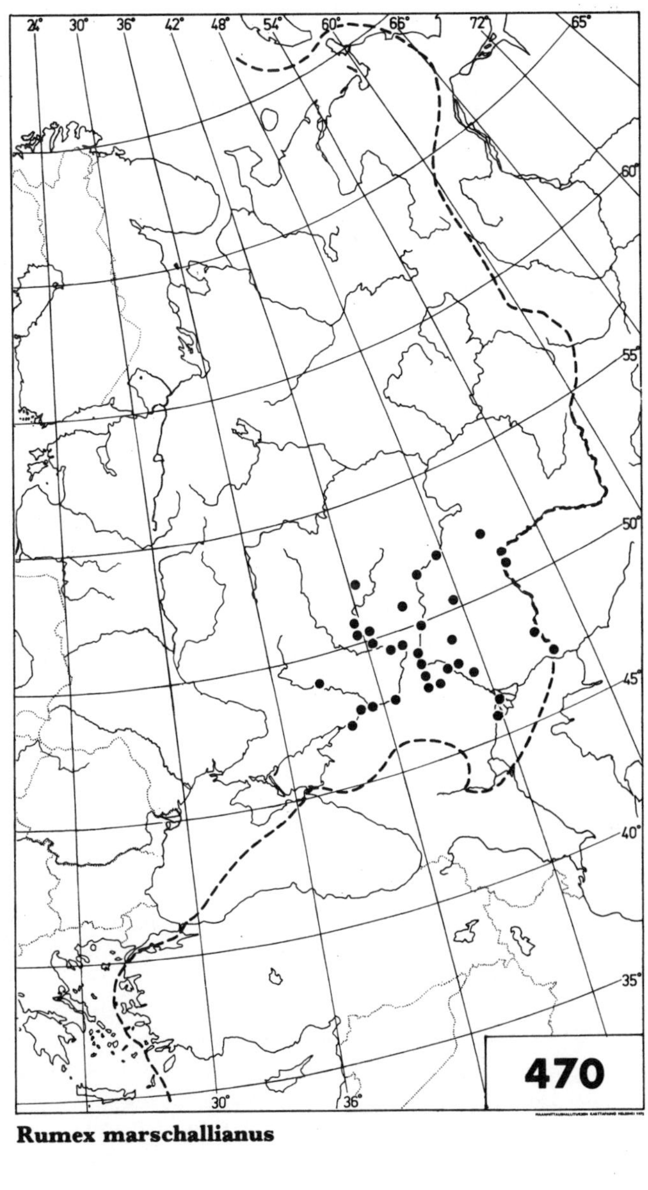

Rumex marschallianus

Rumex ucranicus Besser ex Sprengel — Map 471.

Notes. Rm and Rs(B) omitted (given in Fl. Eur.). [Rs(N)] added (not given in Fl. Eur.): Fl. Bor.-or. Eur. URSS 2: 175, map 251 (1976).

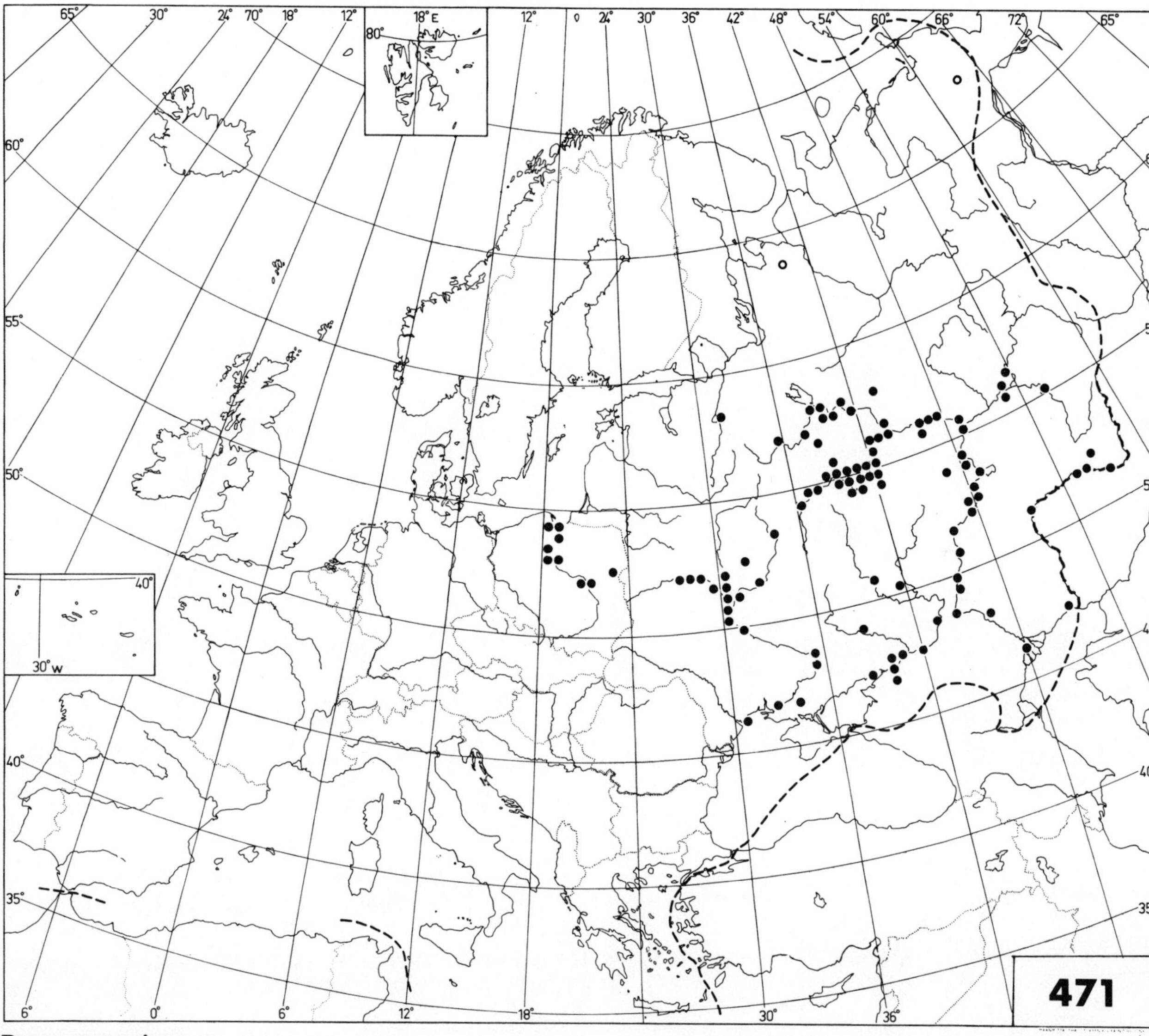

Rumex ucranicus

Rumex bucephalophorus L. — Map 472.

Bucephalophora aculeata (L.) Pau

Taxonomy. J. Cullen in Fl. Turkey 1967: 293.

Nomenclature. Á. Löve & B. M. Kapoor, Taxon 16: 521—522 (1967).

Notes. The data available do not allow separate mapping of the subspecies except for subsp. *aegaeus.* Tu confirmed (?Tu in Fl. Eur.): Fl. Turkey 1967: 292.

R. bucephalophorus subsp. **bucephalophorus**

Bucephalophora aculeata subsp. gallica (Steinh.) Á. Löve & Kapoor; Rumex bucephalophorus subsp. gallicus (Steinh.) Rech. fil.

Notes. »Throughout the range of the species» (Fl. Eur.).

Total range. K. H. Rechinger, Bot. Not. 1939: 497 (1939).

R. bucephalophorus subsp. **hispanicus** (Steinh.) Rech. fil.

Bucephalophora aculeata subsp. hispanica (Steinh.) Á. Löve & Kapoor

Notes. »S. W. Europe» (Fl. Eur.).

Total range. Apparently recorded only for Europe, although it is doubtful whether it can be differentiated from *Rumex bucephalophorus* subsp. *canariensis,* of Islas Canarias and Madeira. K. H. Rechinger, Bot. Not. 1939: 501 (1939).

R. bucephalophorus subsp. **graecus** (Steinh.) Rech. fil.

Notes. »Mediterranean region, from Spain to Greece» (Fl. Eur.).

Total range. K. H. Rechinger, Bot. Not. 1939: 493 (1939).

R. bucephalophorus subsp. **aegaeus** Rech. fil. — Map 473.

Bucephalophora aculeata subsp. aegaea (Rech. fil.) Á. Löve & Kapoor

Notes. Reported from Hs by E. F. Galiano & S. Silvestre, Lagascalia 5: 90 (1975), and from Lu by A. R. Pinto da Silva & A. N. Teles, Mem. Soc. Brot. 21: 136 (1970—71). These Iberian records are not mapped here. »Aegean region» (Fl. Eur.).

Total range. K. H. Rechinger, Bot. Not 1939: 497 (1939).

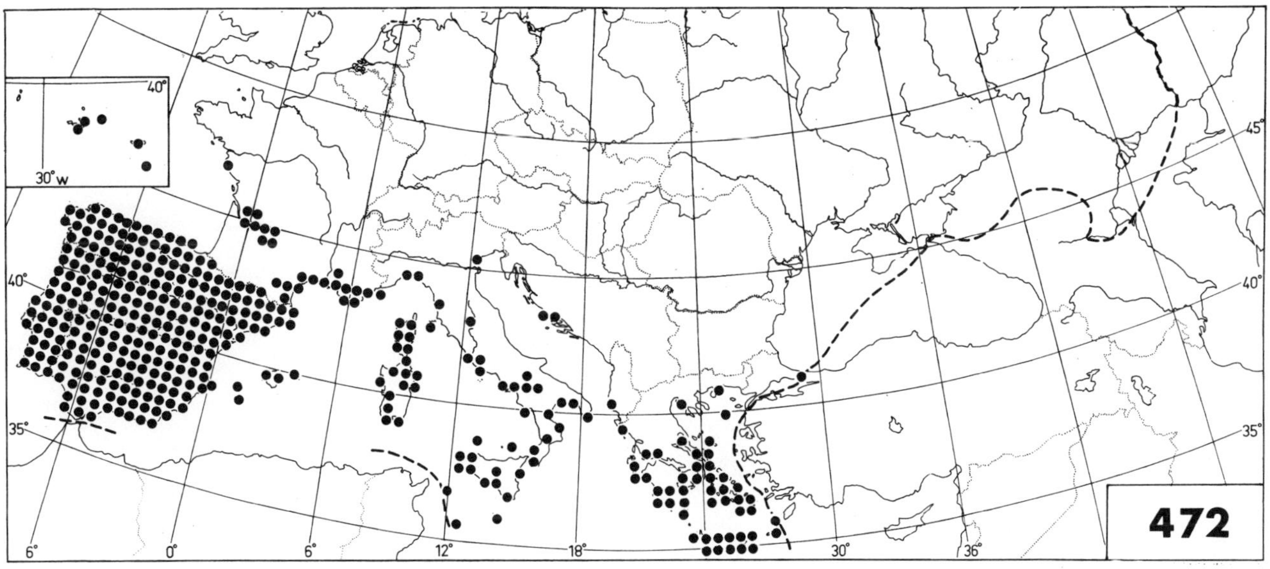

Rumex bucephalophorus

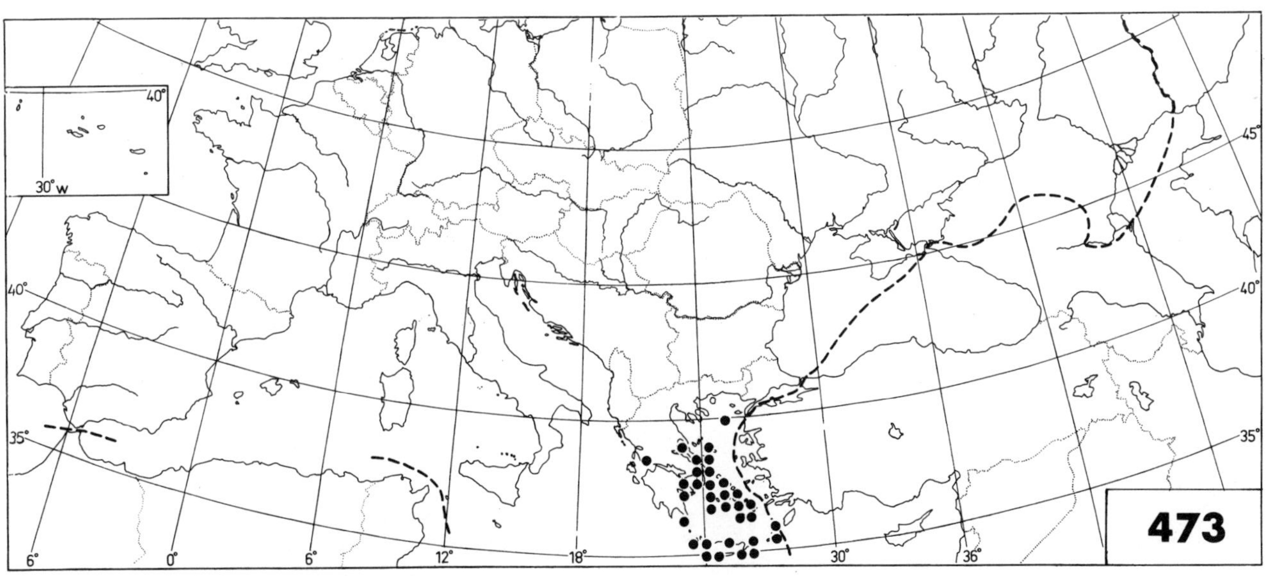

Rumex bucephalophorus subsp. **aegeus**

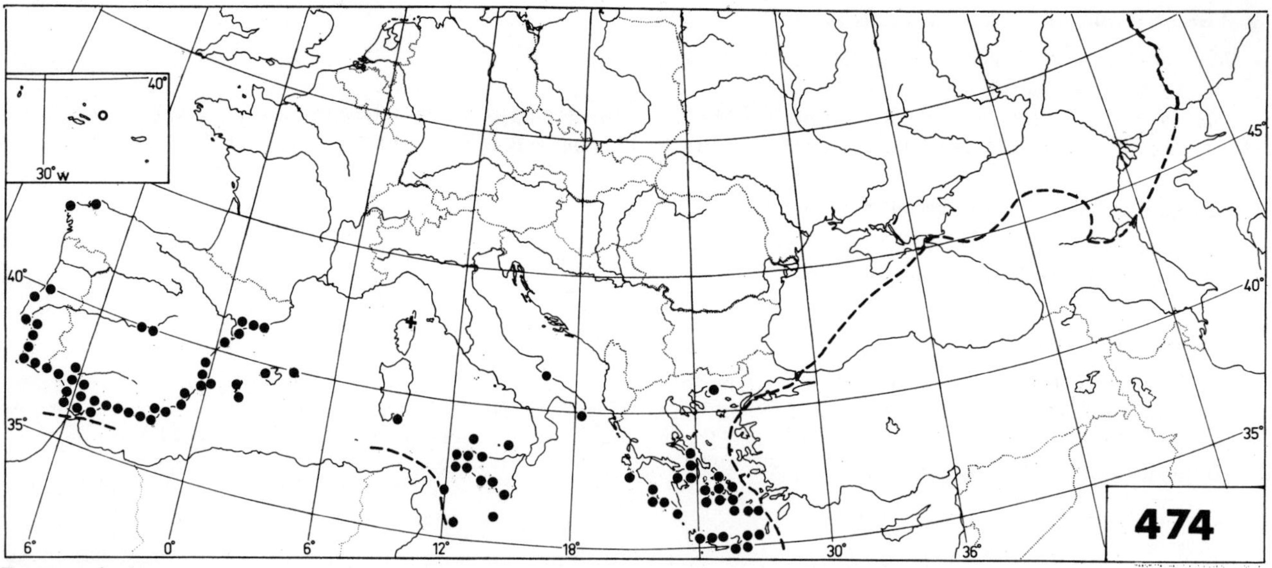

Emex spinosa

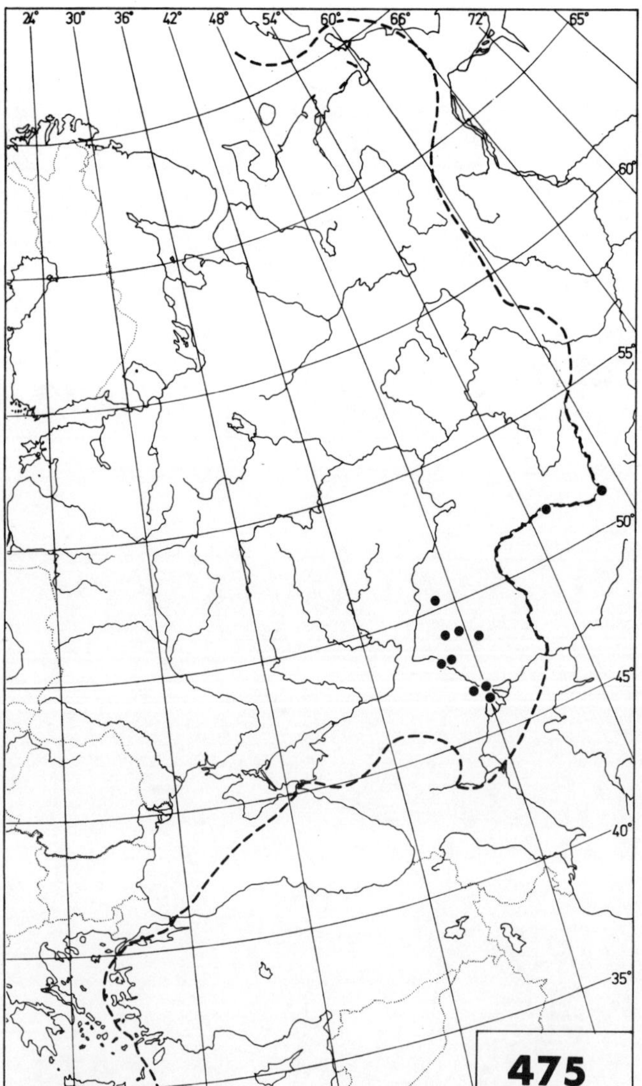

Calligonum aphyllum

Emex spinosa (L.) Campd. — Map 474.

Rumex spinosus L.

Notes. Introduced in Az (given as native in Fl. Eur.), extinct in Co (given as present in Fl. Eur.).

Muehlenbeckia sagittifolia (Ortega) Meissner

Notes. Sometimes found as garden escape, but not truly naturalized ([Az, Lu] in Fl. Eur.). Not mapped.
Native of temperate South America.

Muehlenbeckia complexa Meissner

Notes. [Az], Island of São Miguel, »subspontaneous in a wood»: A. Hansen, Bol. Soc. Brot. 46: 225 (1972); as a garden escape also on the Island of Picos (the species not mentioned in Fl. Eur.). Not mapped.
Native of New Zealand.

Calligonum aphyllum (Pallas) Gürke — Map 475.

Notes. Rs(C) added (not given in Fl. Eur.).

Atraphaxis spinosa L. — Map 476.

Atraphaxis replicata Lam. — Map 477.

Atraphaxis spinosa var. rotundifolia Boiss.

Notes. Rs(W) not confirmed (?Rs(W) in Fl. Eur.).

Atraphaxis frutescens (L.) C. Koch — Map 478.

Atraphaxis lanceolata Meissner

Notes. Rs(C) added (not given in Fl. Eur.), Rs(W) omitted (given in Fl. Eur.).

Atraphaxis billardieri Jaub. & Spach — Map 478.

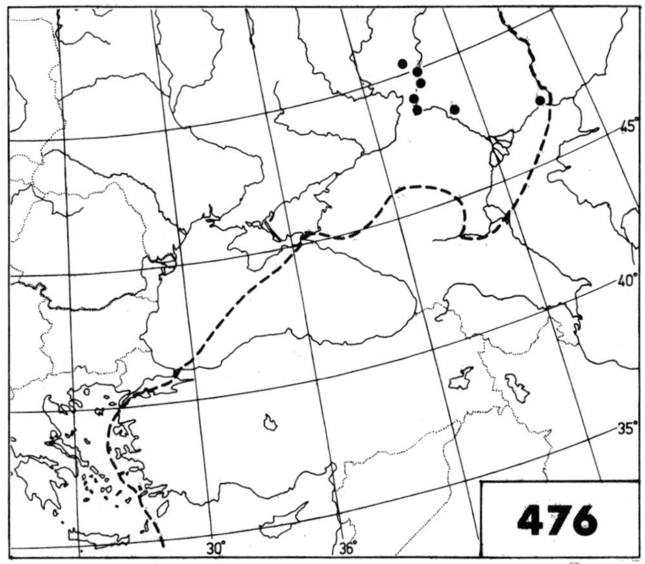

Atraphaxis spinosa

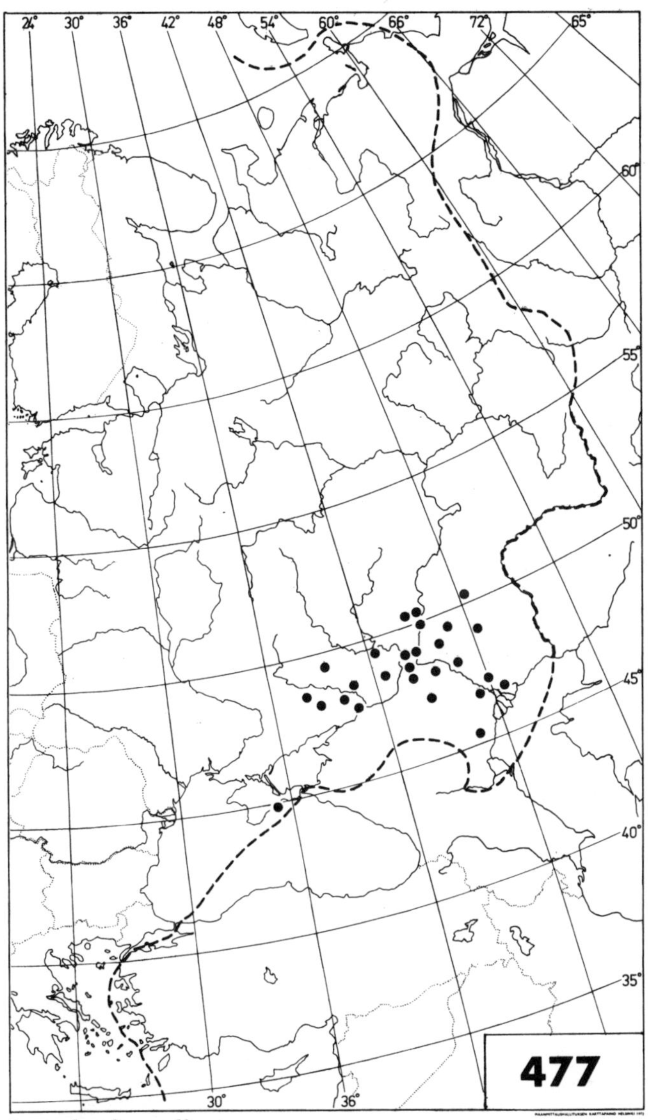

Atraphaxis replicata

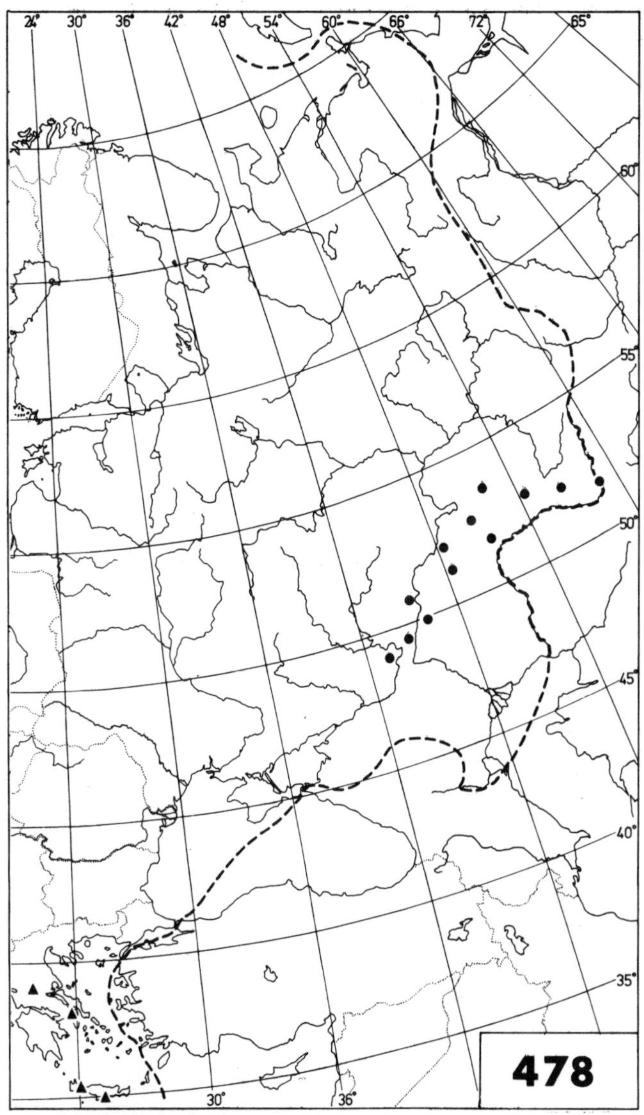

● = **Atraphaxis frutescens**
▲ = **A. billardieri**

5
CHENOPODIACEAE TO BASELLACEAE

CONTENTS

THE COMMITTEE FOR MAPPING THE FLORA OF EUROPE

SECRETARIAT, HELSINKI

J. Jalas (Chairman of the Committee)
J. Suominen (Secretary General)

ADVISERS

T. W. Böcher, København
A. R. Clapham, Arkholme
P. Critopoulos, Athínai
H. Ellenberg, Göttingen
E. Hultén, Stockholm
H. Meusel, Halle
G. Moggi, Firenze
A. Takhtajan, Leningrad
D. A. Webb, Dublin

COMMITTEE MEMBERS
ACTING AS REGIONAL COLLABORATORS

Albania (Al)
I. Mitrushi, Tiranë

Austria (Au)
H. Niklfeld, Wien
Assisted by collaborators in Mapping the Flora of Central Europe

Belgium (Be, excl. Luxembourg)
E. van Rompaey (†), Antwerpen

British Isles (Br, Hb, Channel Islands)
F. H. Perring, Nettleham
D. A. Webb, Dublin (Hb)

Bulgaria (Bu)
S. Kožuharov, Sofia
A. V. Petrova, Sofia

Czechoslovakia (Cz)
J. Futák, Bratislava
J. Holub, Průhonice
Assisted by:
J. Houfek, Průhonice
F. Mladý, Průhonice
V. Skalický, Praha

Denmark (Da, Fa)
A. Hansen, København

Finland (Fe)
J. Suominen, Helsinki
Assisted by:
T. Ulvinen, Oulu
P. Uotila, Helsinki

France (Co, Ga)
P. Dupont, Nantes
Maps checked by several French botanists

German Democratic Republic (Ge: DDR)
H. Meusel, Halle
A. Buhl, Halle
E. Weinert, Halle
Assisted by:
W. Fischer, Potsdam
F. Fukarek, Greifswald
W. Hempel, Dresden
H. D. Knapp, Halle
S. Rauschert, Halle

Federal Republic of Germany (Ge: BRD)
H. Ellenberg, Göttingen
H. Haeupler, Göttingen
J. Krach, München
K. Lewejohann, Göttingen
P. Schönfelder, Regensburg
Assisted by ca. 600 collaborators in Mapping the Flora of Central Europe

Greece (Cr, Gr)
W. Greuter, Berlin
D. Phitos, Patras
Assisted by:
B. Aldén, Göteborg
M. Gustafsson, Lund
L.-Å. Gustavsson, Lund
J. Persson, Göteborg
K.-H. Rechinger, Wien

Hungary (Hu)
R. v. Soó (†), Budapest
A. Terpó, Budapest

Iceland (Is)
E. Einarsson, Reykjavík
Assisted by:
H. Björnsson, Öraefi
I. Davídsson, Reykjavík
B. Jóhannsson, Reykjavík
H. Kristinsson, Akureyri
S. Steindórsson, Akureyri

Italy (It, Sa, Si)
G. Moggi, Firenze
E. Nardi, Firenze
M. Raffaelli, Firenze

Jugoslavia (Ju)
P. Fukarek, Sarajevo
E. Mayer, Ljubljana
Assisted by:
D. Trpin, Ljubljana

Luxembourg (Be, excl. Belgium)
L. Reichling, Luxembourg

Netherlands (Ho)
F. Adema, Leiden
J. Mennema, Leiden

Norway (No, Sb)
R. Y. Berg, Oslo
K. Faegri, Bergen
Assisted by:
R. Elven, Tromsø
E. Fremstad, Bergen
O. I. Rønning, Trondheim (Sb)
K.-E. Sibblund, Oslo

Poland (Po)
J. Kornaś, Kraków
Assisted by:
M. Ciaciura, Wrocław
D. Fijałkowski, Lublin
A. Frey, Kraków
J. Jasnowska, Szczecin
K. Kępczyński, Toruń
M. Kopij, Warszawa
T. Krzaczek, Lublin
J. Mądalski, Wrocław
R. Olaczek, Łódź
L. Olesiński, Olsztyn
A. Pacyna, Kraków
Z. Schwarz, Gdańsk
A. Sokołowski, Białowieża
R. Sowa, Łódź
H. Trzcińska-Tacik, Kraków
A. Zając, Kraków
E. U. Zając, Kraków
W. Żukowski, Poznań

Portugal (Az, Lu)
J. do Amaral Franco, Lisboa
Assisted by:
M. L. da Rocha-Afonso, Lisboa

Romania (Rm)
N. Boşcaiu, Cluj

Spain (Bl, Hs)
E. F. -Galiano, Madrid
R. Folch i Guillèn, Barcelona (Bl, València and Catalonia)
P. Montserrat, Jaca
Assisted by:
B. Cabezudo, Sevilla
B. Valdés, Sevilla

Sweden (Su)
B. Jonsell, Uppsala
Assisted by:
M. Thulin, Uppsala

Switzerland (He)
O. Hegg, Bern
M. Welten, Bern
Assisted by:
R. Sutter, Bern

Turkey (European part; Tu)
H. Demiriz, Istanbul

U.S.S.R. (Rs(N,B,C,W,K,E))
A. I. Tolmachev (†), Leningrad
V. I. Czopik, Kiev
V. N. Tikhomirov, Moskva
Assisted by:
J. E. Alekseev, Moskva
A. Borodina, Leningrad
A. V. Cziczev, Moskva
I. J. Fatare, Riga
Ju. D. Gusev, Leningrad
G. B. Klavina, Riga
N. V. Kozlovskaya, Minsk
L. Laasimer, Tartu
A. N. Laschenkova, Syktyvkar
A. Lekavičius, Vilnius
N. B. Oktyabreva, Moskva
T. Plieva, Leningrad
N. K. Shvedchikova, Moskva
D. I. Tretjakov, Minsk
Z. G. Ulle, Syktyvkar
A. M. Verenko, Kiev
J. A. Vojtjuk, Kiev

PREFACE

Since Volume 4 of Atlas Florae Europaeae was published, the Committee for Mapping the Flora of Europe has suffered the loss of two of its active members and regional collaborators. Professor A.I. Tolmachev died in Leningrad on November 16, 1979, and professor Rezsö von Soó, Academician, died in Budapest on February 10, 1980. They were both born in 1903, and were internationally well-known and highly esteemed specialists in the fields of geobotany and vascular plant taxonomy.

The contents of Volume 5 differ from those of all the earlier volumes in the high proportion of halophytes having a clear concentration in south-eastern Europe (*Chenopodiaceae*) and whose taxonomy and/or area of distribution are apparently in part inadequately known. Temporary difficulties in obtaining data from parts of southern Russia are reflected in a certain underrepresentation of the pertinent taxa in those areas on the maps.

Another feature characterizing the present volume is the inclusion of a high number of non-European species (especially representatives of *Amaranthaceae* and *Aizoaceae*) and also of native European species with extensive secondary areas in Europe (especially many *Chenopodiaceae*). In several countries, especially in the south and east, too little attention has been paid to the occurrence and status of aliens, and so opinions often differ radically from each other, even between neighbouring countries. Differences of this kind often become obvious in the first map drafts, and it has been the special endeavour of the Secretariat to try to level out at least the more disturbing discrepancies and to find a solution resulting in a reasonably uniform and geobotanically plausible total distribution map in Europe. Many of the special problems concerning the status of the aliens have been solved in correspondence between the Secretariat and the individual regional collaborators. Sometimes, however, when marked disharmony between data from neighbouring countries could not be smoothed out by negotiations, the final decision was made by the Secretariat on the basis of all the evidence available.

Of course, all these decisions are open to discussion, and the Secretariat will cordially welcome constructive criticism and suggestions.

DEVIATIONS FROM FLORA EUROPAEA

Included, although not given in Fl. Eur.

1. Native species
   - *Beta adanensis* Pamukç. ex Aellen
   - *Chenopodium exsuccum* (C. Loscos) Uotila
   - *Atriplex recurva* D'Urv.
   - *Salicornia veneta* Pign. & Lausi
   - *Salsola melitensis* Botsch.

2. Non-European species
   - *Chenopodium pumilio* R.Br.
   - *Obione sibirica* (L.) Fischer
   - *Amaranthus spinosus* L.
   - *A. viridis* L.
   - *A. tuberculatus* (Moq.) Sauer (not mapped)
   - *A. rudis* Sauer (not mapped)
   - *Galenia secunda* Sonder
   - *Carpobrotus glaucescens* (Haw.) Schwantes
   - *Drosanthemum floribundum* (Haw.) Schwantes
   - *Oscularia deltoides* (Miller) Schwantes (not mapped)
   - *Ruschia caroli* (L. Bolus) Schwantes (not mapped)
   - *Lampranthus falciformis* (Haw.) N.E. Br. (not mapped)
   - *L. multiradiatus* (Jacq.) N.E. Br.
   - *L. roseus* (Willd.) Schwantes
   - *Erepsia heteropetala* (Haw.) Schwantes
   - *Talinum paniculatum* (Jacq.) Gaertn. (not mapped)

No native or established European records available, although given in Fl. Eur.

1. Native species, according to Fl. Eur.
   Halopeplis pygmaea (Pallas) Bunge ex Ung.-Sternb.

2. Non-European species
   Chenopodium capitatum (L.) Ascherson
   Amaranthus standleyanus Parodi ex Covas
   Mirabilis jalapa L.
   M. longiflora L.
   Phytolaca dioica L.
   Sesuvium portulacastrum L.
   Carpobrotus chilensis (Molina) N.E. Br.
   Lampranthus glaucus (L.) N.E. Br.
   Tetragonia tetragonoides (Pallas) O. Kuntze

Taxonomical deviations from Fl. Eur.

Chenopodium strictum Roth (instead of *C. album* subsp. *striatum* (Krašan) J. Murr) (not mapped)
Atriplex longipes subsp. *praecox* (Hülphers) Turesson (instead of *A. praecox* Hülphers)
Camphorosma monspeliaca subsp. *lessingii* (Litv.) Aellen (instead of *C. lessingii* Litv.)
Amaranthus hybridus L., s. lato, taxonomical treatment and nomenclature according to J.D. Sauer, Ann. Missouri Bot. Garden 54: 103—137 (1967):
A. powellii S. Watson (instead of *A. hybridus* sensu Fl. Eur. + *A. bouchonii* Thell.)
A. patulus Bertol. (instead of *A. cruentus* sensu Fl. Eur.)
A. hypochondriacus L. (instead of *A. hybridus* sensu Fl. Eur., pro p.) (not mapped)
A. paniculatus L. (instead of *A. cruentus* L., s. str.) (not mapped)

Nomenclatural deviations from Fl. Eur.

Atriplex acuminata Waldst. & Kit. (instead of *A. nitens* Schkuhr)
A. prostrata Boucher ex DC. (instead of *A. hastata* L.)
A. prostrata subsp. *polonica* (Zapal.) Uotila (instead of *A. hastata* subsp. *polonica* (Zapal.) Aellen) (not mapped)
Ceratoides latens (J.F. Gmelin) Reveal & Holmgren (instead of *Krascheninnikovia ceratoides* (L.) Gueldenst.)
Arthrocnemum macrostachyum (Moric.) Moris & Delponte (instead of *A. glaucum* (Delile) Ung.-Sternb.)
Anabasis salsa (C.A. Meyer) Bentham & Volkens (corrected author's citation)
Hammada scoparia (Pomel) Iljin (instead of *Haloxylon articulatum* (Moq.) Bunge)
Alternanthera caracasana Humb., Bonpl. & Kunth (instead of *A. peploides* (Humb. & Bonpl.) Urban)
Claytonia perfoliata Donn ex Willd. (instead of *Montia perfoliata* (Donn ex Willd.) Howell)
C. sibirica L. (instead of *Montia sibirica* (L.) Howell)

ABBREVIATIONS AND SYMBOLS

Country and 'territory' abbreviations (in full accordance with Fl. Eur.)

Al Albania
Au Austria, with Liechtenstein
Az Açores (Azores)
Be Belgium, with Luxembourg
Bl Islas Baleares
Br Britain, excluding the Channel Islands and Northern Ireland
Bu Bulgaria
Co Corse
Cr Kriti (*Creta*), with Karpathos, Kasos and Gavdhos
Cz Czechoslovakia
Da Denmark (*Dania*)
Fa Færöer
Fe Finland (*Fennia*)
Ga France (*Gallia*), with the Channel Islands and Monaco; excluding Corse
Ge Germany (both the Federal Republic of Germany and the German Democratic Republic)
Gr Greece, excluding those islands included under Kriti and those which are outside Europe as defined for Fl. Eur.
Hb Ireland (*Hibernia*); both the republic and Northern Ireland
He Switzerland (*Helvetia*)
Ho Netherlands (*Hollandia*)
Hs Spain (*Hispania*), with Gibraltar and Andorra; excluding Islas Baleares
Hu Hungary
Is Iceland (*Islandia*)
It Italy, excluding Sardegna and Sicilia as defined below
Ju Jugoslavia
Lu Portugal (*Lusitania*), excluding Açores
No Norway, excluding Svalbard as defined below
Po Poland
Rm Romania
Rs U.S.S.R. (*Rossia*) (European part)
   Rs(N) *Northern division*: Arctic Europe, Karelo-Lapland, Dvina-Pečora (including the whole of Karelian A.S.S.R.)
   Rs(B) *Baltic division*: Estonia, Latvia, Lithuania, Kaliningradskaya oblast'
   Rs(C) *Central division*: Ladoga-Ilmen, Upper Volga, Volga-Kama, Upper Dnepr, Volga-Don, Ural (including the whole of White Russia, and considered to include the entire Leningradskaya oblast')

Rs(W) *South-western division*: Moldavia, Middle Dnepr, Black Sea, Upper Dnestr (largely including Ukraine)
Rs(K) Krym (*Crimea*)
Rs(E) *South-eastern division*: Lower Don, Lower Volga, Transvolga (including the European part of Kazakhstan)
Sa Sardegna
Sb Svalbard, comprising Spitsbergen, Björnöya (Bear Island) and Jan Mayen
Si Sicilia, with Pantelleria, Isole Pelagie, Isole Lipari and Ustica; including Malta
Su Sweden (*Suecia*)
Tu Turkey (European part), including Imroz

The mapping symbols

● ▲ ■	native occurrence (including archaeophytes)
○ △ □	introduction (established alien)
◒	status unknown or uncertain
+	extinct
×	probably extinct, or, at least, not recorded since 1930
?	record uncertain as regards identification or locality.

If the map gives the areas of two or more taxa, the special symbols (+, ×, ?, etc.) belong to the taxon marked by ●, unless otherwise stated.

Abbreviations for frequently quoted literature

Fl. Arct. URSS 1966 = A.I. Tolmachev (ed.), Flora Arctica URSS. 5. Salicaceae — Portulacaceae. — 208 pp. Mosquae — Leninopoli 1966.

Fl. Eur. = T.G. Tutin, V.H. Heywood et al. (eds.), Flora Europaea. 1. Lycopodiaceae to Platanaceae. — xxxiv + 464 pp. Cambridge 1964.

Fl. France 1973 = M. Guinochet & R. de Vilmorin, Flore de France. 1. — 366 pp. Paris 1973.

Fl. Mallorca 1978 = F. Bonafè Barceló, Flora de Mallorca. 2. — 378 pp. Palma de Mallorca 1978.

Fl. Reipubl. Pop. Bulg. 1966 = D. Jordanov & B. Kuzmanov (eds.), Flora Reipublicae Popularis Bulgaricae. 3. — 638 pp. Serdicae 1966.

Fl. Turkey 1967 = P.H. Davis, Flora of Turkey and the East Aegean Islands. 2. — xii + 581 pp. Edinburgh 1967.

Fl. URSR 1952 = M.I. Kotov (ed.), Flora URSR. 4. — 689 pp. Kiev 1952.

Hegi 1959 = K.H. Rechinger (ed.), G. Hegi, Illustrierte Flora von Mitteleuropa, 2. Aufl. 3(2), Lief. 1, Phytolaccaceae — Amaranthaceae: 453—532. München 1959.

Hegi 1960 = K.H. Rechinger (ed.), G. Hegi, Illustrierte Flora von Mitteleuropa, 2. Aufl. 3(2), Lief. 2—3, Chenopodiaceae: 533—692. München 1960.

Hegi 1961 = K.H. Rechinger (ed.), G. Hegi, Illustrierte Flora von Mitteleuropa, 2. Aufl. 3(2), Lief. 4, Chenopodiaceae — Caryophyllaceae: 693—772. München 1961.

Hegi 1979 = K.H. Rechinger (ed.), G. Hegi, Illustrierte Flora von Mitteleuropa, 2. Aufl. 3(2), Lief. 10, Caryophyllaceae — Portulacaceae: 1173—1265. München 1979.

Hultén AA 1958 = E. Hultén, The Amphi-Atlantic Plants. — Kungl. Svenska Vet.-Akad. Handl., Ser. 4, 7(1): 1—340. 1958.

Hultén Alaska 1968 = E. Hultén, Flora of Alaska and neighboring territories. — 1008 pp. Stanford 1968.

Hultén CP 1971 = E. Hultén, The Circumpolar Plants. II. Dicotyledons. — Kungl. Svenska Vet.-Akad. Handl., Ser. 4, 13(1): 1—463. 1971.

MJW 1965 = H. Meusel, E. Jäger & E. Weinert, Vergleichende Chorologie der zentraleuropäischen Flora. — Text 583 pp., Karten 258 pp. Jena 1965.

Nova Fl. Port. 1971 = J. do Amaral Franco, Nova Flora de Portugal (Continente e Açores). 1. Lycopodiaceae — Umbelliferae. — xxiv + 648 pp. Lisboa 1971.

Rubtsev 1972 = N.I. Rubtsev, Opred. Vysshikh Rast. Kryma. — 552 pp. Leningrad 1972.

ACKNOWLEDGEMENTS

The expenditure of the Secretariat has been covered, as before, by annual allowances from the Finnish National Research Council for Sciences. The Finnish Biological Society Vanamo has continued to act as publisher for Atlas Florae Europaeae. As earlier, working facilities for the Secretariat have been provided by the Department of Botany, University of Helsinki. Technical assistance in preparing this volume has been given by Mrs. Liisa Mäkelä, M.A., Mr. Jari Oksanen, M.A., and Mrs. Paula Oksanen, M.A.

DICOTYLEDONES (cont.)

CHENOPODIACEAE

Polycnemum majus A. Braun — Map 479.

Notes. Established alien in Be and Po (given as native in Fl. Eur.). Tu added (not given in Fl. Eur.): Fl. Turkey 1967: 296.

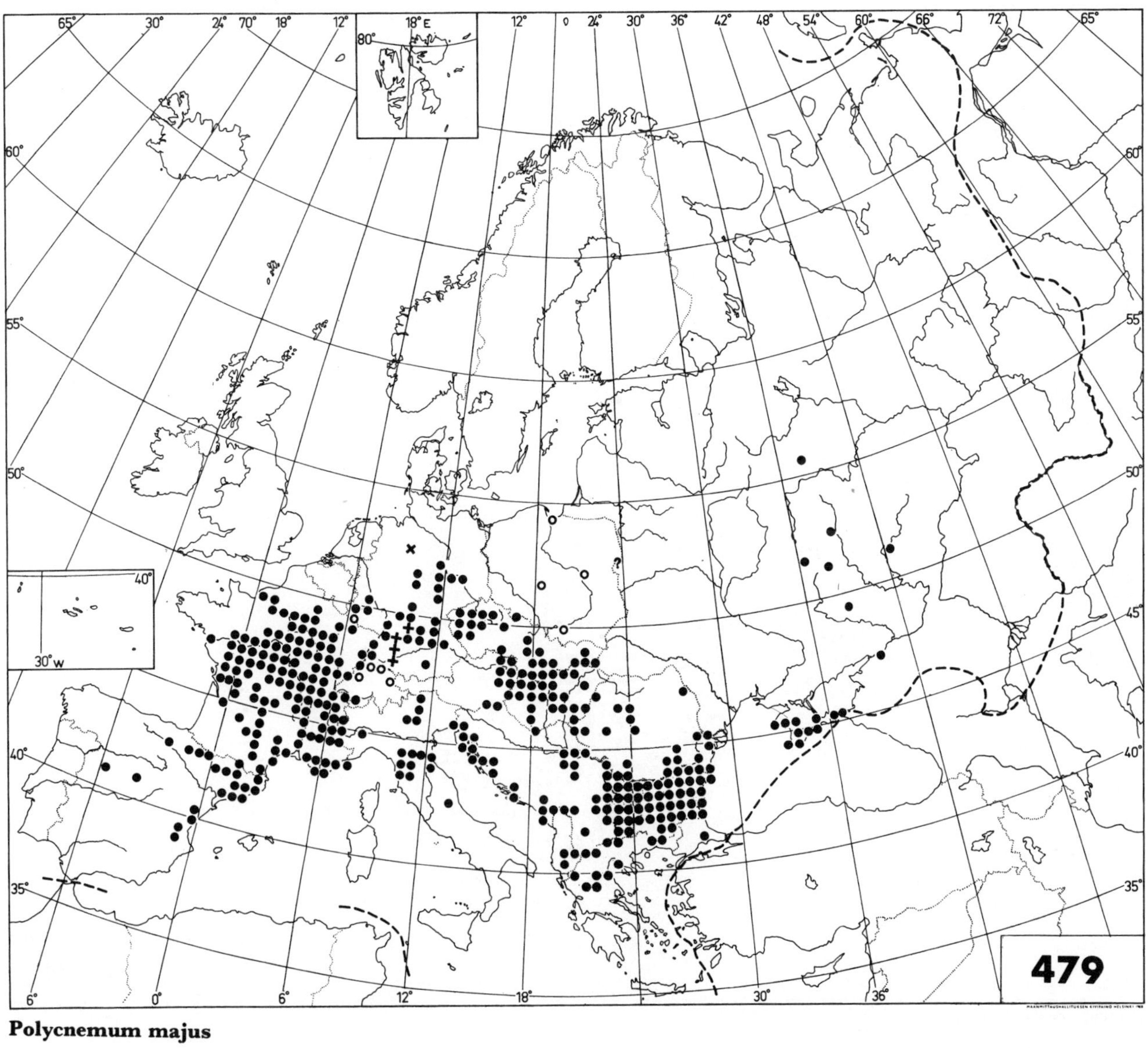

Polycnemum majus

Polycnemum arvense L. — Map 480.

Notes. Al and Lu added (not given in Fl. Eur.): A.R. Pinto da Silva, Agron. Lusit. 30: 196 (1970). Rs(K) and Sa omitted (given in Fl. Eur.); not mentioned for Rs(K) in Rubtsev 1972.

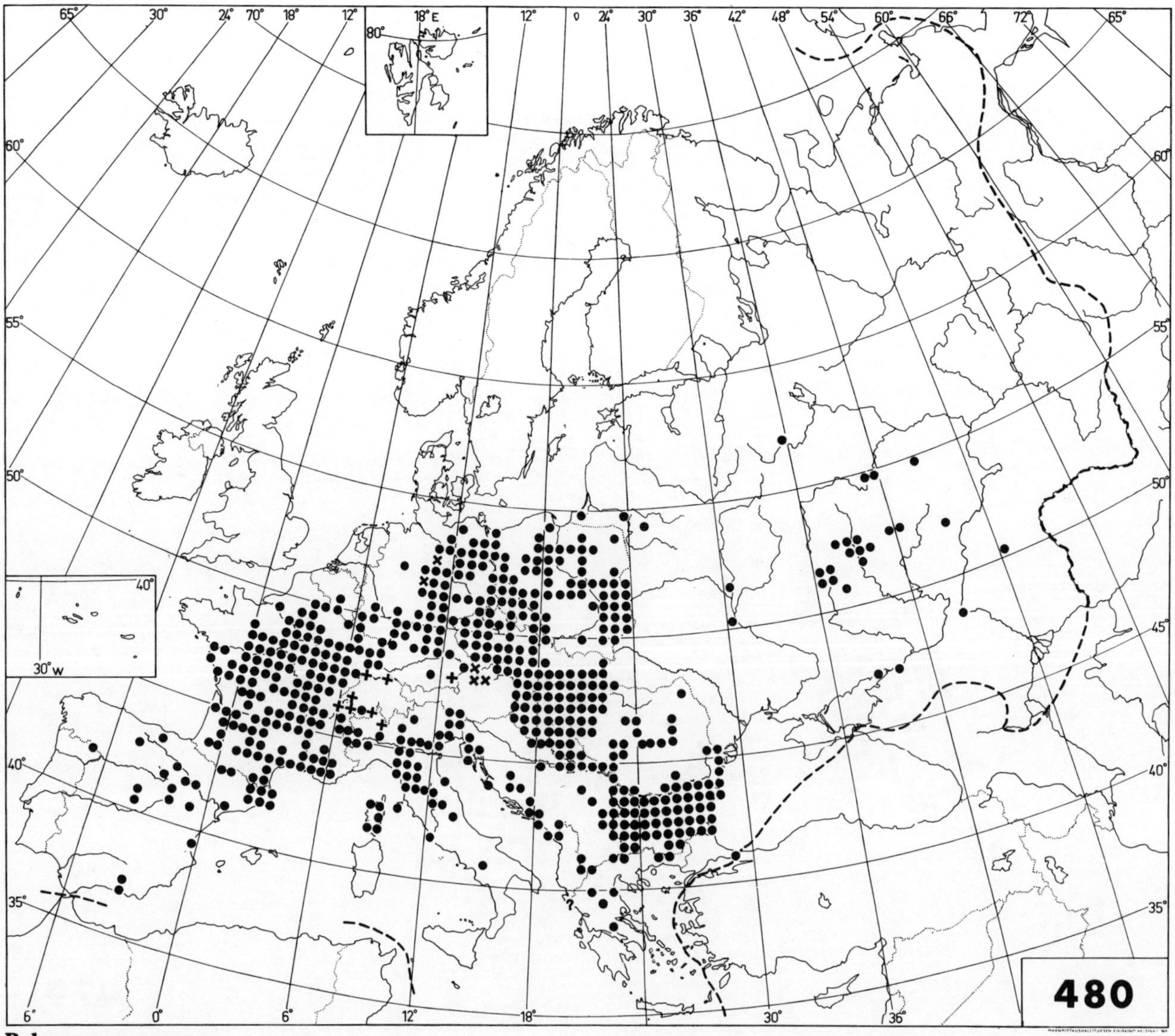

Polycnemum arvense

Polycnemum verrucosum A.F. Láng — Map 481.
Notes. Ju and Tu added (not given in Fl. Eur.): Fl. Turkey 1967: 296.

Polycnemum heuffelii A.F. Láng — Map 482.
Notes. Au added (not given in Fl. Eur.): W. Holzner, Mitt. Bot. Arbeitsgem. Oberösterr. Landesmus. Linz 4: 46 (1973).
Total range. Endemic to Europe.

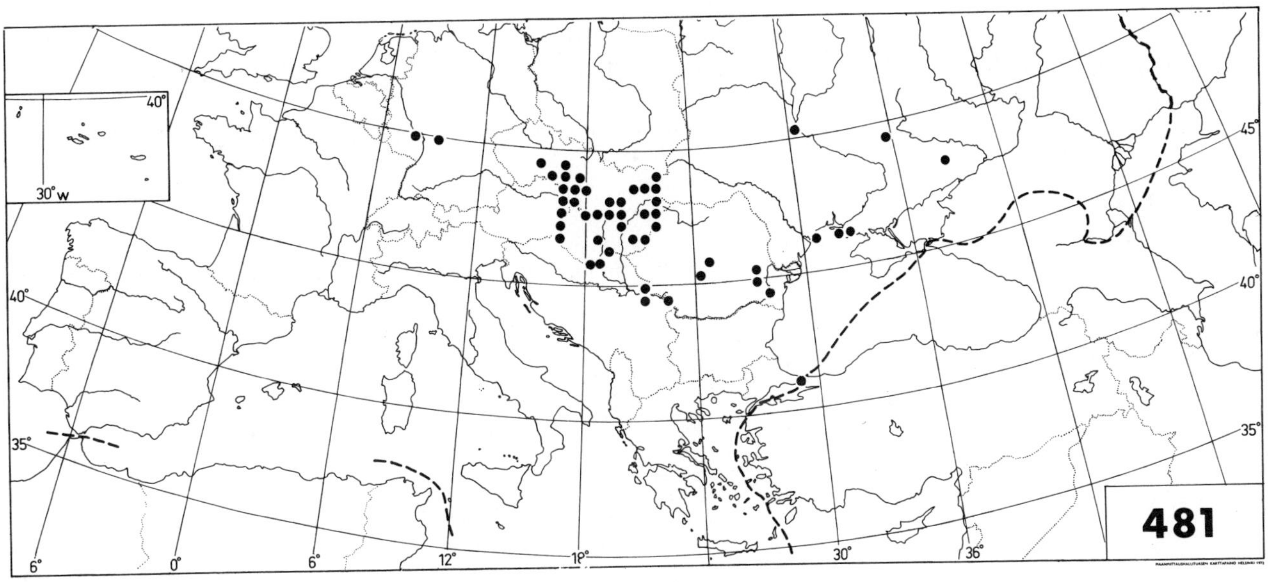

Polycnemum verrucosum

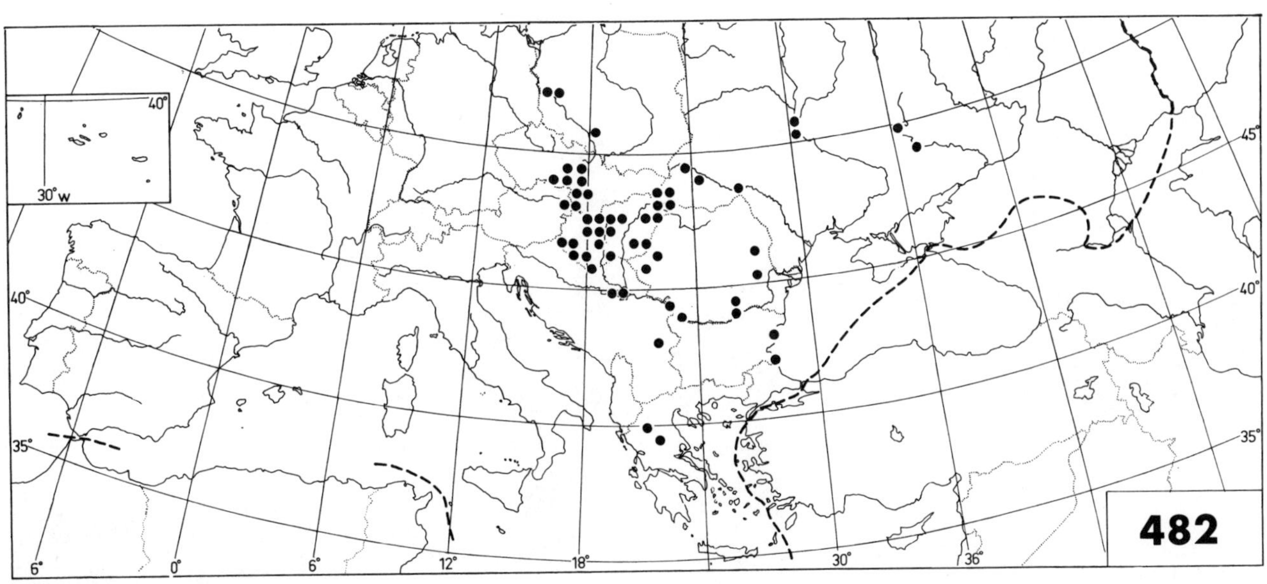

Polycnemum heuffelii

Beta vulgaris L.

Taxonomy. B.V. Ford-Lloyd & J.T. Williams, Bot. Jour. Linn. Soc. 71: 89—102 (1975); K.P. Buttler, Plant Syst. Evol. 128: 123—136 (1977).

Notes. Considered native in Ge (*Ge in Fl. Eur.), Rs(W) not confirmed (?Rs(W) in Fl. Eur.).

Total range. MJW 1965: map 131d.

B. vulgaris subsp. vulgaris

Incl. Beta vulgaris subsp. cicla (L.) Schübler & Martens

Taxonomy. B.V. Ford-Lloyd & J.T. Williams, Bot. Jour. Linn. Soc. 71: 99—100 (1975), recognize the cultivated beets without swollen roots as subsp. *cicla* (L.) Schübler & Martens (»(L.) Koch»), besides subsp. *vulgaris,* which is claimed to include the swollen-rooted beets.

Notes. »Widely cultivated...; often persisting as a casual and occasionally naturalized» (Fl. Eur.). Not mapped.

B. vulgaris subsp. maritima (L.) Arcangeli — Map 483.

? Beta atriplicifolia Rouy; B. maritima L.; B. perennis (L.) Freyn; B. vulgaris subsp. perennis L.

Notes. Al, Az, Be, Bl, Br, Bu, Co, Cr, Da, Ga, Ge, Gr, Hb, Ho, Hs, It, Ju, Lu, Sa, Si, Su, Tu. »Coasts of S. & W. Europe northwards to c. 59° N. in W. Sweden» (Fl. Eur.).

Total range. MJW 1965: map 131d; A.C. Zeven & P.M. Zhukovsky, Dictionary of cultivated plants, p. 92 (Wageningen 1975).

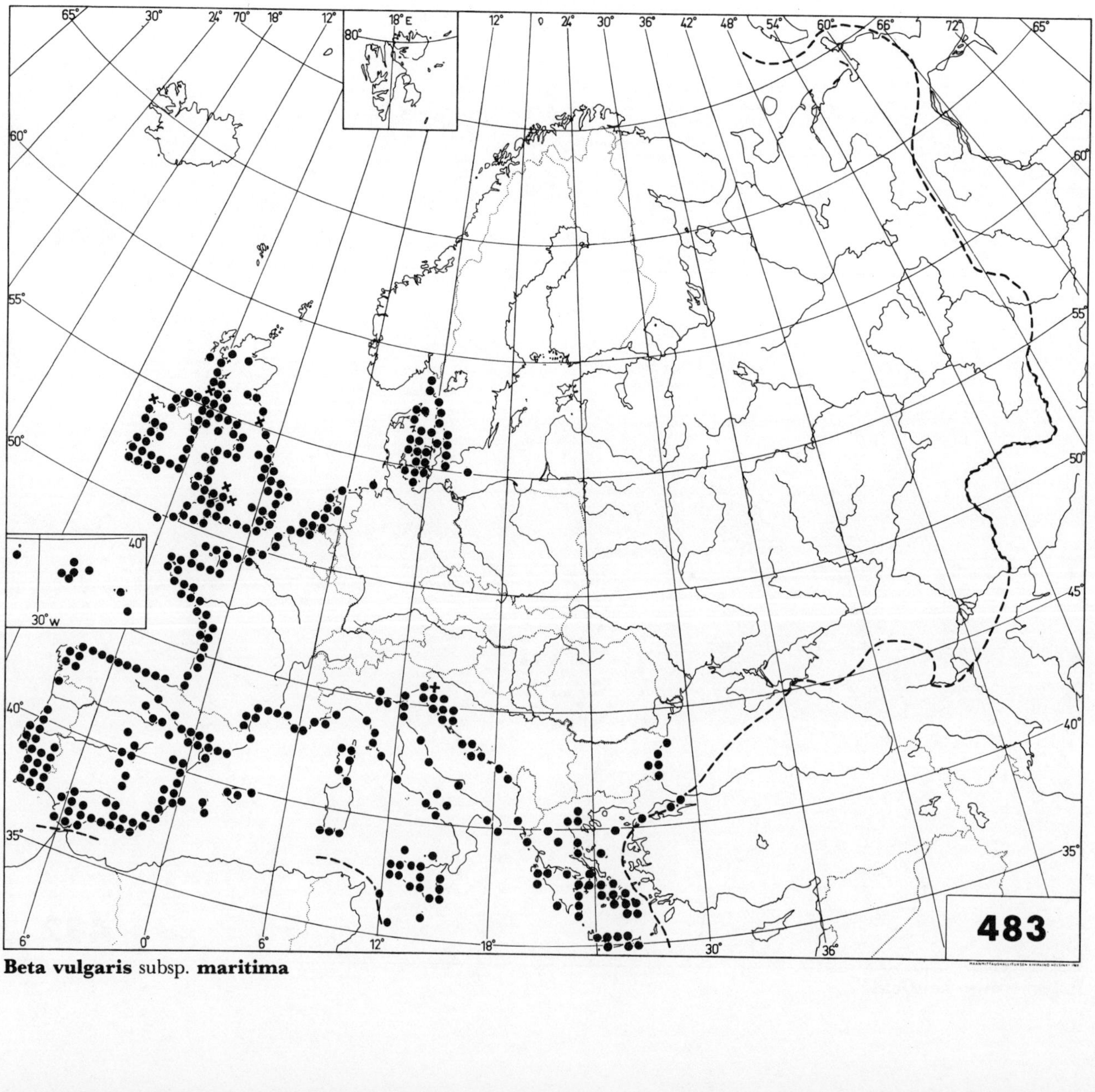

Beta vulgaris subsp. **maritima**

Beta macrocarpa Guss. — Map 484.

Beta bourgaei Cosson; B. vulgaris subsp. macrocarpa (Guss.) Thell.; B. vulgaris var. macrocarpa (Guss.) Moq.

Taxonomy. Included, as a variety, in *Beta vulgaris* subsp. *maritima* by B.V. Ford-Lloyd & J.T. Williams, Bot. Jour. Linn. Soc. 71: 99—100 (1975). The treatment of *B. macrocarpa* as an independent species is preferred by K.P. Buttler, Plant Syst. Evol. 128: 131—132 (1977).

Notes. Bl and Cr added (not given in Fl. Eur.): H. Kuhbier, Veröff. Überseemus. Bremen 5 (Cahiers Optima Leafl. 83): 11 (1978), see also Fl. Mallorca 1978: 92; K.P. Buttler, Plant Syst. Evol. 128: 131 (1977). It omitted (given in Fl. Eur.).

Total range. MJW 1965: map 131d; K.P. Buttler, Plant Syst. Evol. 128: 124 (1977).

Beta adanensis Pamukç. ex Aellen — Map 484.

Beta vulgaris subsp. adanensis (Pamukç. ex Aellen) Ford-Lloyd & Williams

Taxonomy. P. Aellen, Notes Roy. Bot. Garden Edinb. 28: 29 (1967); K.P. Buttler, Plant Syst. Evol. 128: 123—136 (1977).

Notes. Gr. The species not mentioned in Fl. Eur.

Total range. K.P. Buttler, Plant Syst. Evol. 128: 124 (1977).

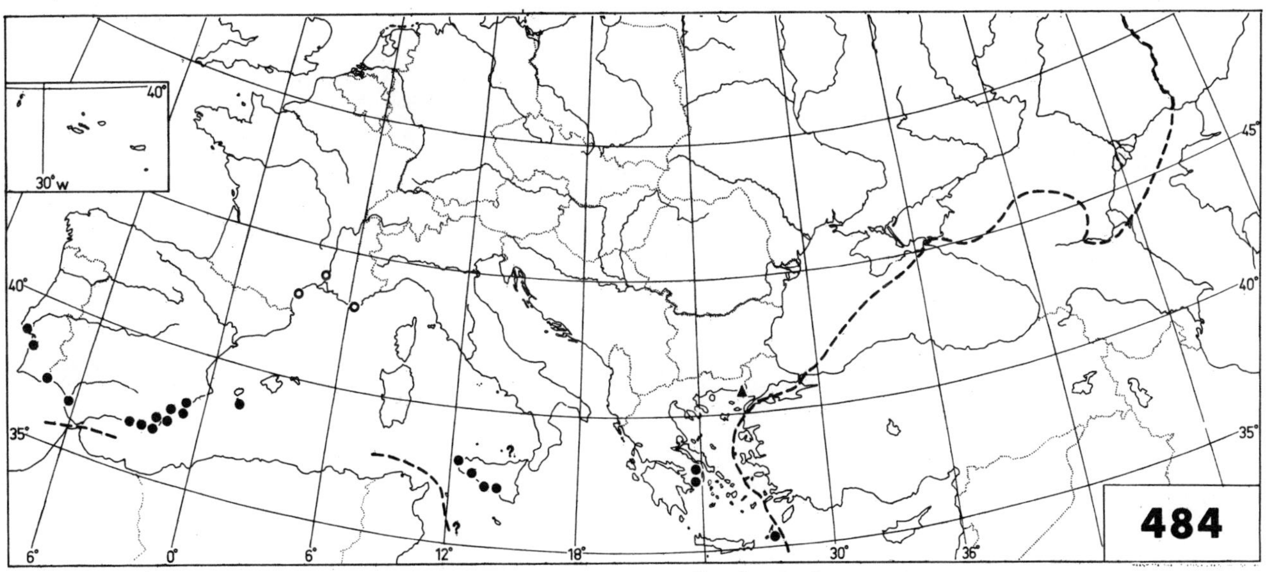

• = **Beta macrocarpa** ▲ = **B. adanensis**

Beta patellaris Moq. — Map 485.

Beta diffusa Cosson; Patellaria cordata Williams, Scott & Ford-Lloyd; Patellifolia patellaris (Moq.) Scott, Ford-Lloyd & Williams

Taxonomy and nomenclature. J.T. Williams, A.J. Scott & B.V. Ford-Lloyd, Feddes Repert. 87: 289—292 (1976); A.J. Scott, B.V. Ford-Lloyd & J.T. Williams, Taxon 26: 284 (1977).

Notes. Bl and Si added (not given in Fl. Eur.): P. Duvigneaud, Soc. Échange Pl. Vasc. Europe Occ. Bassin Médit. 17(suppl.): 15 (1979).

Total range. MJW 1965: 131d; A.C. Zeven & P.M. Zhukovsky, Dictionary of cultivated plants, p. 92 (Wageningen 1975).

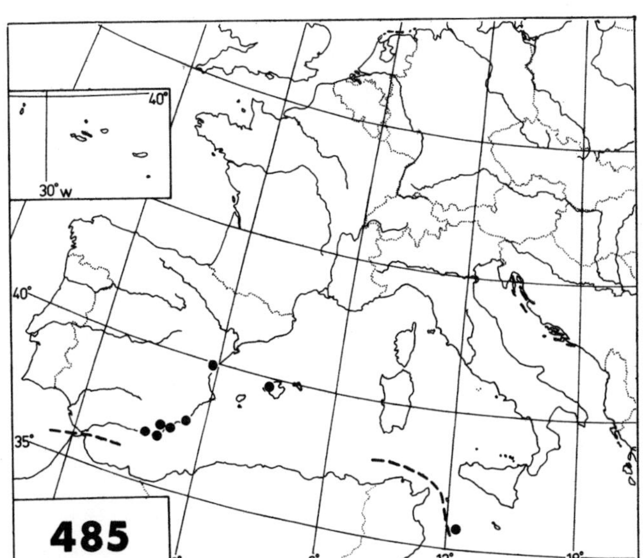

Beta patellaris

Beta trigyna Waldst. & Kit. — Map 486.

Notes. [Br] and *Gr added (not given in Fl. Eur.), [Cz] omitted (given in Fl. Eur.).
Total range. MJW 1965: map 131d; A.C. Zeven & P.M. Zhukovsky, Dictionary of cultivated plants, p. 78 (Wageningen 1975).

Beta nana Boiss. & Heldr. — Map 487.

Notes. See TPC Red Data Sheets Europe (preprint Kew 1977).
Total range. Endemic to Europe.

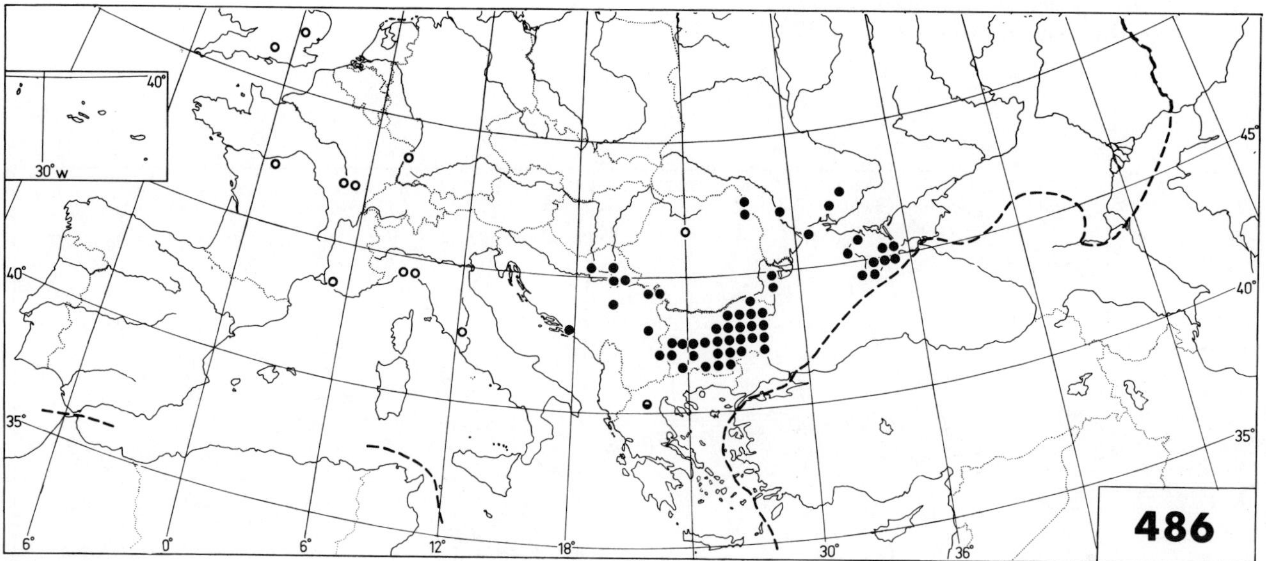

Beta trigyna

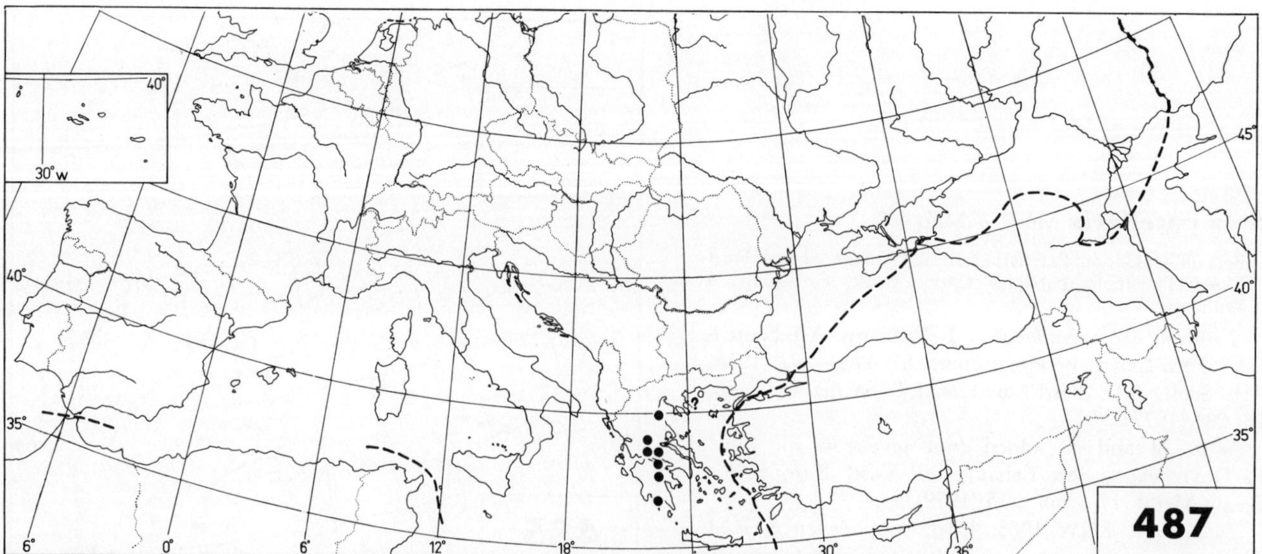

Beta nana

Chenopodium botrys L. — Map 488.

Notes. Established alien in Au, Ge, Po and Rs(C) and doubtful native in He (given as native in Fl. Eur.), [Be], Gr and [Rs(B)] added (not given in Fl. Eur.), Co confirmed (?Co in Fl. Eur.). Map incomplete for Rs(W), and no exact data available from Rs(E) although certainly present there (and given in Fl. Eur.).

Total range. Hultén CP 1971: map 189 (holarctic part); H. Sukopp, Verh. Bot. Ver. Brandenb. 108: 4 (1971).

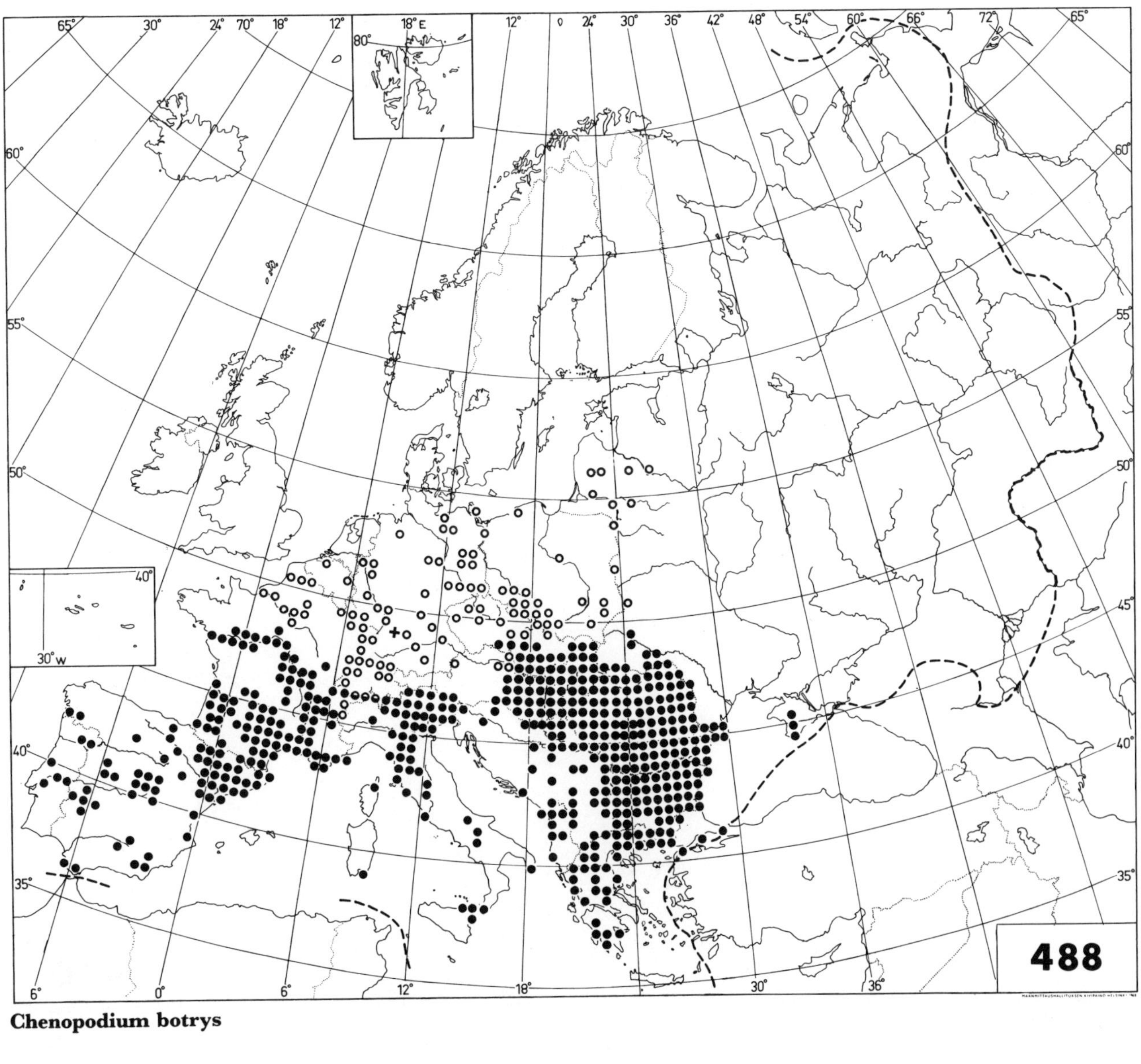

Chenopodium botrys

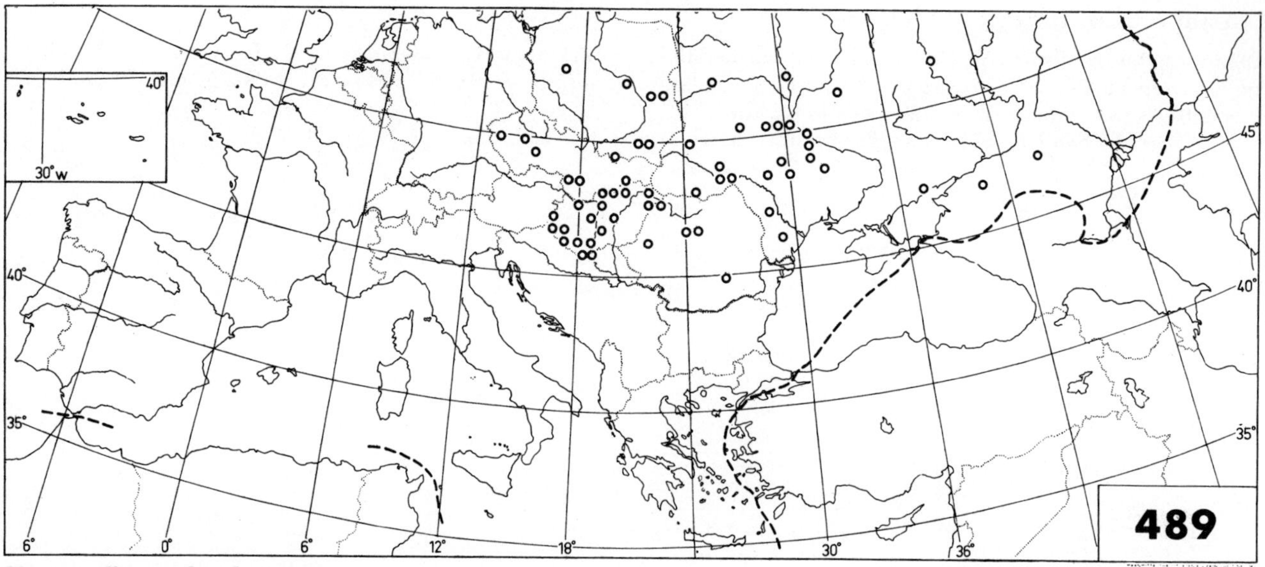

Chenopodium schraderanum

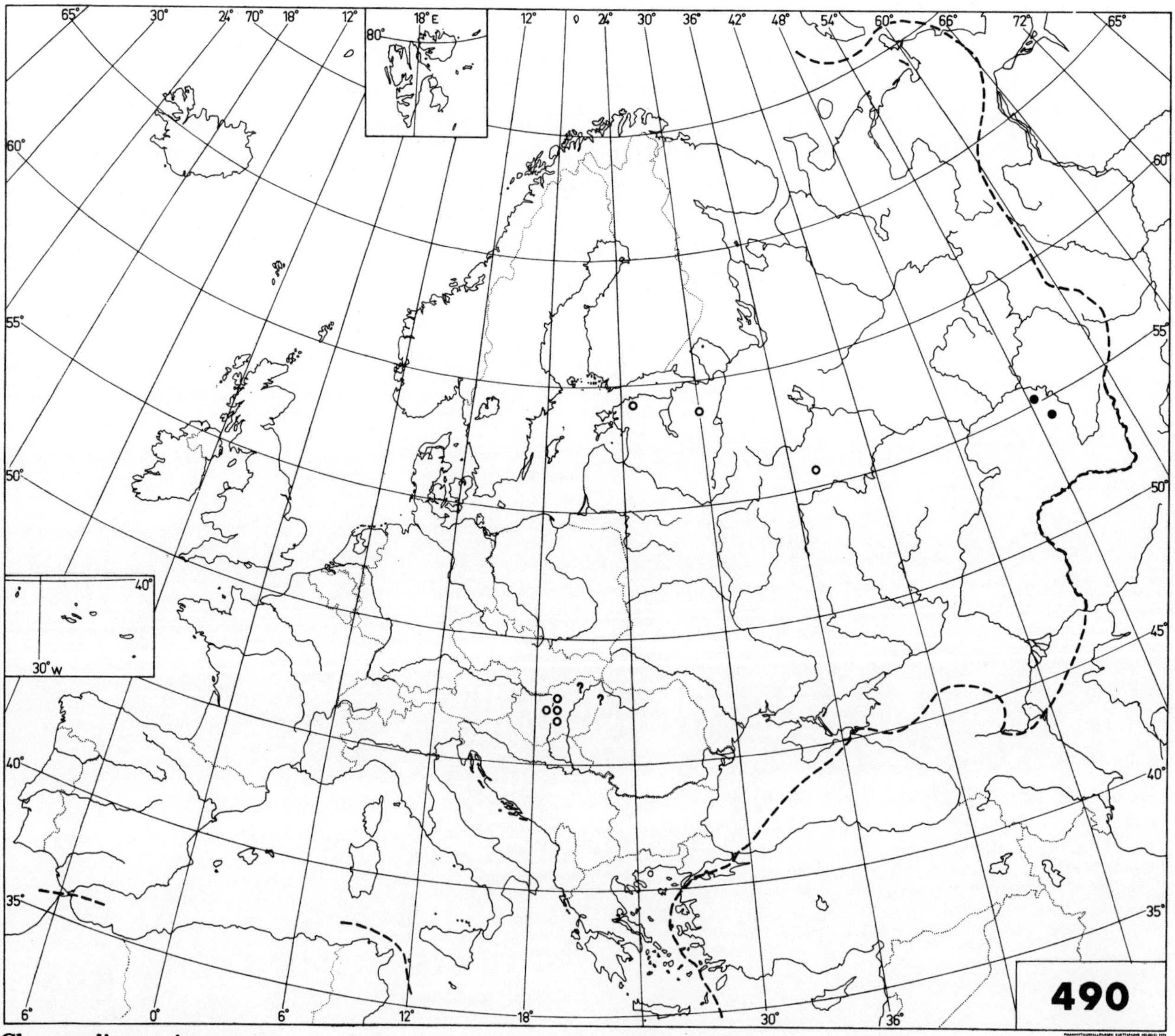

Chenopodium aristatum

Chenopodium schraderanum Schultes — Map 489.

Chenopodium foetidum Schrader non Lam.

Notes. [Cz] and [Rs(E)] added (not given in Fl. Eur.), [Ge] omitted (given in Fl. Eur.).
Native of E. Africa.

Chenopodium aristatum L. — Map 490.

Notes. Considered native in Rs(E) ([Rs(E)] in Fl. Eur.). Apparently immigrant in Rs(C) (given as native in Fl. Eur.) and recorded for Rs(B) but possibly casual (not given in Fl. Eur.).

Chenopodium ambrosioides L. — Map 491.

Chenopodium anthelminticum auct. non L.; C. integrifolium Vorosch.

Notes. [Al] and [Be] added (not given in Fl. Eur.), only casual in Cz ([Cz] in Fl. Eur.).
Native of Tropical America.

Chenopodium multifidum L. — Map 492.

Notes. *Al, *Bl, *Bu, *Gr, *Ju and *Tu added (not given in Fl. Eur.): Fl. Mallorca 1978: 94; Fl. Reipubl. Pop. Bulg. 1966: 528; P. Uotila, Ann. Bot. Fennici 17: 66—67 (1980).
Native of South America (?).

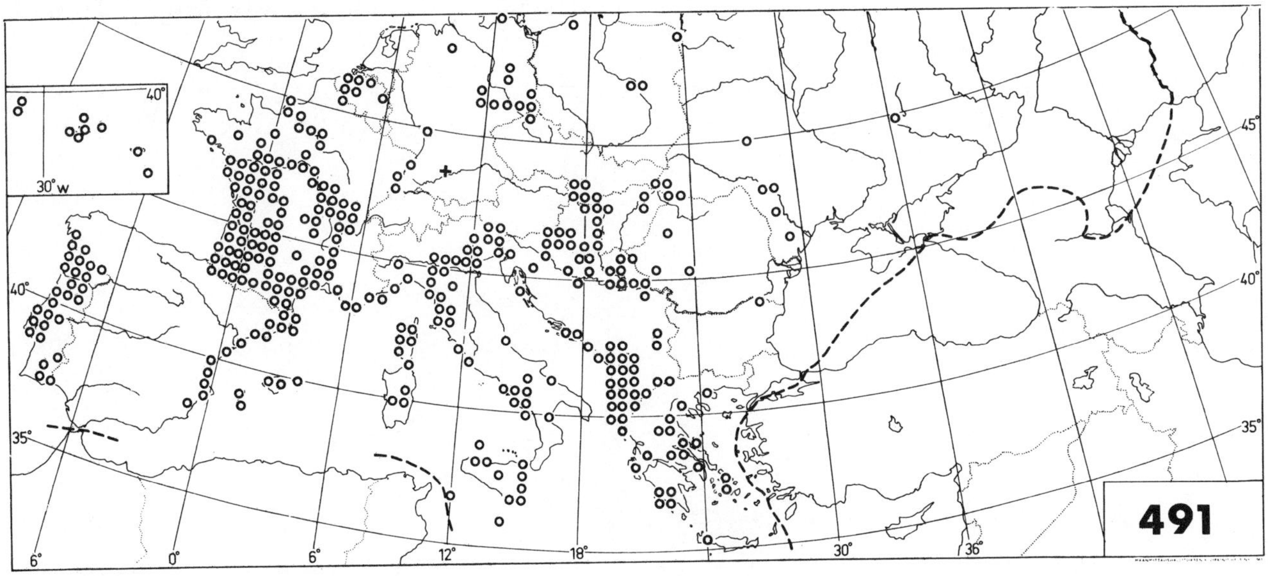

Chenopodium ambrosioides

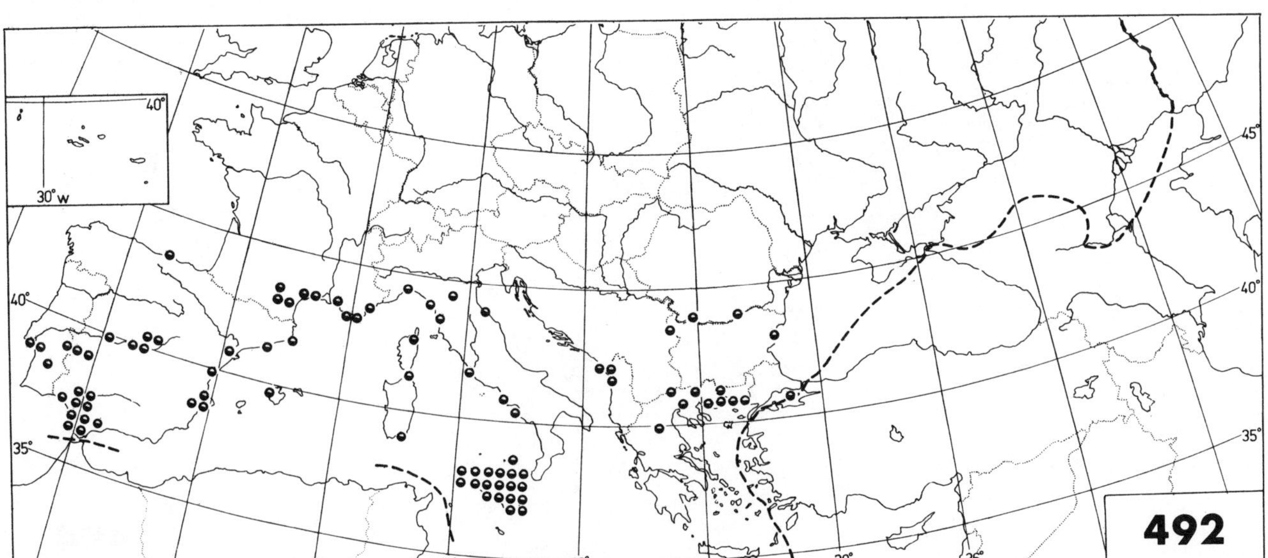

Chenopodium multifidum

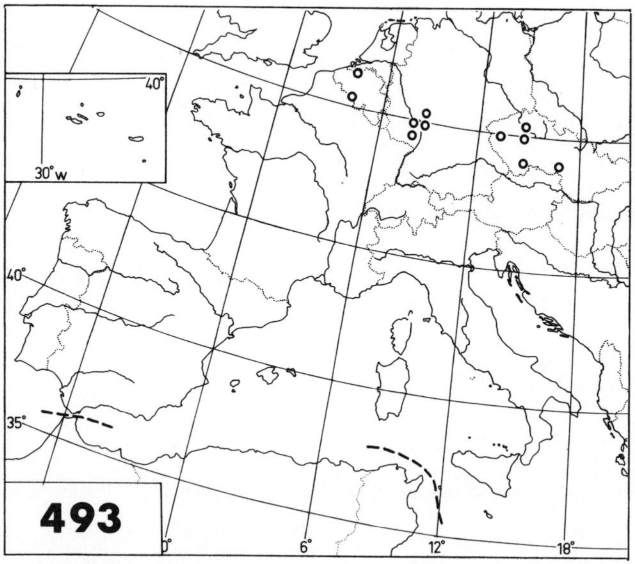

Chenopodium pumilio

Chenopodium pumilio R.Br. — Map 493.

Notes. [Be, Cz, Ge] (the species not mentioned in Fl. Eur.).

Native of Australia and New Zealand.

Chenopodium bonus-henricus L. — Map 494.

Blitum bonus-henricus (L.) C.A. Meyer

Notes. Rs(E) added (not given in Fl. Eur.). Established alien in Br and Hb (given, indirectly, as native in Fl. Eur.). Map incomplete for Rs(W). Sa omitted (given in Fl. Eur.).

Total range. Endemic to Europe (although not mentioned as such in Fl. Eur.): Hegi 1960: 601.

Chenopodium foliosum Ascherson — Map 495.

Blitum virgatum L.; Chenopodium virgatum (L.) Ambrosi non Thunb.

Taxonomy. P. Uotila, Ann. Bot. Fennici 16: 237—240 (1979).

Notes. Considered native in Al, Bu and Gr ([Al, Bu, Gr] in Fl. Eur.), casual in Da and extinct in Su ([Da, Su] in Fl. Eur.), established alien in Ho ([?Ho] in Fl. Eur.),

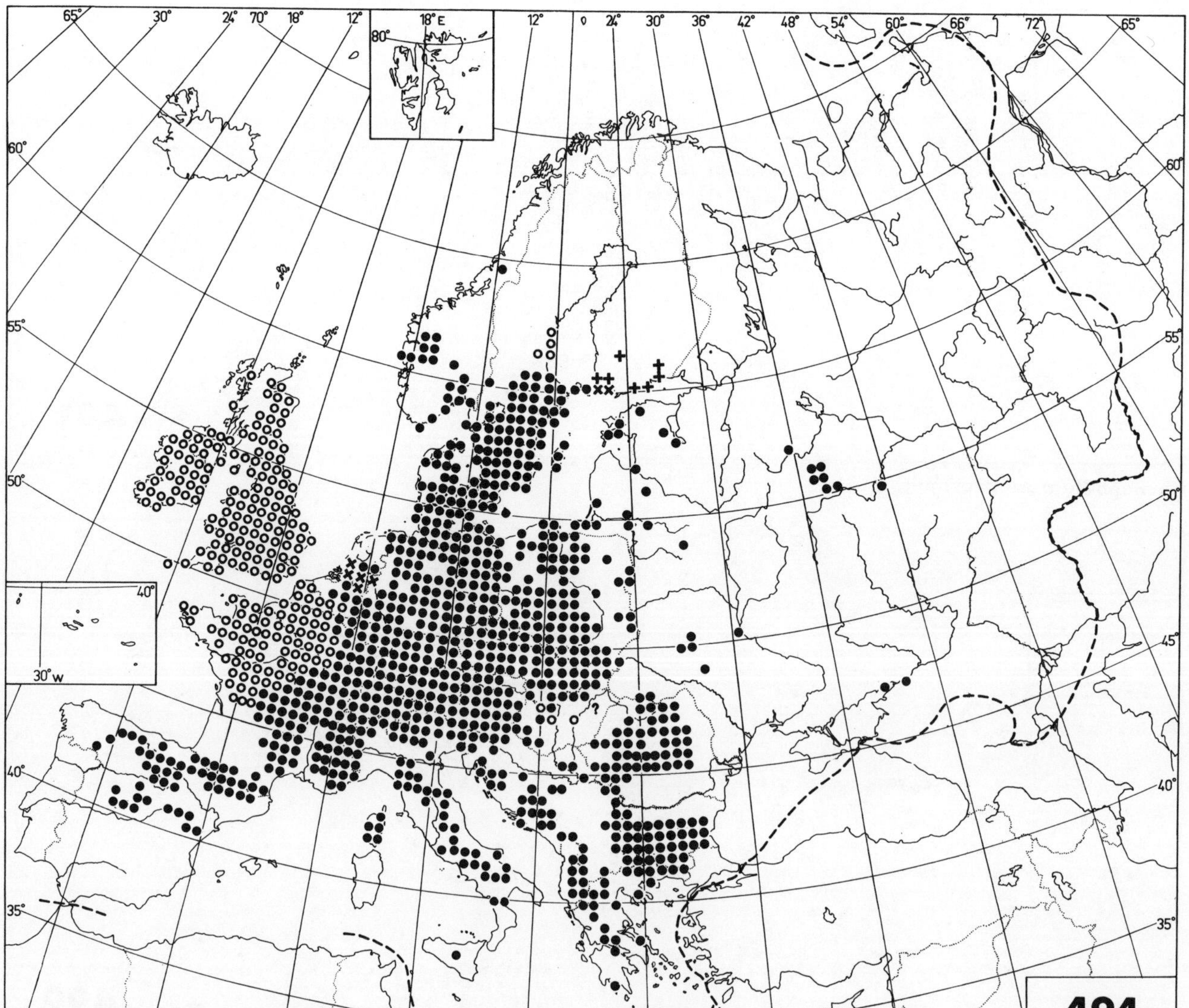

Chenopodium bonus-henricus

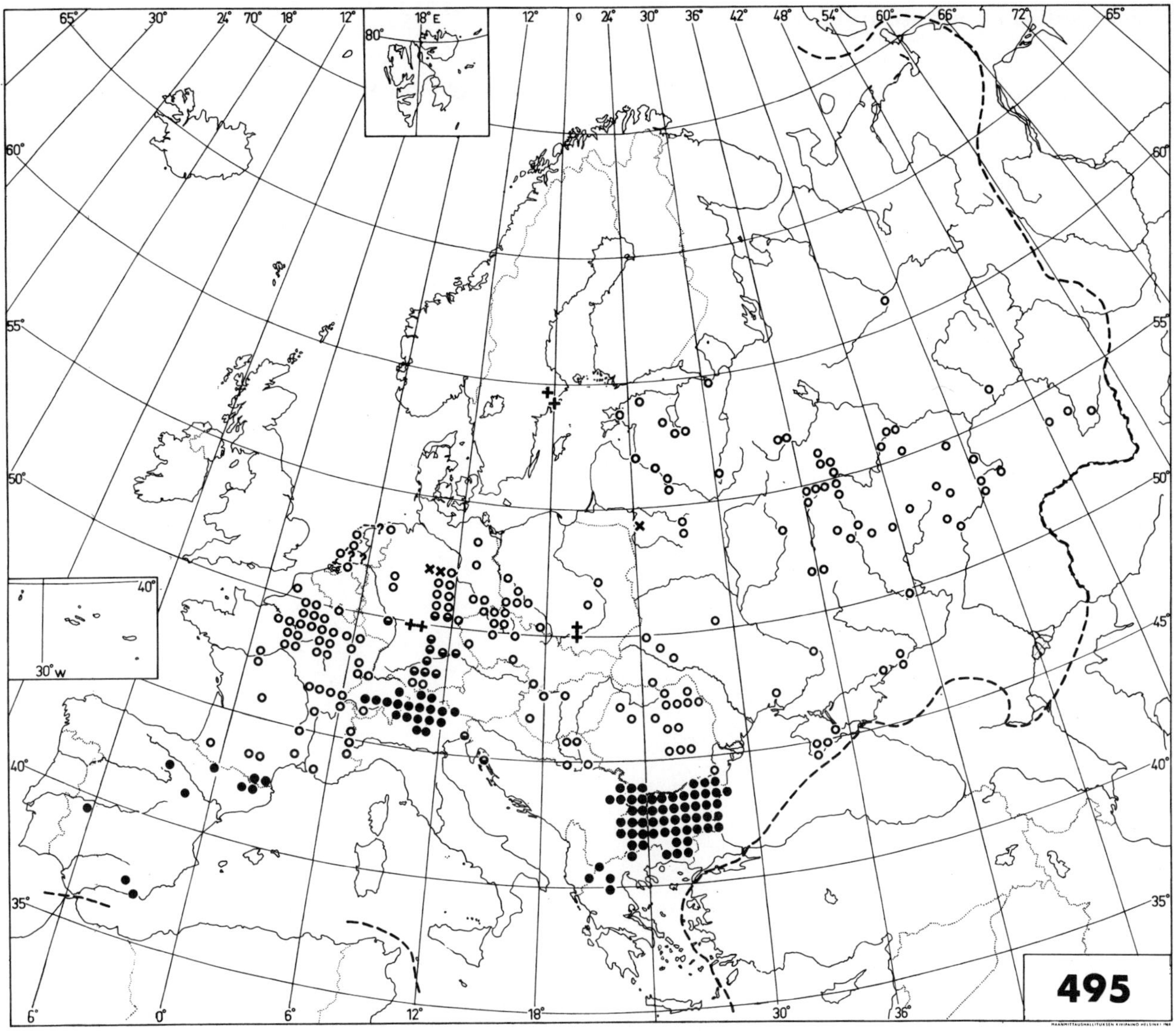

Chenopodium foliosum

doubtful native in Ge (given as native in Fl. Eur.), [Hu] and Ju added (not given in Fl. Eur.). Lu omitted (given in Fl. Eur.), the material belonging to *Chenopodium exsuccum*: P. Uotila, Ann. Bot. Fennici 16: 239 (1979); Hs mapped according to specimens seen by P. Uotila. Probably casual in Rs(N) (not given in Fl. Eur.).

Total range. Hegi 1960: 604 (incl. *Chenopodium exsuccum*).

Chenopodium exsuccum (C. Loscos) Uotila — Map 496.

Blitum exsuccum C. Loscos; Chenopodium virgatum var. exsuccum (C. Loscos) Jahand. & Maire

Taxonomy. P. Uotila, Ann. Bot. Fennici 16: 237—240 (1979).

Notes. Hs, Lu (the species not mentioned in Fl. Eur.): P. Uotila, Ann. Bot. Fennici 16: 239 (1979). The map is based on material seen by P. Uotila.

Total range. P. Uotila, Ann. Bot. Fennici 16: 239 (1979).

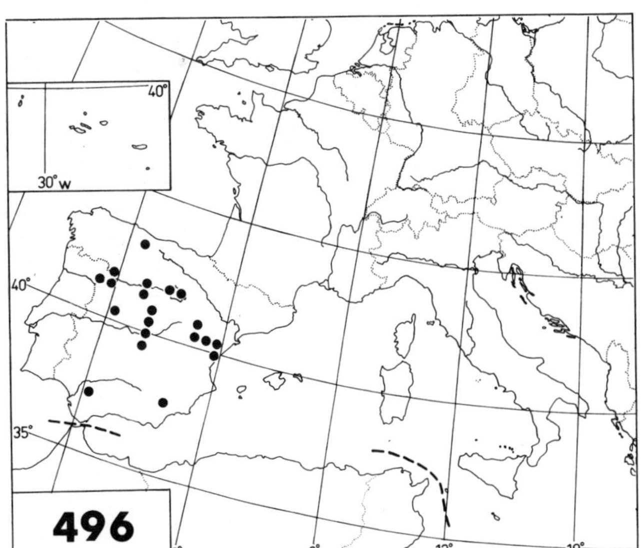

Chenopodium exsuccum

Chenopodium capitatum (L.) Ascherson

Blitum capitatum L.

Notes. Cultivation radically decreased since the 19th century, and evidently not truly naturalized anywhere in Europe. Not mapped.

Native of North America (?). Hultén Alaska 1968: 393.

Chenopodium glaucum L. — Map 497.

Incl. Chenopodium wolffii Simonkai

Notes. Lu added (not given in Fl. Eur.): Nova Fl. Port. 1971: 92. Established alien in No (given as native in Fl. Eur.): P.M. Jörgensen, Norwegian Jour. Bot. 20: 309 (1973). Map incomplete for Rs(W).

Total range. Hegi 1960: 611; MJW 1965: map 133a; Hultén Alaska 1968: 394; Hultén CP 1971: map 133 (northern hemisphere).

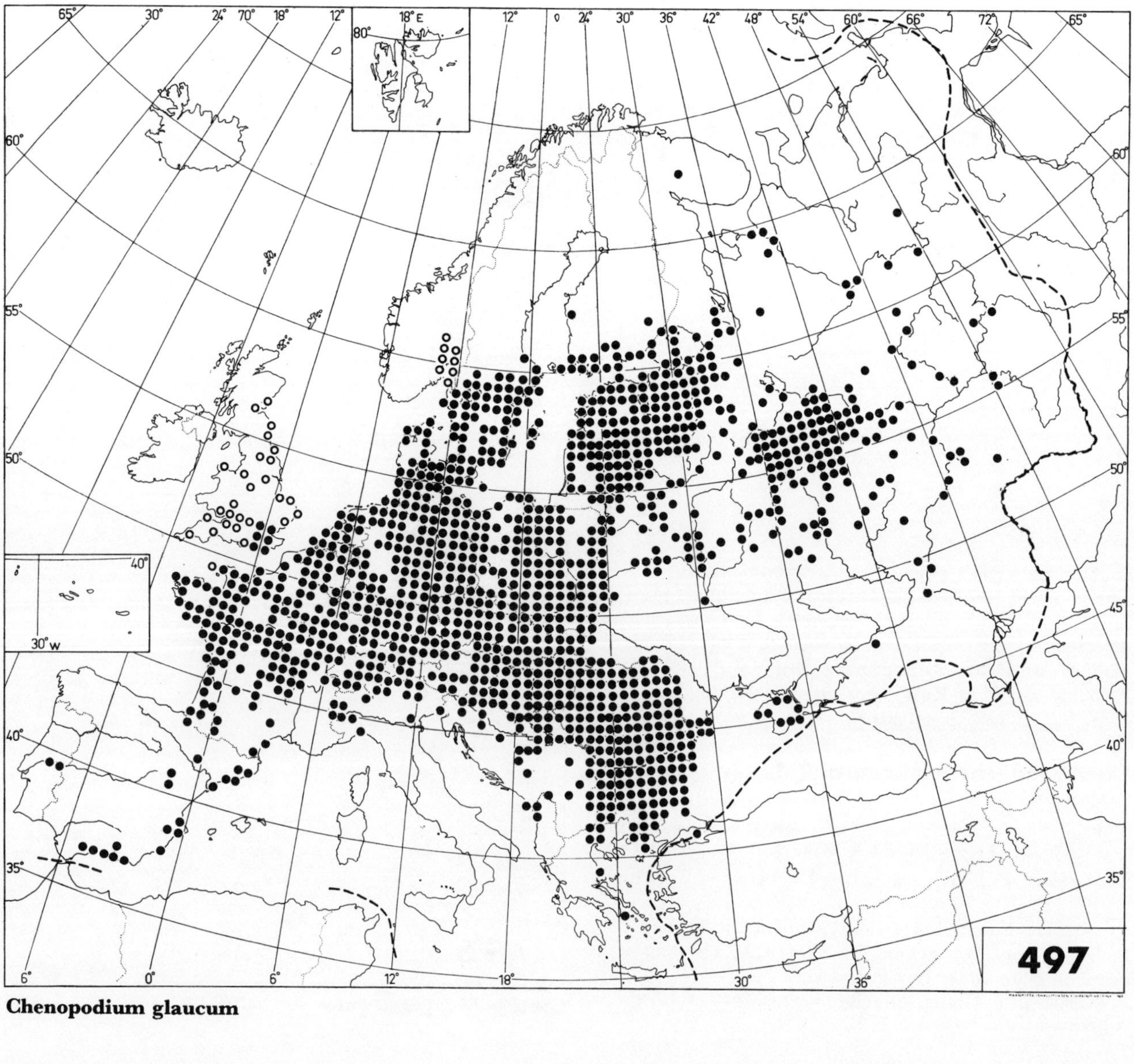

Chenopodium glaucum

Chenopodium rubrum L. — Map 498.

Blitum rubrum (L.) C.A. Meyer

Notes. Lu omitted (given in Fl. Eur.), the records referring to *Chenopodium botryodes*: Nova Fl. Port. 1971: 620. Established alien in No (given as native in Fl. Eur.): P.M. Jörgensen, Norwegian Jour. Bot. 20: 309 (1973). Map incomplete for Rs(W).

Total range. Hultén Alaska 1968: 394; J.T. Williams, Jour. Ecol. 57 (Biol. Fl. British Isles 117): 833 (1969); Hultén CP 1971: map 295 (all three: holarctic part).

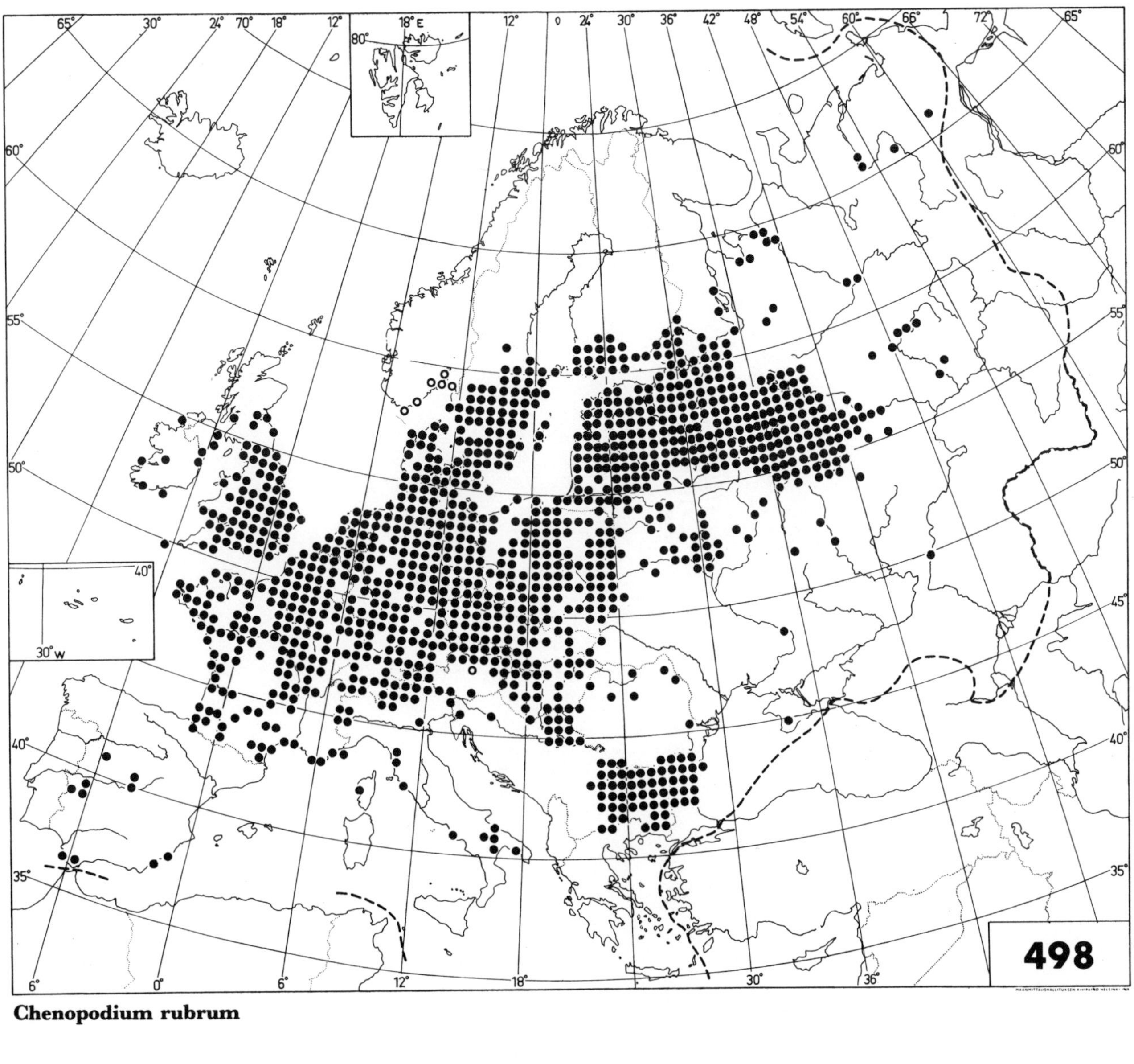

Chenopodium rubrum

Chenopodium botryodes Sm. — Map 499.

Blitum chenopodioides L.; Chenopodium chenopodioides (L.) Aellen; C. crassifolium Hornem.; C. rubrum var. crassifolium (Hornem.) Moq.

Nomenclature. P. Aellen, Ostenia 98 (1933).

Notes. Hb and He omitted (given in Fl. Eur.), Ho (as extinct), It, Ju and Sa added (not given in Fl. Eur.). Recorded for Rs(B) but identity and/or native status doubtful. Rs(K, E) omitted (given in Fl. Eur.): not mentioned from Rs(K) in Rubtsev 1972. Rs(W) mapped according to Fl. URSR 1952: 288.

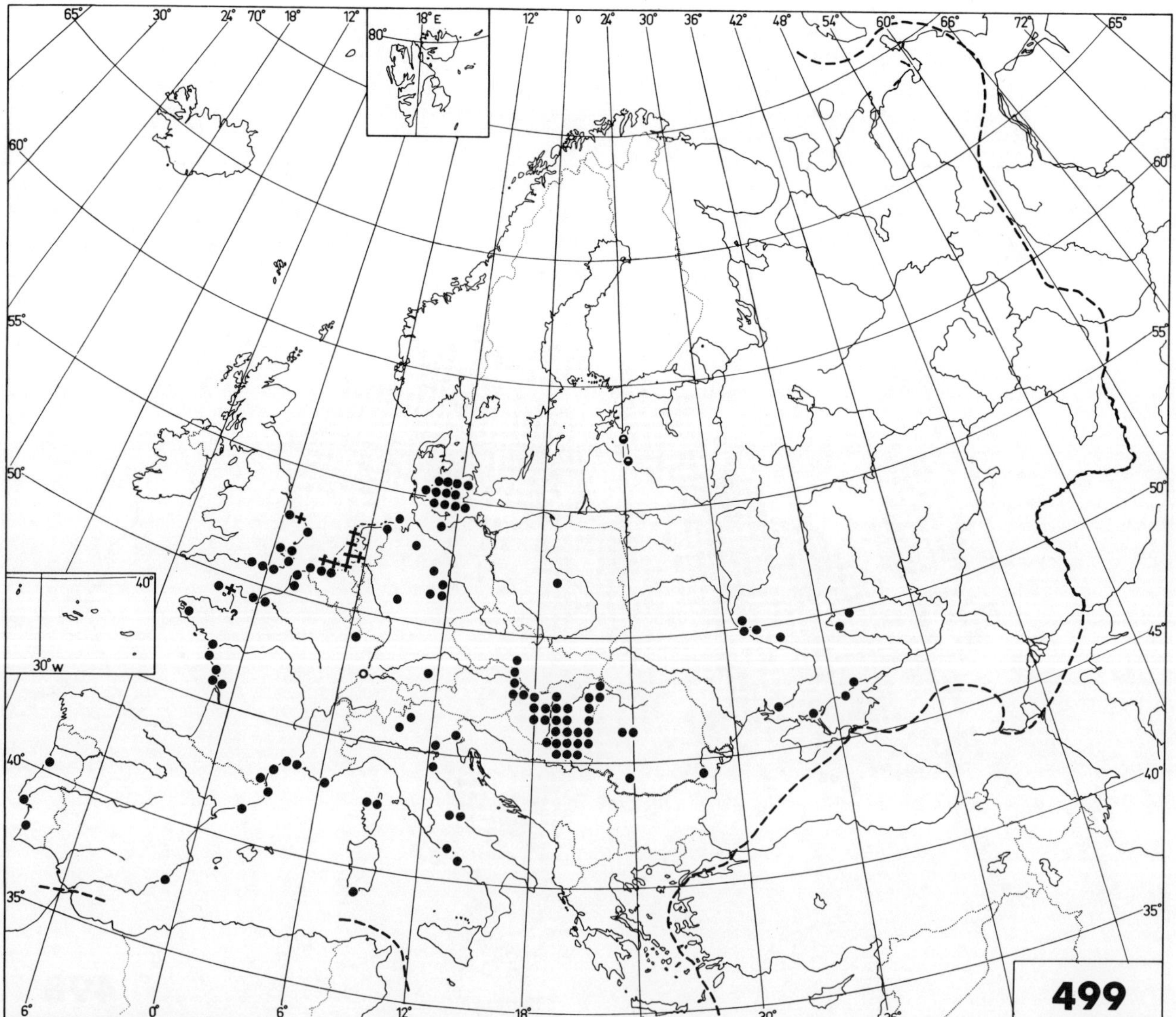

Chenopodium botryodes

Chenopodium hybridum L. — Map 500.

Chenopodium angulosum Lam.

Notes. Established alien in Br and No (given as native in Fl. Eur.). Rs(N) omitted (given in Fl. Eur.). Map incomplete for Rs(W).

Total range. MJW 1965: map 132c; Hultén Alaska 1968: 393; Hultén CP 1971: map 134.

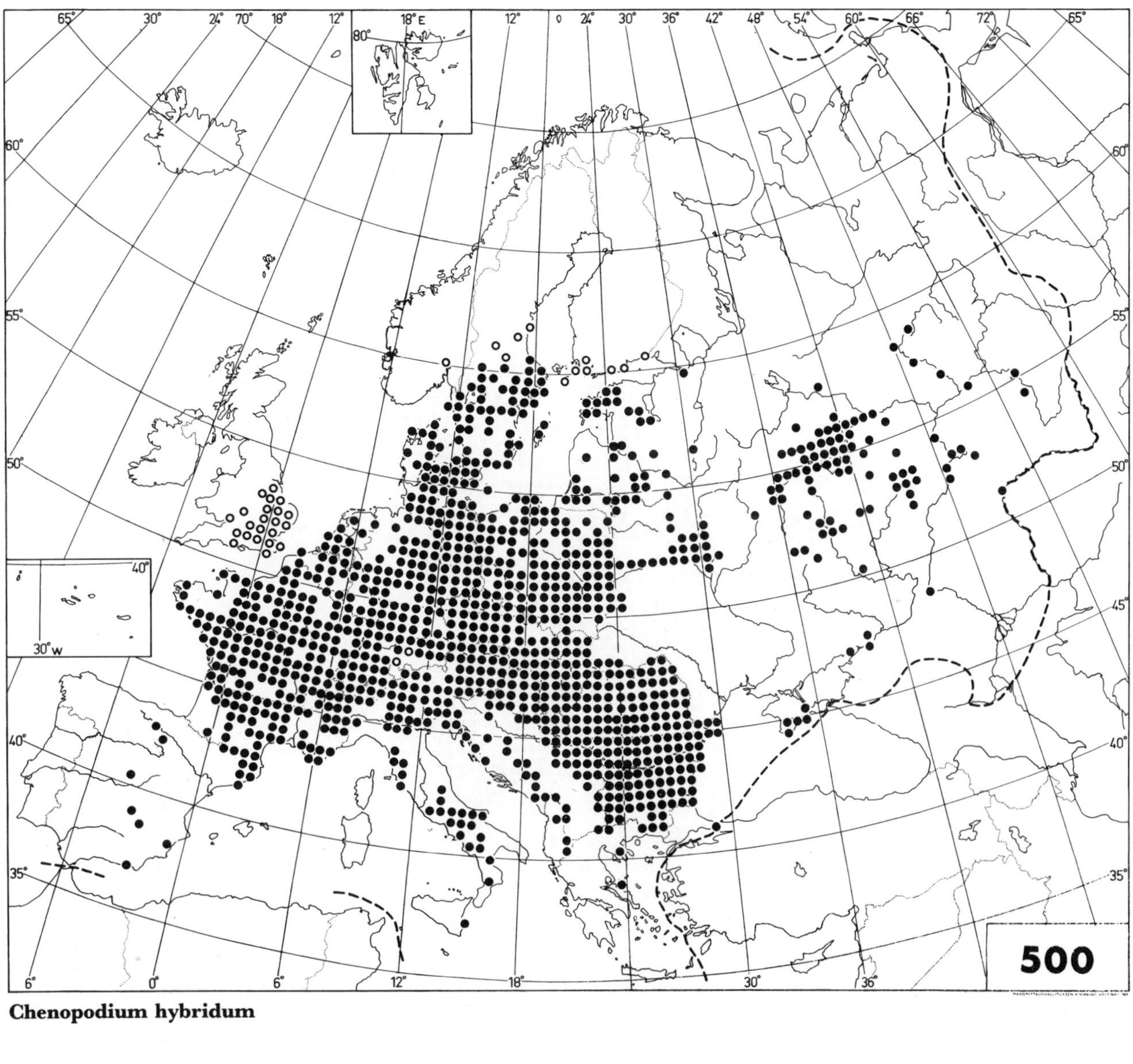

Chenopodium hybridum

Chenopodium polyspermum L. — Map 501.

Notes. Al added (not given in Fl. Eur.), Cr and Sa omitted (given in Fl. Eur.). Map incomplete for Rs(W).

Total range. Hegi 1960: 619; MJW 1965: map 132b.

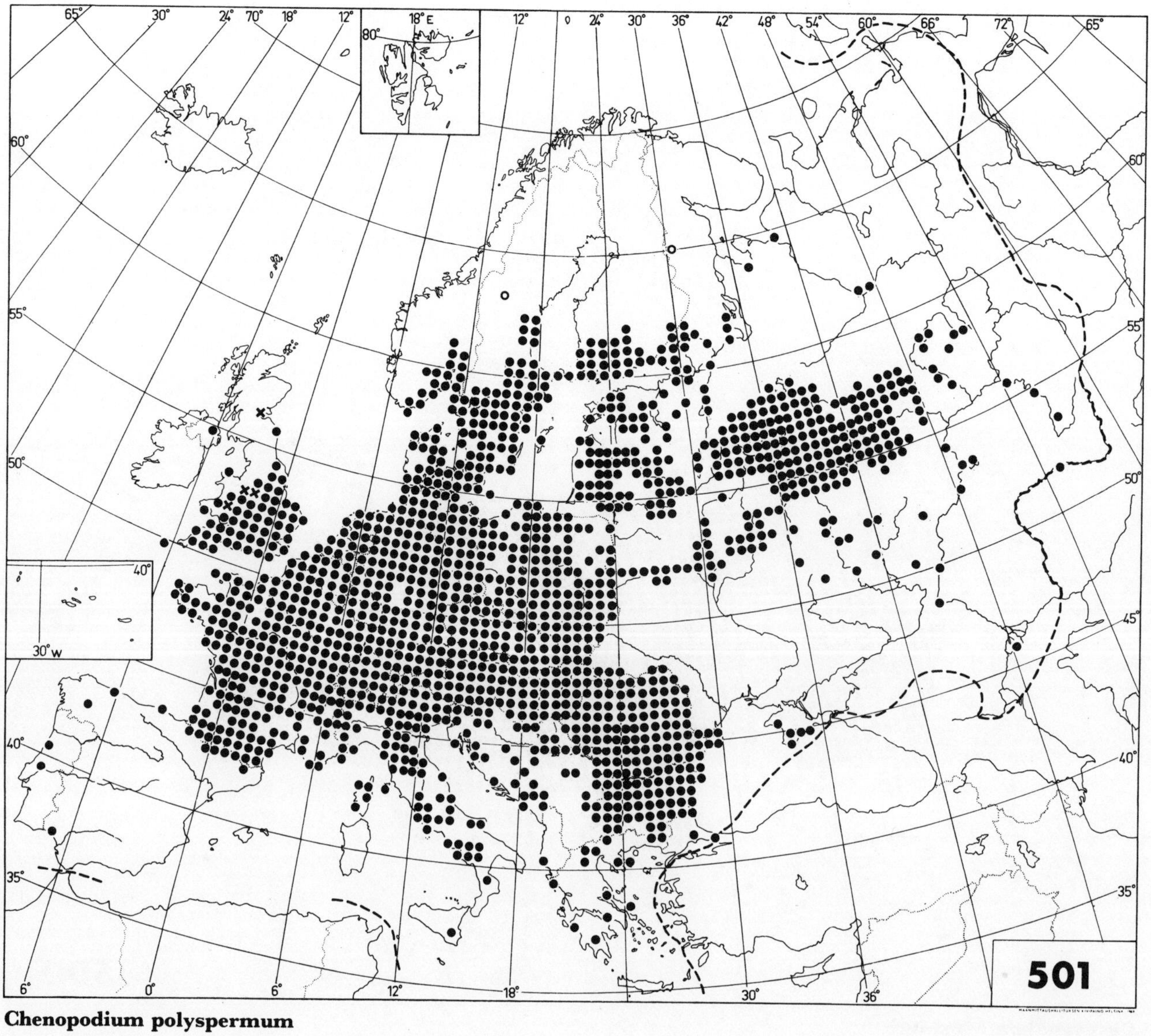

Chenopodium polyspermum

Chenopodium vulvaria L. — Map 502.

Chenopodium foetidum Lam.; C. olidum Curtis

Notes. Only casual in Fe and Da (given, indirectly, as native in Fl. Eur.): P. Uotila & J. Suominen, Ann. Bot. Fennici 13: 16 (1976), A. Hansen & A. Pedersen, Bot. Tidsskr. 63: 252 (1968). Extinct in Ho (given as present in Fl. Eur.): J. Mennema, Gorteria 8: 162—163 (1977). Map incomplete for Rs(W).

Total range. MJW 1965: map 132d.

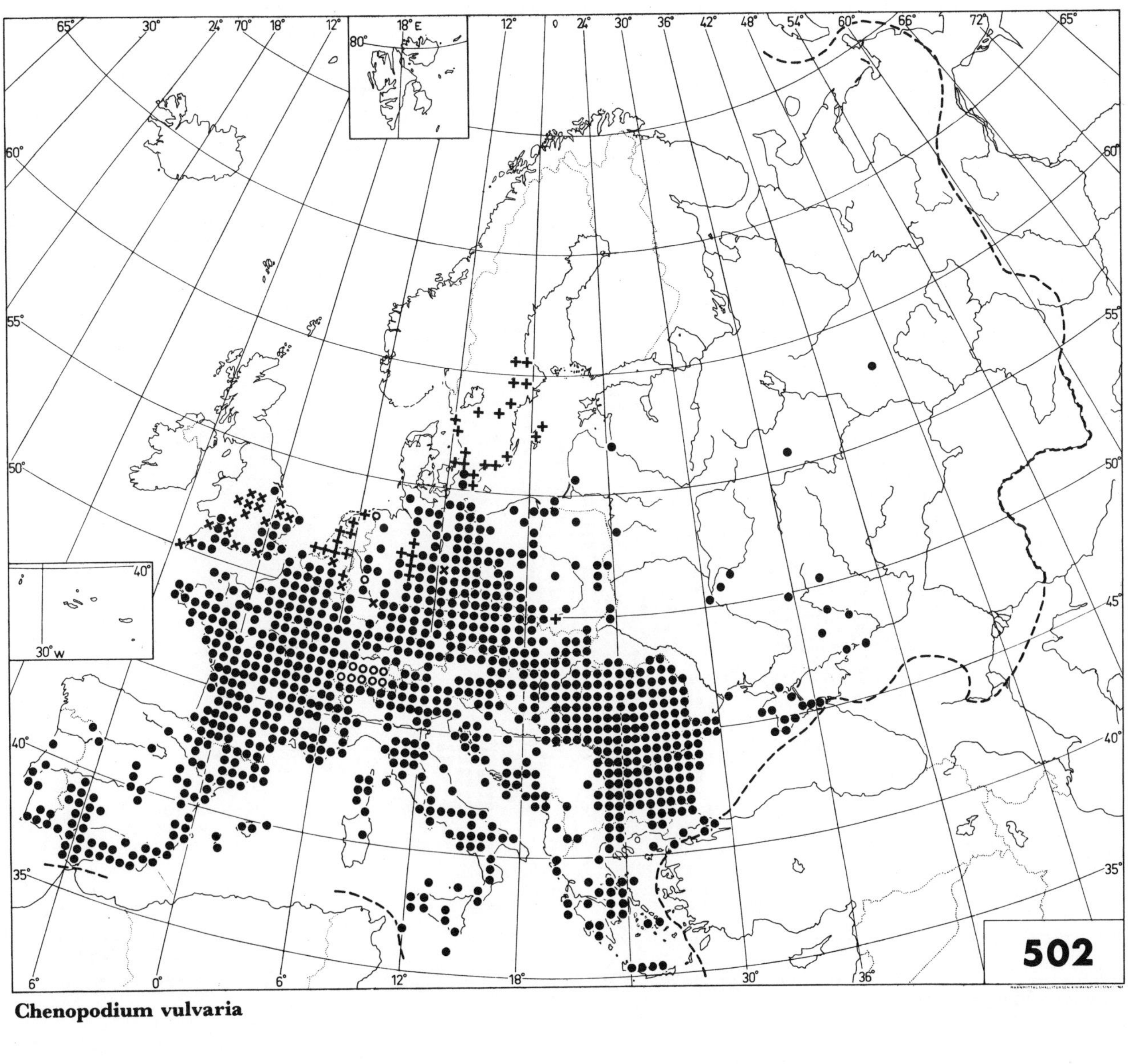

Chenopodium vulvaria

Chenopodium urbicum L. — Map 503.

Incl. Chenopodium rhombifolium Muhl.

Notes. Formerly well established and common in ruderal places throughout much of Europe, now strongly declining. Established alien in Br (given, indirectly, as native in Fl. Eur.). In Da mainly collected before 1900, established occurrences of present date not known (given as native in Fl. Eur.): A. Hansen & A. Pedersen, Bot. Tidsskr. 63: 252 (1968). Probably extinct in Fe and only casual in No (given as present in Fl. Eur.): P. Uotila & J. Suominen, Ann. Bot. Fennici 13: 15 (1976), P.M. Jörgensen, Norwegian Jour. Bot. 20: 318 (1973). He and Lu added (not given in Fl. Eur.): Nova Fl. Port. 1971: 93. Map incomplete for Rs(W). Becoming extinct in Su: Ö. Nilsson & L.-Å. Gustafsson, Svensk Bot. Tidskr. 73: 356—357 (1979).

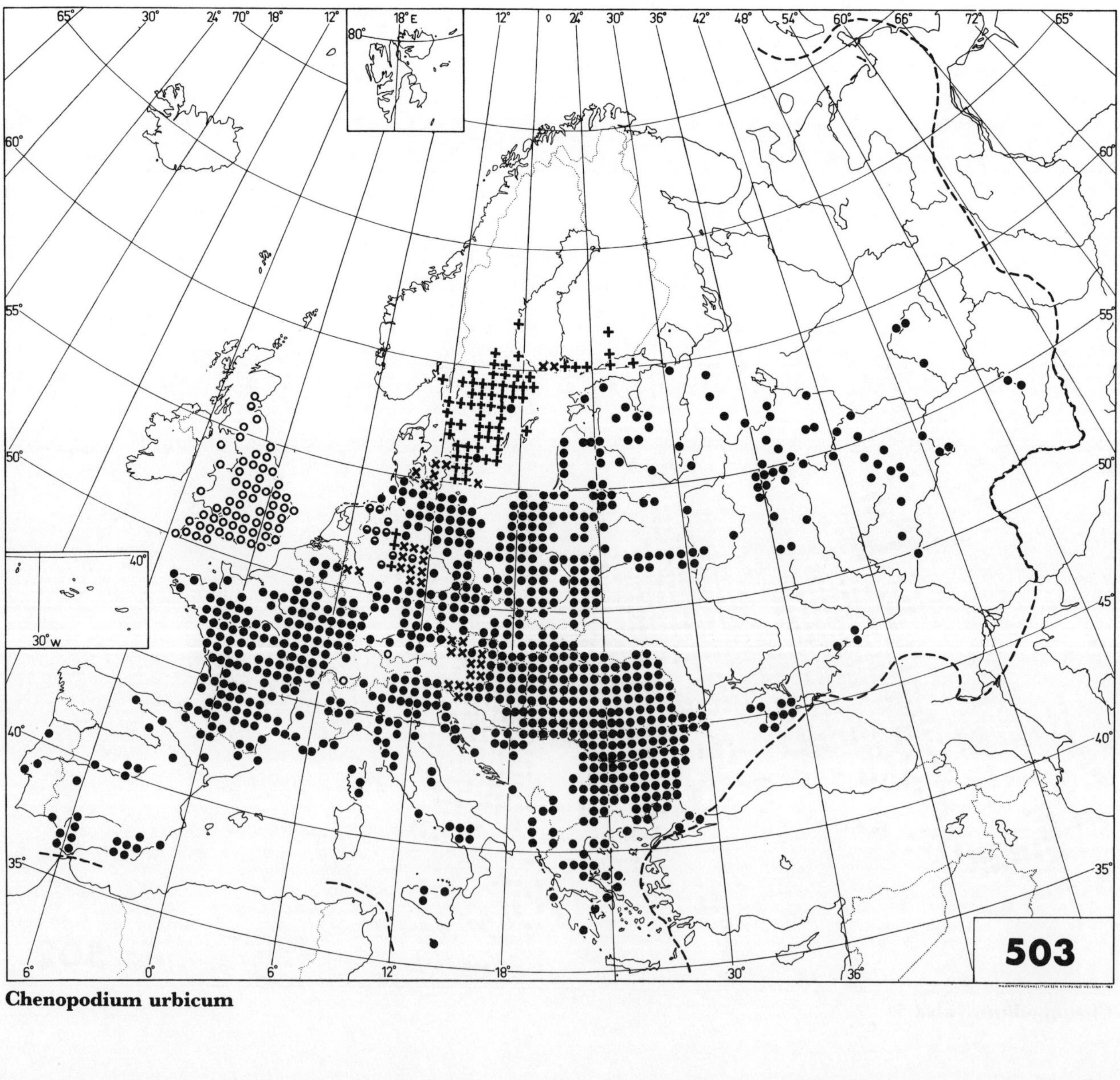

Chenopodium urbicum

Chenopodium murale L. — Map 504.

Notes. Largely a recent immigrant or even casual towards the north of the range. Only casual in Fe (given, indirectly, as native in Fl. Eur.): P. Uotila & J. Suominen, Ann. Bot. Fennici 13: 9 (1976). Rs(W) mapped according to Fl. URSR 1952: 296. Tu added (not given in Fl. Eur.): Fl. Turkey 1967: 302.

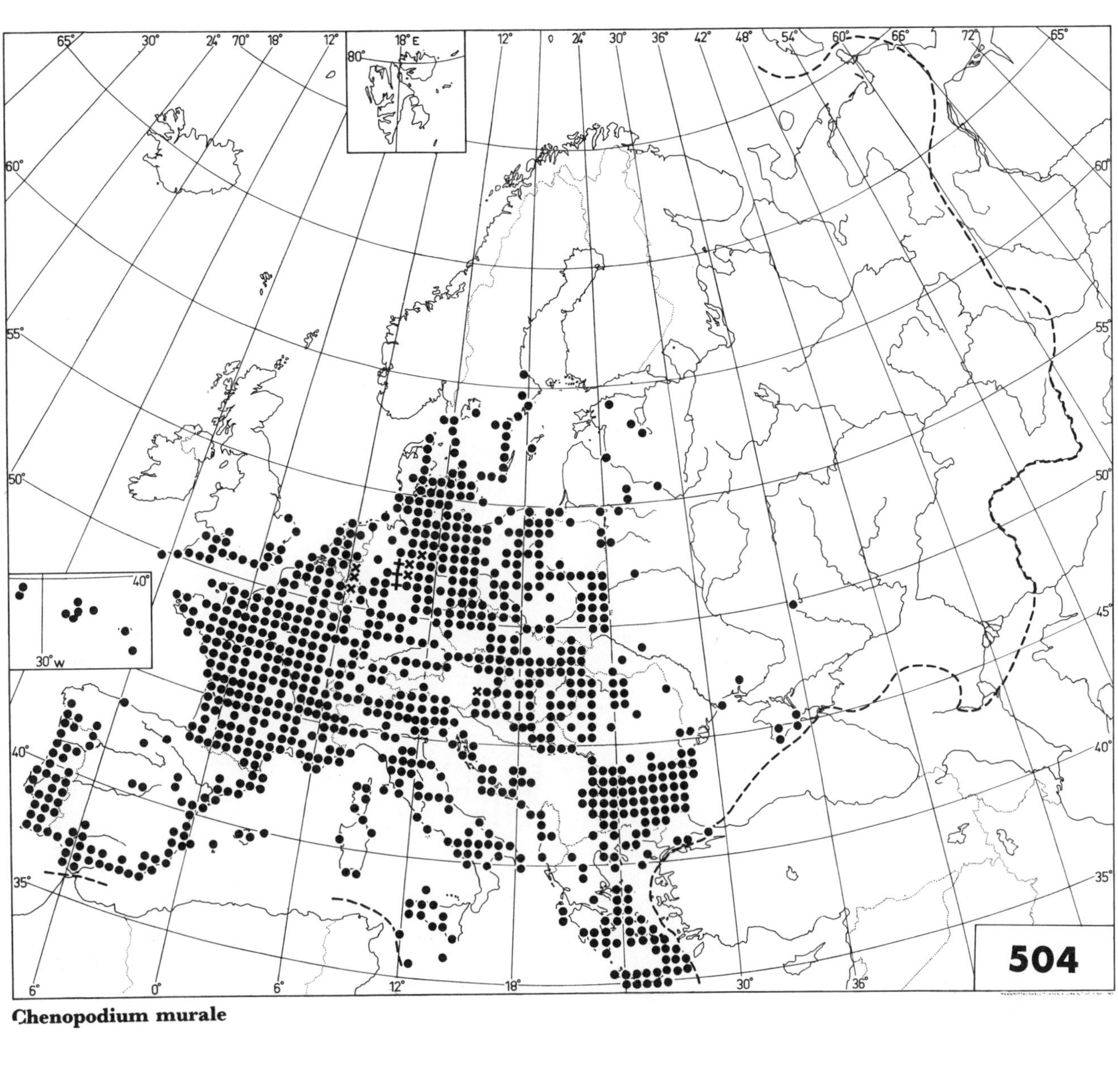

Chenopodium murale

Chenopodium ficifolium Sm. — Map 505.

Chenopodium serotinum auct. non L.

Taxonomy. Concerning the typification of *Chenopodium serotinum* L., see P. Uotila, Ann. Bot. Fennici 14: 197—198 (1977).
Notes. Established alien in Da, Rs(B) and Su, and only casual in Fe and No (given as native in Fl. Eur.), Ga, Ge and Hu added (not given in Fl. Eur.), Hs confirmed (?Hs in Fl. Eur.), Sa and Si omitted (given in Fl. Eur.). Map incomplete since data are lacking from Rs(N, K, E) (given in Fl. Eur.); not mentioned for Rs(K) in Rutbsev 1972. Rs(W) mapped according to Fl. URSR 1952: 312.

Chenopodium acerifolium Andrz. — Map 506.

Chenopodium album subsp. hastatum (Klinggr.) J. Murr; C. klinggraeffii (Abromeit) Aellen

Notes. The available data are contradictory and cover mainly the western fringe of the probable area in Europe. »Poland and U.S.S.R. Po Rs(N, C, W, E)» (Fl. Eur.).
Total range. Hegi 1960: 633.

Chenopodium jenissejense Aellen & Iljin — Map 507.

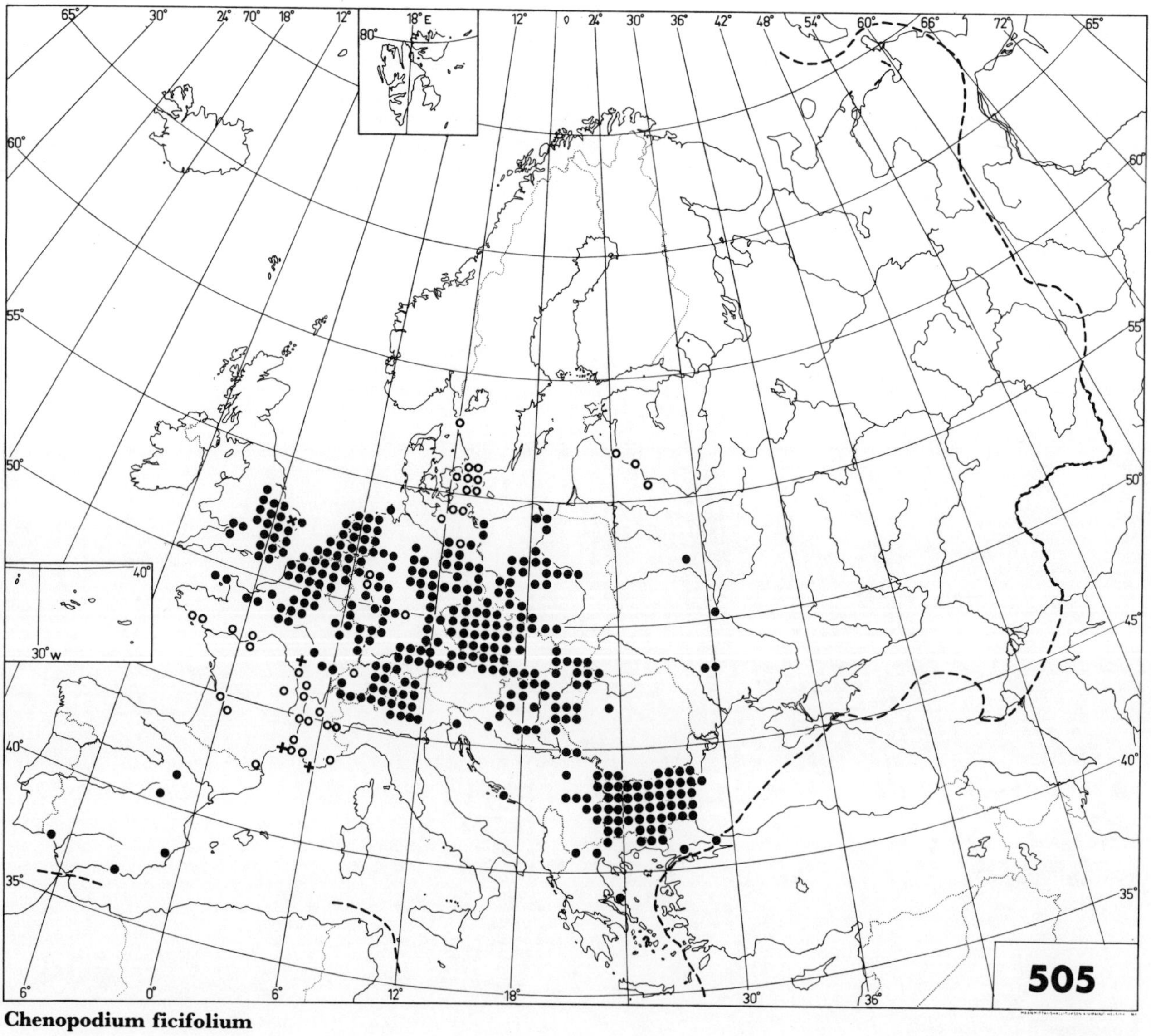

Chenopodium ficifolium

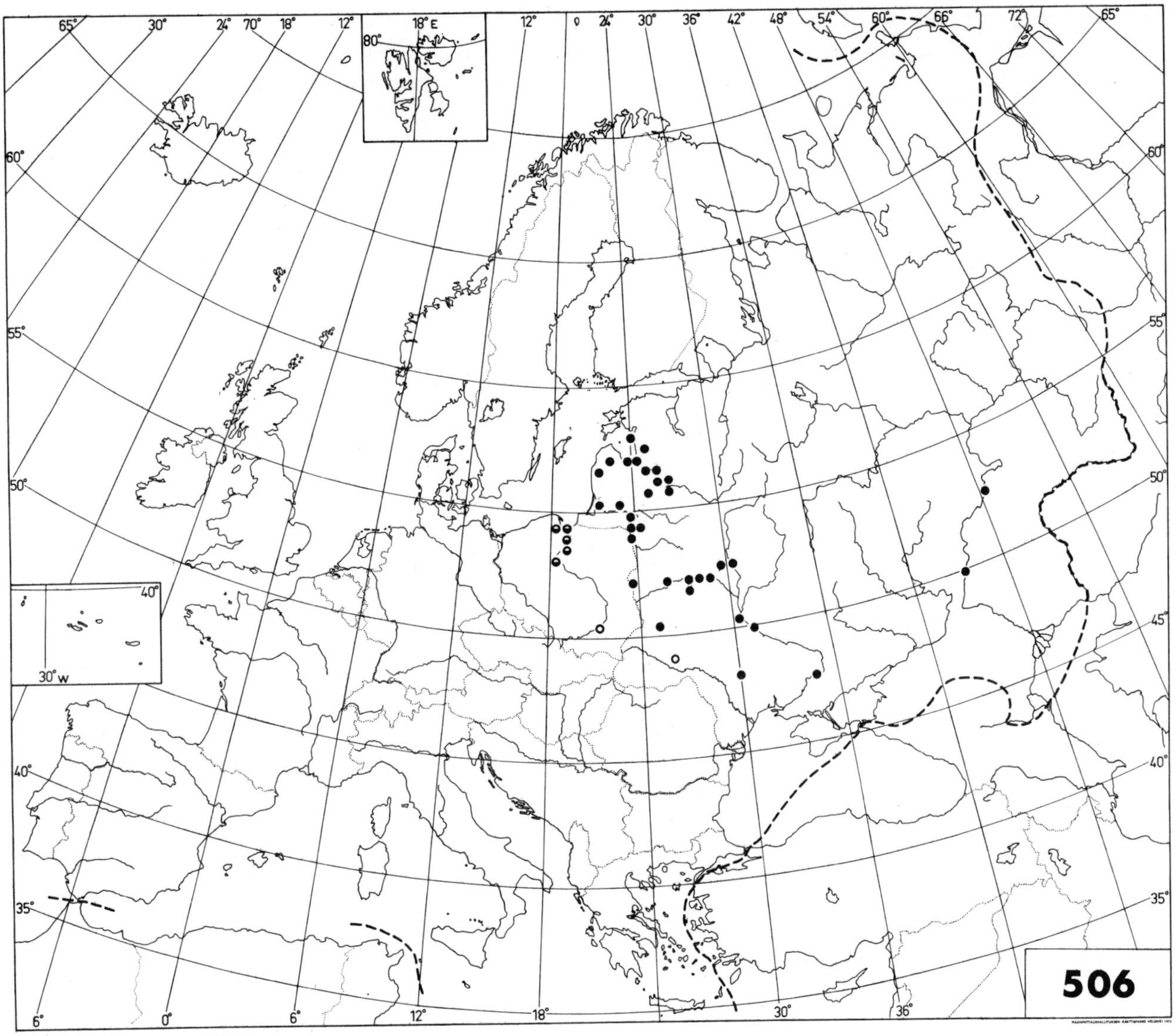

Chenopodium acerifolium

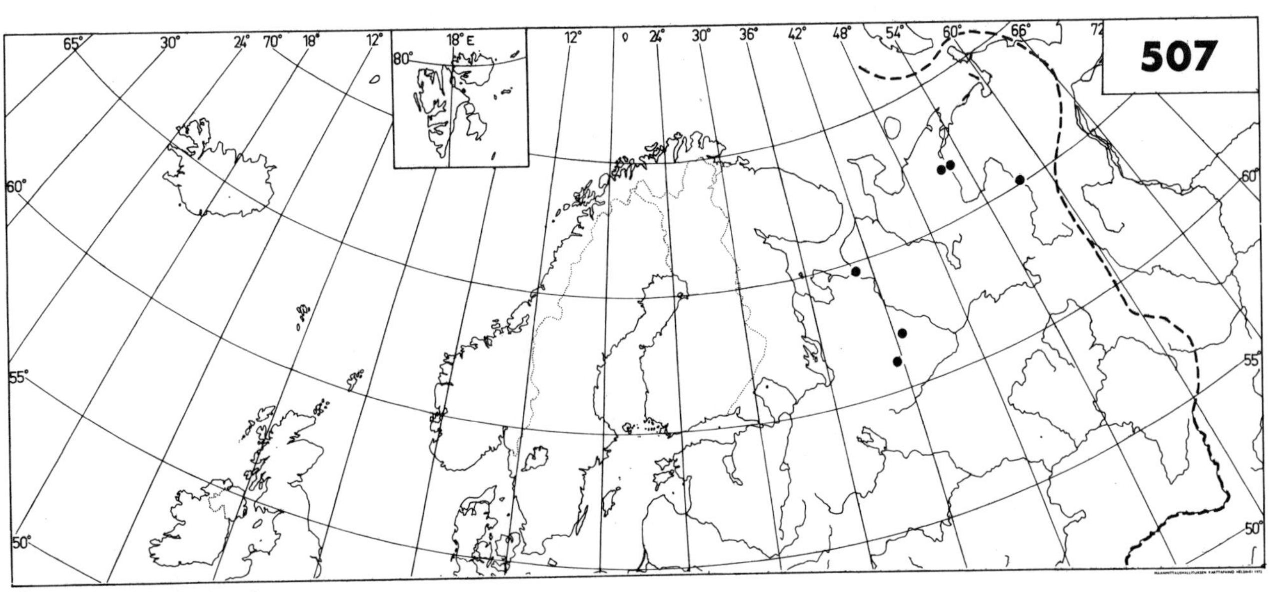

Chenopodium jenissejense

Chenopodium opulifolium Schrader ex Koch & Ziz — Map 508.

Notes. [Az] and Tu added (not given in Fl. Eur.): A. Hansen, Anuario Soc. Brot. 37: 89 and 99 (1971), J. Jalas *et al.*, Notes Roy. Bot. Garden Edinb. 35: 71 (1976). Casual in Br and Da, an established alien in Be and Su and doubtful native in He (given as native in Fl. Eur.). Only a few scattered records available from Rs (given, indirectly, for all Rs territories in Fl. Eur.), and, consequently, the map incomplete. Probably an established alien (or only casual) in Rs(B), Rs(N) omitted.

Chenopodium album L., s. lato — Map 509.

Chenopodium album s. str. + C. reticulatum + C. strictum + C. suecicum + C. zerovii

Taxonomy. A. Beaugé, Chenopodium album et espèces affines. Étude historique et statistique. 447 pp. (Paris 1974); P. Uotila, Ann. Bot. Fennici 11: 44—58 (1974), 14: 199—205 (1977), Acta Bot. Fennica 108: 1—35 (1978).

Notes. Fa omitted ([Fa] in Fl. Eur.). The occurrence in Rs is seriously underrepresented.

Chenopodium album L., s. str.

Incl. Chenopodium reticulatum Aellen. Excl. C. album subsp. striatum (Krašan) J. Murr and C. striatiforme J. Murr

Notes. Not mapped separately.

Total range. Almost cosmopolitan. J.T. Williams, Jour. Ecol. 51 (Biol. Fl. British Isles 87): 713 (1963); Hultén Alaska 1968: 395 (both: holartic part); Hultén CP 1971: map 246 (northern boundary).

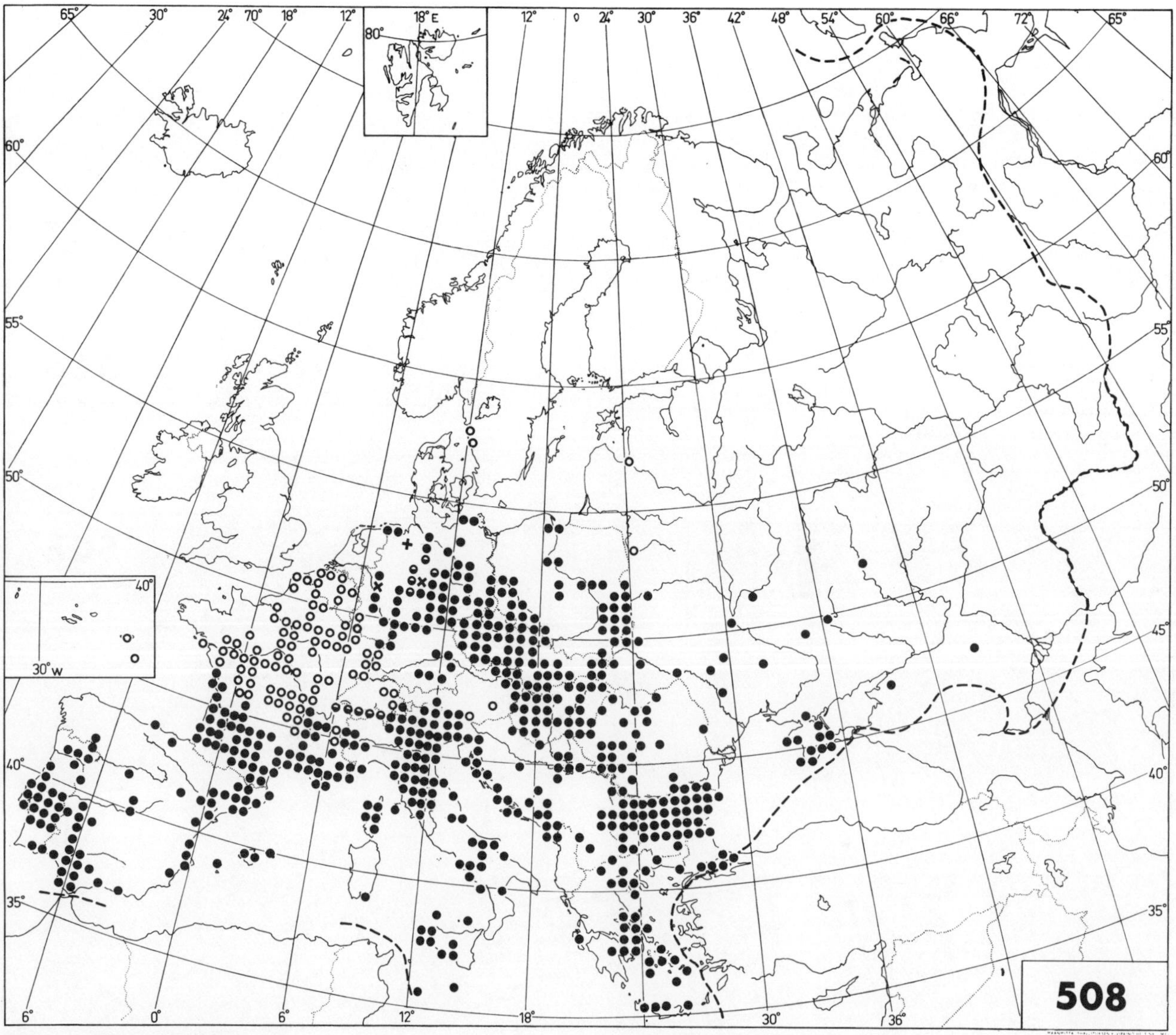

Chenopodium opulifolium

Chenopodium strictum Roth

Chenopodium album subsp. striatum (Krašan) J. Murr; C. betaceum Andrz.; C. concatenatum auct. non Thuill.; C. striatum (Krašan) J. Murr

Taxonomy. P. Uotila, Ann. Bot. Fennici 10: 338—339 (1973), 14: 199—205 (1977), Acta Bot. Fennica 108: 26 (1978), P. Uotila & J. Suominen, Ann. Bot. Fennici 13: 14—15 (1976).

Notes. Data only sporadically available, and the species not consistently kept apart from *Chenopodium album* s. str. Hence not mapped.

C. strictum subsp. **strictum**

Chenopodium album subsp. striatum (Krašan) J. Murr

Notes. Not native in C. Europe but locally established. »C. & E. Europe, extending westwards to Belgium; introduced in Spain» (Fl. Eur.).

C. strictum subsp. **striatiforme** (J. Murr) Uotila

Chenopodium album subsp. microphyllum (Boenningh.) Sterner; C. concatenatum subsp. striatiforme (J. Murr) J. Murr; C. strictum subsp. microphyllum (Boenningh.) Jörgensen; C. striatiforme J. Murr

Taxonomy. P.M. Jörgensen, Norwegian Jour. Bot. 20: 317 (1973); P. Uotila, Ann. Bot. Fennici 14: 199—205 (1977).

Notes. Possibly to be considered native in S.E. Europe but exact data not available. Finds from the Baltic Sea area mapped in P. Uotila, Ann. Bot. Fennici 14: 203 (1977) (the taxon not recognized in Fl. Eur.).

Chenopodium zerovii Iljin

Notes. »C. & S. Ukraine» (Fl. Eur.). Not mapped separately.

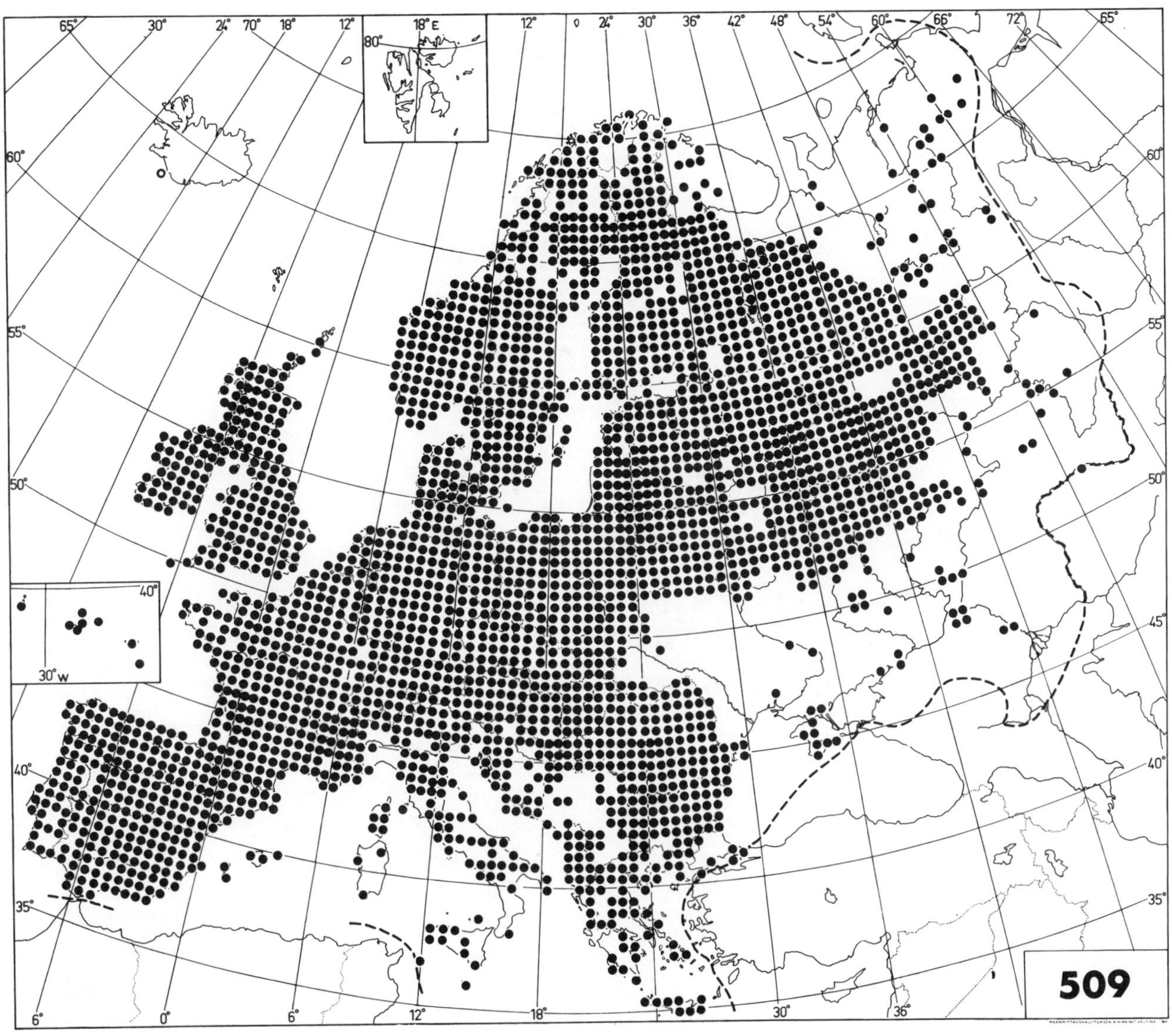

Chenopodium album sensu lato

Chenopodium suecicum J. Murr — Map 510.

Chenopodium album subsp. pseudopulifolium (J.B. Scholz) J. Murr; C. neumanii J. Murr; C. pseudomurale J. Murr; C. pseudopulifolium (J.B. Scholz) J. Murr; C. viride sensu Aellen, non L.

Taxonomy. A. Pedersen, Flora og Fauna 73: 127—131 (1967); P. Uotila, Ann. Bot. Fennici 9: 29—32 (1972), 11: 44—58 (1974), Acta Bot. Fennica 108: 1—35 (1978).

Notes. Au, Br, Cz, Da, Fe, Ga, Ge, Hb, He, Ho, Is, It, No, Po, Rs(N, B, C, W, E), Su. The map was prepared by P. Uotila mainly on the basis of material seen by himself (published in Acta Bot. Fennica 108: 23 (1978), slightly complemented) and does not entirely agree with the data provided by the individual countries (also reported from Hu and Rm). It was not possible to distinguish between native occurrences and established aliens. Casual towards the south of the range. See also Hegi 1960: 634—635, and P. Aellen, Phyton (Austria) 10: 259—260 (1963).

Total range. Hegi 1960: 635 (Eurasia); P. Uotila, Acta Bot. Fennica 108: 22 (1978).

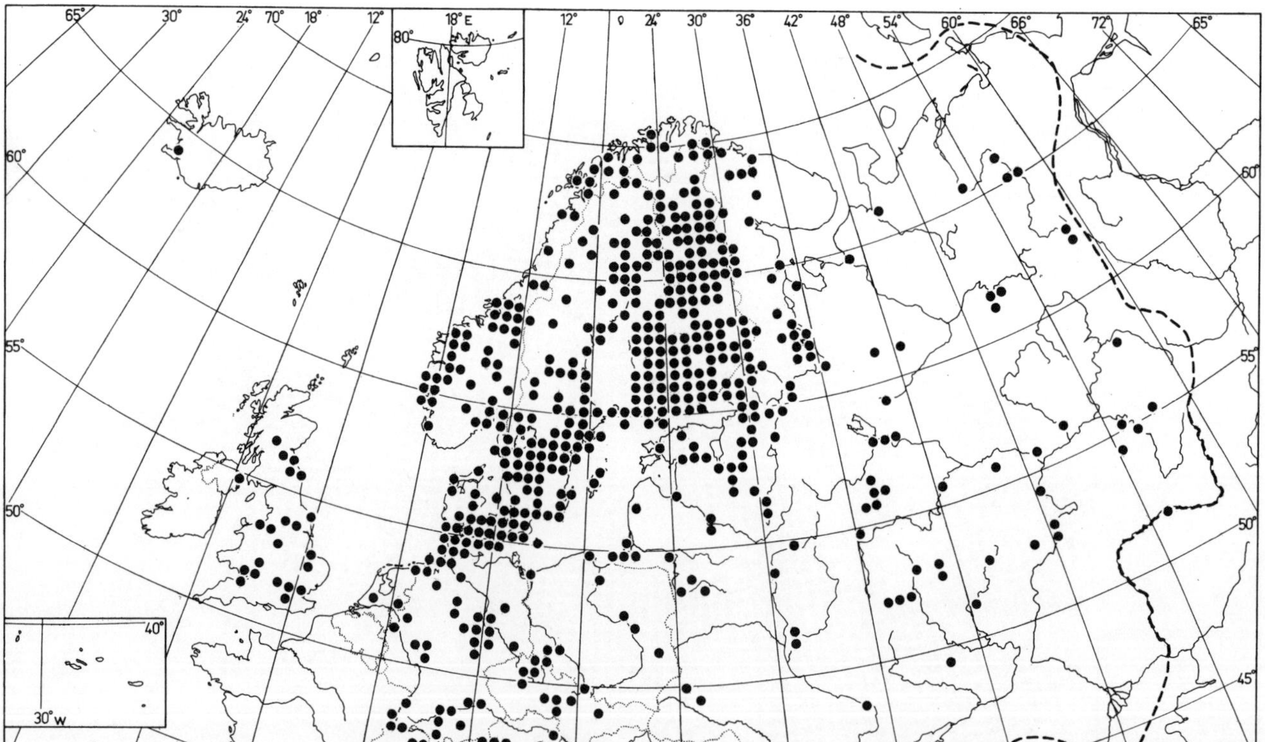

Chenopodium suecicum

Chenopodium giganteum D. Don — Map 511.

Chenopodium album subsp. amaranticolor Coste & Reyn.; C. amaranticolor (Coste & Reyn.) Coste & Reyn.

Notes. [Cr, Gr, Si] added (not given in Fl. Eur.).
Native of N. India.

Cycloloma atriplicifolia (Sprengel) Coulter — Map 512.

Cycloloma platyphylla (Michx) Moq.

Notes. [Al] added (not given in Fl. Eur.).
Native of C. & W. North America.

Spinacia oleracea L.

Spinacia glabra Miller; S. inermis Moench; S. spinosa Moench

Notes. »A widely cultivated vegetable, occurring throughout most of Europe as an escape from cultivation, but rarely persisting» (Fl. Eur.). Not mapped.
Native origin unknown (?W. Asia).

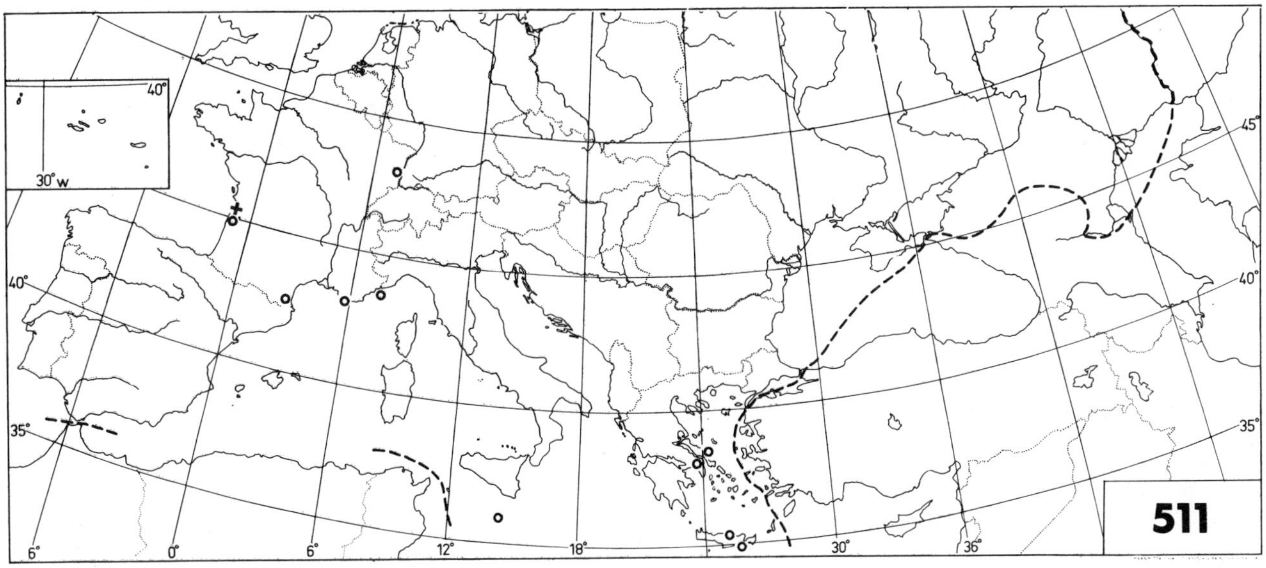

Chenopodium giganteum

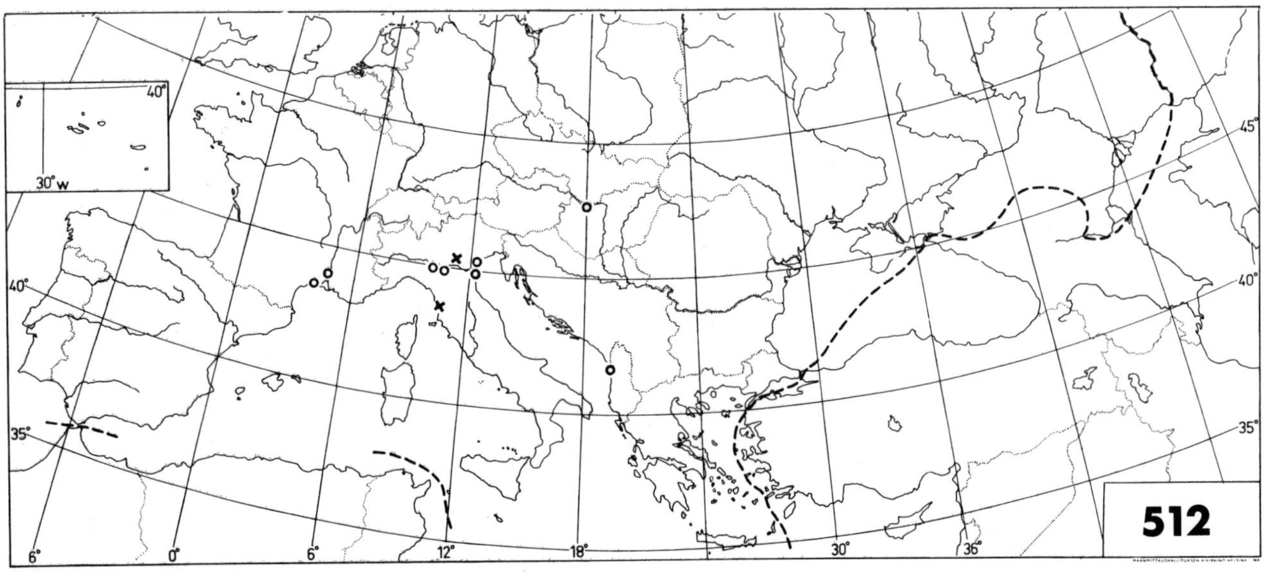

Cycloloma atriplicifolia

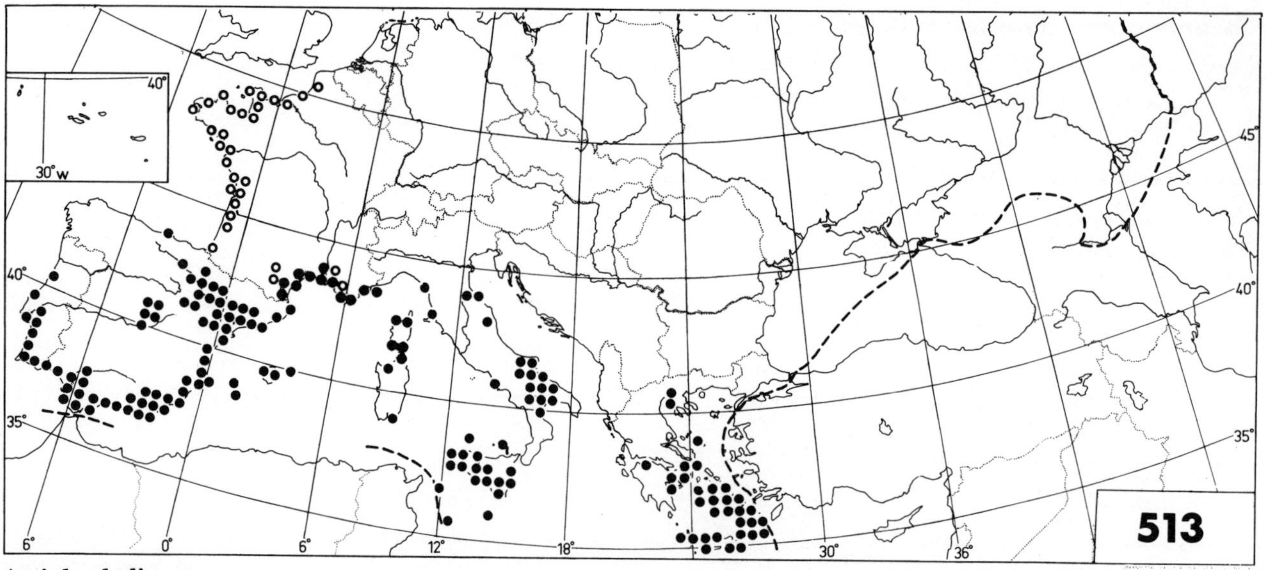

Atriplex halimus

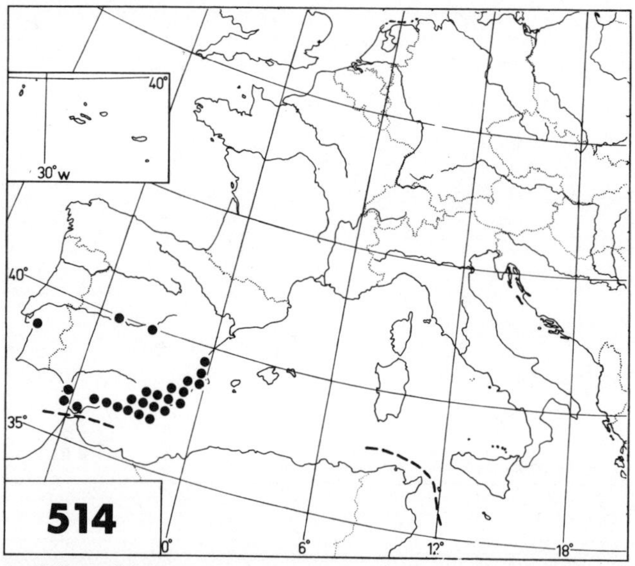

Atriplex glauca

Atriplex halimus L. — Map 513.

Notes. [Be] and Bu omitted (given in Fl. Eur.)

Atriplex glauca L. — Map 514.

Atriplex cana Ledeb. — Map 515.

Atriplex mollis Desf. — Map 515.

Notes. »Recorded from Malta but not seen for many years and possibly only introduced» (Fl. Eur.).
Native of E. Mediterranean and N. Africa.

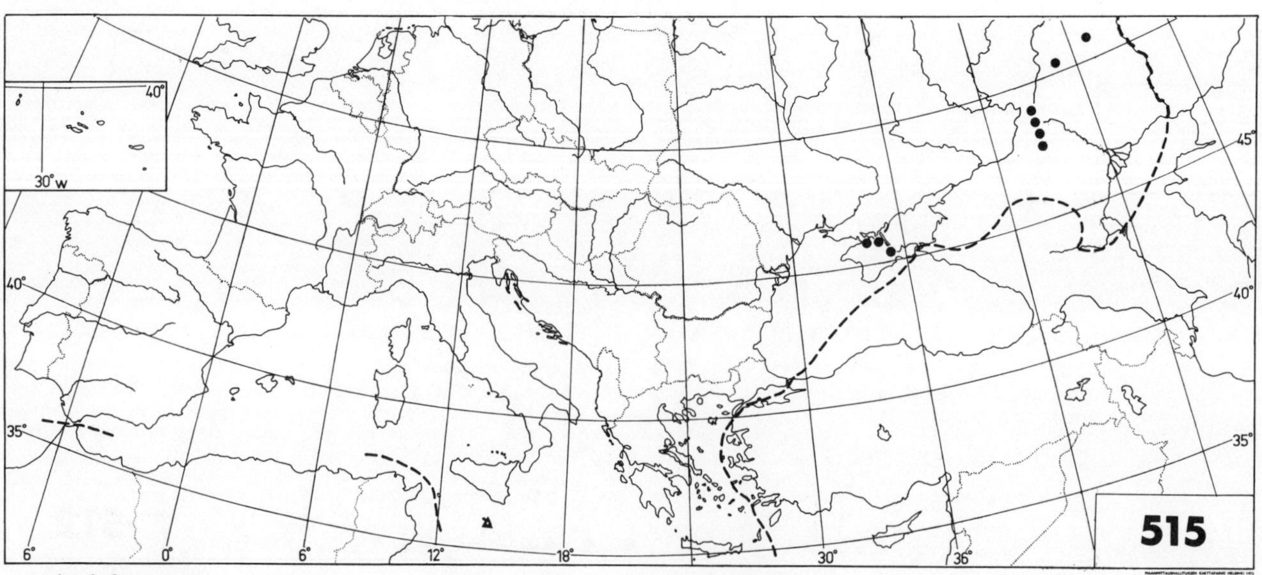

• = **Atriplex cana** ▲ = **A. mollis**

Atriplex hortensis L.

Notes. »Cultivated throughout most of Europe as a vegetable and naturalized in the centre and south» (Fl. Eur.). Not mapped. *Native* of C. Asia. Hultén Alaska 1968: 399.

Atriplex acuminata Waldst. & Kit. — Map 516.

Atriplex nitens Schkuhr; A. hortensis subsp. nitens (Schkuhr) E. Pons

Nomenclature. S. Priszter, Bot. Közl. 56: 213—215 (1969); R. Soó, Feddes Repert. 83: 142 (1972); S. Rauschert, Feddes Repert. 85: 642—643 (1974).

Notes. [Ga] and Rs(B) added (not given in Fl. Eur.). Doubtful native in It (given as native in Fl. Eur.). Data lacking for most of Rs(W).

Total range. Hegi 1960: 672; H. Meusel & A. Buhl, Wiss. Zeitschr. Univ. Halle-Wittenb. Math.-Nat. 11: 1275 (1962); MJW 1965: map 133d.

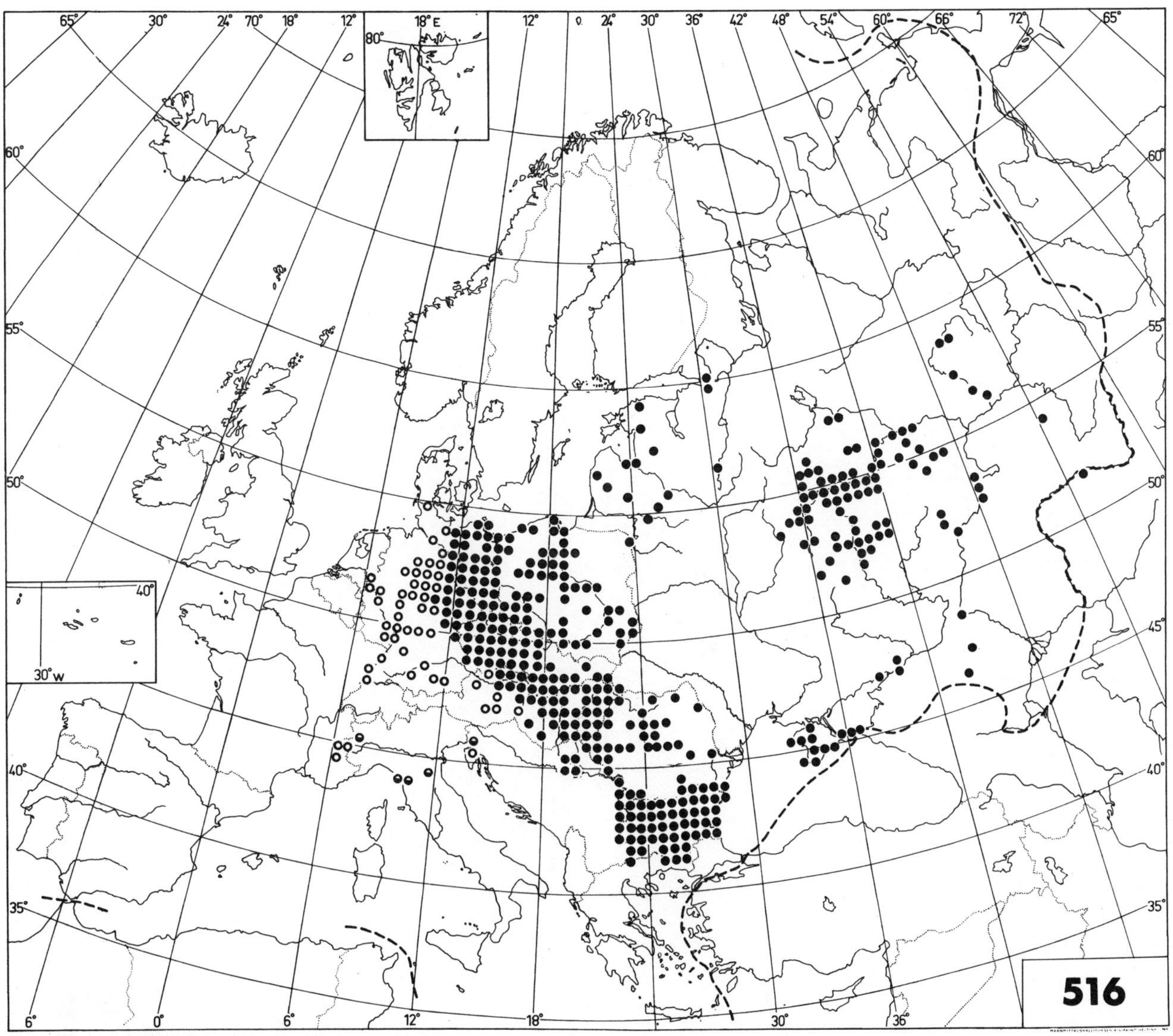

Atriplex acuminata

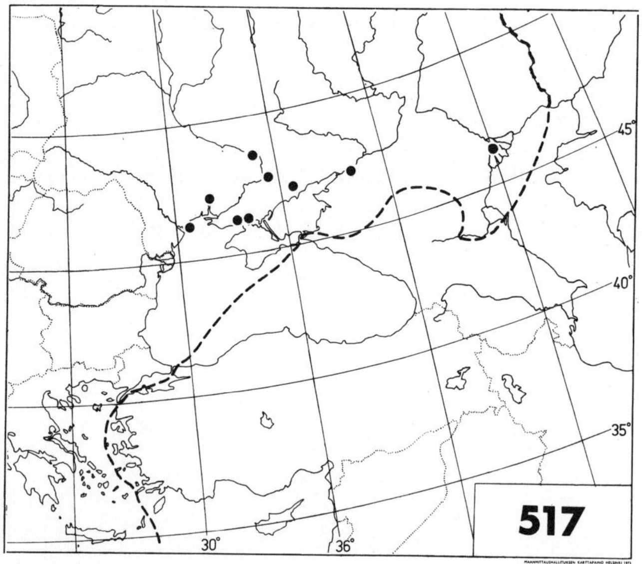

Atriplex aucheri

Atriplex aucheri Moq. — Map 517.

Atriplex amblyostegia Turcz.; A. desertorum D. Sosn.

Taxonomy. It has been claimed (V.I. Czopik *in litt.* 1978) that the Ukrainean *Atriplex aucheri* is not identical with *A. amblyostegia* of Kazahstan and that they should possibly be treated as two separate taxa.

Notes. Rs(K) omitted (given in Fl. Eur.), not mentioned for Rs(K) in Rubtsev 1972.

Total range. H. Meusel & A. Buhl, Wiss. Zeitschr. Univ. Halle-Wittenb. Math.-Nat. 11: 1275 (1962); MJW 1965: map 133d.

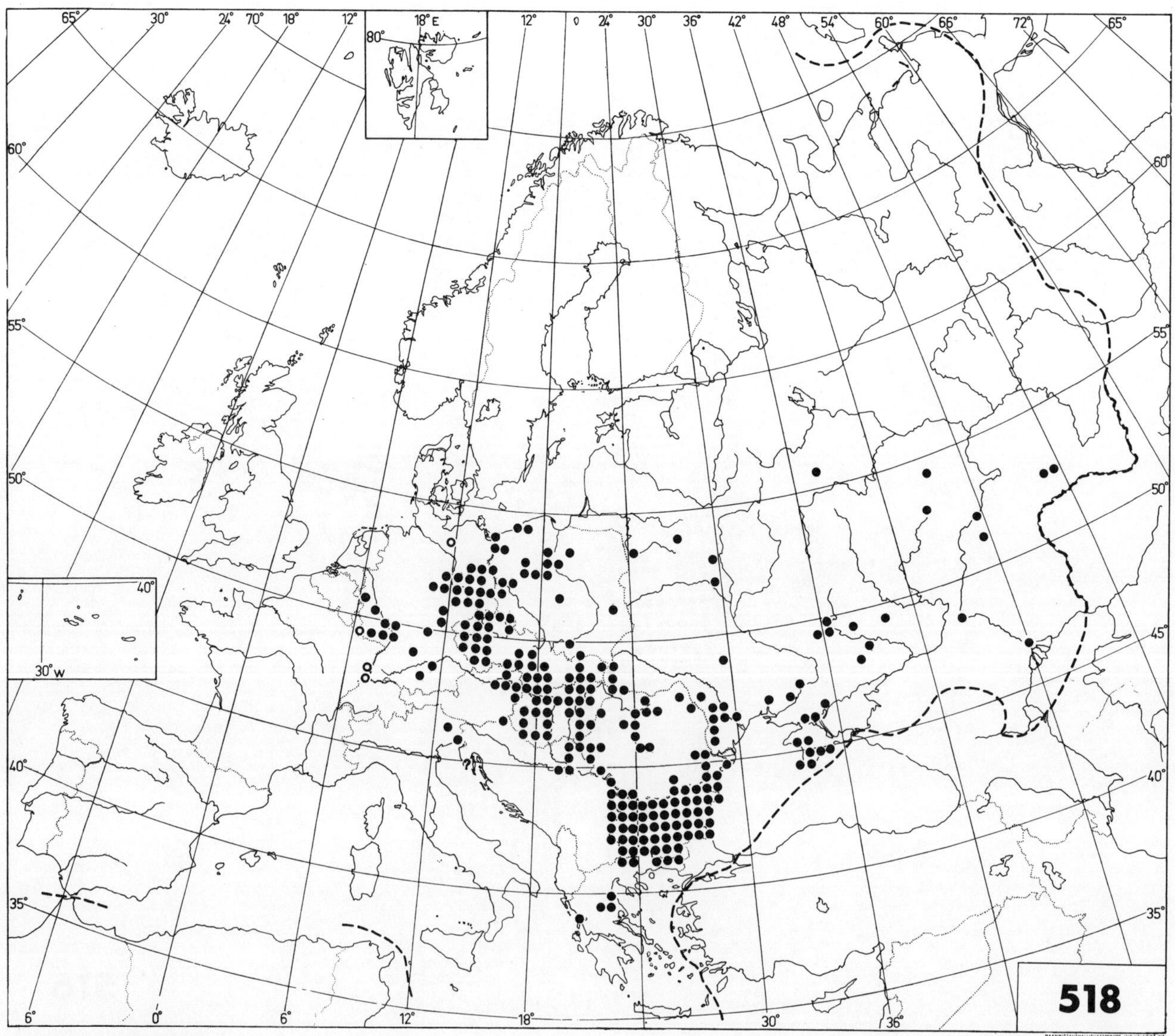

Atriplex oblongifolia

Atriplex oblongifolia Waldst. & Kit. — Map 518.

Notes. Naturalized alien in Ga (given as native in Fl. Eur.).
Total range. MJW 1965: map 134a.

Atriplex heterosperma Bunge — Map 519.

Atriplex hortensis subsp. heterosperma (Bunge) Meijden; A. micrantha C.A. Meyer

Taxonomy. R. van der Meijden, Gorteria 4: 103—108 (1968).
Notes. No records available from Au, Ga, Po and Rm ([?Au, Ga, Po, ?Rm] in Fl. Eur.). [Hs] and Rs(W) added (not given in Fl. Eur.): J. Vigo, Butll. Inst. Catalana Hist. Nat. 38 (Sec. Bot. 1): 50—51 (1974).

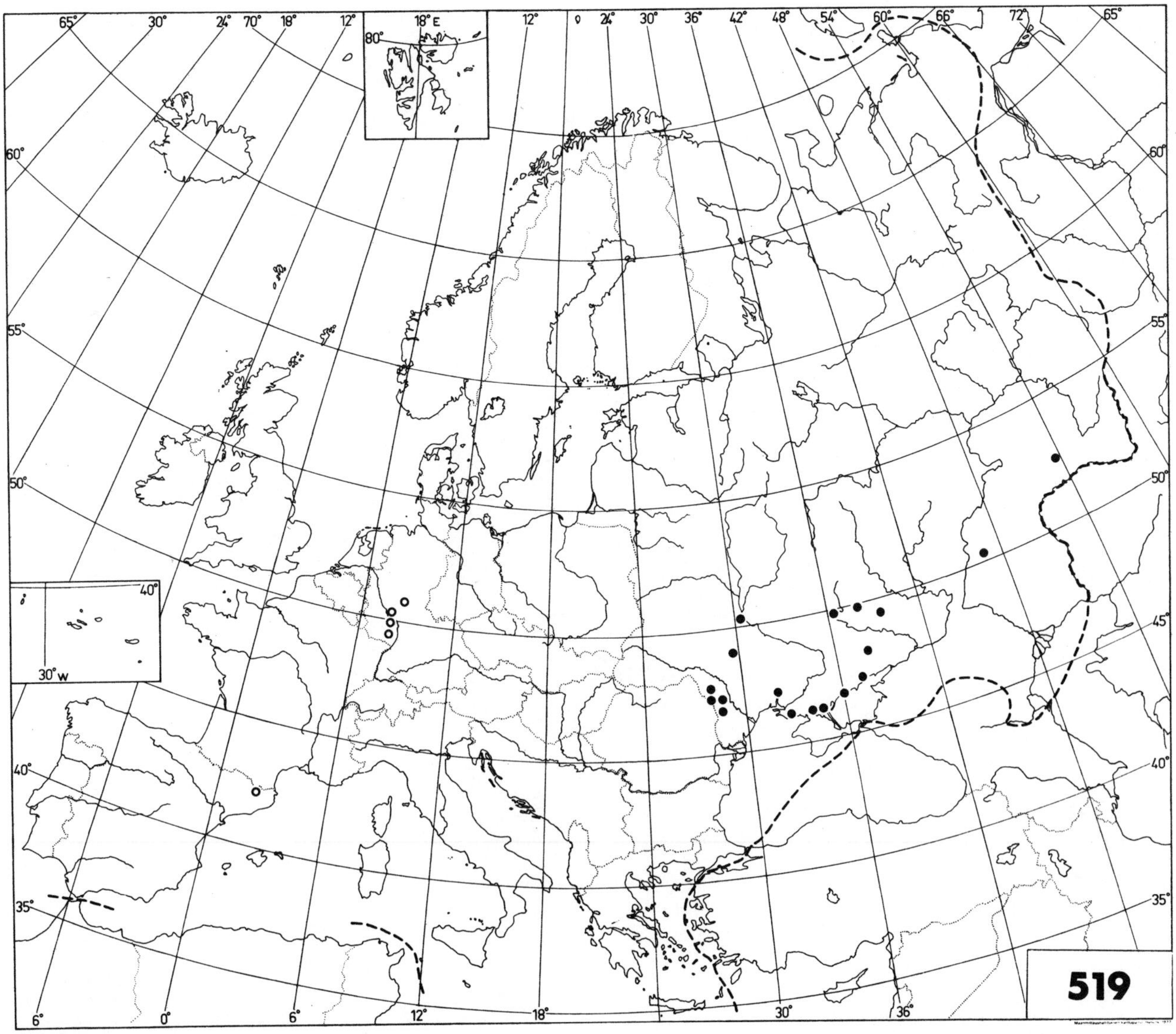

Atriplex heterosperma

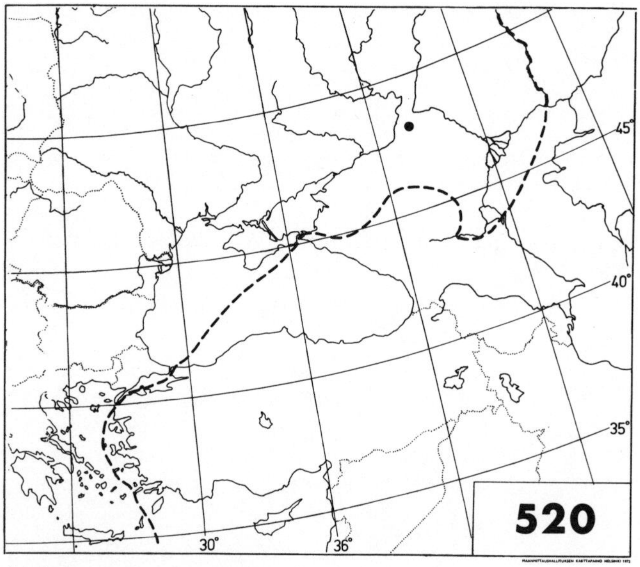

Atriplex sphaeromorpha

Atriplex sphaeromorpha Iljin — Map 520.

Notes. Rs(W) omitted (given in Fl. Eur.).

Atriplex rosea L. — Map 521.

Taxonomy. The taxonomical status of *Atriplex rosea* subsp. *tarraconensis* (Sennen) O. Bolòs & Vigo, Butll. Inst. Catalana Hist. Nat. 38 (Sec. Bot. 1): 88 (1974), needs further study.

Notes. Al, [Be, Br, He, Ho] omitted (given in Fl. Eur.), Rs(B, C, W, K, E) added (not given in Fl. Eur.); map incomplete for Rs(W).

Total range. Hegi 1960: 684; MJW 1965: map 134d.

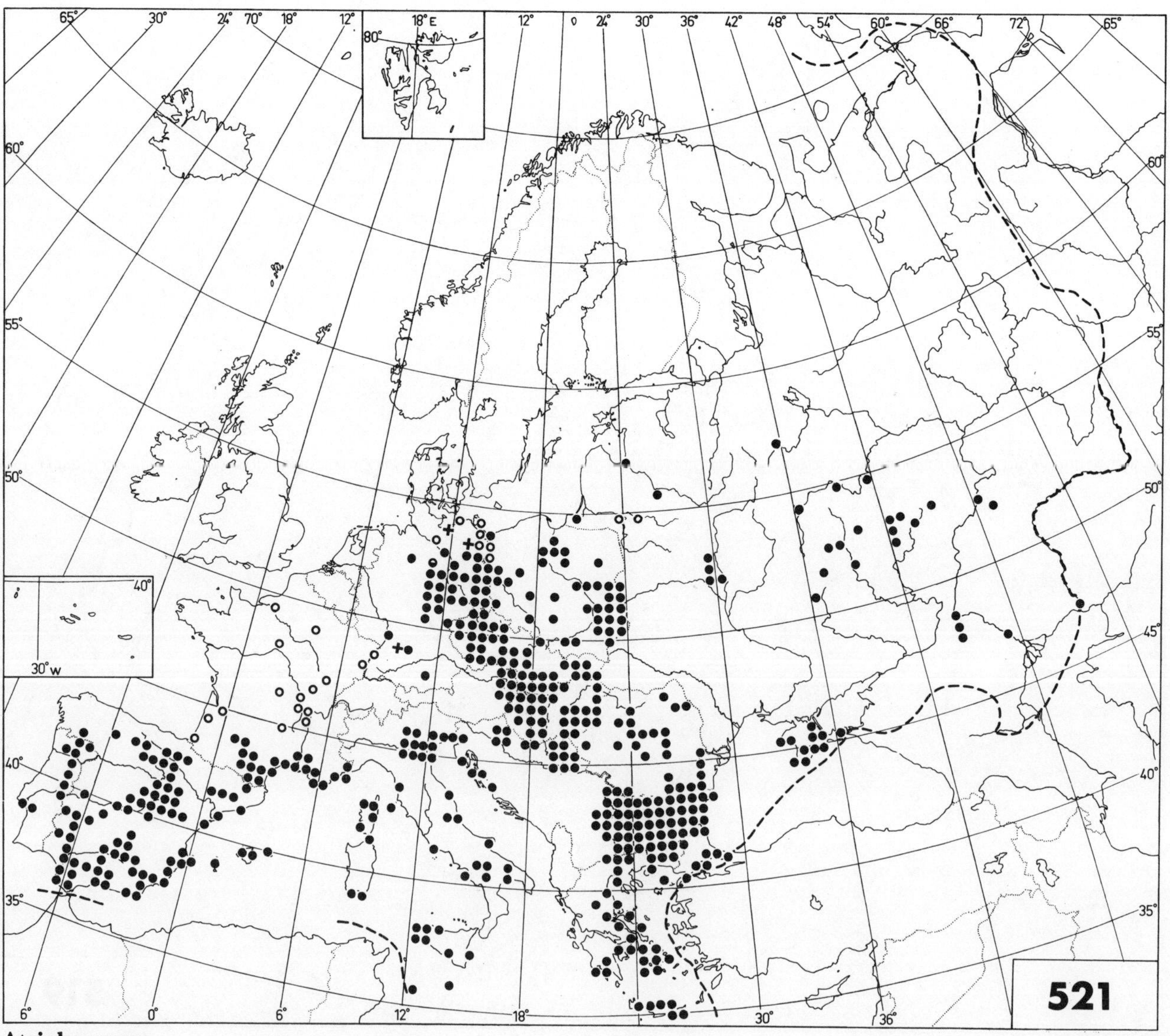

Atriplex rosea

Atriplex laciniata L. — Map 522.

Atriplex arenaria J. Woods; A. maritima L.; A. sabulosa Rouy

Taxonomy and nomenclature. P.M. Taschereau, Canad. Jour. Bot. 50: 1591—1592 (1972).

Notes. A. Hansen & A. Pedersen, Bot. Tidsskr. 63: 219 (1968), among others, have drawn attention to the northward expansion of the species into Scotland, No and Su, during the last few decades.

Total range. Hultén AA 1958: map 267.

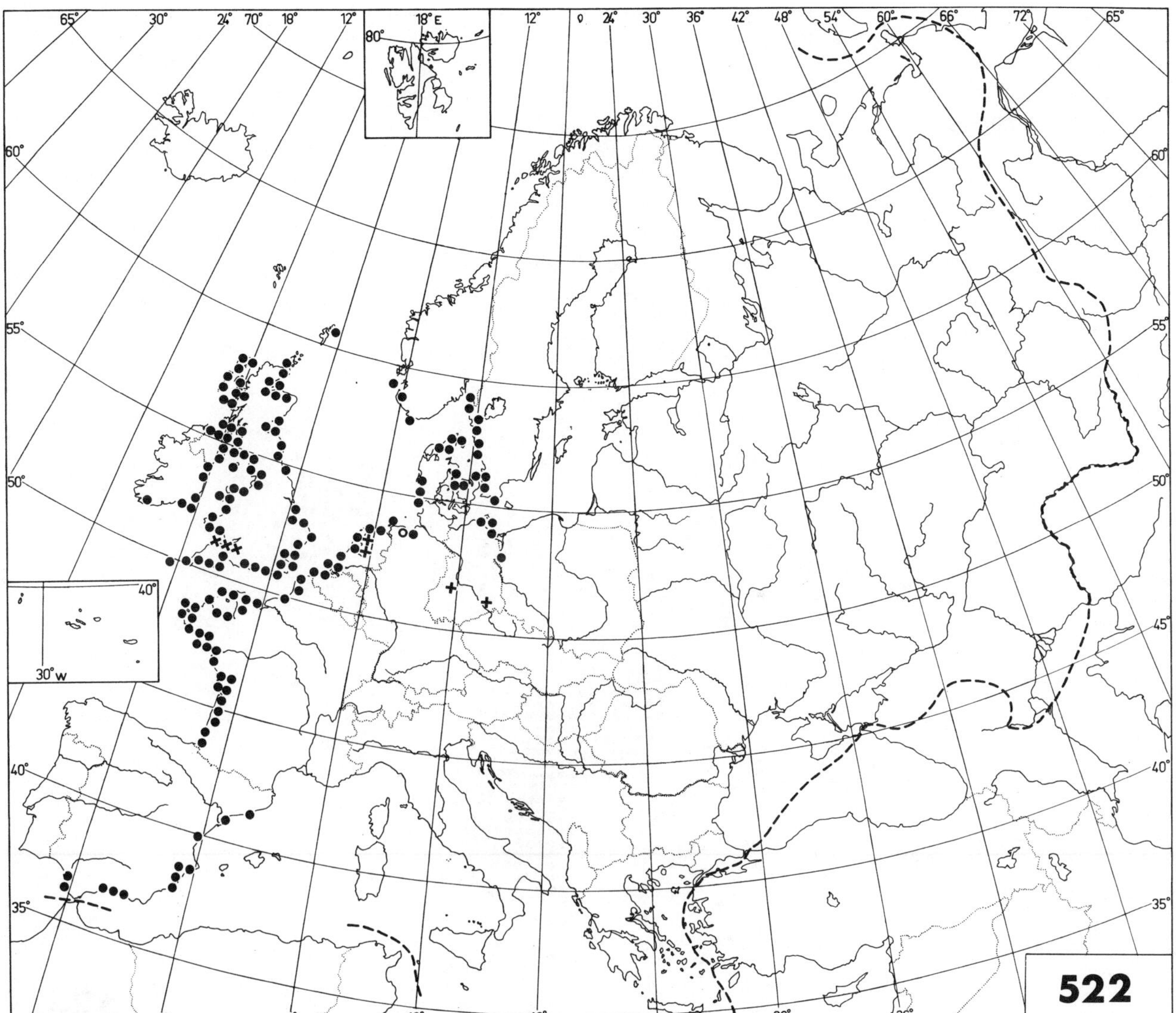

Atriplex laciniata

Atriplex tatarica L. — Map 523.

Atriplex laciniata auct. non L., ? Chenopodium serotinum L. Incl. A. tornabenii Tineo, excl. A. recurva D'Urv.

Nomenclature. Concerning the typification of *Chenopodium serotinum* L., see P. Uotila, Ann. Bot. Fennici 14: 197—198 (1977).

Notes. Al, [Be] and Sa omitted (given in Fl. Eur.). Bl added (not given in Fl. Eur.): Fl. Mallorca 1978: 101, P. Duvigneaud, Soc. Échange Pl. Vasc. Europe Occ. Bassin Médit. 17 (suppl.): 15 (1979). Ga and Rs([N, B], C, W, K, E) added (not given in Fl. Eur.); map incomplete for Rs(W). Established alien in Ge and Po (given as native in Fl. Eur.).

Atriplex recurva D'Urv. — Map 524.

Atriplex tatarica auct. non L.; A. tatarica subsp. recurva (D'Urv.) Rech. fil.

Taxonomy. M. Gustafsson, Bot. Not. 123: 371—383 (1970).

Notes. Cr, Gr (included in *Atriplex tatarica* in Fl. Eur.).

Total range. Endemic to Europe.

Atriplex littoralis L. — Map 525.

Atriplex laciniata auct. non L.

Taxonomy. P.M. Taschereau, Canad. Jour. Bot. 50: 1581—1582 (1972).

Notes. Sa omitted (given in Fl. Eur.), Rs(B, C, W, E) added (not given in Fl. Eur.); map incomplete for Rs(W).

Total range. Hegi 1960: 677; MJW 1965: map 134b; F. Fukarek, Natur Naturschutz Mecklenburg 5: 167 (1967); N.A. Miniaev, Vestn. Leningrad Univ. (Biol.) 1969: 57 (1969).

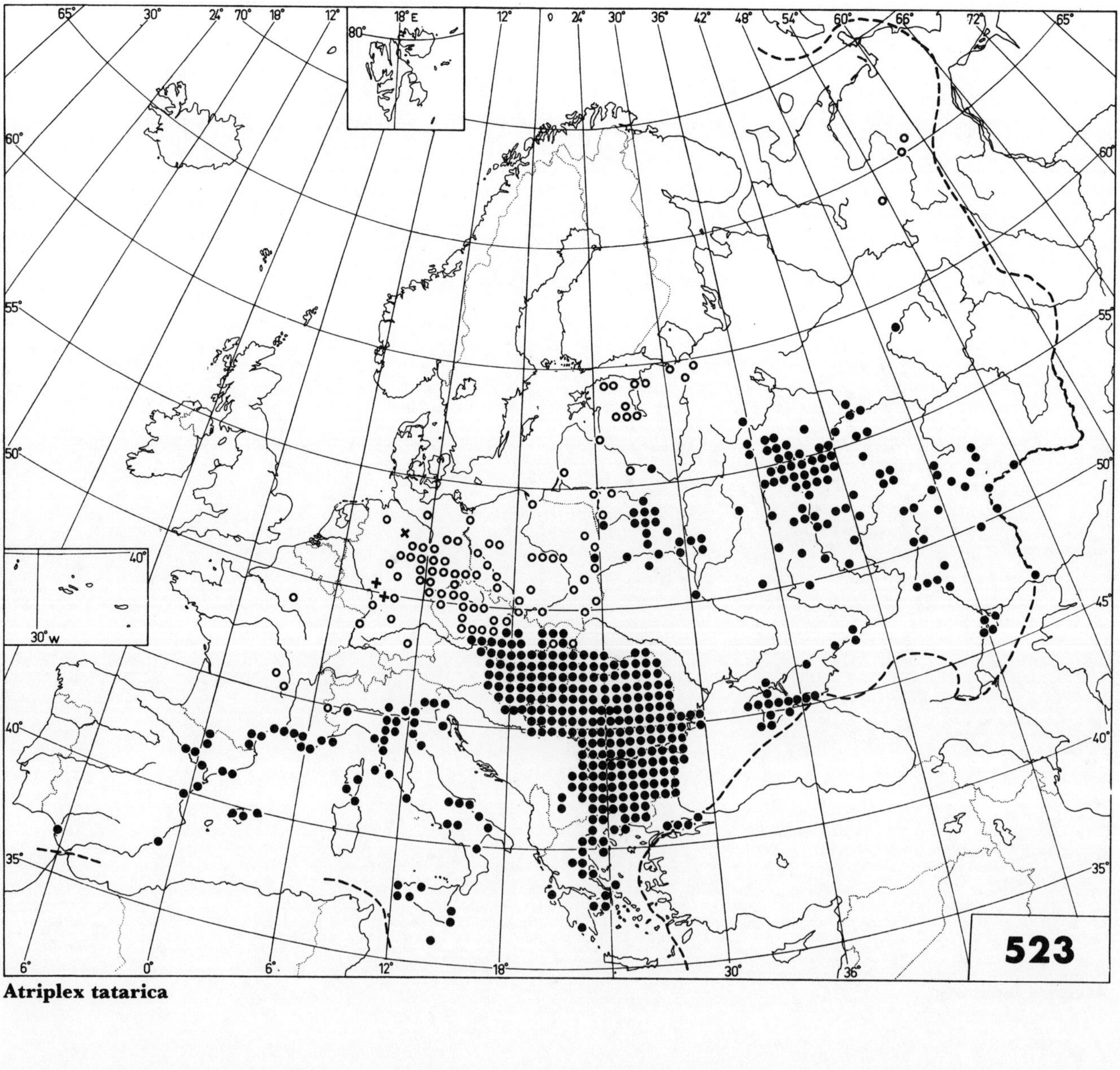

Atriplex tatarica

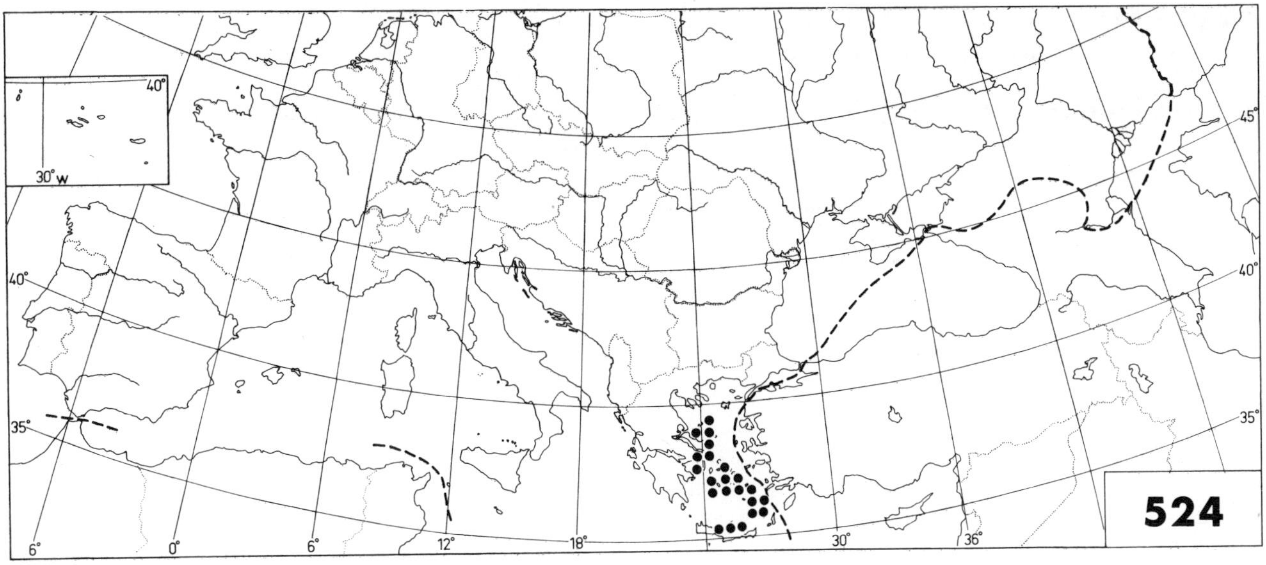

Atriplex recurva

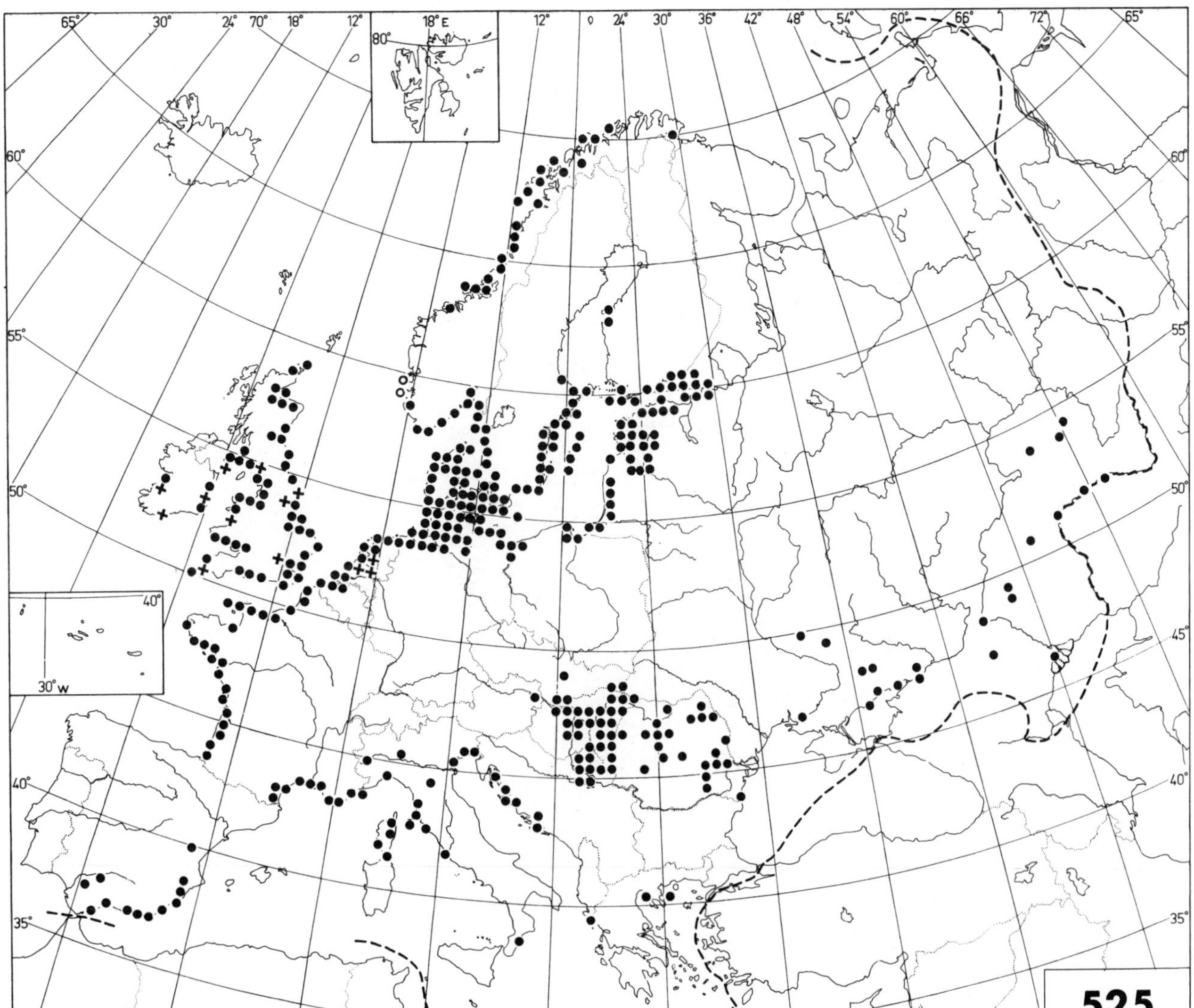

Atriplex littoralis

Atriplex patens (Litv.) Iljin — Map 526.

Notes. Rs(C) added (not given in Fl. Eur.).

Atriplex patula L. — Map 526.

Taxonomy. P.M. Taschereau, Canad. Jour. Bot. 50: 1574—1576 (1972).
Notes. Al, Az and Cr added (not given in Fl. Eur.), Fa and Is omitted (given in Fl. Eur.). Map incomplete for Rs(W).
Total range. Hultén Alaska 1968: 398 (northern hemisphere).

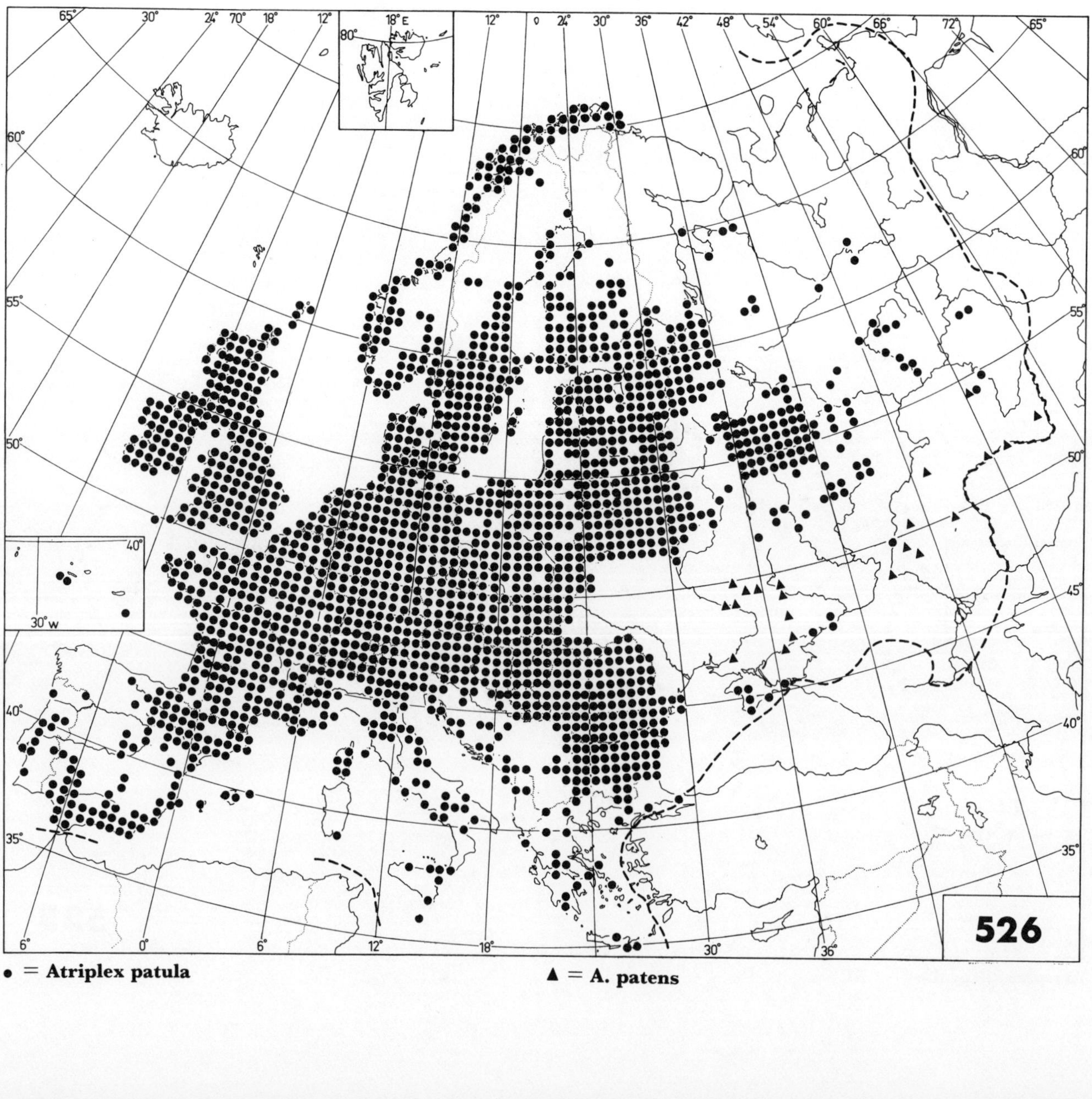

• = **Atriplex patula** ▲ = **A. patens**

Atriplex calotheca (Rafn) Fries — Map 527.

Atriplex hastata L., nomen ambig.; A. hastifolia Salisb., nomen ambig.; A. prostrata subsp. calotheca (Rafn) M. Gust.

Taxonomy. M. Gustafsson, Op. Bot. (Lund) 39: 1—63 (1976).
Nomenclature. S. Rauschert, Feddes Repert. 85: 643 (1974); E.M. Jones, Watsonia 10: 248 (1975).
Total range. Endemic to Europe (although not mentioned as such in Fl. Eur.).

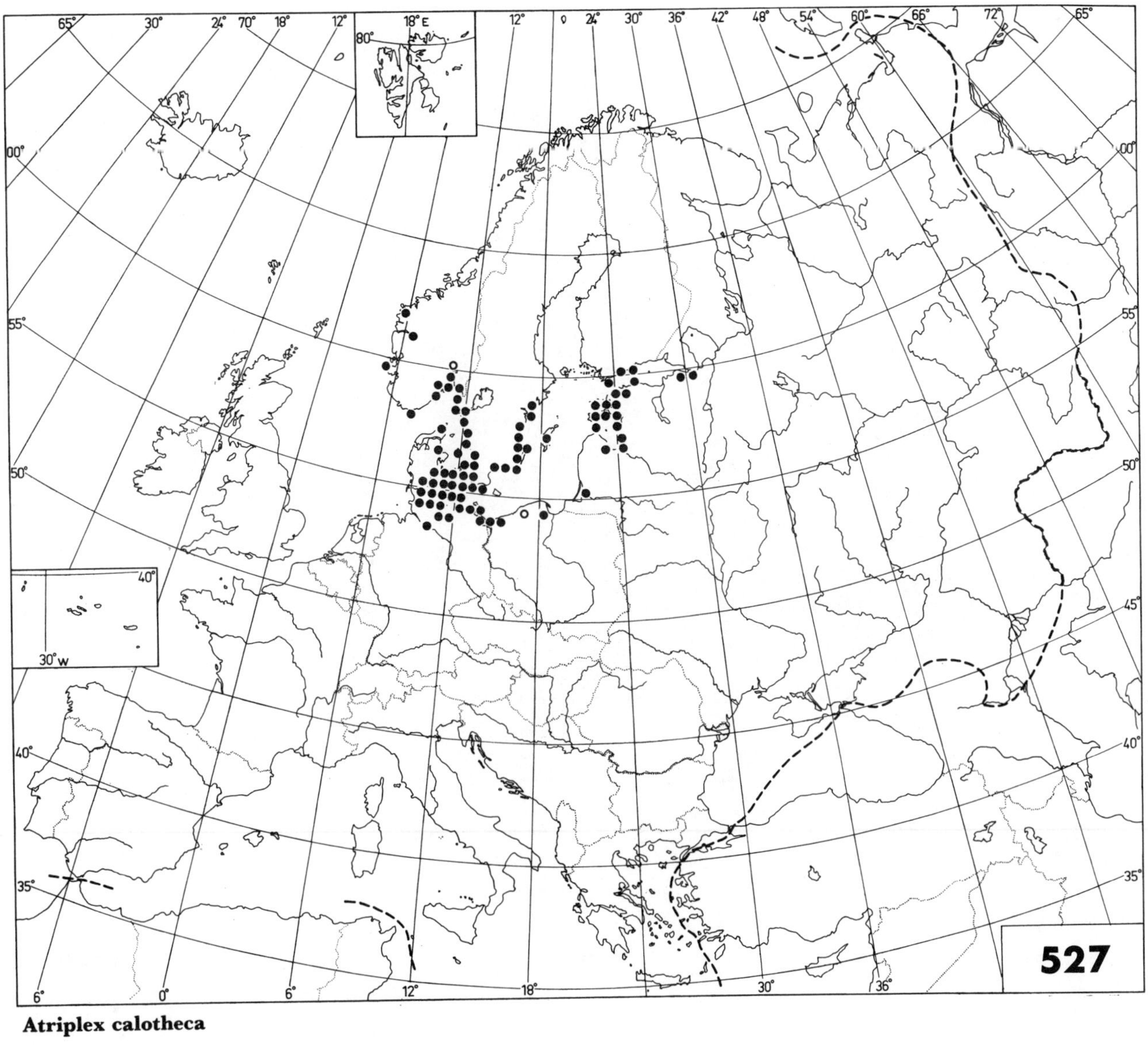

Atriplex calotheca

Atriplex prostrata Boucher ex DC. — Map 528.

Atriplex deltoidea Bab.; A. hastata auct non L.; A. latifolia Wahlenb.; A. oppositifolia DC.; A. ruderalis Wallr.; A. triangularis Willd.

Taxonomy and nomenclature. A. Pedersen, Bot. Tidsskr. 63: 289—303 (1968); P.M. Taschereau, Canad. Jour. Bot. 50: 1583—1586 (1972); S. Rauschert, Feddes Repert. 85: 643—644 (1974); E.M. Jones, Watsonia 10: 248—249 (1975); M. Gustafsson, Op. Bot. (Lund) 39: 1—63 (1976; preprint 1975). The more distinct variants are listed below as five subspecies, although they are not recognized in Fl. Eur.

Notes. Cr added (not given in Fl. Eur.): W. Greuter, Ann. Mus. Goulandris 1: 28 (1973). Map incomplete for Rs(W). The subspecies are not mapped separately.

Total range. Hultén CP 1971: map 190.

A. prostrata subsp. **prostrata**

Atriplex latifolia subsp. prostrata (Boucher ex DC.) Hiitonen

Notes. Da, Fe, Ga, Ge, No, Po, Rs(B, C), Su.

A. prostrata subsp. **polonica** (Zapał.) Uotila

Atriplex hastata subsp. polonica (Zapał.) Aellen

Taxonomy. P. Aellen, Feddes Repert. 69: 144 (1964); P. Uotila, Ann. Bot. Fennici 14: 197 (1977).
Notes. ?Cz (no material traced), Po, Rs(W). »S.E. Poland, E. Czechoslovakia and W. Ukraine» (Fl. Eur.).

A. prostrata subsp. **triangularis** (Willd.) Rauschert

Atriplex latifolia subsp. triangularis (Willd.) Hiitonen

Nomenclature. S. Rauschert, Feddes Repert. 85: 643—644 (1974).
Notes. Au, Da, ?Fe, Ge, Hu, It.

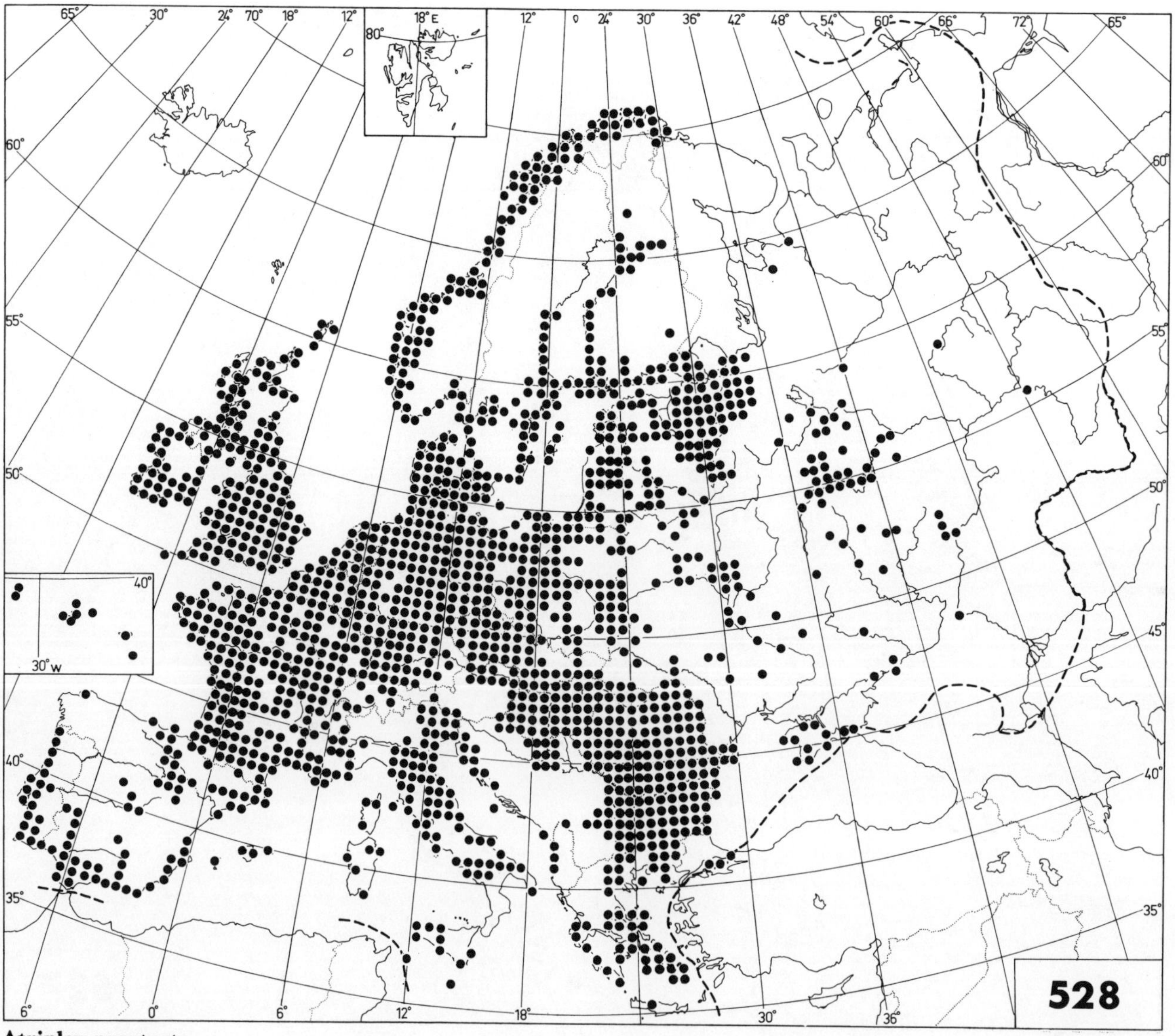

Atriplex prostrata

A. prostrata subsp. **latifolia** (Wahlenb.) Rauschert

Atriplex latifolia subsp. hastifolia Hiitonen, non A. hastifolia Salisb.

Nomenclature. S. Rauschert, Feddes Repert. 85: 644 (1974).
Notes. Au, Cz, Da, Ge, [He], Hu, It, Ju, Su, and certainly elsewhere.

A. prostrata subsp. **deltoidea** (Bab.) Rauschert

Atriplex hastata subsp. deltoidea (Bab.) Nyman; A. latifolia subsp. deltoidea (Bab.) Hiitonen

Nomenclature. S. Rauschert, Feddes Repert. 85: 644 (1974).
Notes. Br, Da, Fe, Ga, Ge, Su, and certainly elsewhere.

Atriplex glabriuscula Edmondston — Map 529.

Atriplex babingtonii J. Woods; A. patula subsp. glabriuscula (Edmondston) Hall & Clements

Taxonomy. P.M. Taschereau, Canad. Jour. Bot. 50: 1589—1591 (1972).
Notes. Be omitted (given in Fl. Eur.), the locality given in Atlas Fl. Belge Luxemb., text p. 55 (1978) being in Ho. Hs not confirmed (?Hs in Fl. Eur.), Rs(N, B) added (not given in Fl. Eur.).
Total range. Hultén AA 1958: map 266; MJW 1965: map 134c.

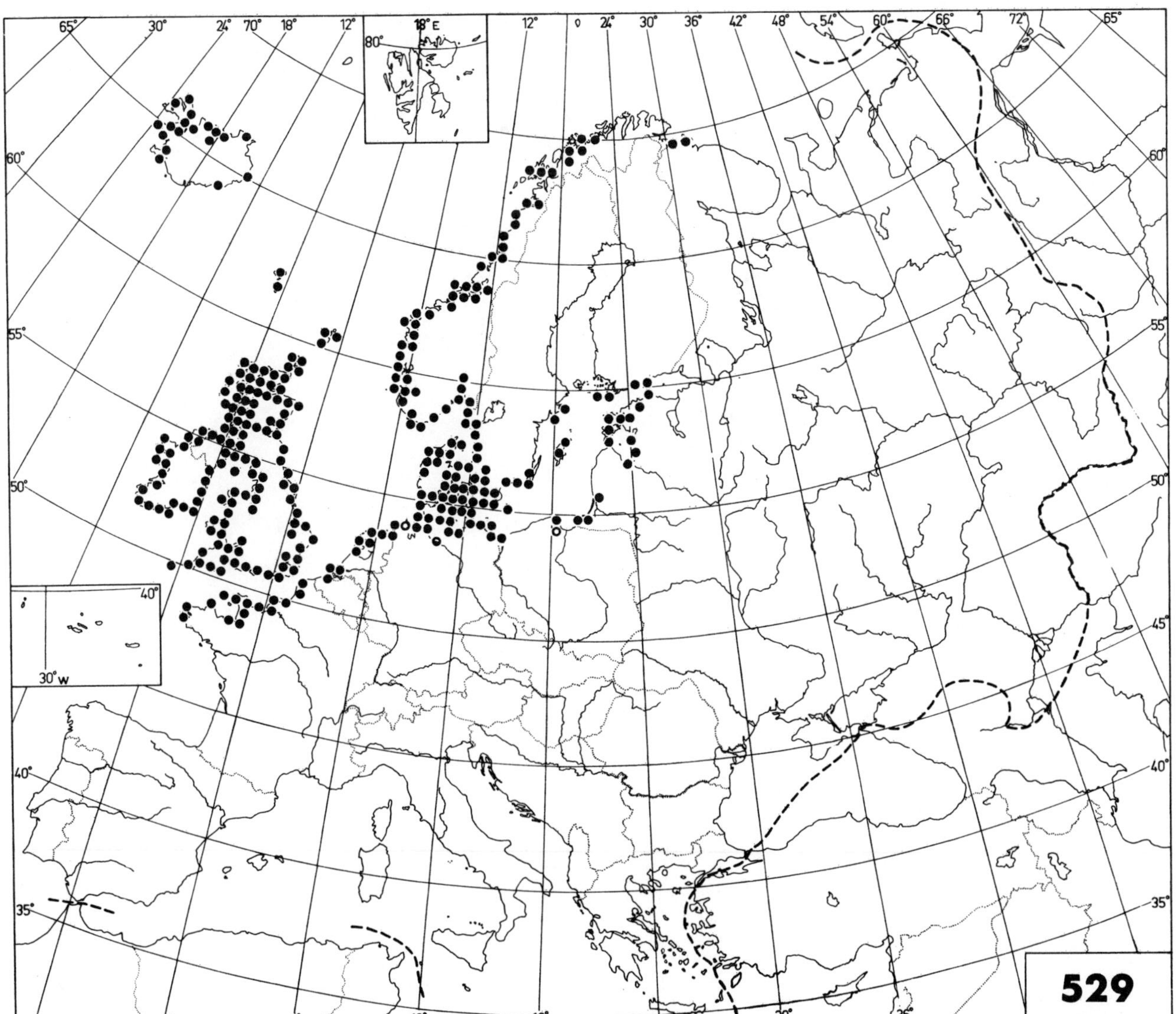

Atriplex glabriuscula

Atriplex longipes Drejer — Map 530.

Incl. Atriplex kuzenevae N. Semen., A. lapponica Pojark., A. nudicaulis Boguslaw and A. praecox Hülphers

Taxonomy. M. Gustafsson, Hereditas 71: 173—194 (1972), Op. Bot. (Lund) 39: 1—63 (1976), recognizes the two subspecies given below. The plant known as *Atriplex longipes* subsp. *kattegatensis* Turesson is very likely of hybrid origin, derived from crosses between *A. longipes* and *A. prostrata*: M. Gustafsson, Bot. Not. 126: 398—416 (1973). See also E.M. Jones, Watsonia 10: 249—250 (1975) and P.M. Taschereau, Watsonia 11: 195—198 (1977).

Notes. Is and Rs(B, C) added (not given in Fl. Eur.): M. Gustafsson, Op. Bot. (Lund) 39: 20 (1976). Native in No ([No] in Fl. Eur.).

A. longipes subsp. **longipes** — Map 530.

Notes. Br, Da, Fe, No, ?Po, Rs(B), Su. See P.M. Taschereau, Watsonia 11: 197 (1977), and also M. Gustafsson, Op. Bot. (Lund) 39: 19—20 (1976).

Total range. ? Endemic to Europe.

A. longipes subsp. **praecox** (Hülphers) Turesson — Map 530.

? Atriplex nudicaulis Boguslaw; A. praecox Hülphers

Notes. Br, Fe, Is, No, Rs(N, B), Su: P.M. Taschereau, Watsonia 11: 195—198 (1977); M. Gustafsson, Op. Bot. (Lund) 39: 20 and 22 (1976); Fl. Arct. URSS 1966: 182. Since the taxonomical status and affinities of those species described from the Arctic coasts of Rs(N) are still unsettled, the map of subsp. *praecox* must be considered tentative for that part of the range.

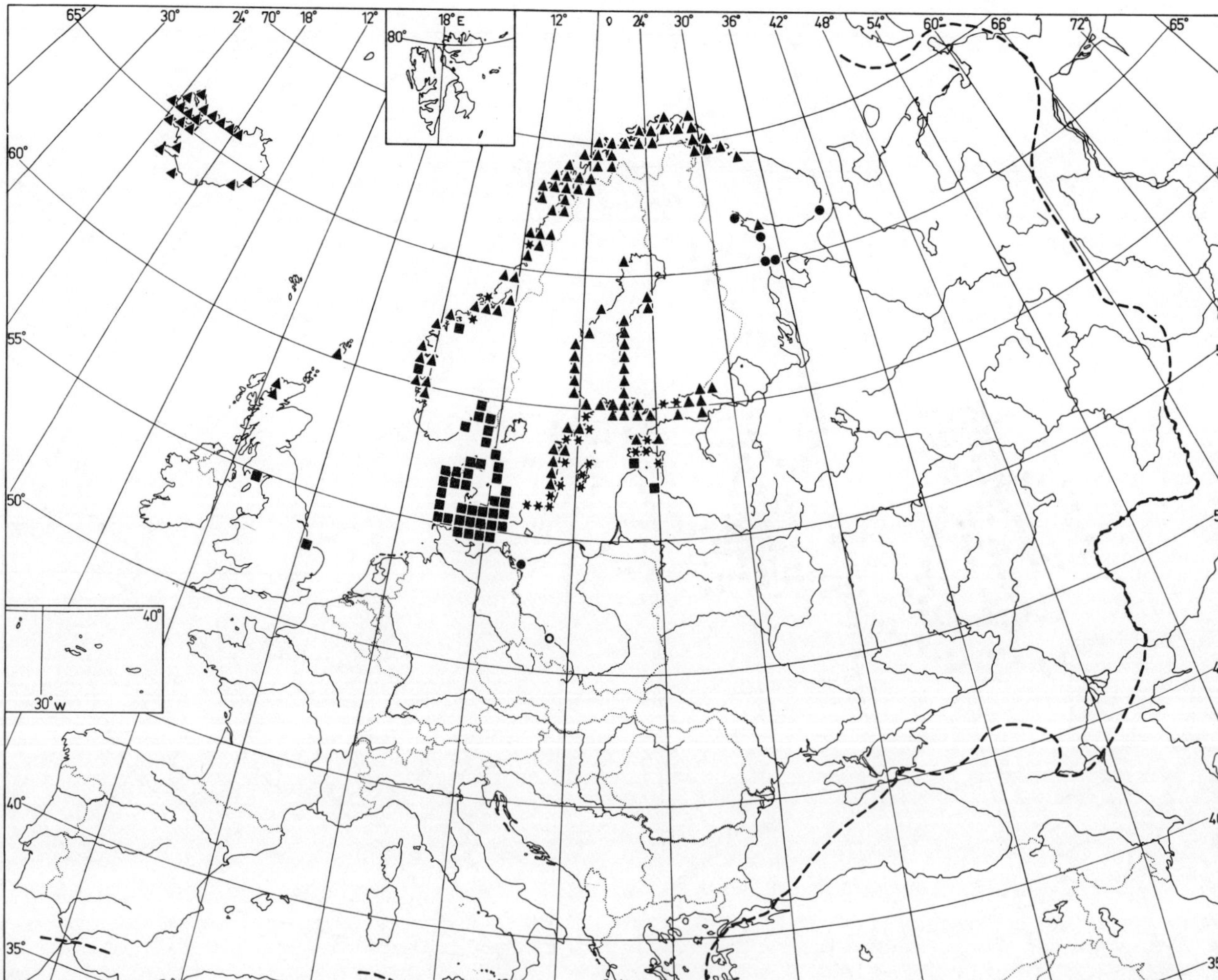

● = **Atriplex longipes**

■ = **A. longipes** subsp. **longipes**

▲ = **A. longipes** subsp. **praecox**

★ = subsp. **longipes** + subsp. **praecox**

Halimione portulacoides (L.) Aellen — Map 531.

Atriplex portulacoides L.; Obione portulacoides (L.) Moq.

Generic delimitation. Nova Fl. Port. 1971: 94, 550.

Notes. Cr added (not given in Fl. Eur.): W. Greuter, Ann. Mus. Goulandris 1: 29 (1973). Extinct in Su (not given in Fl. Eur.).

Total range. MJW 1965: map 133b.

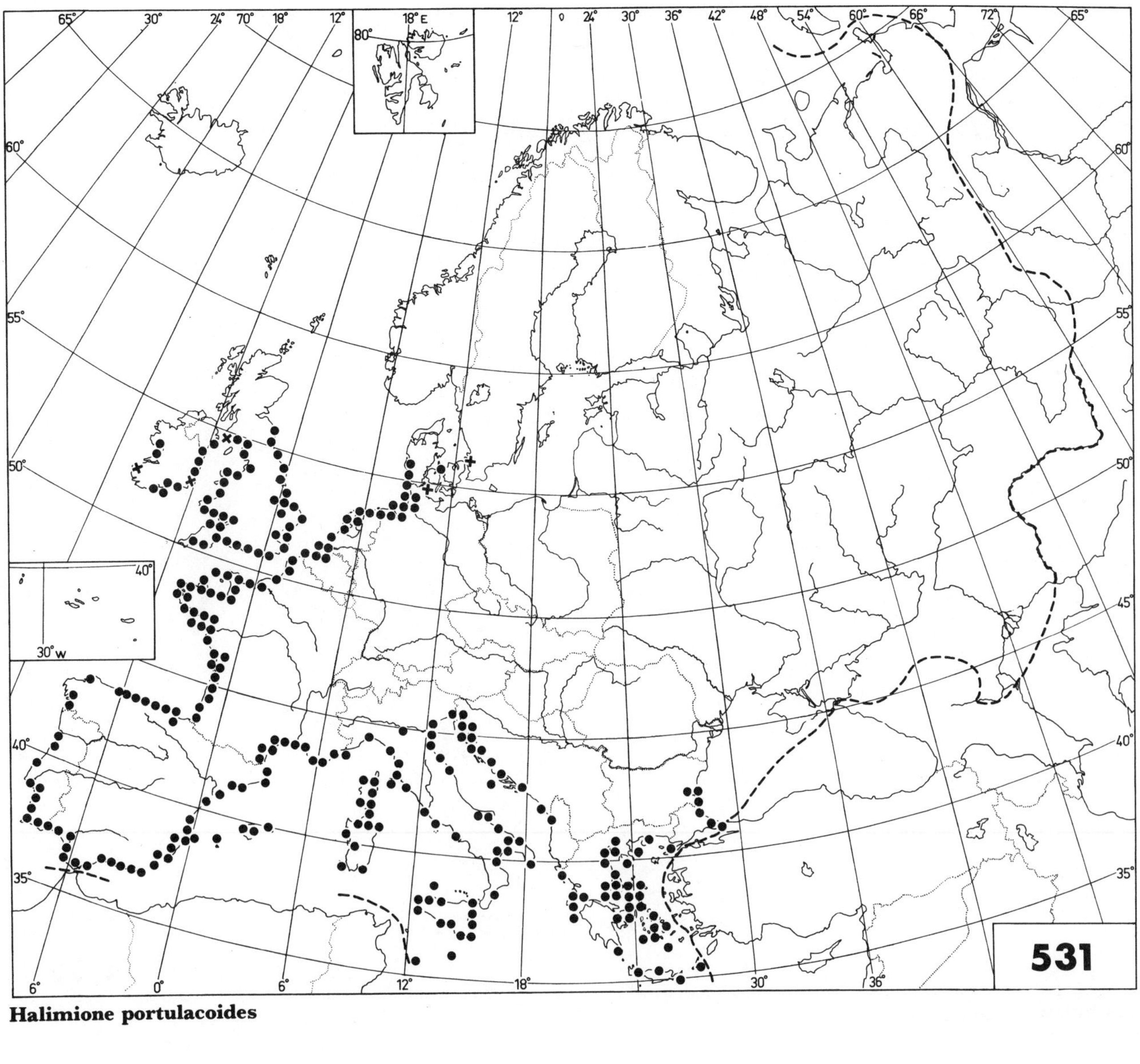

Halimione portulacoides

Halimione verrucifera (Bieb.) Aellen — Map 532.

Atriplex verrucifera Bieb.; Obione verrucifera (Bieb.) Moq.

Notes. Rs(C) and Tu added (not given in Fl. Eur.): Fl. Turkey 1967: 313.
Total range. MJW 1965: map 133b.

Halimione pedunculata (L.) Aellen — Map 533.

Atriplex pedunculata L.; Obione pedunculata (L.) Moq.

Notes. Rs(C) added (not given in Fl. Eur.). Map incomplete for Rs(W).
Total range. MJW 1965: map 133c; F. Fukarek, Natur Naturschutz Mecklenburg 5: 157 (1967).

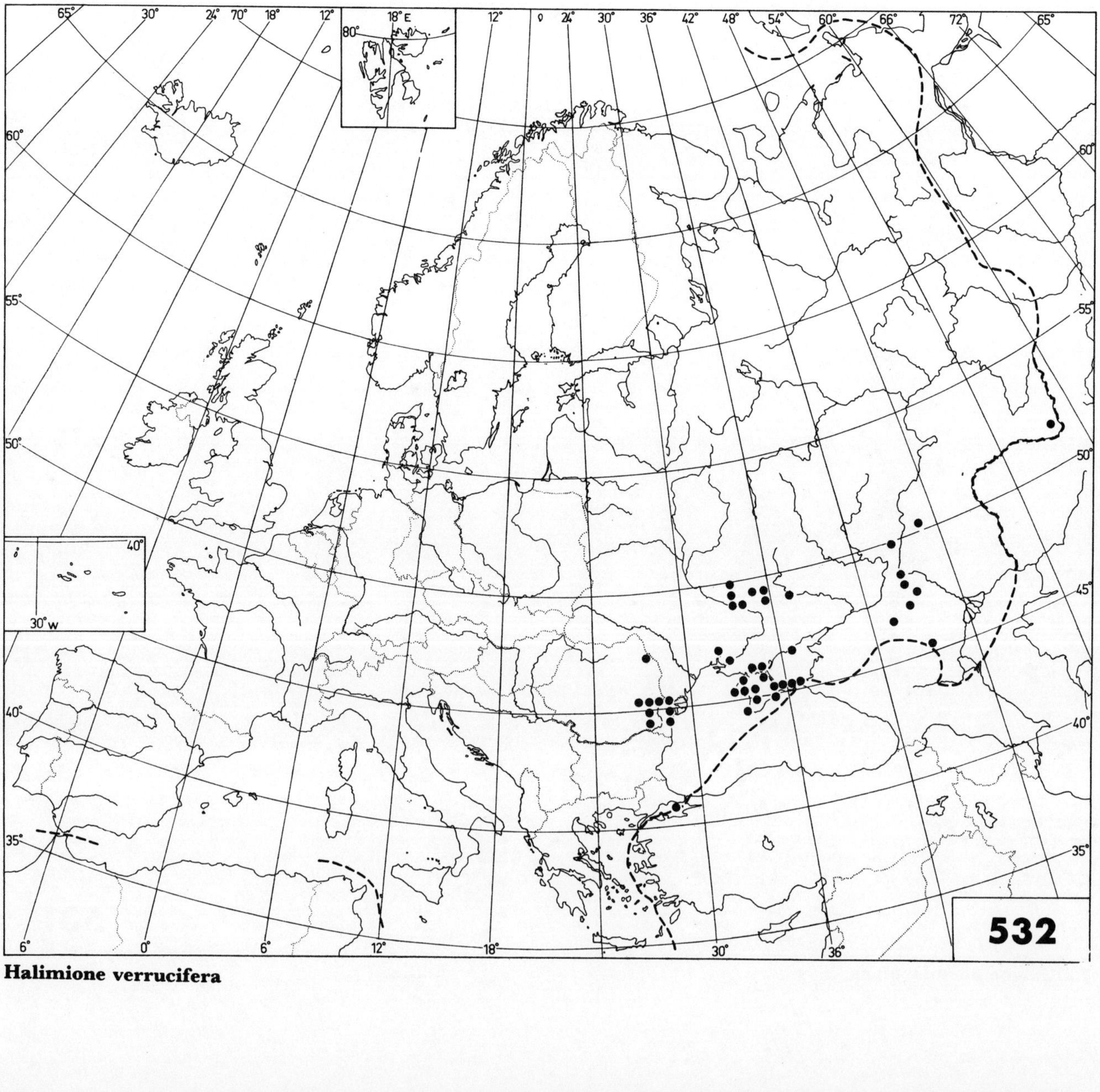

Halimione verrucifera

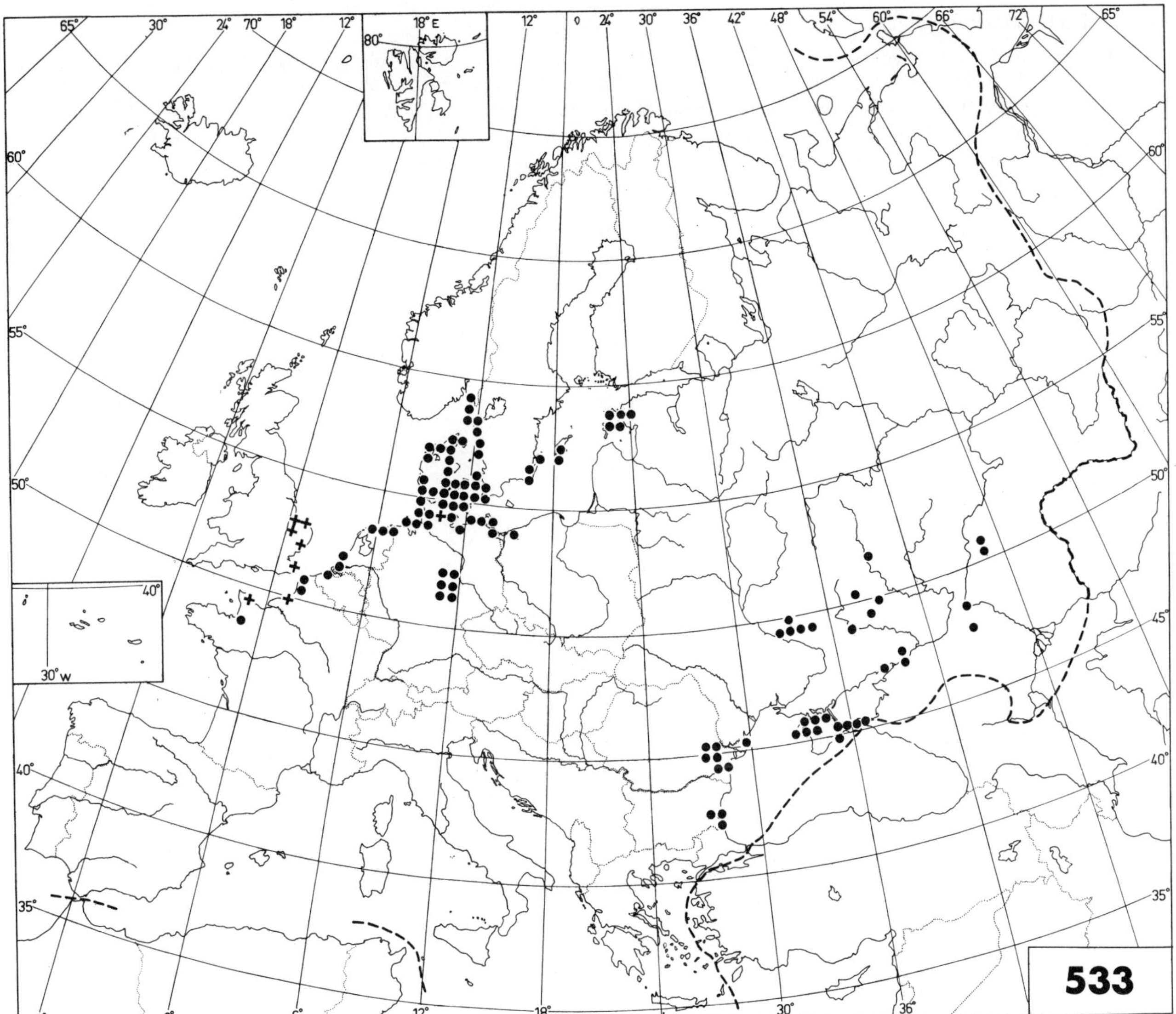

Halimione pedunculata

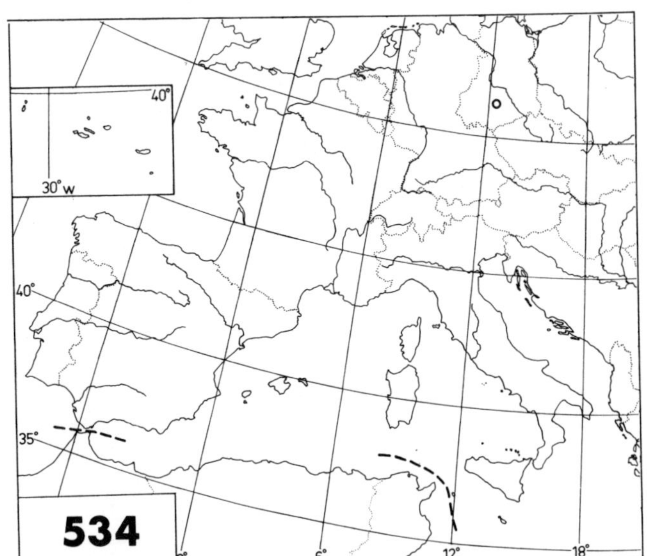

Obione sibirica

Obione sibirica (L.) Fischer — Map 534.

Atriplex sibirica L.; Obione muricata Gaertner

Taxonomy. Hegi 1961: 696.

Notes. [Ge] (Leipzig) (the species not mentioned in Fl. Eur.): P. Gutte, Mitt. Flor. Kart. Halle 4(2): 33—35 (1978).

Native of Siberia, C. Asia and China.

Ceratoides latens (J.F. Gmelin) Reveal & Holmgren — Map 535.

Axyris ceratoides L.; Ceratoides papposa (Pers.) Botsch. & Ikonn.; Ceratospermum papposum Pers.; Eurotia ceratoides (L.) C.A. Meyer; E. ferruginea (Nees) Boiss.; Krascheninnikovia ceratoides (L.) Gueldenst.; K. latens J.F. Gmelin

Generic delimitation. A.D.J. Meeuse & A. Smit, Taxon 20: 644 (1971).

Nomenclature. V.P. Botschantzev & S. Ikonnikov, Nov. Syst. Pl. Vasc. (Leningrad) 6: 267 (1970); J.L. Reveal & N.H. Holmgren, Taxon 21: 209 (1972); V.I. Grubov, Taxon 25: 362 (1976); see also Taxon 27: 288—289 (1978).

Notes. Extinct in Cz (?Cz in Fl. Eur.), Ju added (not given in Fl. Eur.). Records lacking from Rs(C); Rs(W) mapped according to Fl. URSR 1952: 340.

Total range. Hegi 1961: 701; E.J. Jäger, Flora 160: 234 (1971).

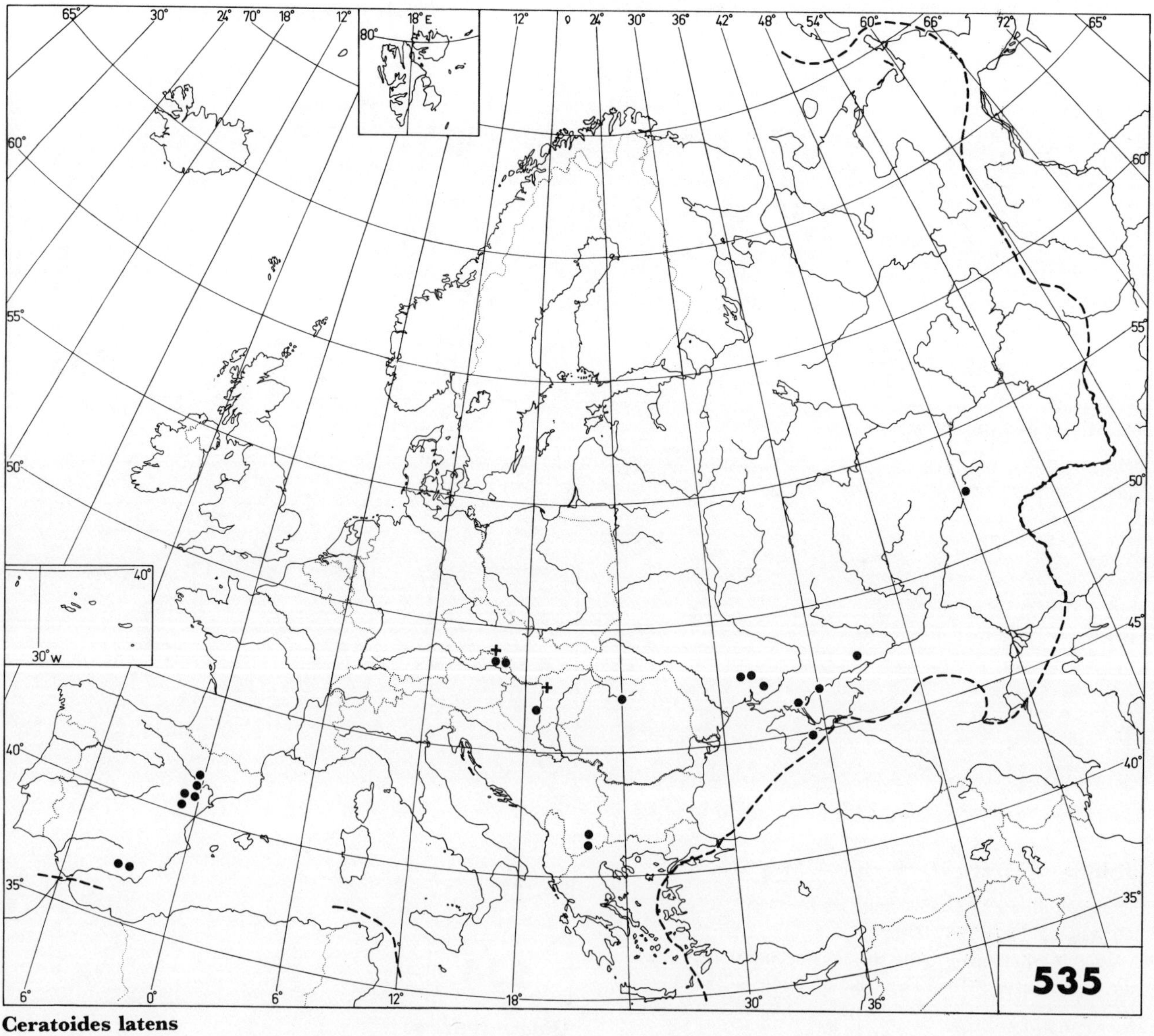

Ceratoides latens

Ceratocarpus arenarius L. — Map 536.

Axyris amaranthoides L. — Map 537.

Notes. Only casual in Cz ([Cz] in Fl. Eur.), possibly only casual in Rs(N, B) (not given in Fl. Eur.).

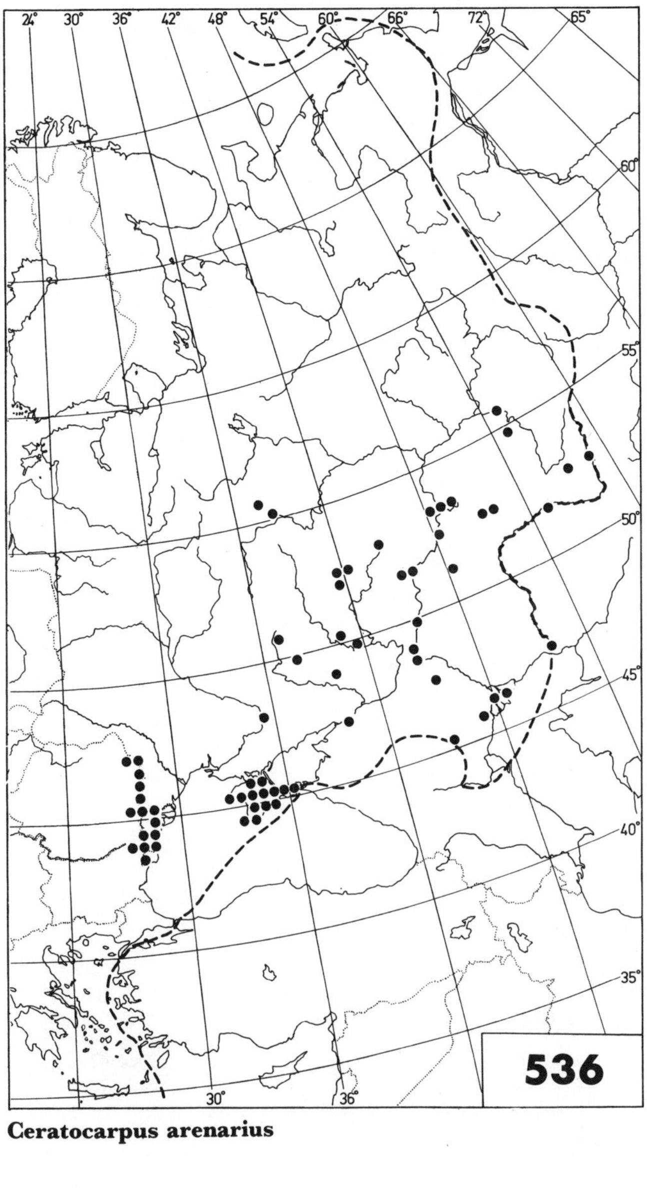

Ceratocarpus arenarius

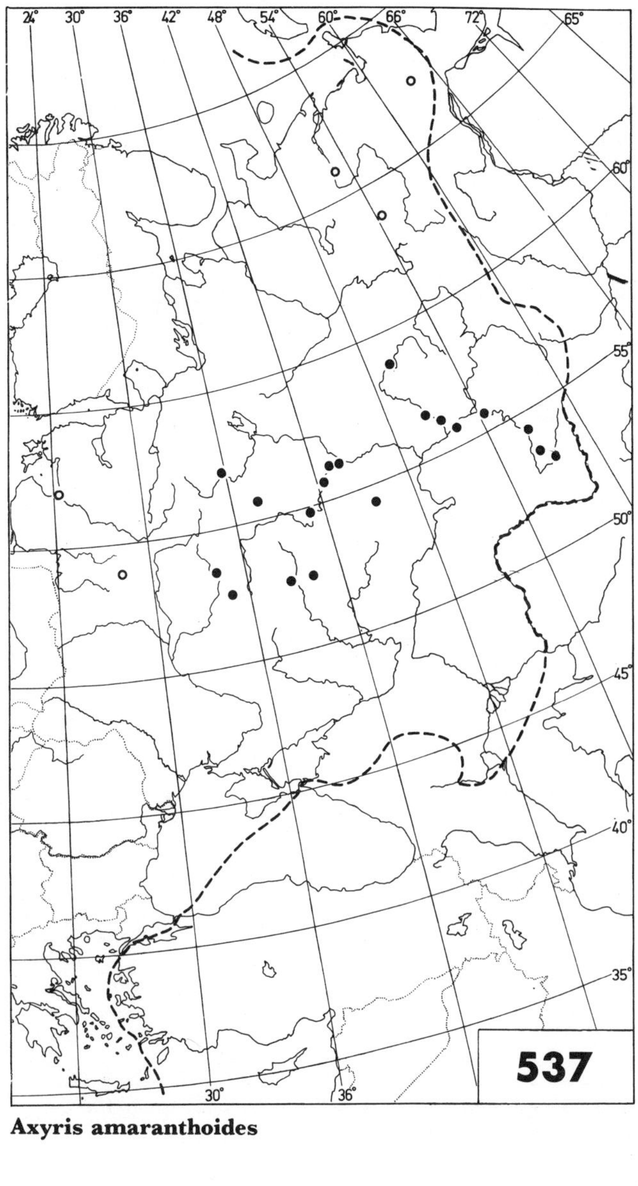

Axyris amaranthoides

Camphorosma monspeliaca L. — Map 538.

Camphorosma ruthenica Bieb. Incl. C. lessingii Litv. and C. nestensis Turrill

Taxonomy. P. Aellen, Notes Roy. Bot. Garden Edinb. 28: 31 (1967), and in Fl. Turkey 1967: 314—315. *Camphorosma nestensis*, of northern Greece, is probably of but little taxonomical value.

Notes. Rm omitted (given in Fl. Eur.), data not available from Rs(C). Rs(W) mapped according to URSR 1952: 345.

C. monspeliaca subsp. **monspeliaca** — Map 538.

Camphorosma ruthenica Bieb.

Notes. Al, Bu, Co, Ga, Hs, It, Ju, Rs(?C, ?W, K, E), Sa, Si.

C. monspeliaca subsp. **lessingii** (Litv.) Aellen — Map 538.

Camphorosma lessingii Litv.; C. ruthenica auct. non Bieb.

Notes. Rs(E).

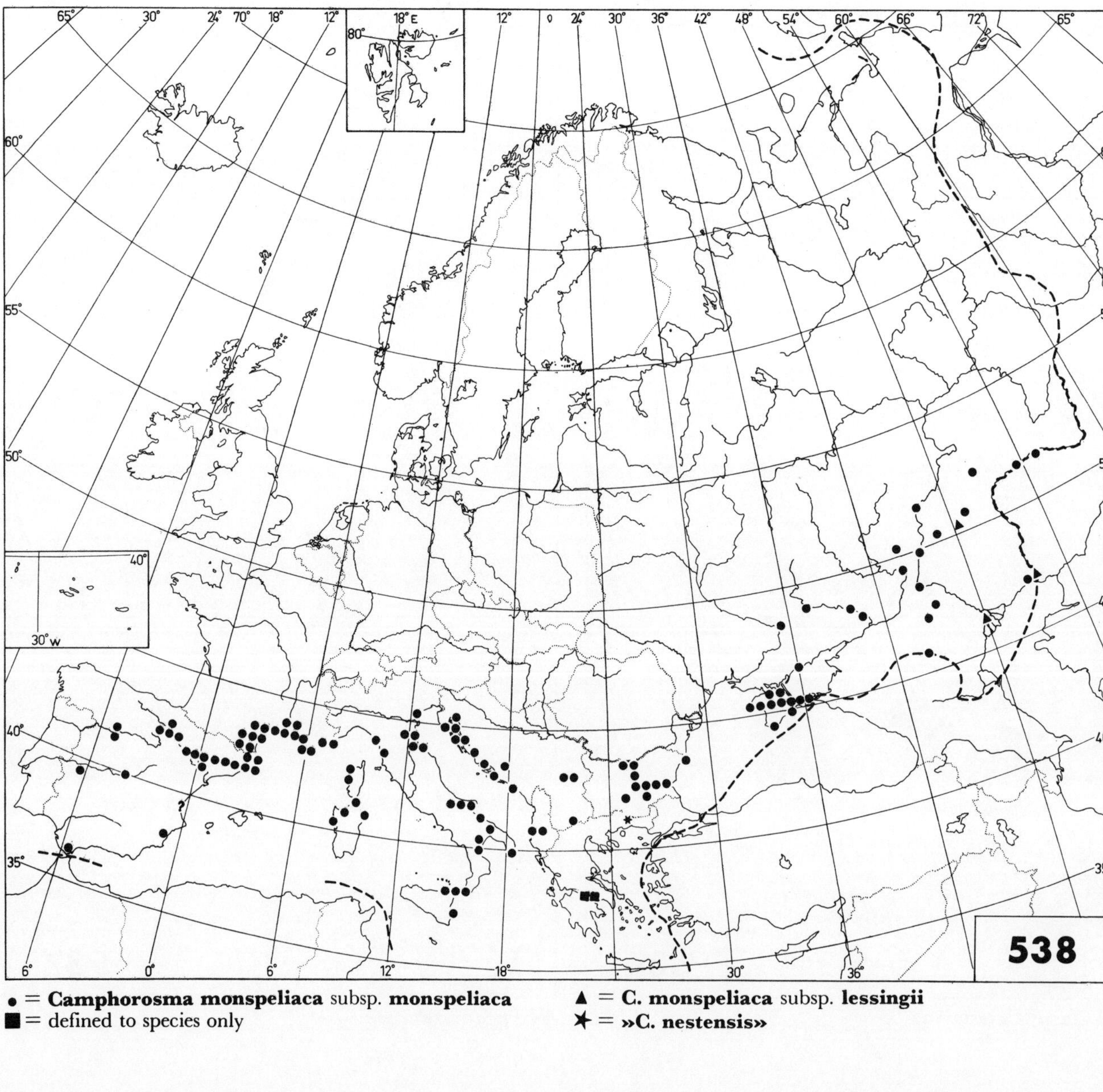

• = **Camphorosma monspeliaca** subsp. **monspeliaca** ▲ = **C. monspeliaca** subsp. **lessingii**
■ = defined to species only ✶ = **»C. nestensis»**

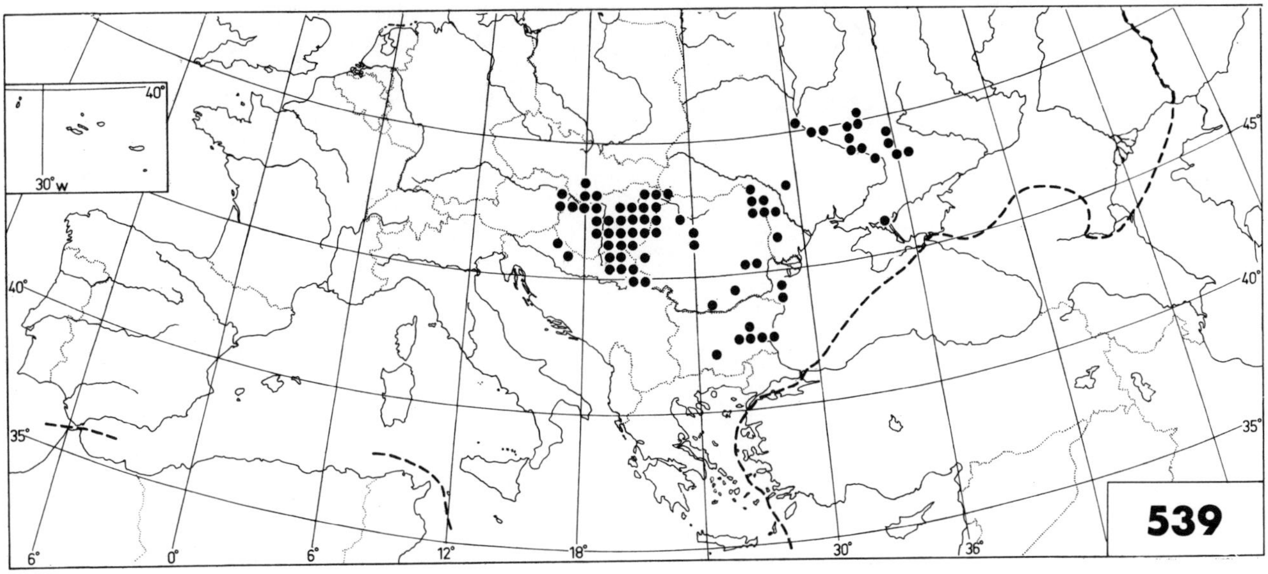

Camphorosma annua

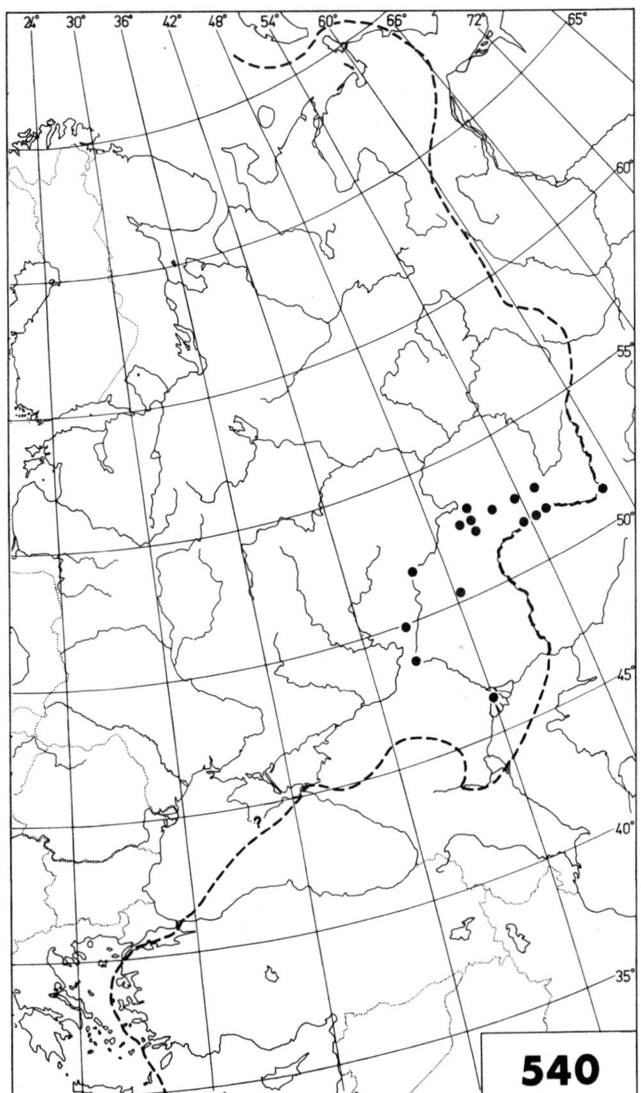

Camphorosma songorica

Camphorosma annua Pallas — Map 539.

Camphorosma ovata Waldst. & Kit.

Total range. Endemic to Europe.

Camphorosma songorica Bunge — Map 540.

Camphorosma annua auct. non Pallas

Notes. Rs(C) added (not given in Fl. Eur.), Rs(W) omitted (given in Fl. Eur.); not reported from Rs(K) in Rubtsev 1972.

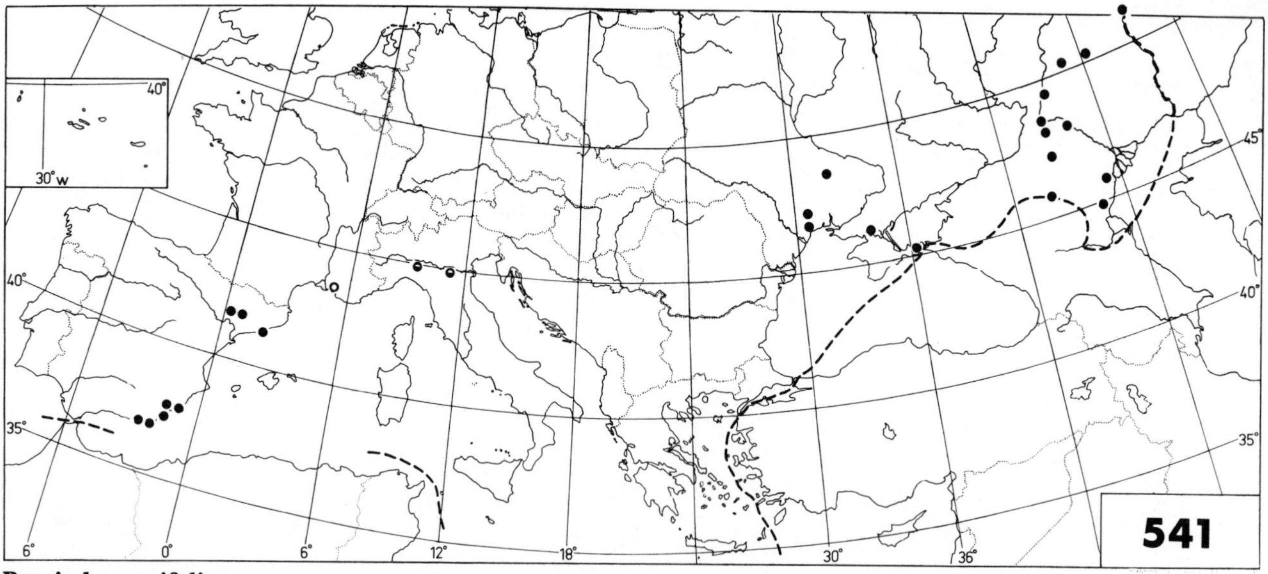

Bassia hyssopifolia

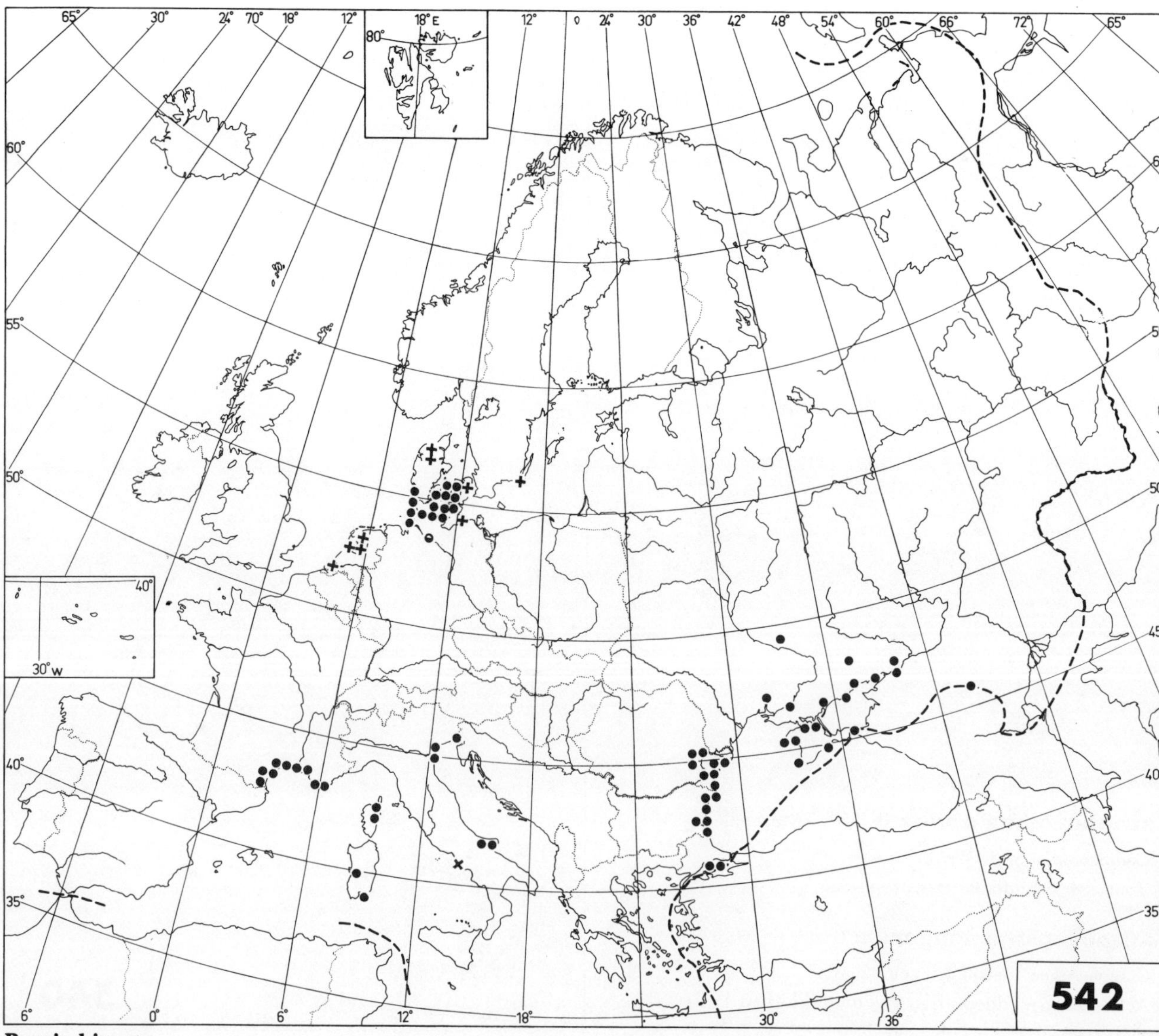

Bassia hirsuta

Bassia hyssopifolia (Pallas) Volk. — Map 541.

Echinopsilon hyssopifolia (Pallas) Moq.; E. reuterianus Boiss.; Kochia hyssopifolia (Pallas) Schrader

Generic delimitation of *Bassia* All. (incl. *Kochia* Roth): A.J. Scott, Feddes Repert. 89: 101—109 (1978).

Notes. *It added (not given in Fl. Eur.). Presence in Rs(K) doubtful according to Rubtsev 1972: 131 (given in Fl. Eur.), Rs(W) mapped according to Fl. URSR 1952: 348.

Bassia hirsuta (L.) Ascherson — Map 542.

Echinopsilon hirsutum (L.) Moq.; Kochia hirsuta (L.) Nolte

Notes. Be not confirmed (?Be in Fl. Eur.). Decreasing alarmingly in Da and extinct in Ho and Su (given as present in Fl. Eur.): NU A 1978(9): 124—125 (Hotade djur och växter i Norden). Ju omitted (given in Fl. Eur.), Tu added (not given in Fl. Eur.): Fl. Turkey 1967: 315. Map incomplete for Rs(W).

Bassia sedoides (Pallas) Ascherson — Map 543.

Echinopsilon sedoides (Pallas) Moq.; Kochia sedoides (Pallas) Schrader; Salsola cinerea Waldst. & Kit.; S. sedoides Pallas

Notes. Not confirmed for Bu (given as present in Fl. Eur.): B.A. Kuzmanov, Candollea 34: 16 (1979). Cz omitted (given in Fl. Eur.). Map incomplete for Rs(W).

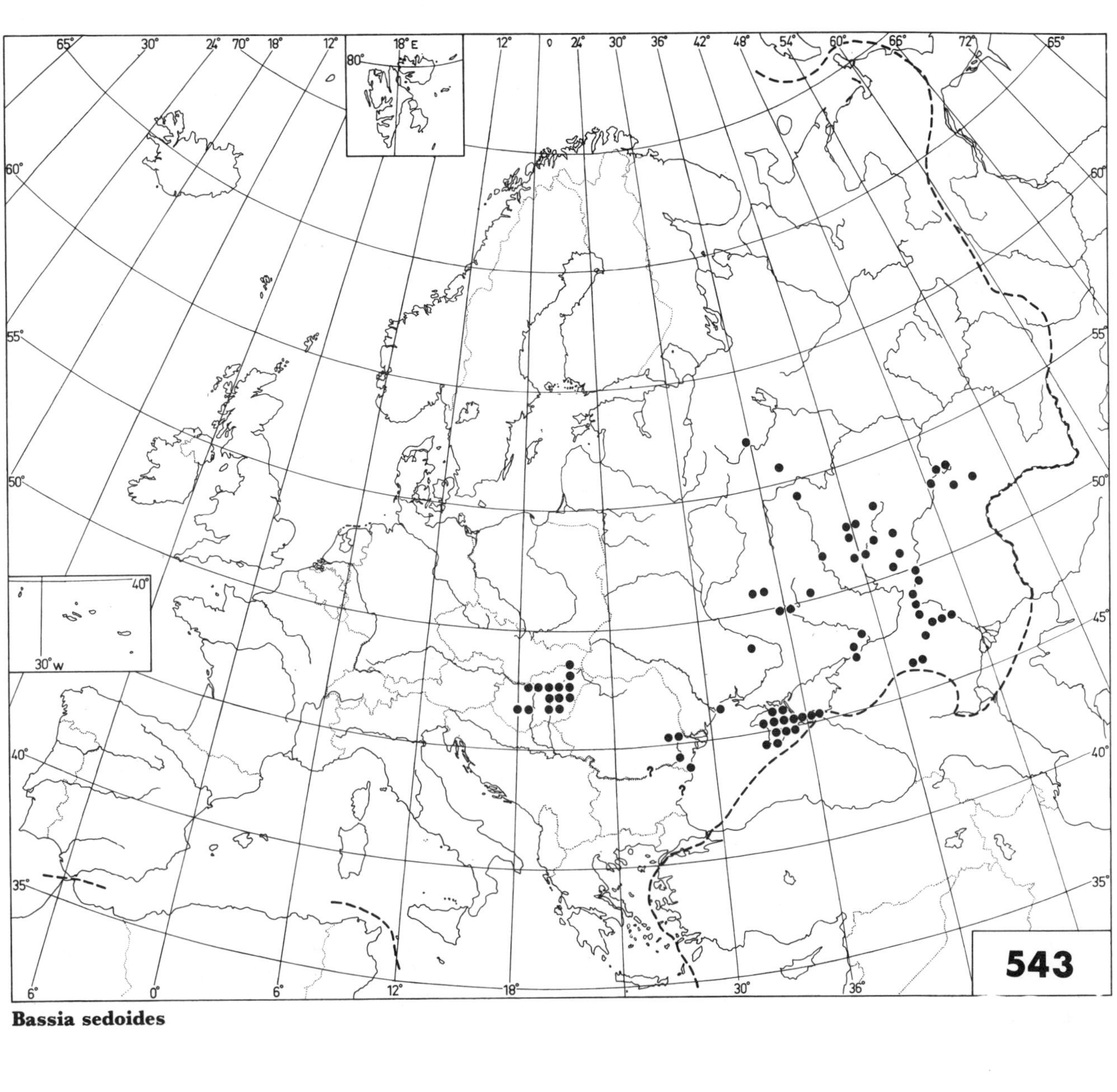

Bassia sedoides

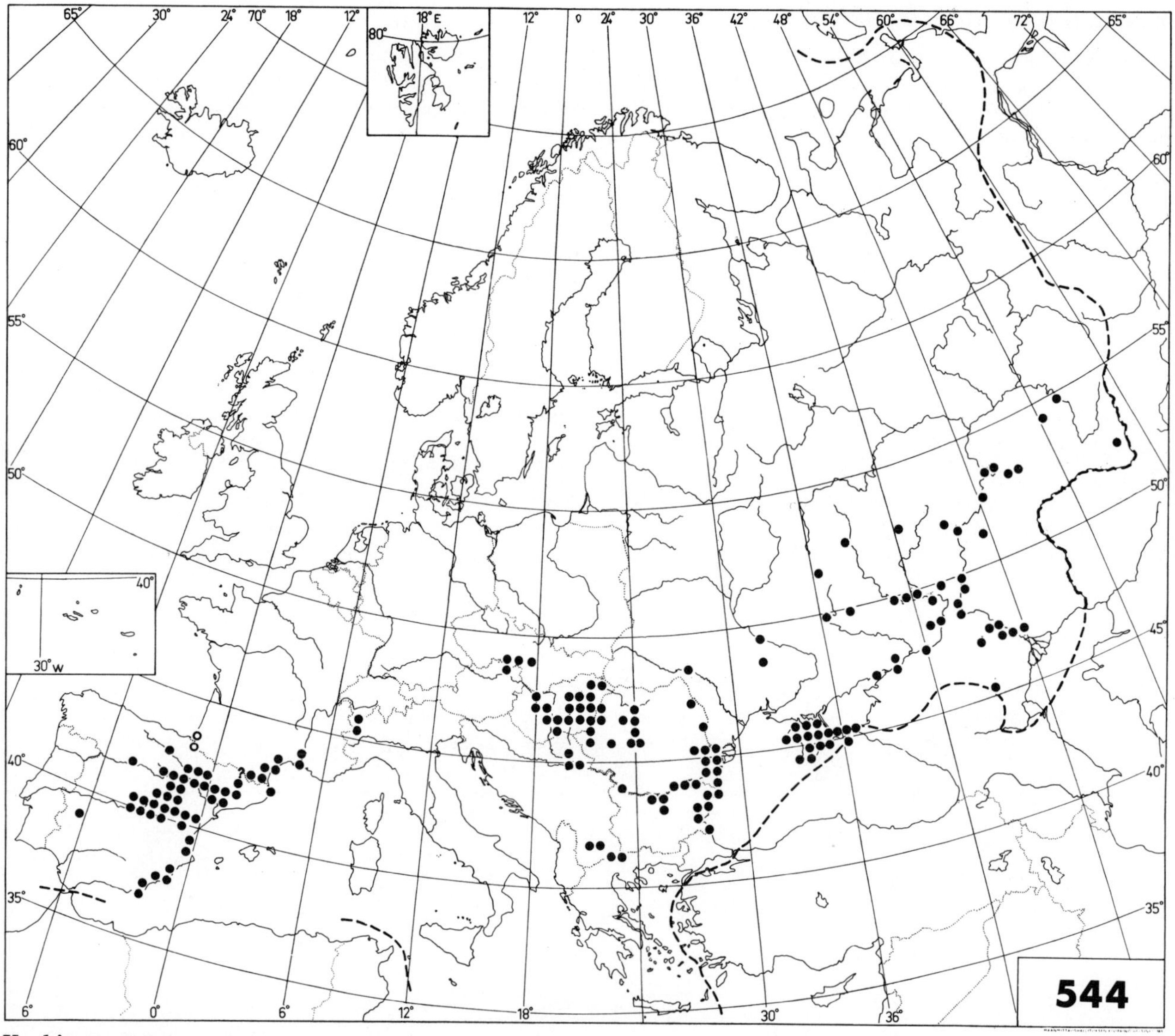

Kochia prostrata

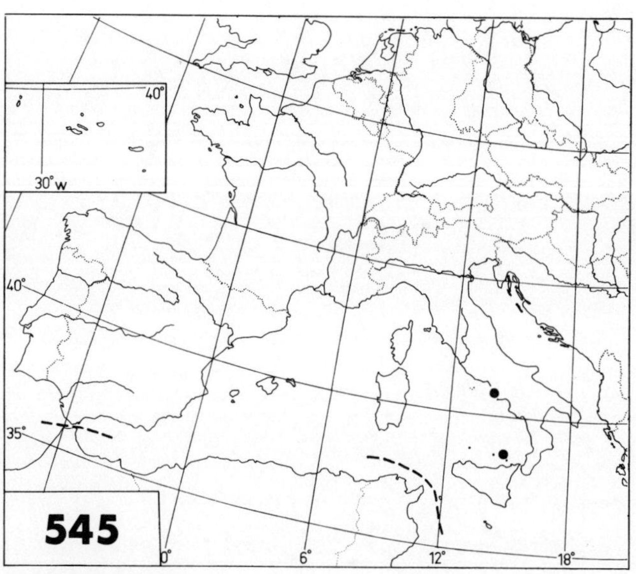

Kochia saxicola

Kochia prostrata (L.) Schrader — Map 544.

Bassia prostrata (L.) A.J. Scott

Generic delimitation of *Kochia* Roth. To be included in *Bassia* according to A.J. Scott, Feddes Repert. 89: 101—109 (1978).

Taxonomy and nomenclature. A.J. Scott, Feddes Repert. 89: 108 (1978).

Notes. Al omitted (given in Fl Eur.). Map incomplete for Rs(W).

Kochia saxicola Guss. — Map 545.

Bassia saxicola (Guss.) A.J. Scott

Taxonomy and nomenclature. A.J. Scott, Feddes Repert. 89: 108 (1978).

Notes. It, Si (Fl. Eur. gives only It, although Isole Lipari and Stromboli are part of Si).

Total range. Endemic to the islands of Ischia, Capri and Stromboli.

Kochia laniflora (S.G. Gmelin) Borbás — Map 546.

Bassia laniflora (S.G. Gmelin) A.J. Scott; Kochia arenaria (Gaertner, Meyer & Scherber) Roth; ? K. sicorica O. Bolòs & Masclans; Salsola laniflora S.G. Gmelin

Taxonomy and nomenclature. From the formal description of *Kochia sicorica* O. Bolòs & Masclans, Butll. Inst. Catalana Hist. Nat. 38 (Sec. Bot. 1): 89 (1974), which is the only information available, it is not clear whether the new species is sufficiently different from *K. laniflora* to be treated as a separate taxon. A.J. Scott, Feddes Repert. 89: 108 (1978).

Notes. Al and Gr added (not given in Fl. Eur.), [He] omitted (given in Fl. Eur.). Map incomplete for Rs(W). ?Si added (not given in Fl. Eur.): A. Ponzo, La Flora Trapanese, p. 112 (Palermo 1900), as *Kochia arenaria.*

Total range. Hegi 1961: 707; MJW 1965: map 135a.

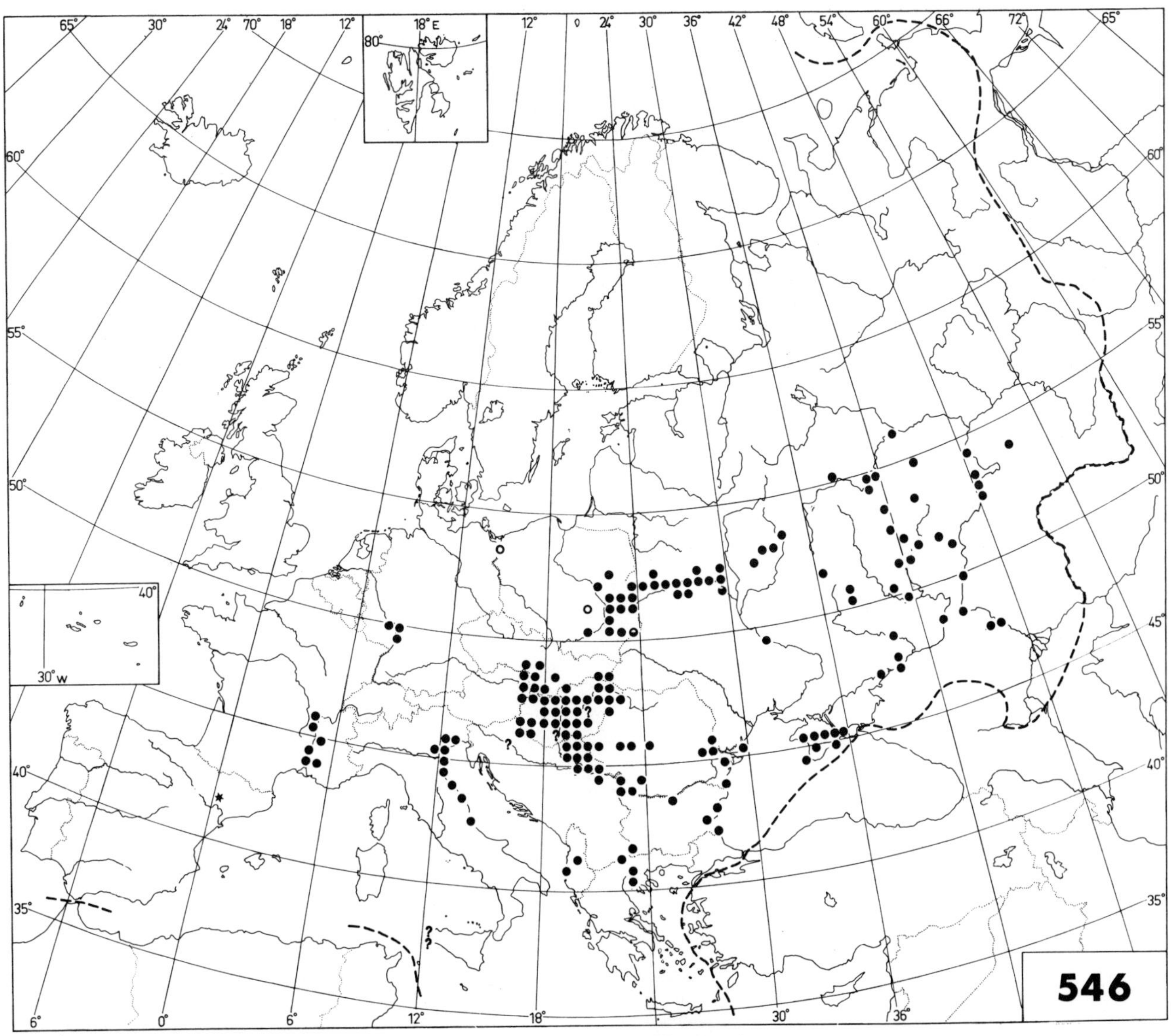

• = **Kochia laniflora** ✶ = **»K. sicorica«**

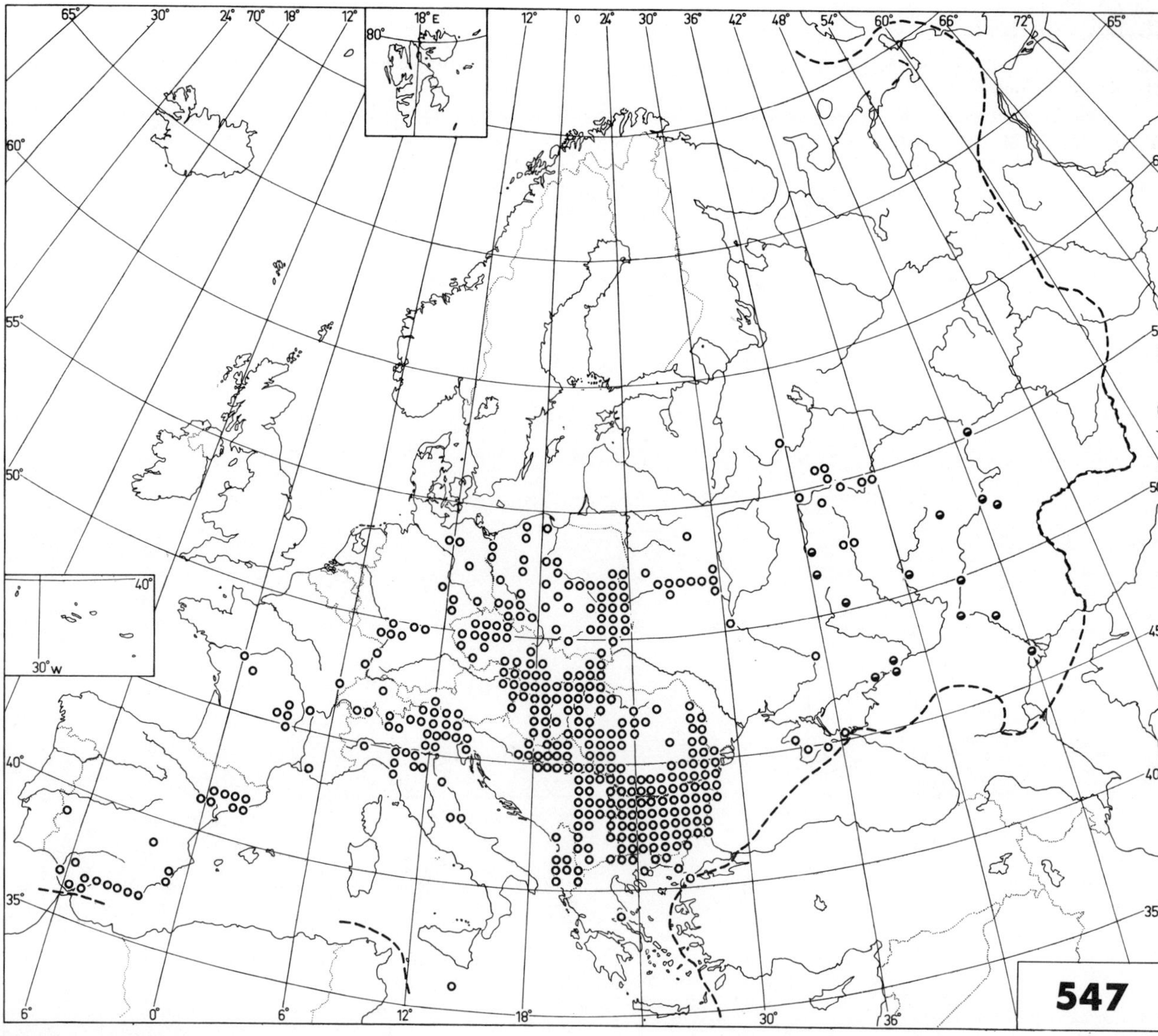

Kochia scoparia

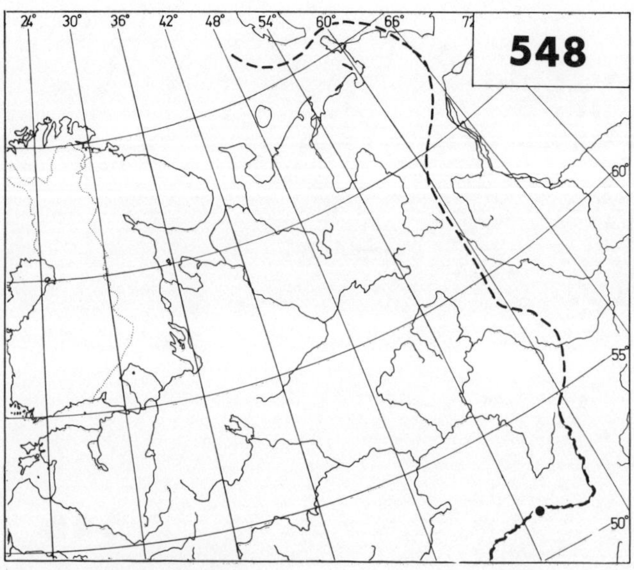

Corispermum uralense

Kochia scoparia (L.) Schrader — Map 547.

Bassia scoparia (L.) A.J. Scott. Incl. Kochia densiflora Turcz.; K. scoparia subsp. densiflora (Turcz.) Aellen

Taxonomy and nomenclature. A.J. Scott, Feddes Repert. 89: 108 (1978).

Notes. [Bu, Gr, Si] added (not given in Fl. Eur.), [Da, Ho, Lu] omitted (given in Fl. Eur.). Rs(K) mapped according to Rubtsev 1972: 133, who gives it as an escape from cultivation.

Total range. Hultén CP 1971: map 207.

Corispermum uralense (Iljin) Aellen — Map 548.

Corispermum squarrosum auct. non L.; C. squarrosum subsp. uralense Iljin

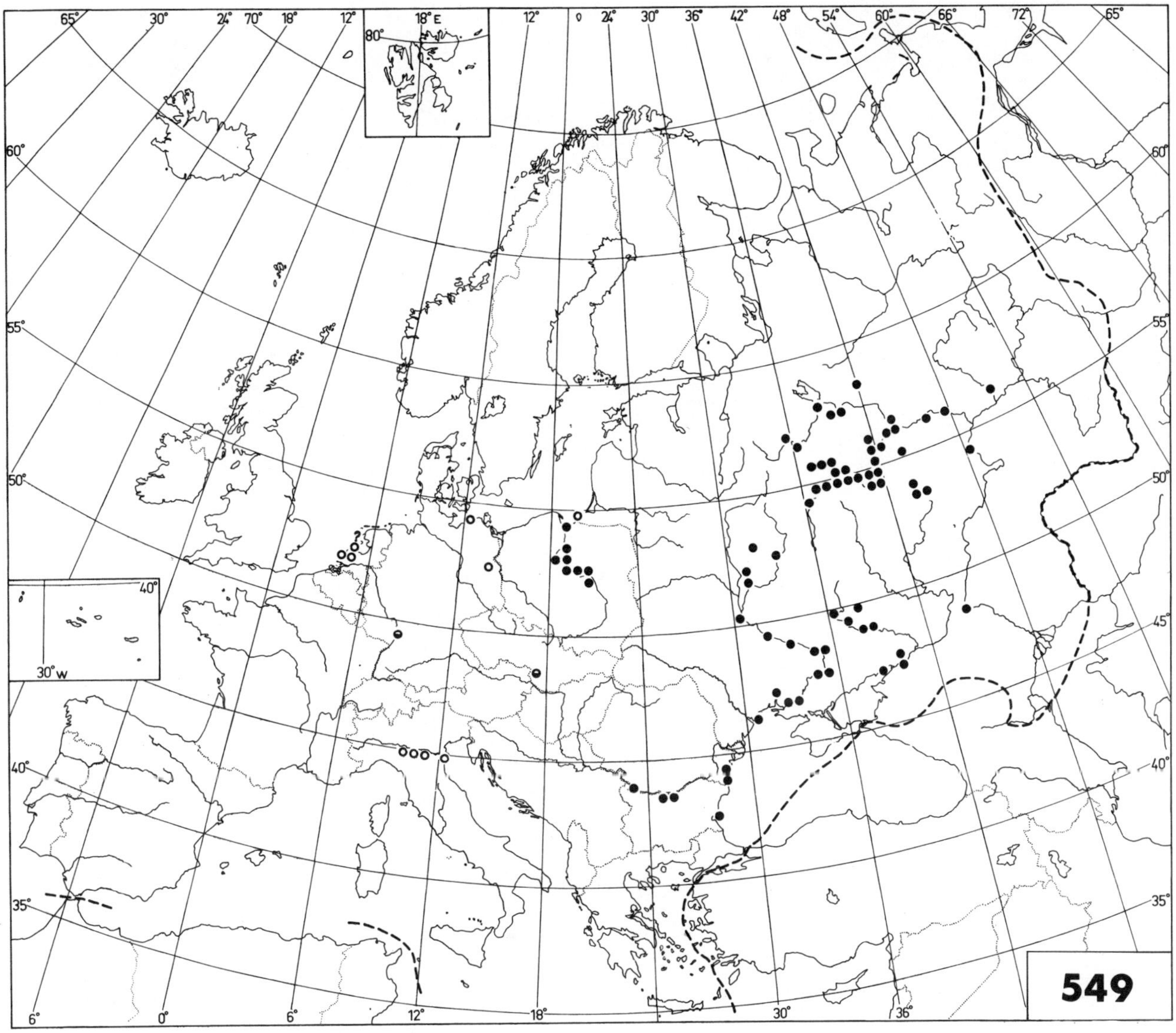

Corispermum marschallii

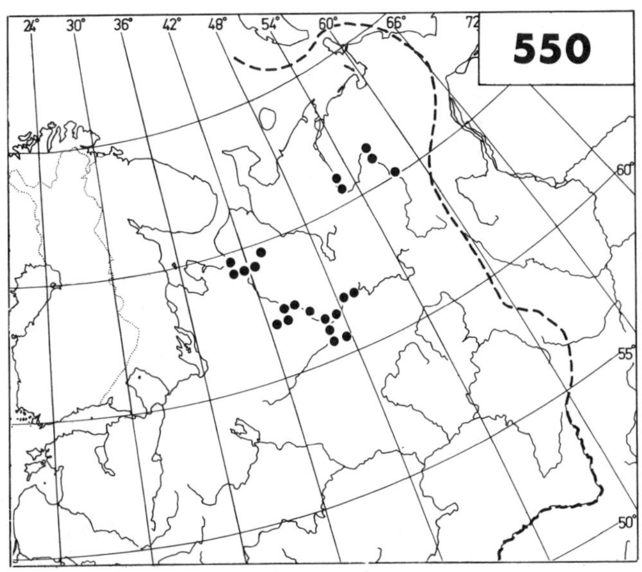

Corispermum algidum

Corispermum marschallii Steven — Map 549.

Notes. Bu, [Ho] and [Rs(B)] added (not given in Fl. Eur.): Fl. Reipubl. Pop. Bulg. 1966: 572—573 (where, however, *Corispermum canescens* Kit. is cited in the synonymy); B.A. Kuzmanov, Candollea 34: 15 (1979). Doubtful native in Cz and Ge (given as native in Fl. Eur.): Hegi 1961: 718. Si not confirmed ([?Si] in Fl. Eur.).

Corispermum algidum Iljin — Map 550.

Total range. The range extends far east of the Ural Mountains (given as endemic to Europe in Fl. Eur.): M.M. Iljin, Bull. Jard. Bot. URSS 28: 642 (1929).

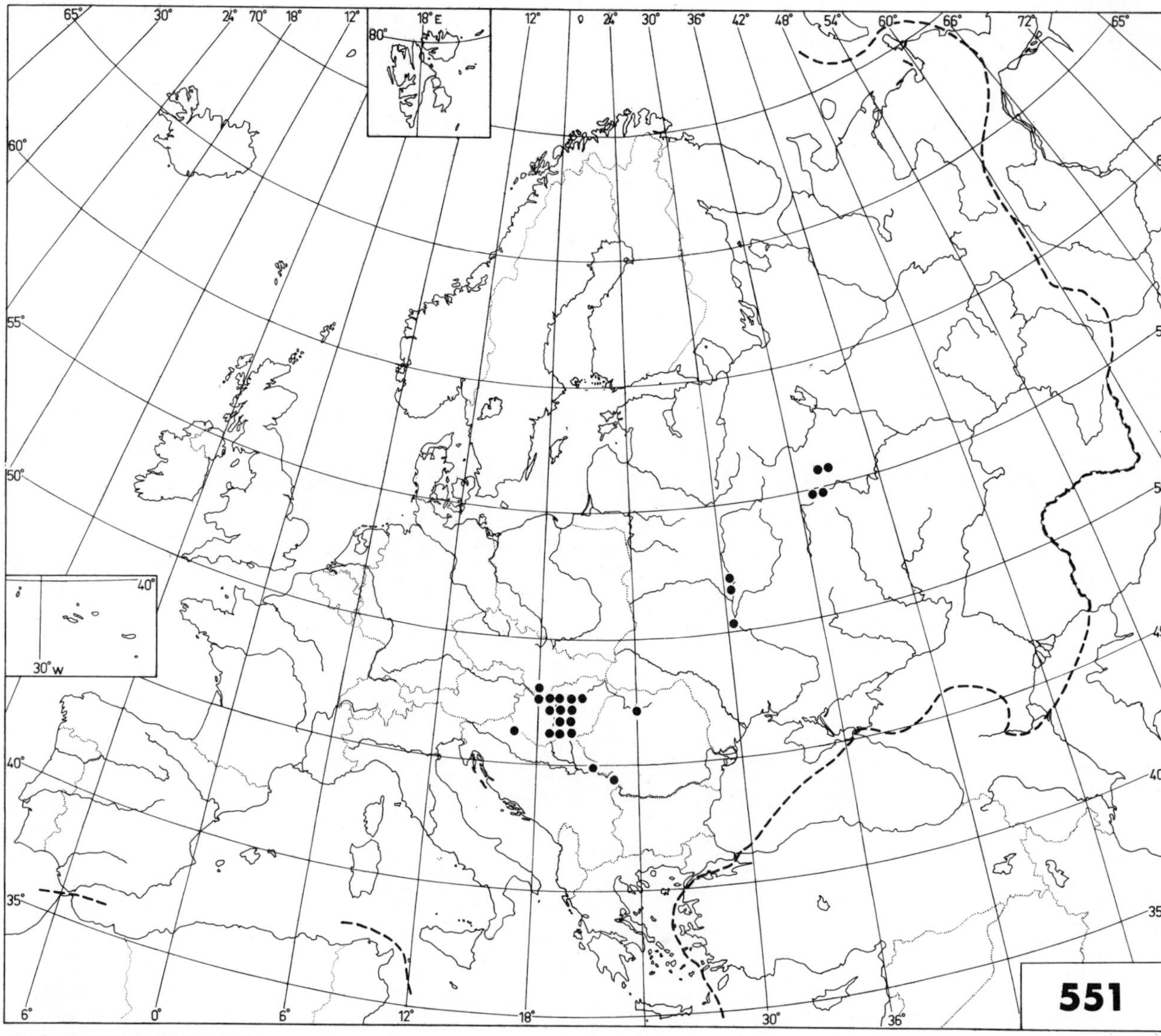

Corispermum canescens

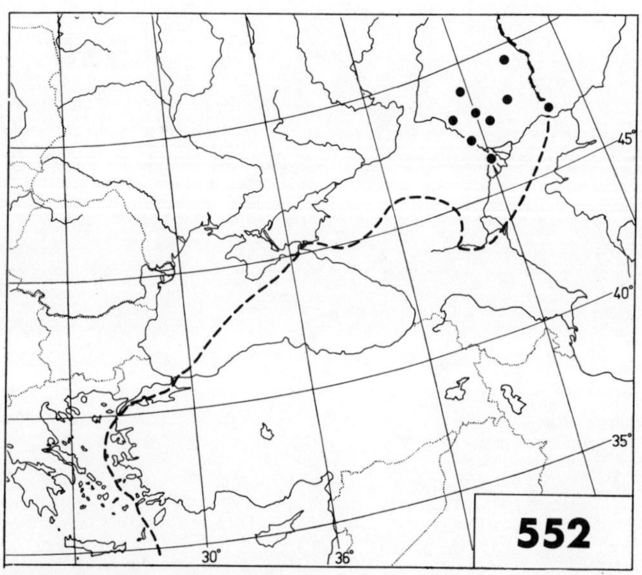

Corispermum aralocaspicum

Corispermum canescens Kit. ex Schultes — Map 551.

Incl. Corispermum borysthenicum Andrz., C. stenopterum Klokov, C. ucrainicum Iljin and C. volgicum Klokov

Notes. Ju added (not given in Fl. Eur.), no data available from Rs(W, E).

Corispermum aralocaspicum Iljin — Map 552.

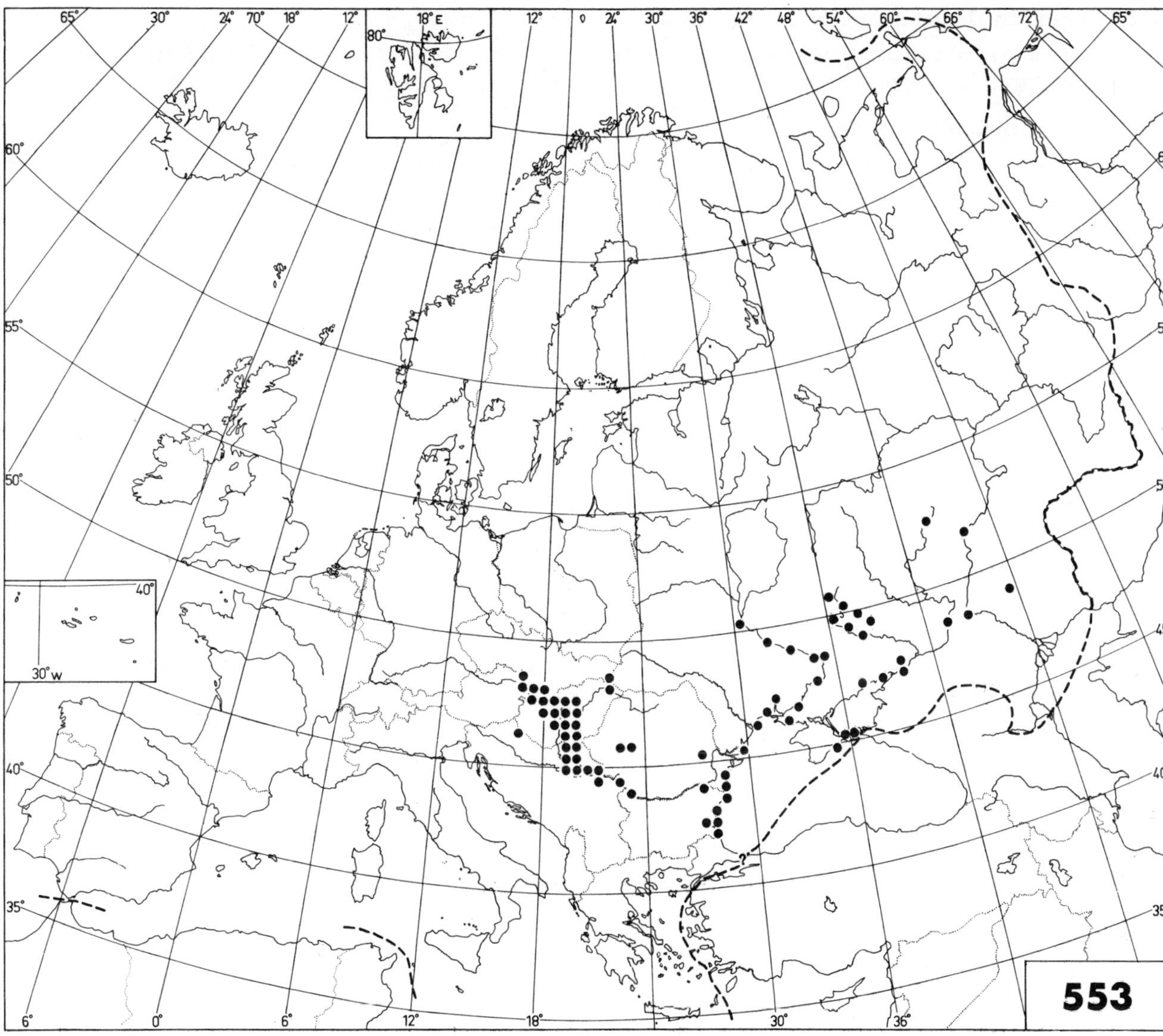

Corispermum nitidum

Corispermum nitidum Kit. ex Schultes — Map 553.

Incl. Corispermum calvum and C. nitidulum Klokov

Notes. Doubtful native in Au, and now extinct (given as native and present in Fl. Eur.). Gr omitted (given in Fl. Eur.); note, however, the remark in Fl. Turkey 1967: 318 (under *Corispermum filifolium*): »Specimens from Greek Thrace (e.g. *Tedd 777*) apparently belong to *C. nitidum* Kit.». Rs(C) omitted (given in Fl. Eur.). ?Tu added (not given in Fl. Eur.), probably referable to *C. filifolium*, recorded there in Fl. Turkey 1967: 318.

Corispermum filifolium C.A. Meyer — Map 554.

Notes. Possibly only casual in Ge ([Ge] in Fl. Eur.). Given for Tu in Fl. Turkey 1967: 318; see also under *Corispermum nitidum* (not given in Fl. Eur.).

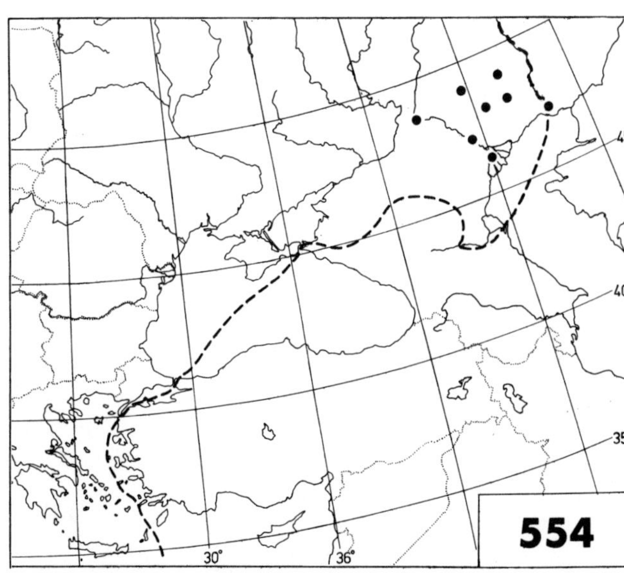

Corispermum filifolium

Corispermum declinatum Stephan ex Steven

Notes. Data not available from Rs(E) (given in Fl. Eur. as the only territory). Recorded from five squares in Rs(C), near Moscow, but doubtful native. Not mapped.

Corispermum hyssopifolium L. — Map 555.

Incl. Corispermum czernjaevii Klokov, C. glabratum Klokov, C. hybridum Besser ex Andrz. and C. insulare Klokov

Notes. [Cz], Rs(*B) and Rs(C) added (not given in Fl. Eur.). Doubtful native in Po (given as native in Fl. Eur.), Rm and Rs(K) omitted (given in Fl. Eur.).

Total range. Hultén Alaska 1968: 400.

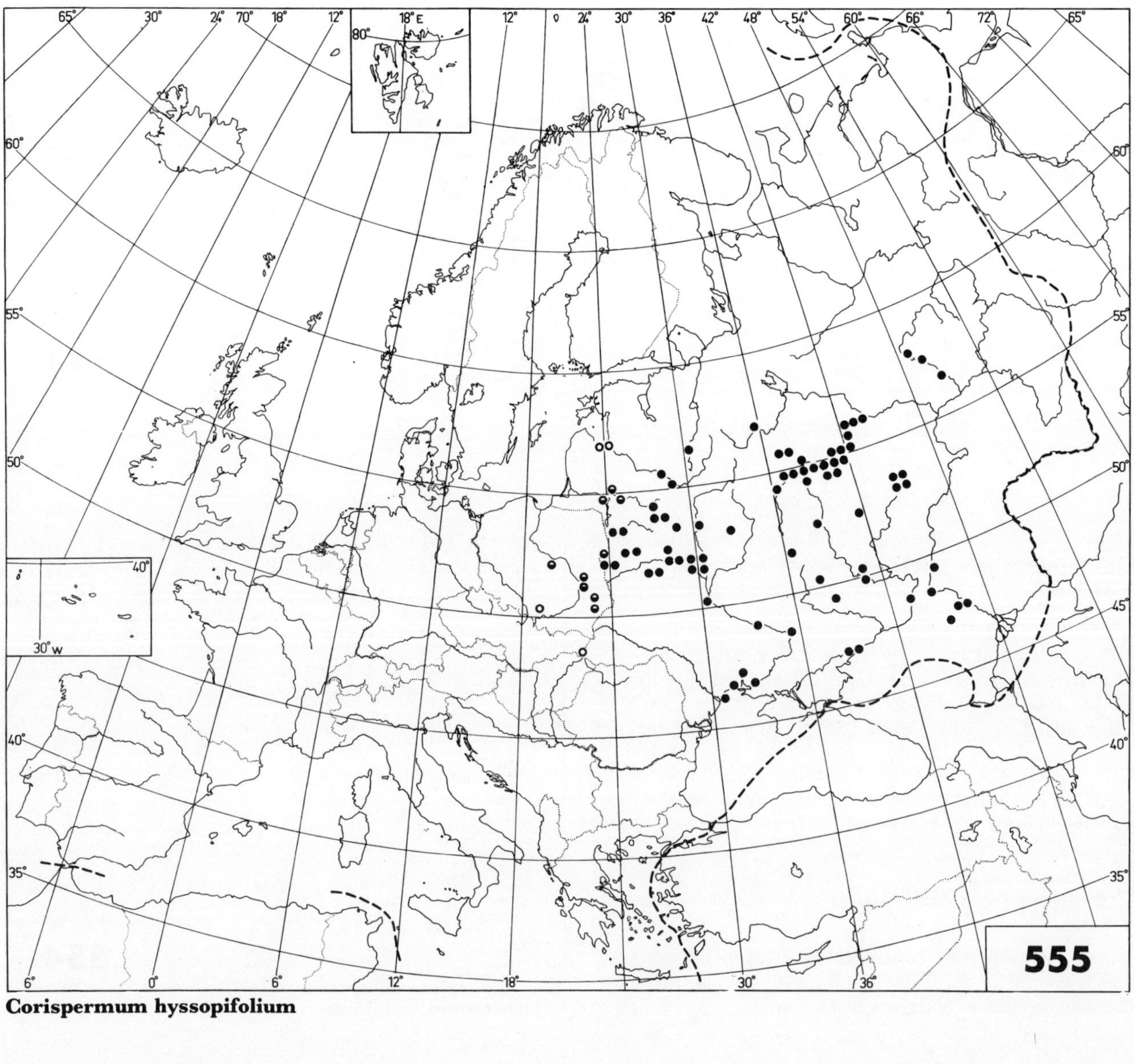

Corispermum hyssopifolium

Corispermum intermedium Schweigger, s. lato — Map 556.

Corispermum hyssopifolium auct. non L., C. membranaceum (Bischoff) Iljin. Incl. C. leptopterum (Ascherson) Iljin, sensu Fl. Eur.

Taxonomy. The only taxonomically sound solution here seems to be similar to that anticipated by P. Aellen in Hegi 1961: 719—722, i.e. to consider *Corispermum intermedium* s. str. just another variant or weakly differentiated ecotype within the aggregate including *C. leptopterum* sensu (Hegi and) Fl. Eur., with its numerous subordinate taxa. The present variation, ± connecting two »basic» taxa the native ranges of which were perhaps originally well separated, the southwestern *C. leptopterum* and the eastern *C. intermedium,* may possibly be due to hemerochorous dispersal and subsequent hybridization. But further studies are needed to elucidate this point.

Notes. [Be], [Cz], Ga, [Ge], [Ho], It, [Po], Rs([B], C, E).

Corispermum intermedium Schweigger, s. str. — Map 556.

Taxonomy. It was not possible to check whether the Russian records in fact refer to this taxon, considered a plant of sand dunes along the southeastern shores of the Baltic by P. Aellen in Hegi 1961: 720.

Notes. »C. Europe. Ge, Po» (Fl. Eur.).

Total range. When circumscribed as in Hegi 1961 and Fl. Eur., certainly endemic to Europe (not mentioned as such in Fl. Eur.).

Corispermum leptopterum (Ascherson) Iljin — Map 556.

Corispermum gallicum Iljin

Notes. »S. France, Italy, Sicilia; casual elsewhere and locally naturalized. Ga, It, Si, [Be, Ge, Ho]» (Fl. Eur.).

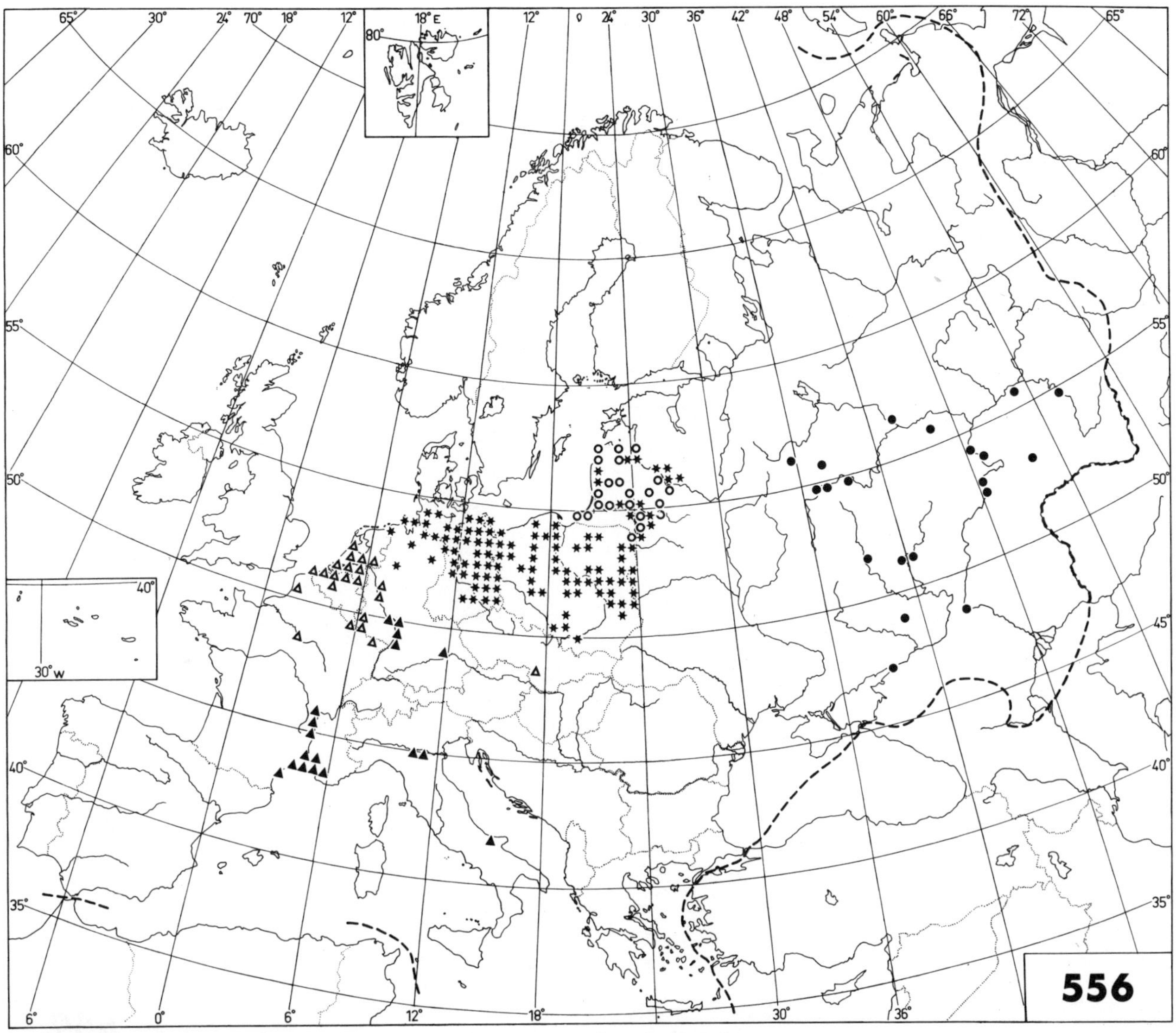

Corispermum intermedium sensu lato
• = **C. intermedium**
▲ = **C. leptopterum**
✱ = both species or collective

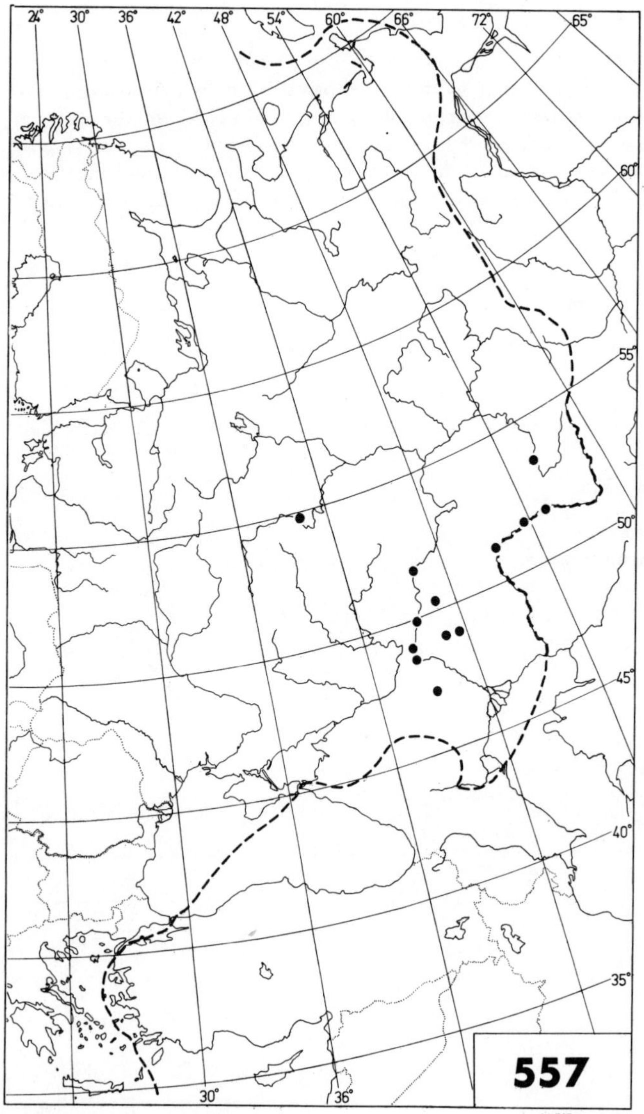

Corispermum orientale

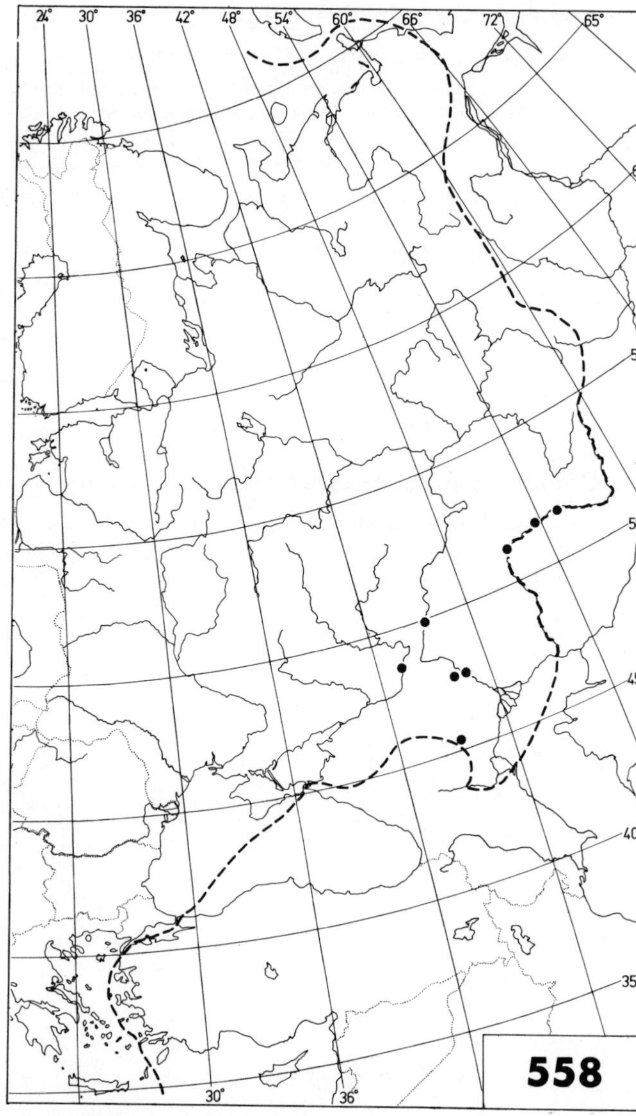

Agriophyllum squarrosum

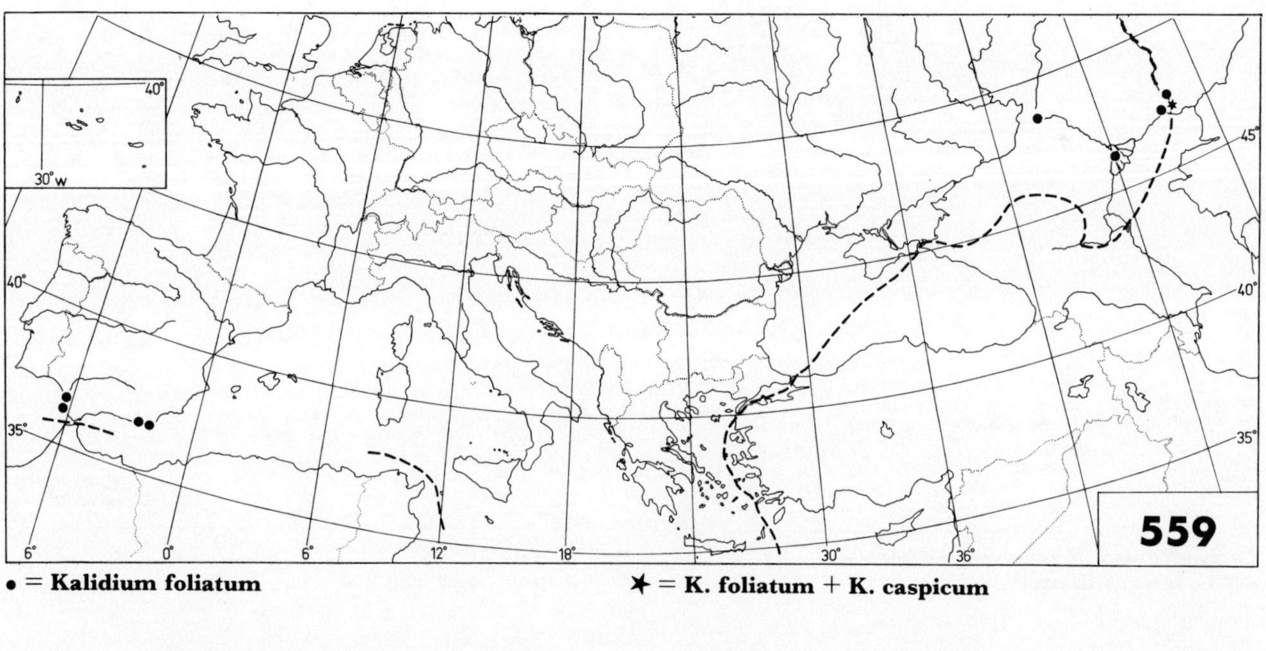

• = **Kalidium foliatum** ✶ = **K. foliatum + K. caspicum**

Corispermum orientale Lam. — Map 557.

Notes. Rs(C) added (not given in Fl. Eur.).

Agriophyllum squarrosum (L.) Moq. — Map 558.

Agriophyllum arenarium Bieb. ex C.A. Meyer; Corispermum squarrosum L.

Total range. I.F. Musaev, Vestn. Leningrad. Univ. (Biol.) 1966(3): 23 (1966).

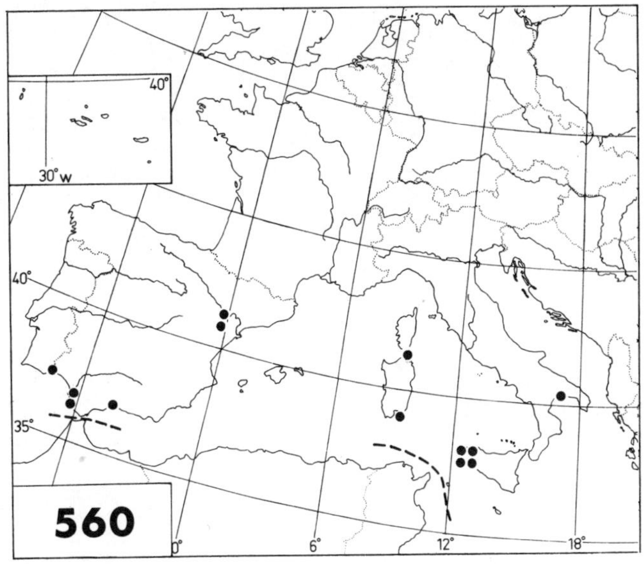

Halopeplis amplexicaulis

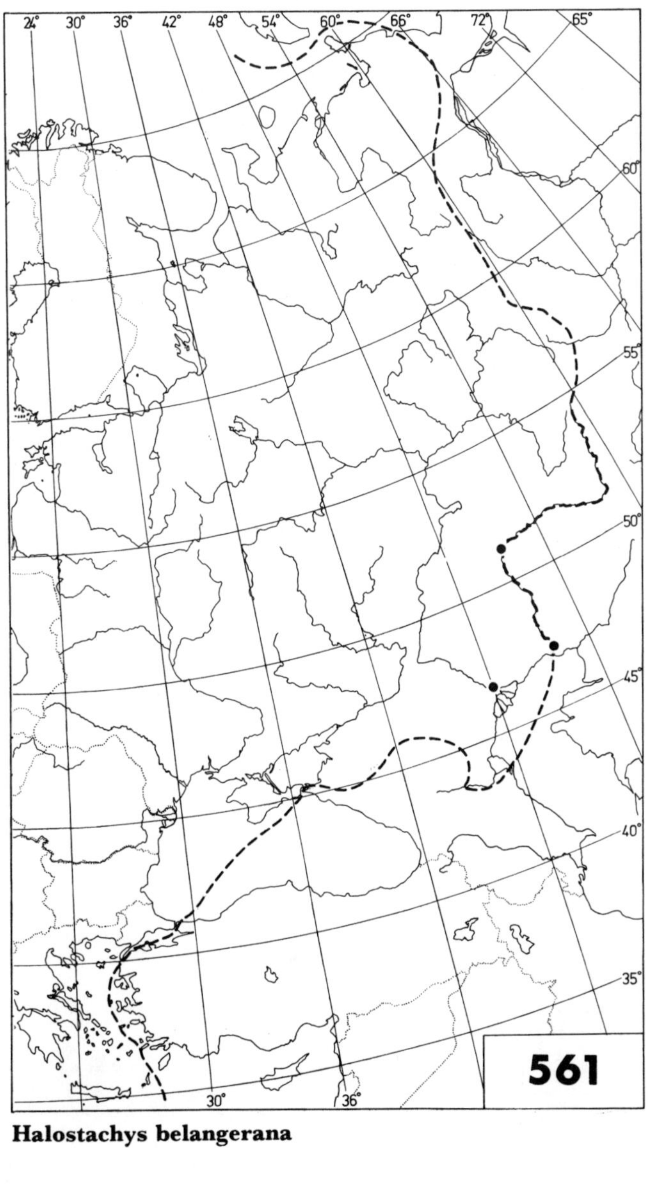

Halostachys belangerana

Kalidium foliatum (Pallas) Moq. — Map 559.

Kalidiopsis wagenitzii Aellen; Salicornia foliata Pallas

Generic delimitation. A.J. Scott, Bot. Jour. Linn. Soc. 75: 361, 372 (1977).

Total range. N.I. Nikolskaya, Bot. Žurn. 57: 302 (1972).

Kalidium caspicum (L.) Ung.-Sternb. — Map 559.

Salicornia caspica L.

Total range. N.I. Nikolskaya, Bot. Žurn. 57: 303 (1972).

Halopeplis amplexicaulis (Vahl) Ung.-Sternb. ex Cesati, Passer. & Gibelli — Map 560.

Halostachys perfoliata sensu Willk., non (Forskål) Moq.

Halopeplis pygmaea (Pallas) Bunge ex Ung.-Sternb.

Notes. Rs(E) omitted (given in Fl. Eur.); the species thus unknown from Europe and not mapped.

Halostachys belangerana (Moq.) Botsch. — Map 561.

Arthrocnemum belangeranum Moq.; Halocnemum caspicum Bieb. pro p.; Halostachys caspia C.A. Meyer; H. caspica auct.; Salicornia caspica Pallas non L.

Taxonomy. Halostachys is considered a monotypic genus with a single species, *Halostachys caspica* (Bieb.) C.A. Meyer ex Schrenk, by A.J. Scott, Bot. Jour. Linn. Soc. 75: 371 (1977).

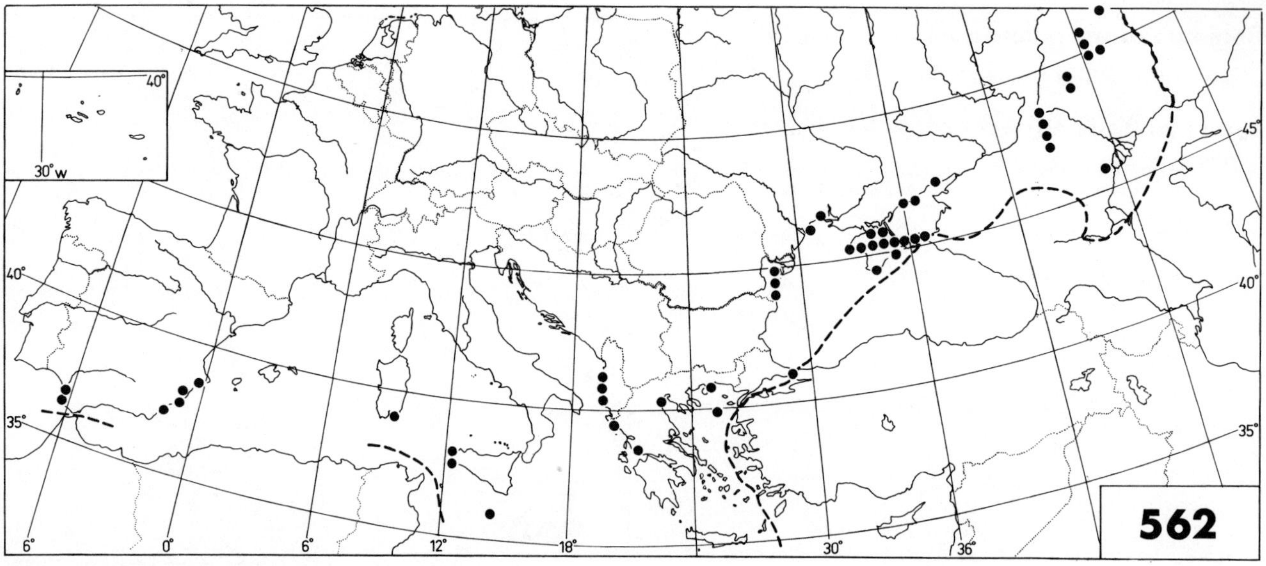

Halocnemum strobilaceum

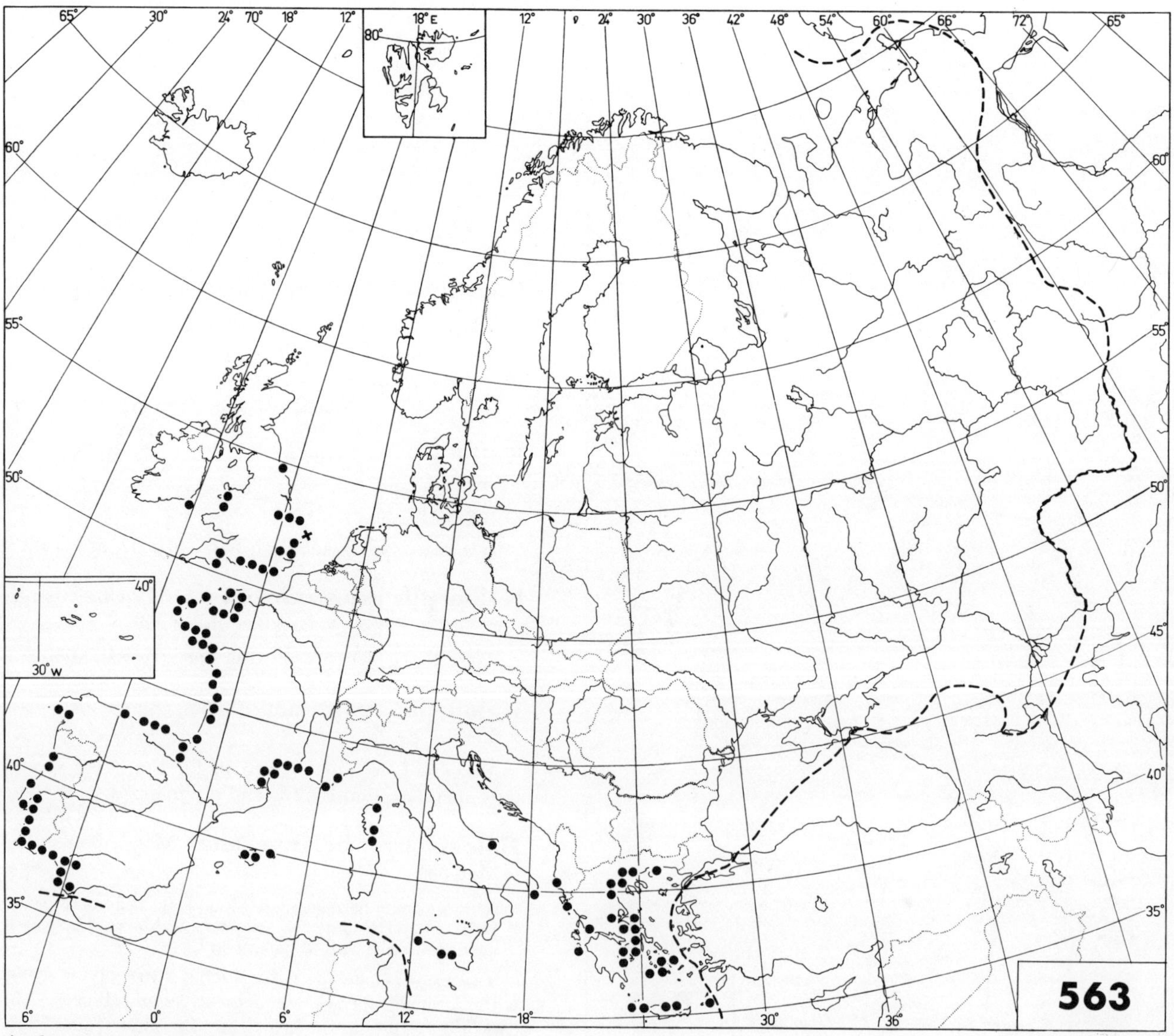

Arthrocnemum perenne

Halocnemum strobilaceum (Pallas) Bieb. — Map 562.

Halostachys perfoliata (Forskål) Moq.

Notes. Rs(W) mapped according to Fl. URSR 1952: 368.

Arthrocnemum perenne (Miller) Moss — Map 563.

Salicornia alpini Lag.; S. ambigua Michx; S. perennis Miller; S. radicans Sm.; Sarcocornia perennis (Miller) A.J. Scott

Generic delimitation. P.W. Ball, Feddes Repert. 69 (Fl. Eur. Notulae Syst. 3): 1—2 (1964); A.J. Scott, Bot. Jour. Linn. Soc. 75: 357—374 (1977).

Notes. Bl mapped according to Fl. Mallorca 1978: 104 and P. Duvigneaud, Soc. Échange Pl. Vasc. Europe Occ. Bassin Médit. 17(suppl.): 15 (1979). Ju and Sa omitted (given in Fl. Eur.).

Arthrocnemum fruticosum (L.) Moq. — Map 564.

Salicornia arabica auct.; S. fruticosa (L.) L.; Sarcocornia fruticosa (L.) A.J. Scott

Nomenclature. A.J. Scott, Bot. Jour. Linn. Soc. 75: 367 (1977).

Notes. Presence in Cr doubtful, because of extensive confusion with the other two species (given as present in Fl. Eur.): W. Greuter, Ann. Mus. Goulandris 1: 29 (1973). Cf. also E.F. Galiano & S. Silvestre, Lagascalia 5: 102 (1975).

Arthrocnemum macrostachyum (Moric.) Moris & Delponte — Map 565.

Arthrocnemum glaucum (Delile) Ung.-Sternb.; Salicornia fruticosa Lag. non (L.) L.; S. glauca Delile non Stokes; S. macrostachya Moric.

Generic delimitation and nomenclature. M. Zohary, Fl. Palaest. 1: 156 (1966); A.J. Scott, Bot. Jour. Linn. Soc. 75: 357—374 (1977).

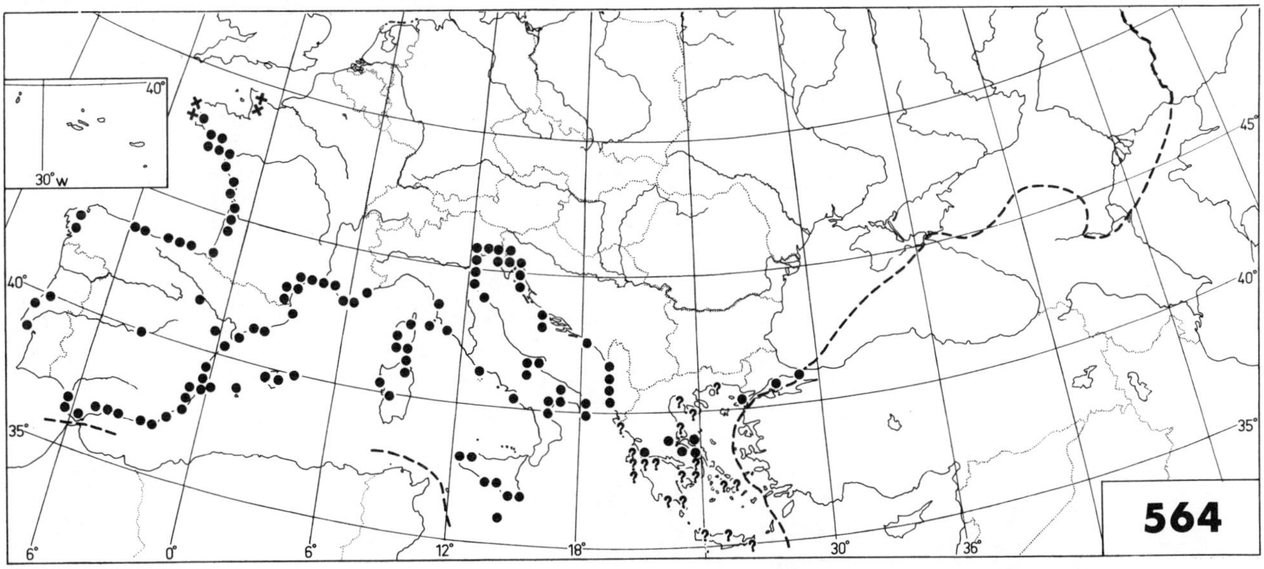

Arthrocnemum fruticosum

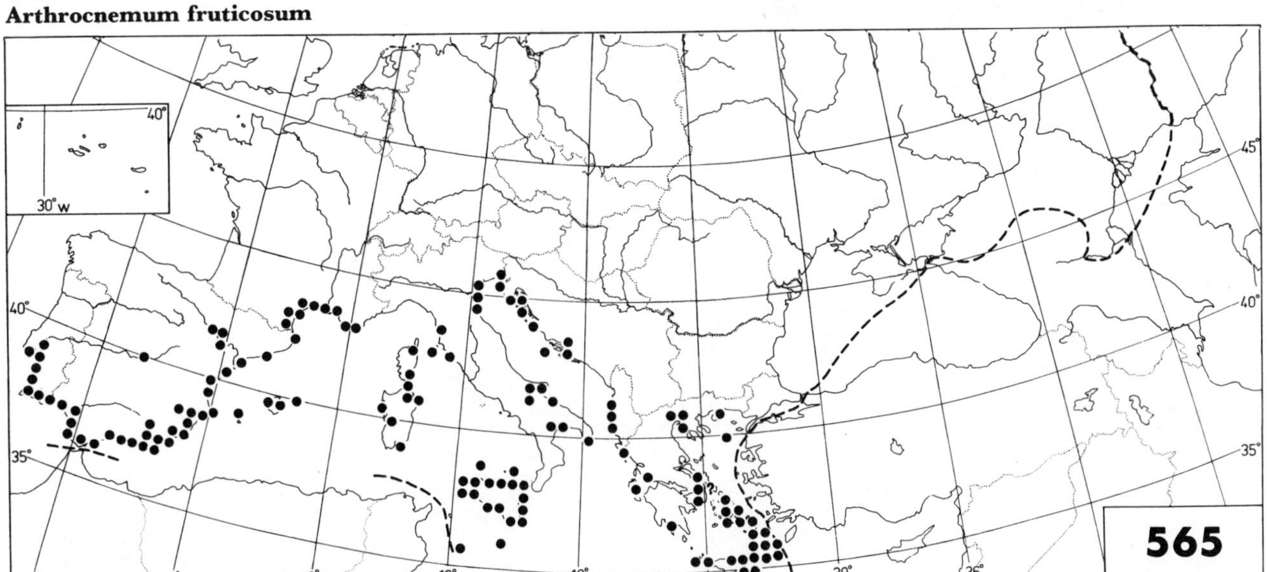

Arthrocnemum macrostachyum

Salicornia L. (the genus) — Map 566.

Taxonomy. P.W. Ball, Feddes Repert. 69 (Fl. Eur. Notulae Syst. 3): 1—8 (1964); A.J. Scott, Bot. Jour. Linn. Soc. 75: 357—374 (1977).

Notes. Rs(C) and Sa omitted (given as present in Fl. Eur.).

Total range. Hegi 1961: 727; MJW 1965: map 135b; Hultén Alaska 1968: 401; Hultén CP 1971: map 181.

Salicornia europaea group — Map 567.

Salicornia europaea + S. prostrata + S. ramosissima

Taxonomy. Diploids with 2n = 18.

Notes. As above for the genus.

Salicornia europaea L.

Salicornia herbacea (L.) L. Incl. S. brachystachya (G.F.W. Meyer) D. König, S. obscura P.W. Ball & Tutin, S. patula Duval-Jouve and S. stricta Dumort.

Notes. Hs omitted, the records in question being referable to *Salicornia ramosissima.* Hu omitted, the material having been identified as *S. prostrata*: R. Soó, Synopsis Syst.-Geobot. Fl. Veget. Hungariae 4: 411—412 (Budapest 1970). Lu omitted, the specimens in question belonging to *S. nitens,* according to Nova Fl. Port. 1971: 98. Rs(C) and Sa omitted; see above for the genus. »N.W. Europe; possibly this or closely related spp. throughout Europe... All except Au, Az, Fa, He, Is, Sb» (Fl. Eur.). Not mapped separately.

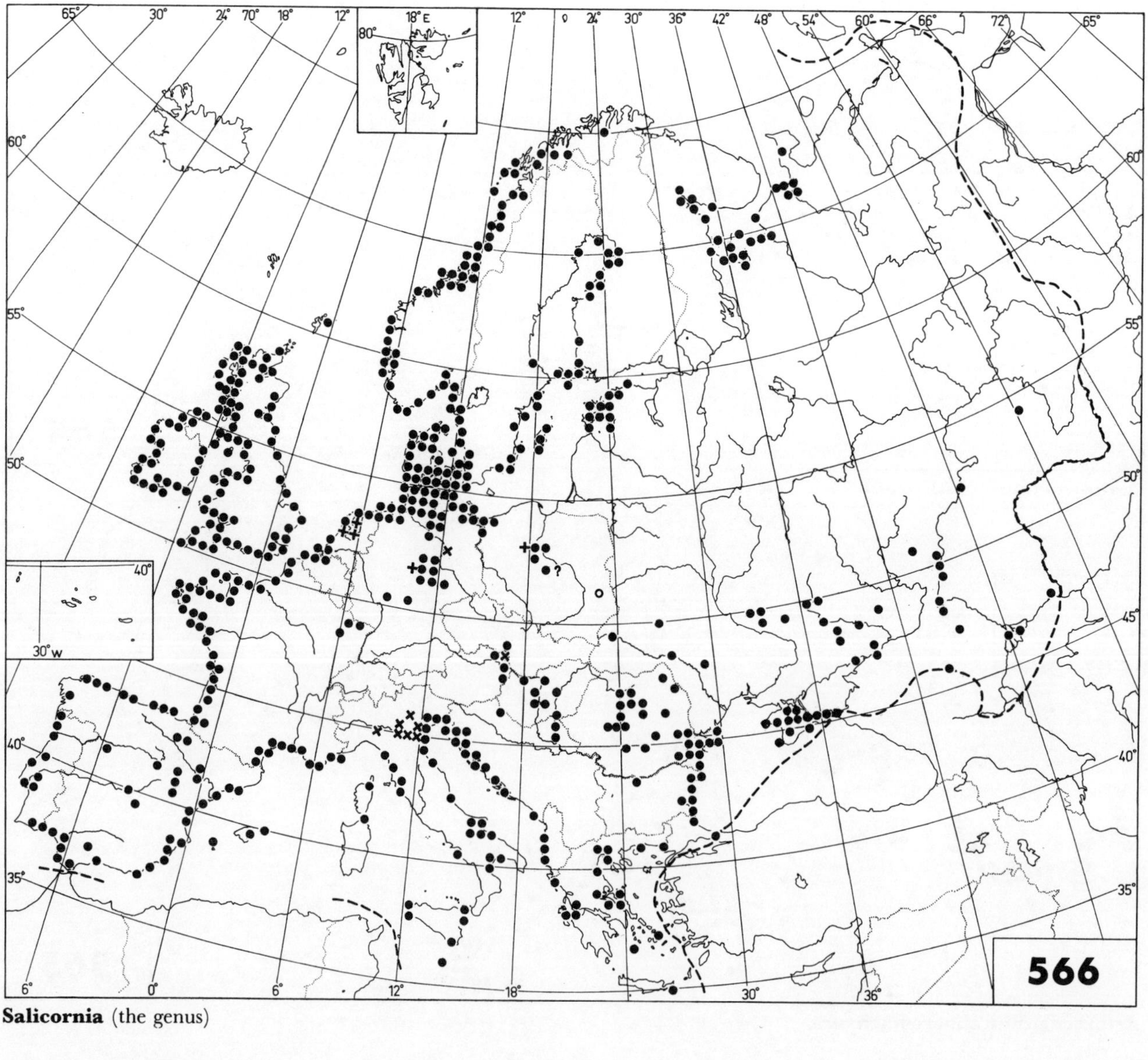

Salicornia (the genus)

Salicornia ramosissima J. Woods

Salicornia appressa Dumort.; S. prostrata auct. non Pallas; S. smithiana Moss

Taxonomy. P.W. Ball, Feddes Repert. 69 (Fl. Eur. Notulae Syst. 3): 4—6 (1964).

Notes. ?Bl added (as »*Salicornia* cf. *ramosissima* J. Woods») (not given in Fl. Eur.): H. Kuhbier, Veröff. Überseemus. Bremen 5 (Cahiers Optima Leafl. 83): 11 (1978); see, however, J. Duvigneaud, Soc. Échange Pl. Vasc. Europe Occ. Bassin Médit. 17(suppl.): 15 (1979). Hs added (not given in Fl. Eur.): E.F. Galiano & B. Valdés, Lagascalia 2: 204 (1972), E.F. Galiano & S. Silvestre, Lagascalia 5: 104 (1975). »N.W. Europe and ?W. Mediterranean region» (Fl. Eur.). Not mapped separately.

Total range. Given as endemic to Europe in Fl. Eur.

Salicornia prostrata Pallas

Salicornia europaea subsp. prostrata (Pallas) O. Schwarz; S. ramosissima auct. non J. Woods; S. simonkaiana (Soó) Soó

Taxonomy. P.W. Ball, Feddes Repert. 69 (Fl. Eur. Notulae Syst. 3): 4—6 (1964).

Notes. Bu added (not given in Fl. Eur.): D. Jordanov & B. Kuzmanov in Fl. Reipubl. Pop. Bulg. 1966: 575, and B.A. Kuzmanov, Candollea 34: 16 (1979), as »*Salicornia ramosissima* Woods», syn. *S. simonkaiana.* »E. & E.C. Europe» (Fl. Eur.). Not mapped separately.

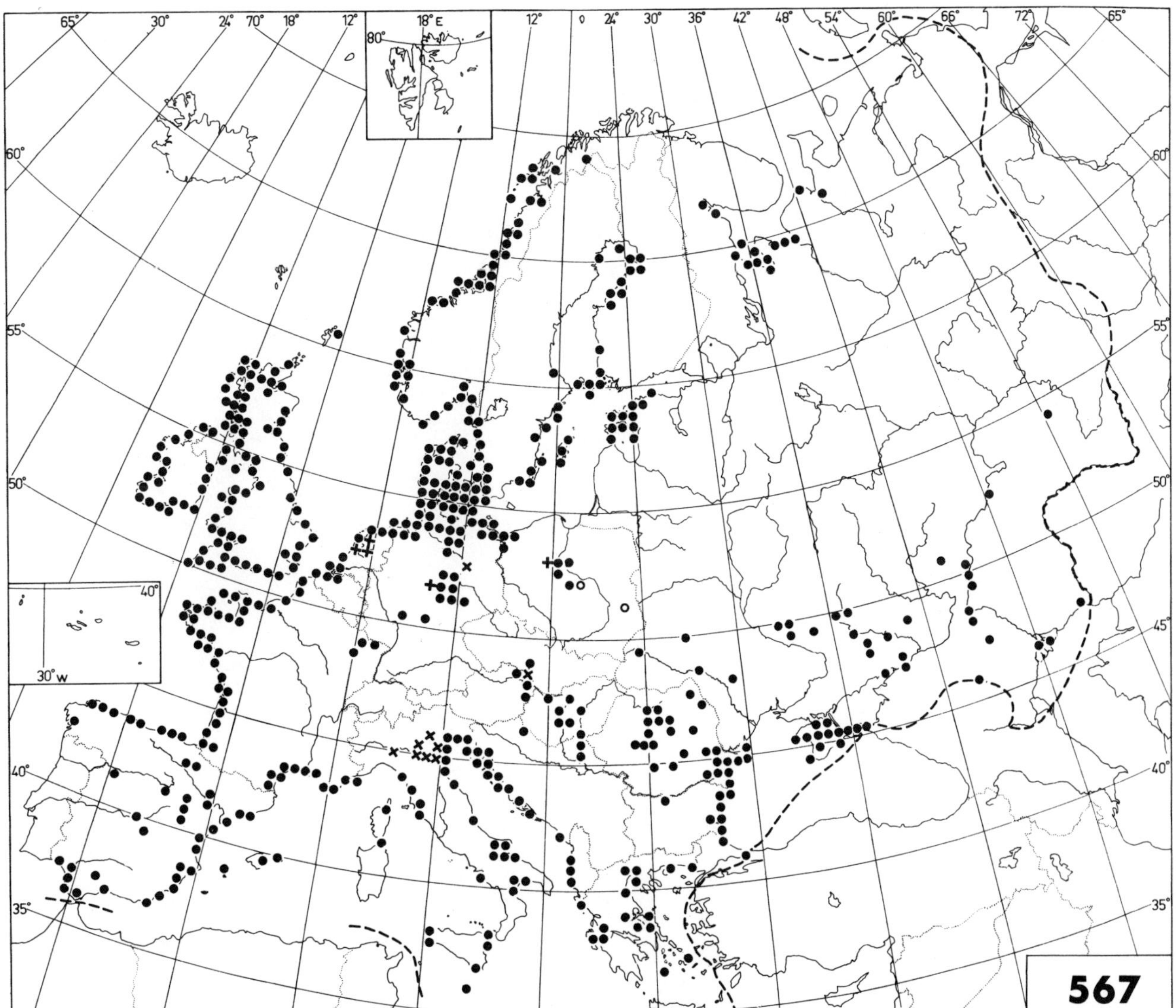

Salicornia europaea group

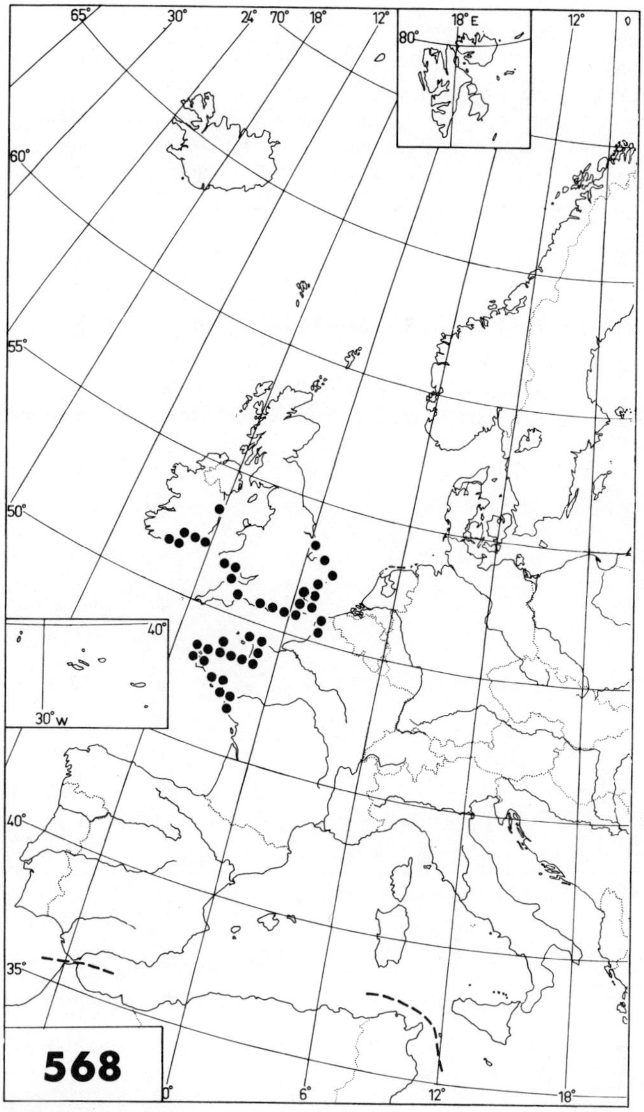

Salicornia pusilla

Salicornia pusilla J. Woods — Map 568.

Salicornia disarticulata Moss

Taxonomy. Diploid with 2n = 18.

Total range. Endemic to Europe.

Salicornia procumbens group — Map 569.

Salicornia dolichostachya + S. fragilis + S. nitens + S. veneta

Taxonomy. Tetraploids with 2n = 36.

Notes. In addition to the information in Fl. Eur., members of the *Salicornia procumbens* group have also been recorded from S. Europe (Ga, It and Lu). However, the map must still be considered quite provisional.

Salicornia nitens P.W. Ball & Tutin

Salicornia ramosissima auct. non J. Woods

Taxonomy. P.W. Ball, Feddes Repert. 69 (Fl. Eur. Notulae Syst. 3): 6—7 (1964).

Notes. Lu added (not given in Fl. Eur.): Nova Fl. Port. 1971: 98. »England, S. Ireland; probably elsewhere in W. and S. Europe» (Fl. Eur.). Not mapped separately.

Salicornia fragilis P.W. Ball & Tutin

? Salicornia procumbens Sm.; S. stricta auct. non Dumort. Incl. S. emerici Duval-Jouve, S. lutescens P.W. Ball & Tutin and S. stricta subsp. decumbens Aellen

Taxonomy. P.W. Ball, Feddes Repert. 69 (Fl. Eur. Notulae Syst. 3): 6—7 (1964).

Notes. ?Tu added (not given in Fl. Eur.): Fl. Turkey 1967: 323. »Known certainly only from C. & S. England and Ireland but probably widespread throughout W. & S. Europe» (Fl. Eur.). Not mapped separately.

Total range. Outside Europe apparently in Anatolia (given as endemic to Europe in Fl. Eur.): Fl. Turkey 1967: 323.

Salicornia veneta Pign. & Lausi — Map 569.

Taxonomy. D. Lausi, Giorn. Bot. Ital. 103: 183—188 (1969).

Notes. It (the species not mentioned in Fl. Eur.).

Total range. Endemic to Europe.

Salicornia dolichostachya Moss

? Salicornia procumbens Sm.; S. stricta sensu D. König, pro max. parte, non Dumort. Incl. S. pojarkovae N. Semen. and S. strictissima K. Gram

Taxonomy. P.W. Ball, Feddes Repert. 69 (Fl. Eur. Notulae Syst. 3): 6—8 (1964).

Notes. Probably also present in Fe, according to V. Johansen & R. Elven, Blyttia 37: 65—66 (1979) (not given in Fl. Eur.). »N.W. Europe» (Fl. Eur.). Not mapped separately.

Total range. Endemic to Europe (Fl. Eur.).

S. dolichostachya subsp. dolichostachya

? Salicornia leiosperma K. Gram; ? S. oliveri Moss; S. pojarkovae N. Semen.

Taxonomy. N.Z. Semenova-Tian-Shanskaya in Fl. Murmanskoy Oblasti 3: 190, 192—193, 367—368 (1956); V. Johansen & R. Elven, Blyttia 37: 61—64 (1979) (*Salicornia pojarkovae*).

Notes. Mentioned, as *Salicornia* cf. *pojarkovae*, from the northern shores of the Gulf of Bothnia (Baltic Sea area; Fe) by V. Johansen & R. Elven, Blyttia 37: 65—66 (1979). »Throughout the range of the species, except the Baltic region» (Fl. Eur.).

S. dolichostachya subsp. strictissima (K. Gram) P.W. Ball

Salicornia strictissima K. Gram

Notes. »Baltic region and possibly parts of the North Sea coasts» (Fl. Eur.).

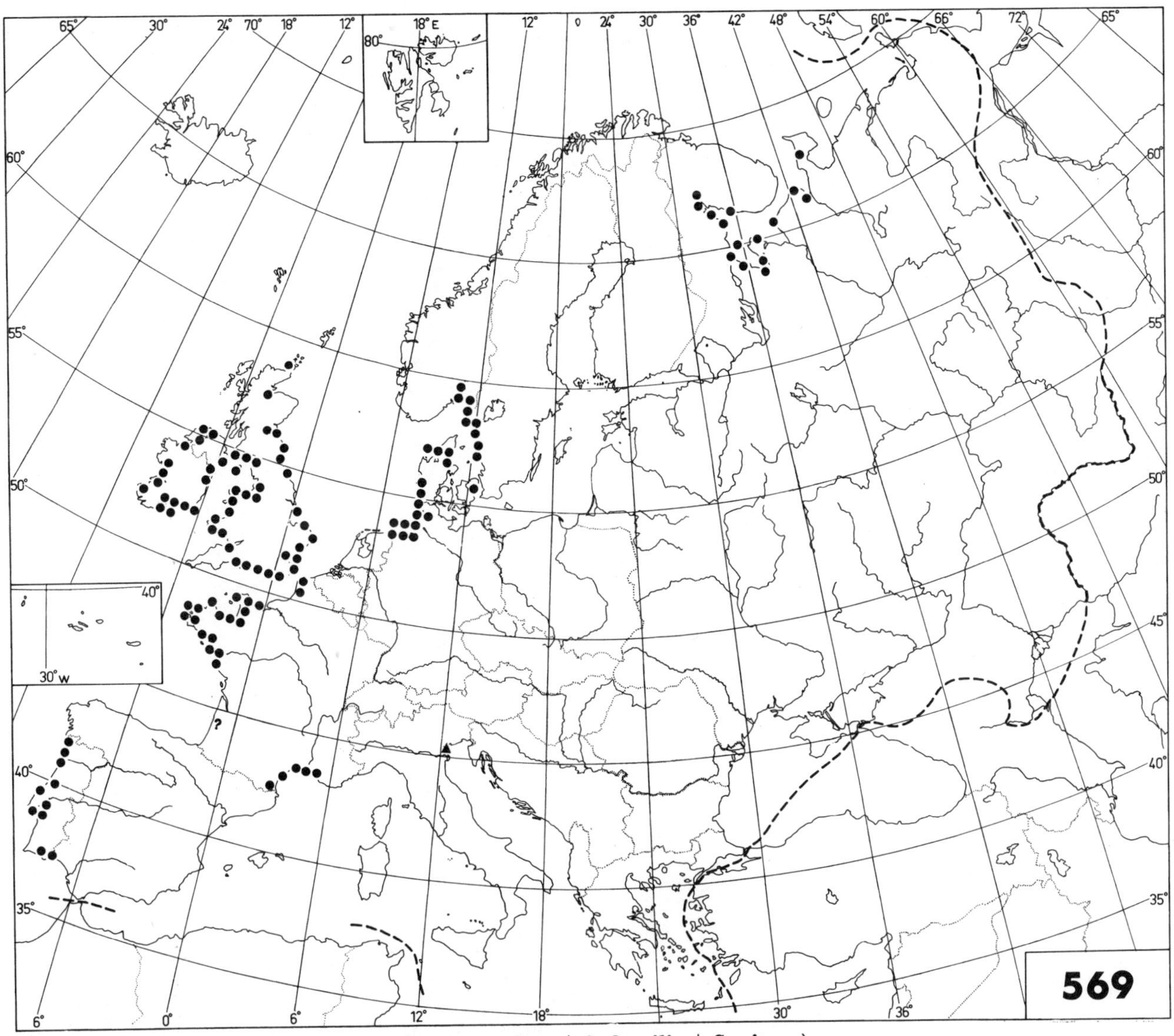

• = **Salicornia procumbens** group (**S. dolichostachya** + **S. fragilis** + **S. nitens**)
▲ = **S. veneta**

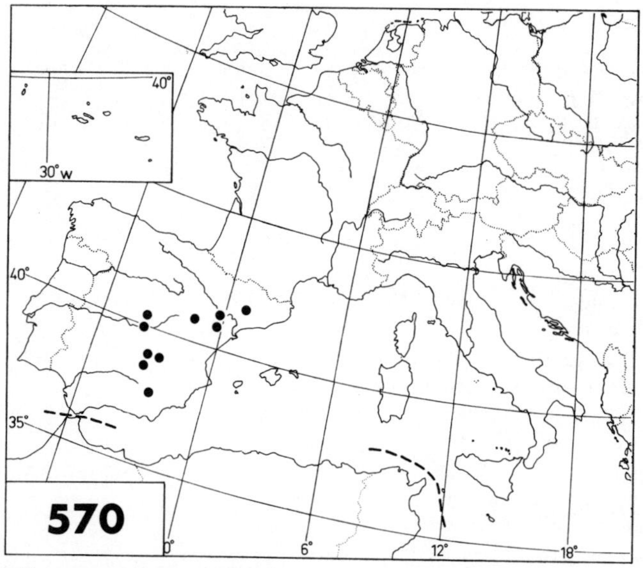

Microcnemum coralloides

Microcnemum coralloides (Loscos & Pardo) Font Quer — Map 570.

Arthrocnenum coralloides Loscos & Pardo; Microcnemum fastigiatum Ung.-Sternb.

Taxonomy. G. Wagenitz, Ber. Deutsch. Bot. Ges. 72: 151—158 (1959).

Total range. Subsp. *coralloides* is endemic to Europe, another subsp. being confined to C. Anatolia; see G. Wagenitz, Ber. Deutsch. Bot. Ges. 72: 151—158 (1959).

Suaeda vera J.F. Gmelin — Map 571.

Suaeda fruticosa auct. non Forskål

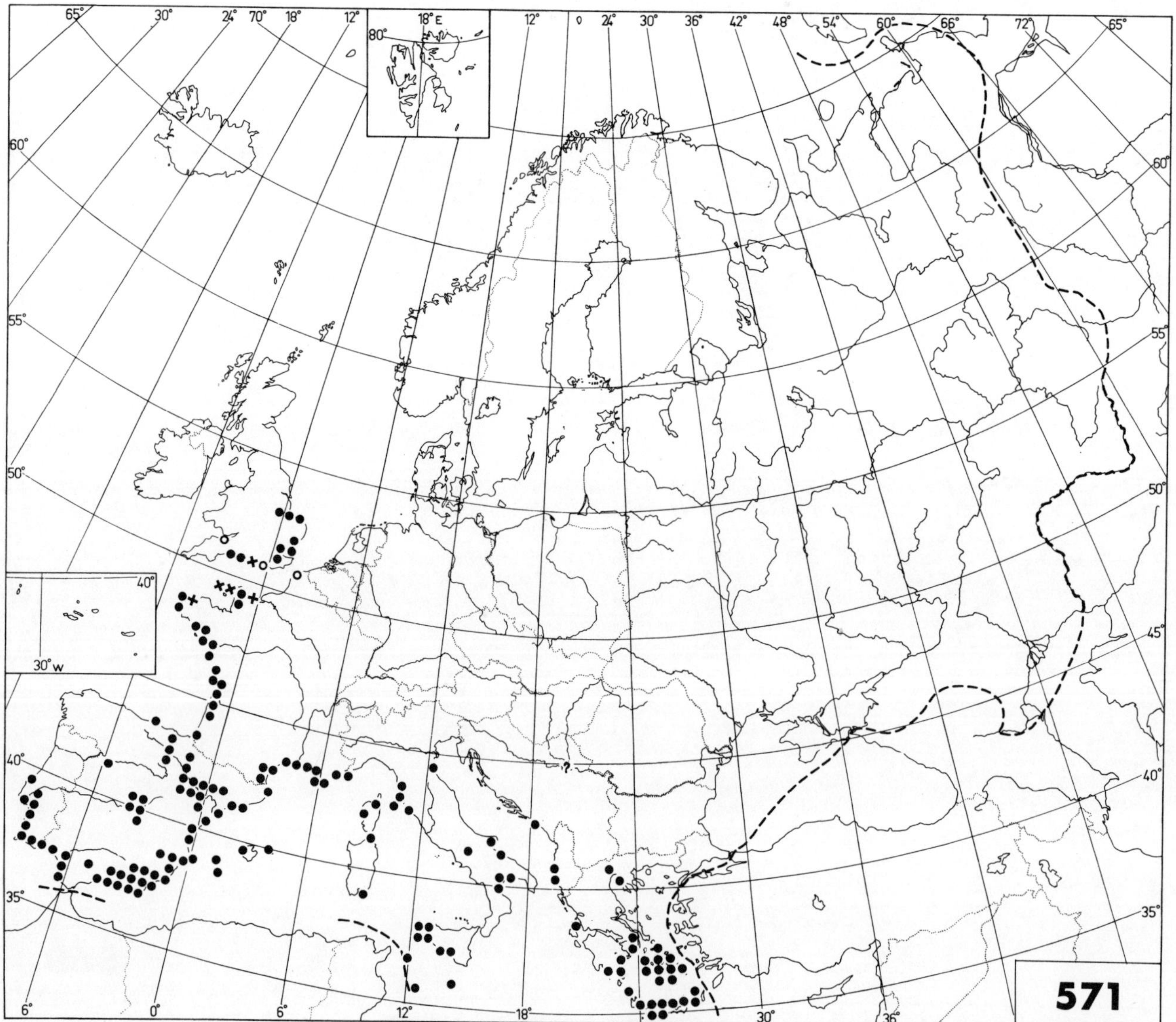

Suaeda vera

Suaeda pruinosa Lange — Map 572.

Suaeda fruticosa var. brevifolia Moq., pro p.; S. vera auct. non J.F. Gmelin; S. vera subsp. pruinosa (Lange) O. Bolòs & Vigo

Nomenclature. O. Bolòs & J. Vigo, Butll. Inst. Catalana Hist. Nat. 38 (Sec. Bot. 1): 89 (1974).

Suaeda physophora Pallas — Map 573.

Suaeda dendroides (C.A. Meyer) Moq. — Map 573.

Suaeda confusa Iljin — Map 574.

Notes. Rs(W) mapped according to Fl. URSR 1952: 376, 378.

Suaeda baccifera Pallas — Map 574.

Notes. Data not available from Rs(W) (given in Fl. Eur.).

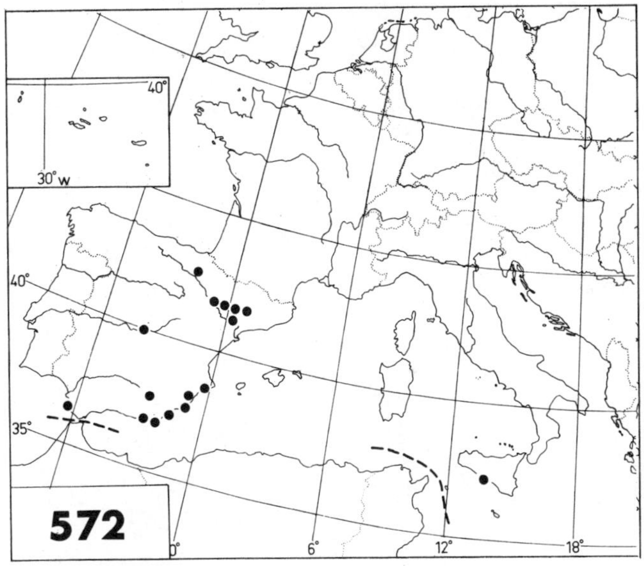

Suaeda pruinosa

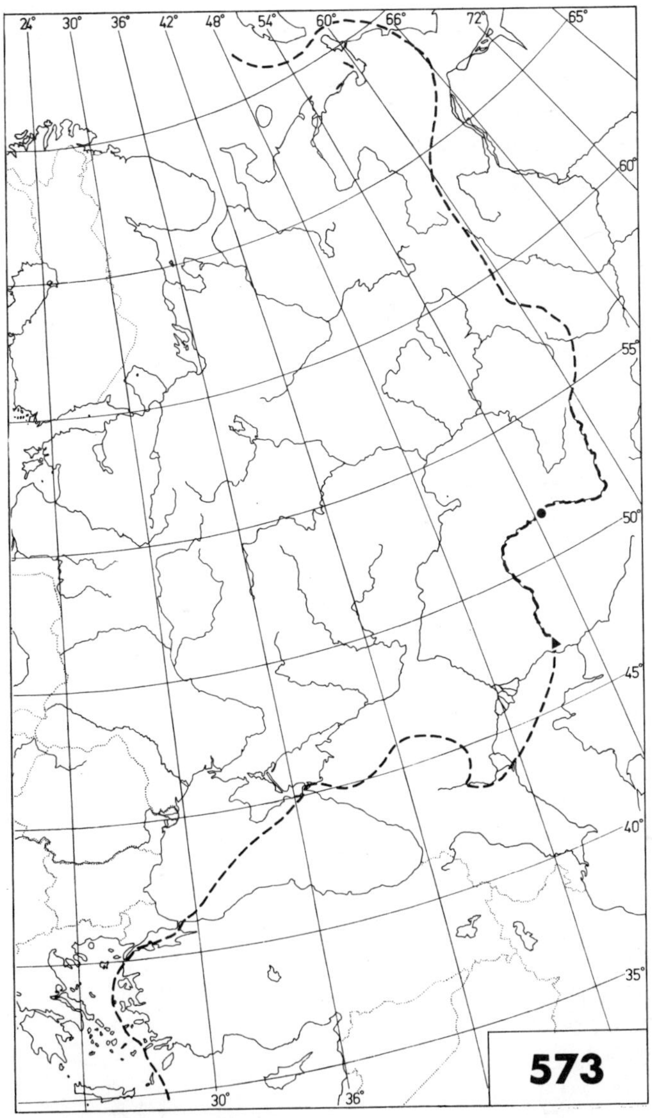

• = **Suaeda physophora**
▲ = **S. dendroides**

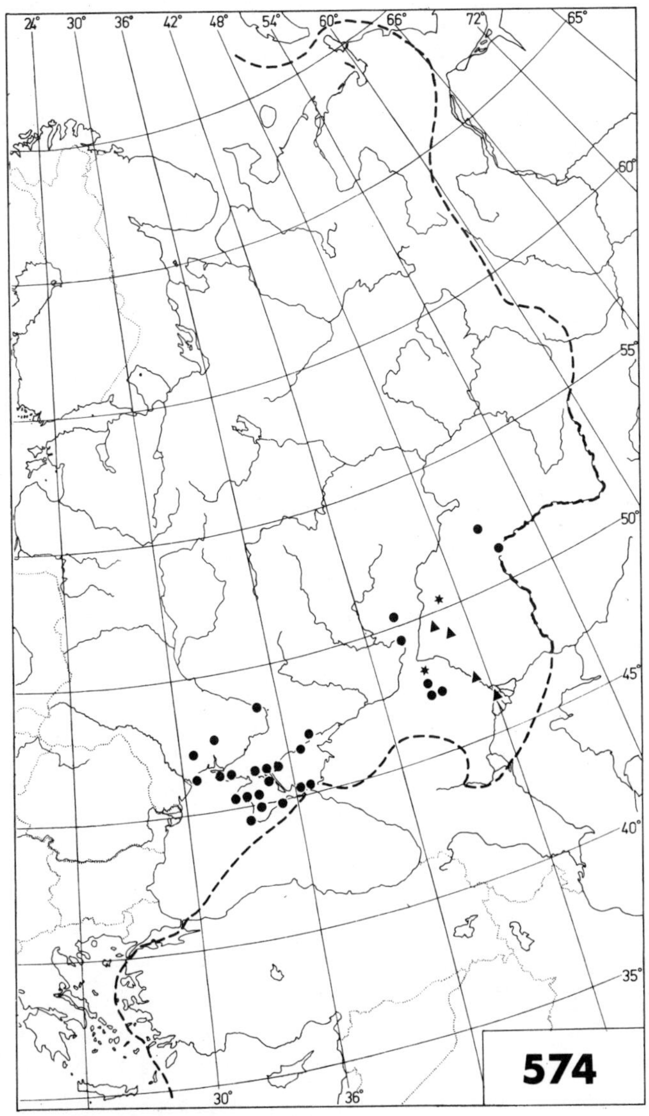

• = **Suaeda confusa**
▲ = **S. baccifera**
★ = **S. confusa + S. baccifera**

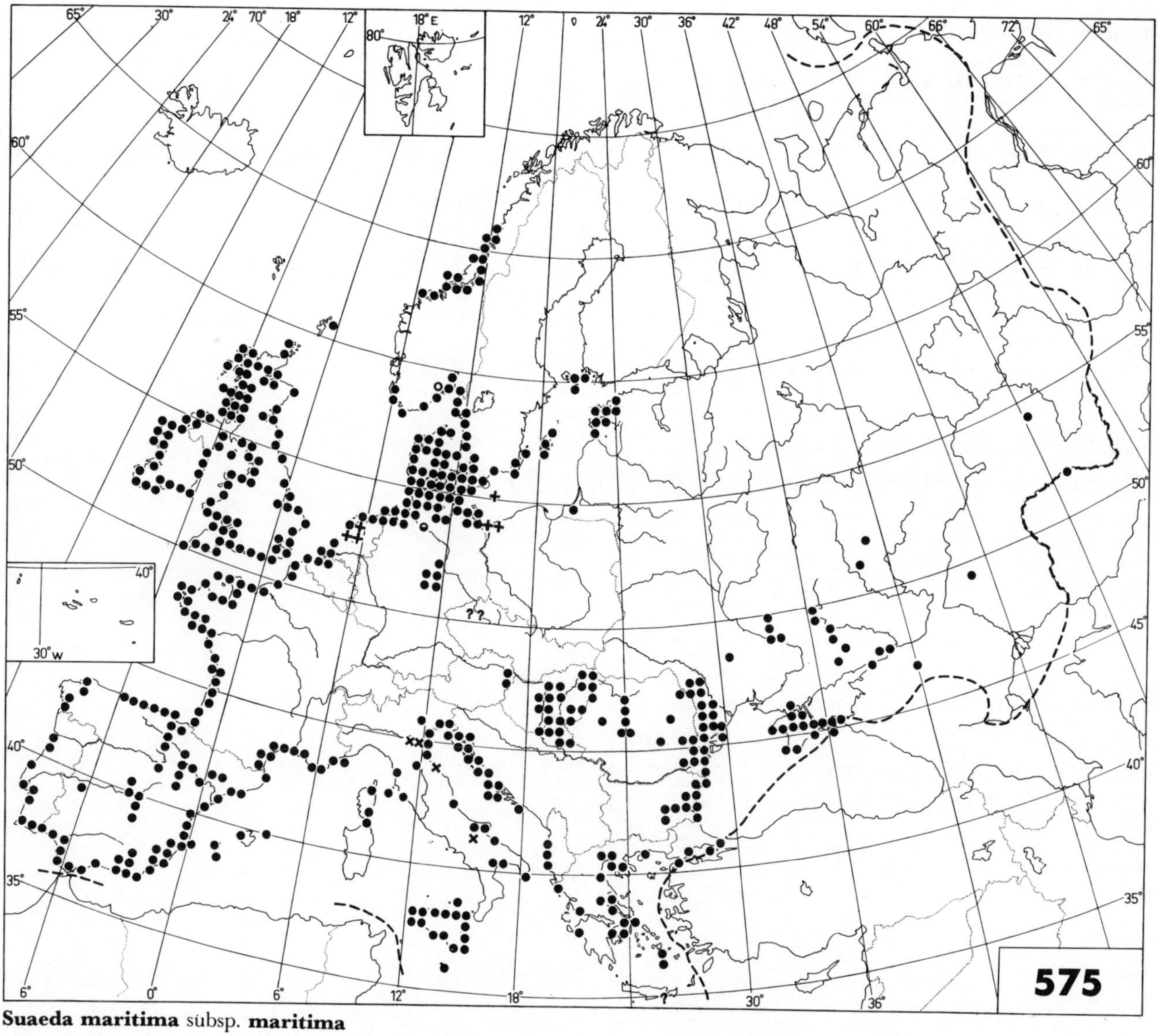

Suaeda maritima subsp. **maritima**

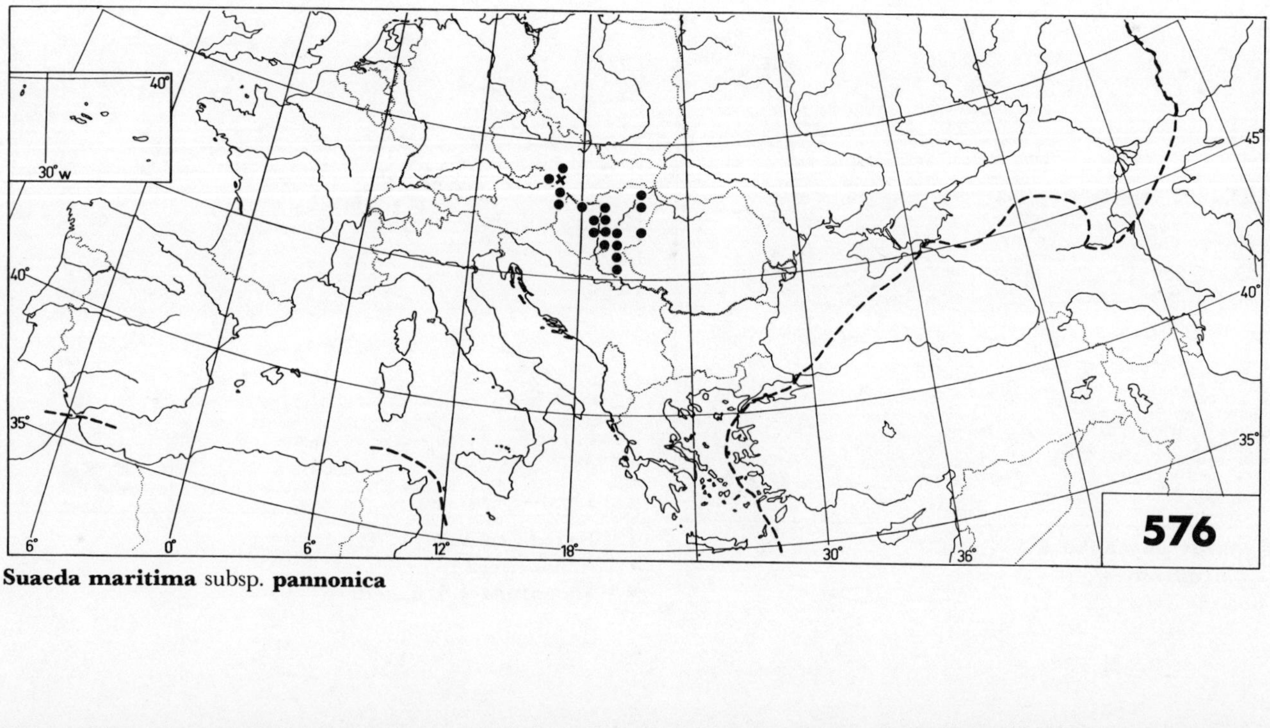

Suaeda maritima subsp. **pannonica**

Suaeda maritima (L.) Dumort.

Suaeda albescens Láz.-Ibiza; S. cavanillesiana (Láz.-Ibiza) Coutinho; S. depressa (Pursh) S. Watson; S. pannonica G. Beck; S. prostrata Pallas; S. salsa (L.) Pallas

Taxonomy and nomenclature. P. Aellen in Fl. Turkey 1967: 325—326; J. McNeill, I.J. Bassett & C.W. Crompton, Rhodora 79: 133—138 (1977).

Notes. Presence in Cr doubtful (given as present in Fl. Eur.), extinct in Po (given as present in Fl. Eur.), Sa omitted (given in Fl. Eur.). Owing to differences in the interpretation and delimitation of the subspecies recognized in Fl. Eur., only subsp. *pannonica* could be mapped separately.

Total range. Hultén CP 1971: map 179 (northern hemisphere).

S. maritima subsp. **maritima** — Map 575.

Incl. Suaeda maritima subsp. prostrata (Pallas) Soó and S. maritima subsp. salsa (L.) Soó

Notes. Al, Au, Be, Bl, Br, Bu, Co, ?Cr, ?Cz, Da, Fe, Ga, Ge, Gr, Hb, Ho, Hs, Hu, It, Ju, Lu, No, †Po, Rm, Rs(B, C, W, K, E), Si, Su, Tu. »Recorded throughout the range of the species, but many of the records for E.C. & E. Europe may be referable to (b) [= subsp. *salsa*]» (Fl. Eur.).

S. maritima subsp. **pannonica** (G. Beck) Soó ex P.W. Ball — Map 576.

Taxonomy. Subsp. *pannonica* seems fairly well differentiated from the rest of *Suaeda maritima* and has recently even been treated as a separate species: P. Aellen in Hegi 1961: 737.

Notes. Au, Cz, Hu, Ju. For Cz, see J. Soják, Preslia (Praha) 34: 408 and 413 (1962). »... ostwärts bis in die Dobrudscha» (Hegi 1961: 737), »E. Austria, Hungary, N.W. Romania» (Fl. Eur.).

Total range. Evidently endemic to Europe (although this is not mentioned in Fl. Eur.).

For the text of maps 577 and 578, see overleaf.

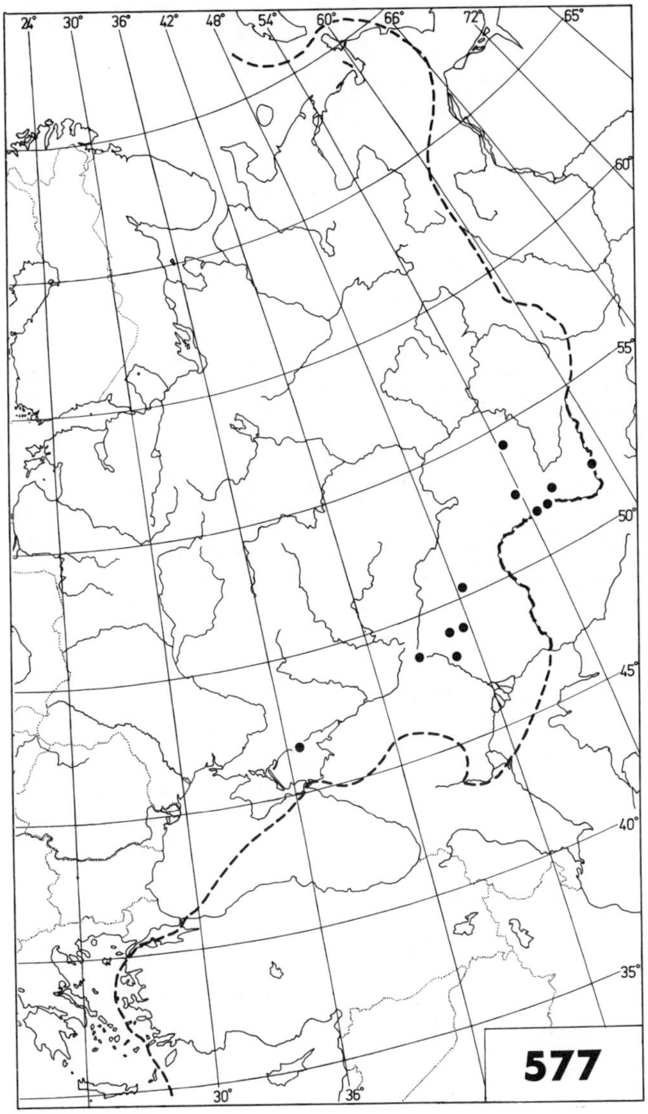

Suaeda corniculata

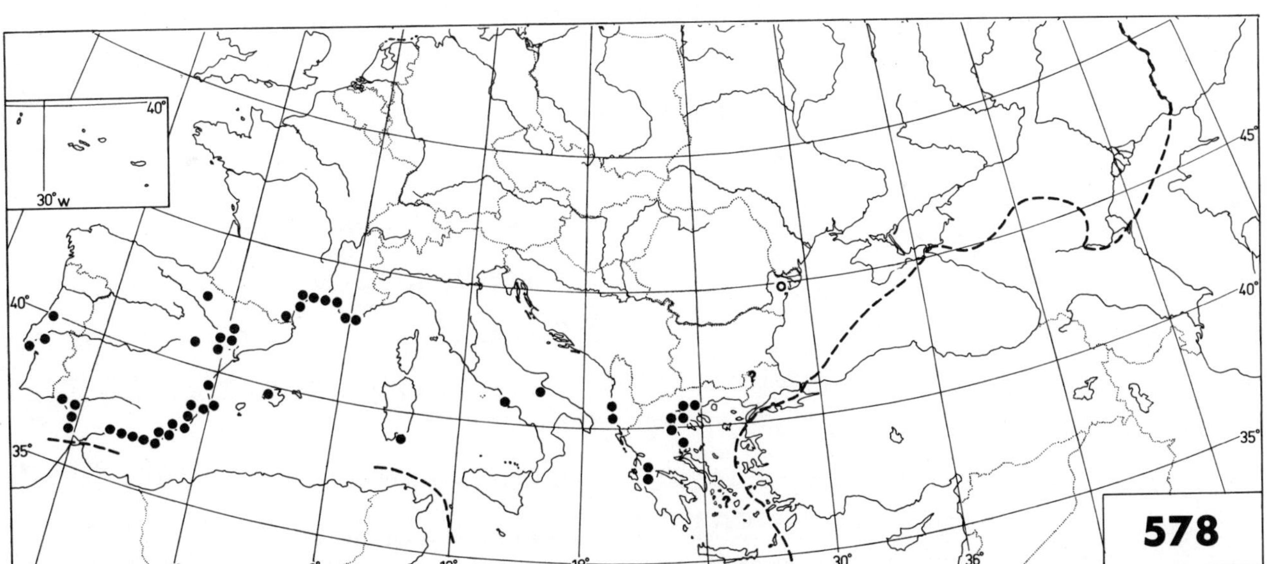

Suaeda splendens

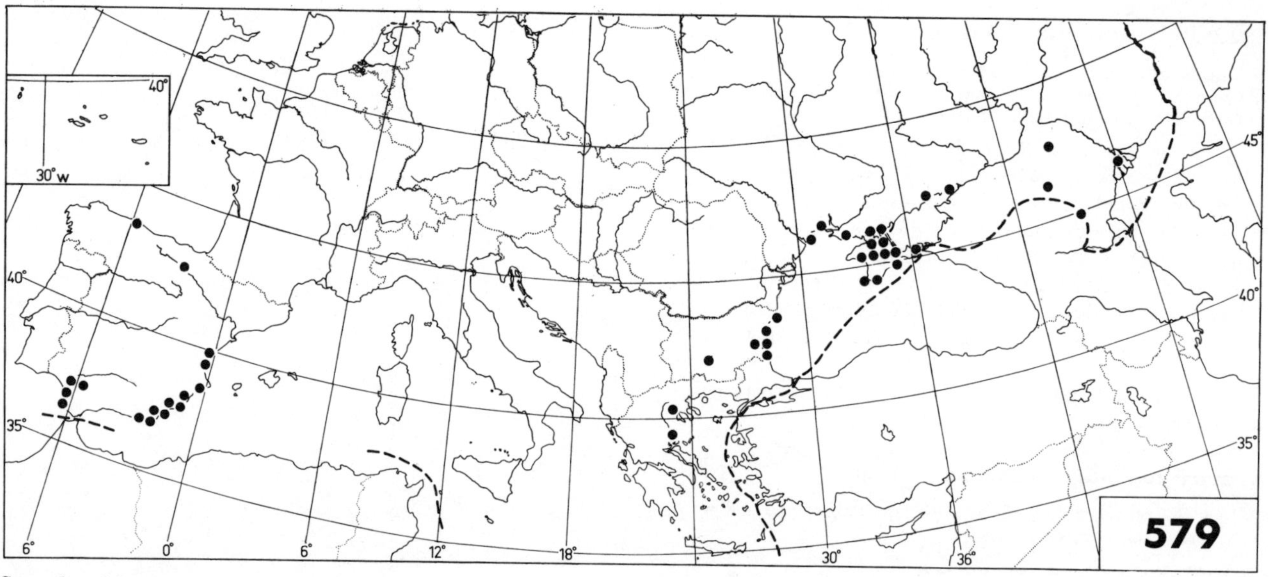

Suaeda altissima

Suaeda corniculata (C.A. Meyer) Bunge — Map 577.

Notes. Rs(C, W) added (not given in Fl. Eur.). Recorded for Au (not given in Fl. Eur.) by P. Aellen, Ann. Naturhist. Mus. Wien 75: 17—20 (1971), but the precise taxonomic identity of the plant is not fully established (not mapped).

Suaeda splendens (Pourret) Gren. & Godron — Map 578.

Suaeda setigera (DC.) Moq.

Notes. Bl added (not given in Fl. Eur.): Fl. Mallorca 1978: 109. Co omitted (given in Fl. Eur.). Reported for Rm as introduced before 1900 (?Rm in Fl. Eur.). Presence in Tu doubtful (given as present in Fl. Eur., not recognized in Fl. Turkey 1967: 324—327).

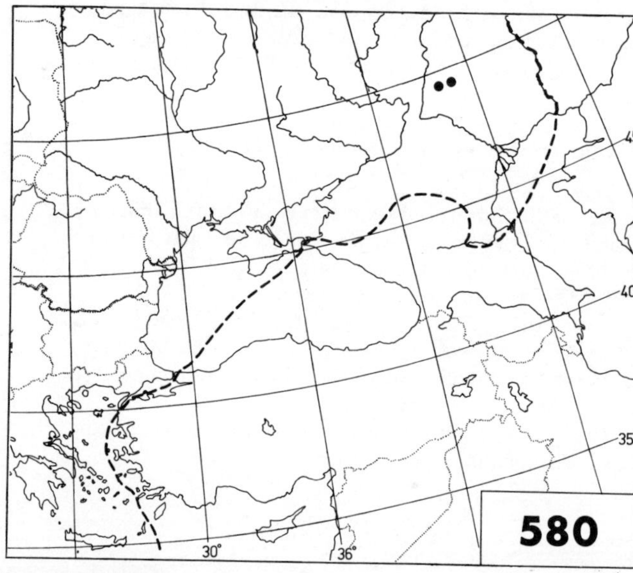

Suaeda linifolia

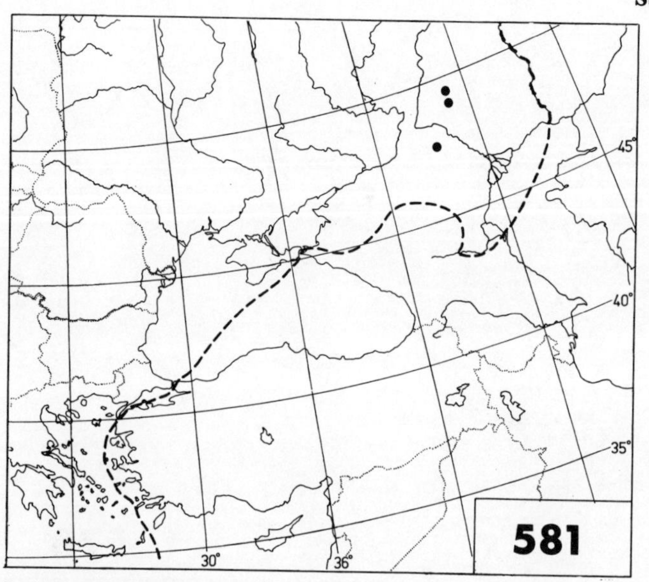

Suaeda eltonica

Suaeda altissima (L.) Pallas — Map 579.

Suaeda linifolia Pallas — Map 580.

Suaeda eltonica Iljin — Map 581.

Suaeda heterophylla (Kar. & Kir.) Bunge ex Boiss. — Map 582.

Notes. Bu added (not given in Fl. Eur.): D. Jordanov & B. Kuzmanov in Fl. Reipubl. Pop. Bulg. 1966: 569—570.

Suaeda kossinskyi Iljin — Map 582.

Bienertia cycloptera Bunge ex Boiss. — Map 583.

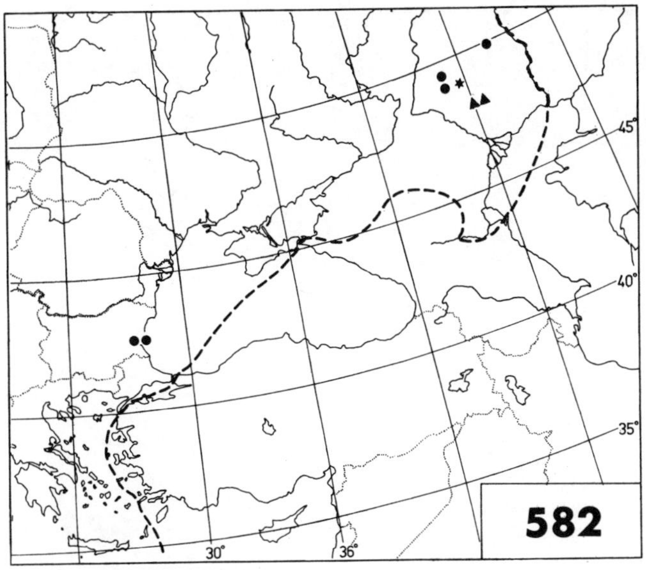

• = **Suaeda heterophylla**
▲ = **S. kossinskyi**
✶ = **S. heterophylla + S. kossinskyi**

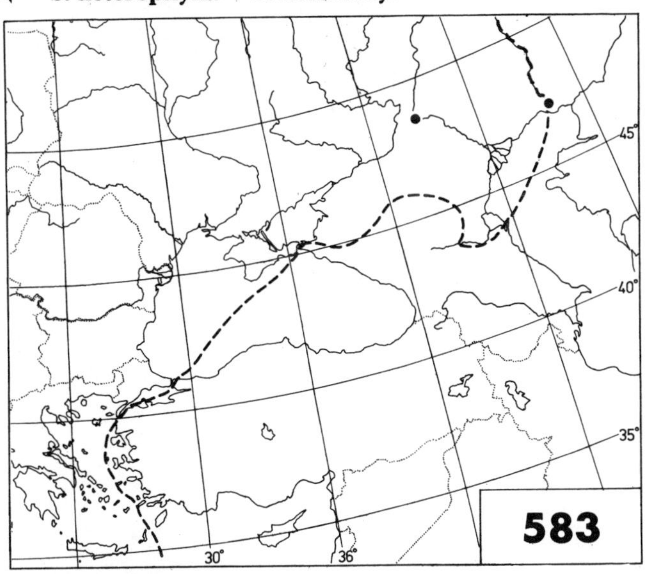

Bienertia cycloptera

Horaninovia minor Fischer & Meyer

Notes. No European data available. »... occurs at the mouth of the Ural river in W. Kazakhstan, and has also been doubtfully recorded from S.E. Russia (Astrakhan)» (Fl. Eur.).

Salsola foliosa (L.) Schrader — Map 584.

Caspia foliosa (L.) Galushko; Salsola clavifolia Pallas

Generic delimitation. V. Botschantzev, Nov. Syst. Pl. Vasc. (Leningrad) 13: 94, 99 (1976); A. Galushko, Nov. Syst. Pl. Vasc. (Leningrad) 13: 250 (1976).

Salsola brachiata Pallas — Map 585.

Climacoptera brachiata (Pallas) Botsch.

Generic delimitation. V. P. Botschantzev, Sbornik rabot akad. V.N. Sukaczevy 75-let.: 111 (1956); see also S.K. Czerepanov, Additam. Corr. Fl. URSS: 186 (Leningrad 1973).

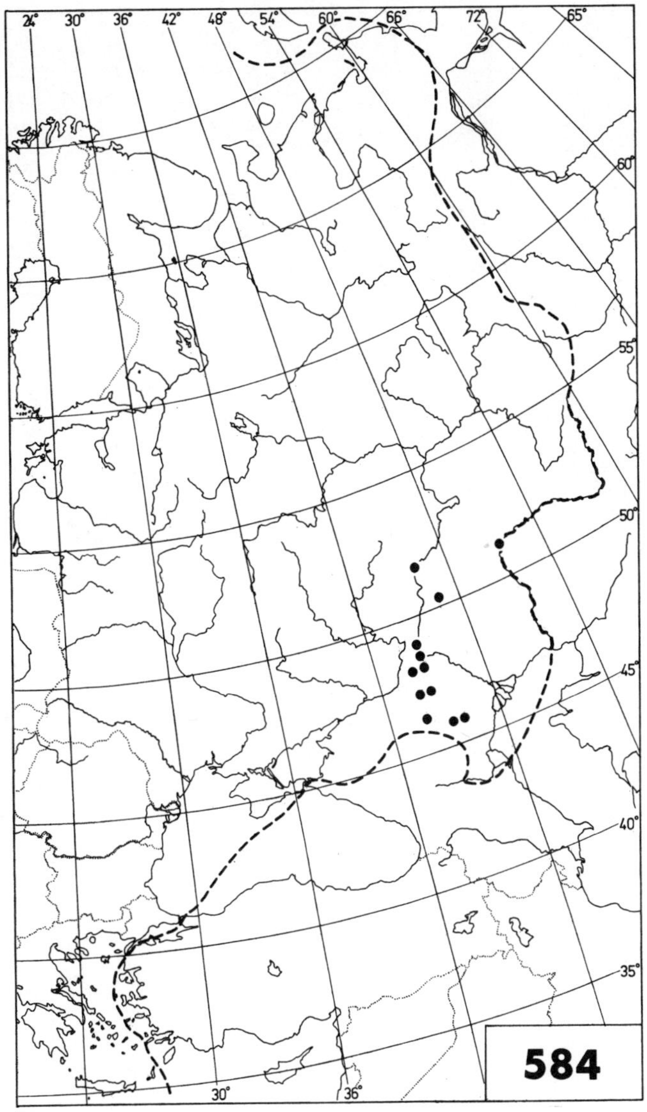

Salsola foliosa

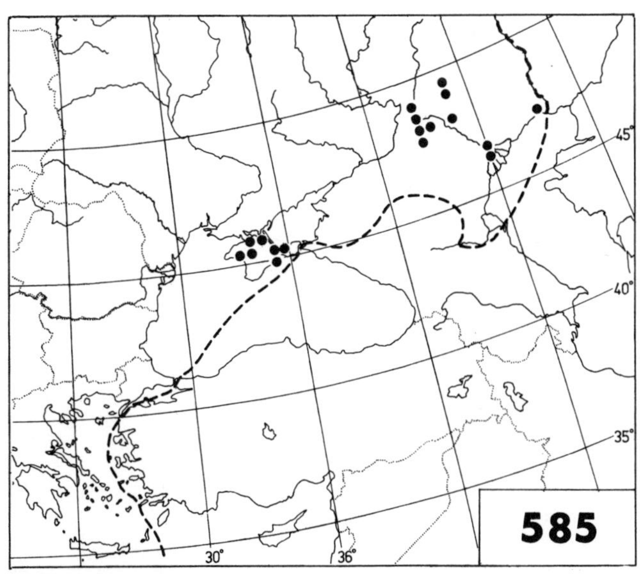

Salsola brachiata

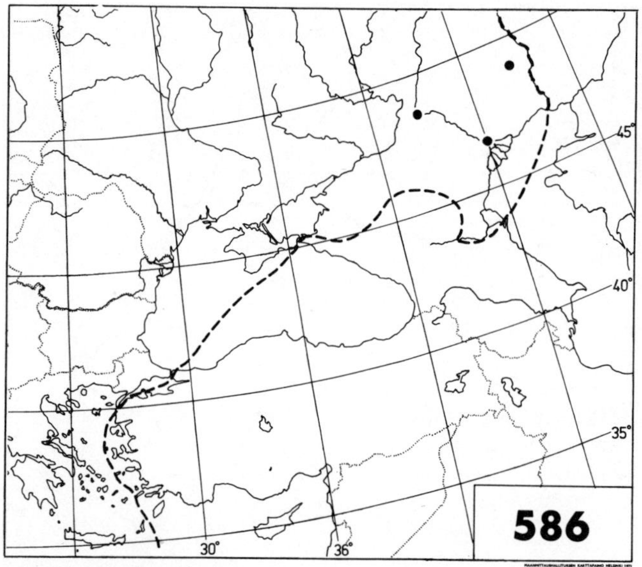

Salsola paulsenii

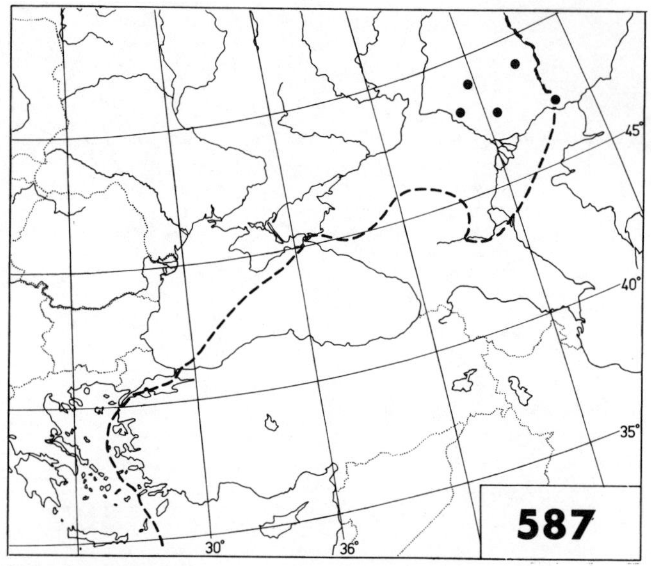

Salsola pellucida

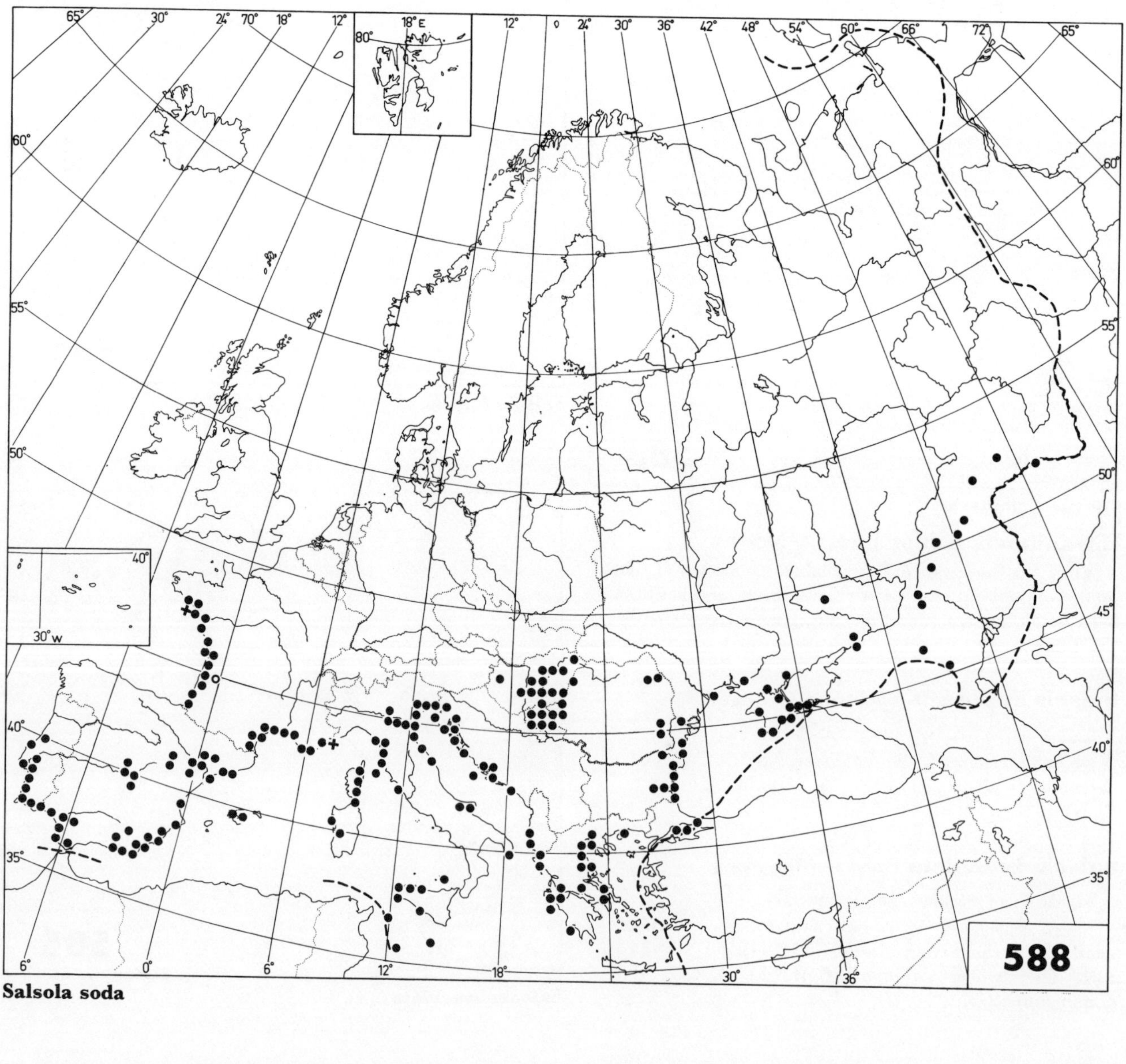

Salsola soda

Salsola paulsenii Litv. — Map 586.

Salsola pellucida Litv. — Map 587.

Salsola soda L. — Map 588.

Notes. Co, Sa and Si added (not given in Fl. Eur.). Rs(W) mapped according to Fl. URSR 1952: 390.

Salsola acutifolia (Bunge) Botsch. — Map 589.

Salsola mutica C.A. Meyer

Notes. Rs(W) added (not given in Fl. Eur.): Fl. URSR 1952: 390.

Salsola tamariscina Pallas — Map 590.

Notes. Data not available from Rs(W) (given in Fl. Eur.).

Total range. MJW 1965: map 135c.

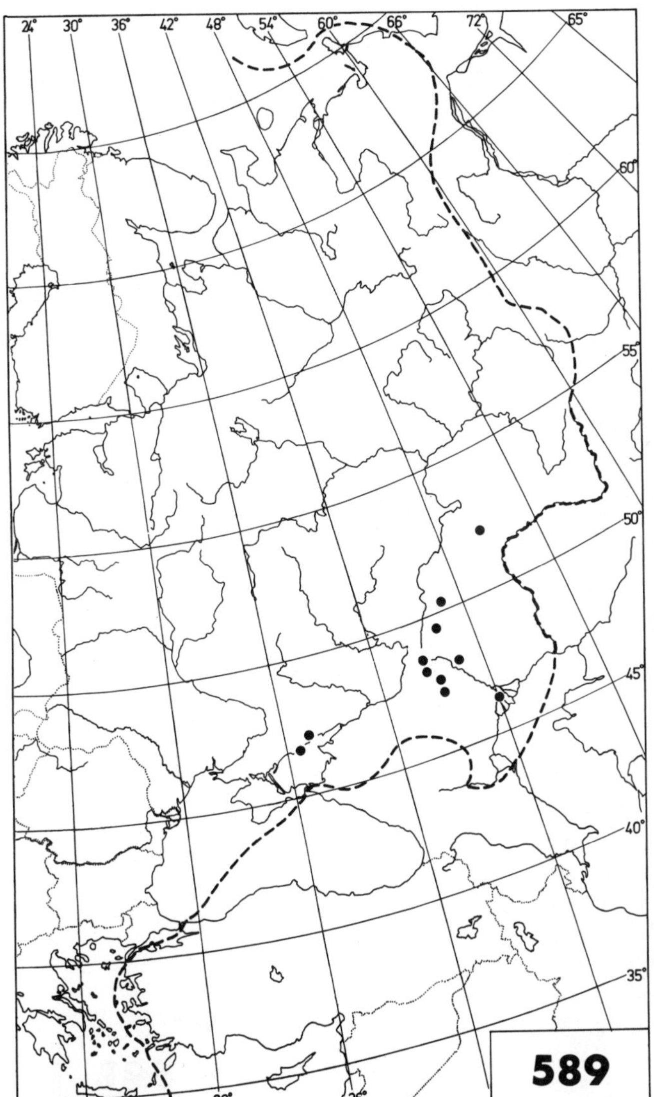

Salsola acutifolia

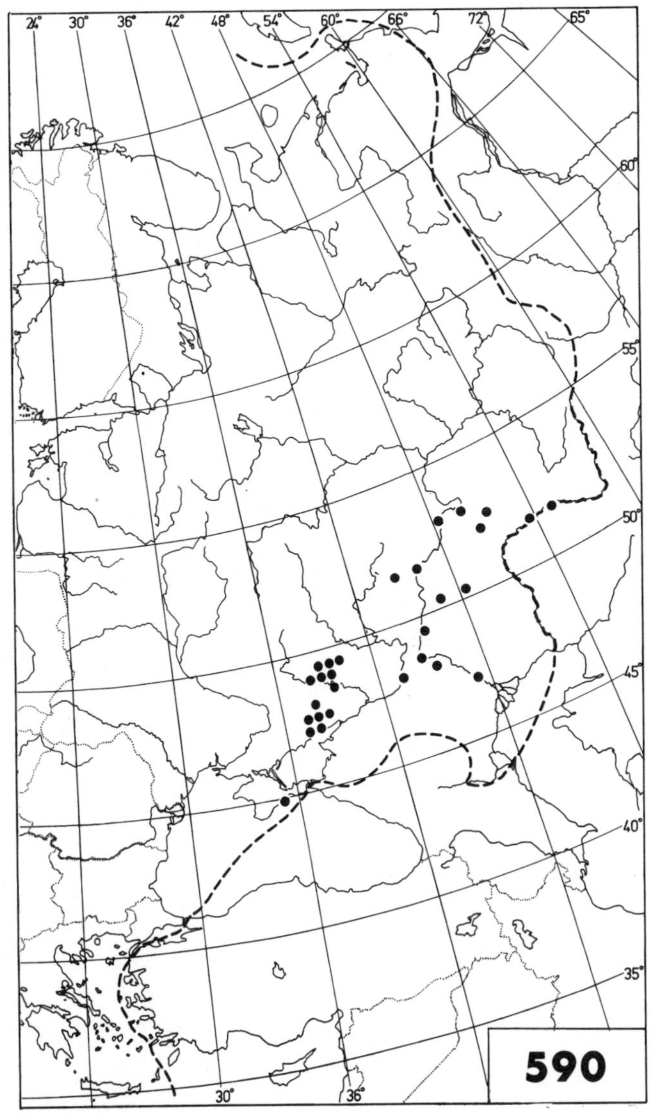

Salsola tamariscina

Salsola kali L. — Map 591.

Salsola aptera Iljin; S. pestifer A. Nelson; S. pontica Iljin; S. praecox Litv.; S. ruthenica Iljin; S. tragus L.

Notes. Established alien in He (the status not defined in Fl. Eur.). Rs(W) mapped according to Fl. URSR 1952: 386, 387 (as *S. ruthenica* and *S. tragus*). The data available do not allow separate mapping of the subspecies.

Total range. H. Walter, Einführung in die Phytologie III(2): 187 (1954); Hegi 1961: 744; MJW 1965: map 135c; F. Fukarek, Natur Naturschutz Mecklenburg 5: 169 (1967).

S. kali subsp. **kali**

Notes. »Coasts of Europe, northwards to 63°30′ N.» (Fl. Eur.).

S. kali subsp. **tragus** (L.) Nyman

Notes. »Coasts of S. Europe» (Fl. Eur.).

S. kali subsp. **ruthenica** (Iljin) Soó

Incl. Salsola pestifer A. Nelson

Notes. »Throughout the range of the species, but often only casual in the north» (Fl. Eur.).

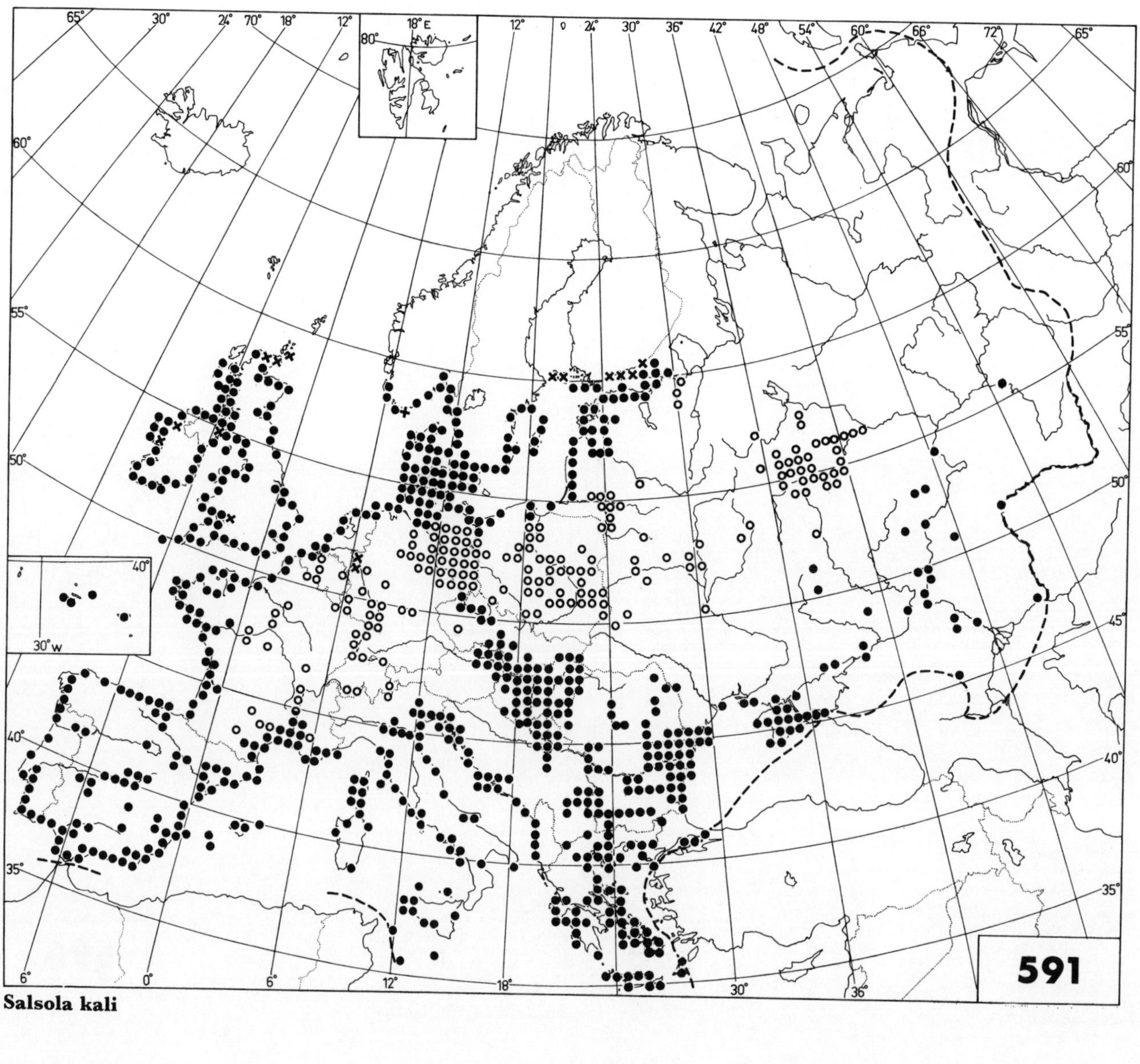

Salsola kali

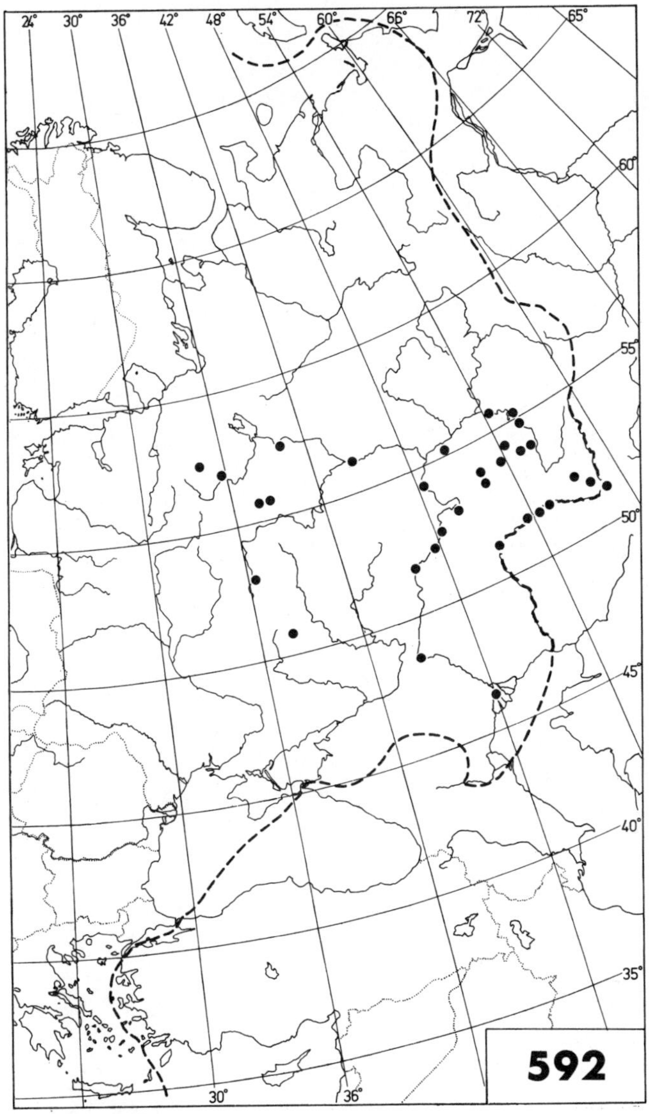

Salsola collina

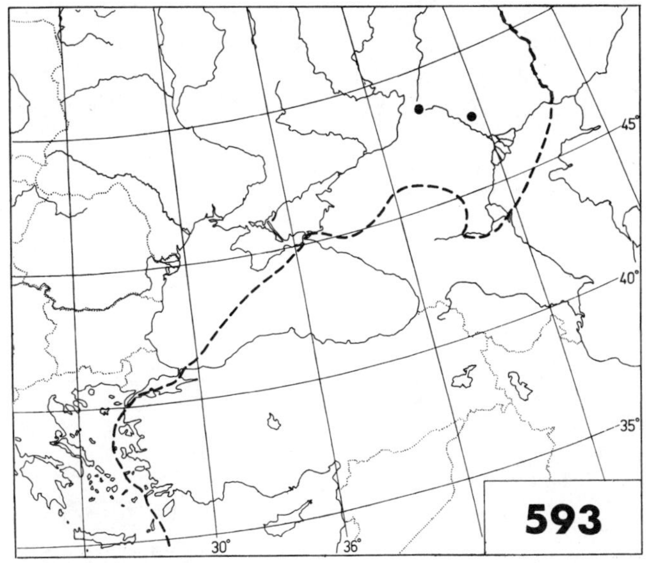

Salsola lanata

Salsola collina Pallas — Map 592.

Salsola kali subsp. collina (Pallas) O. Bolòs & Vigo

Nomenclature. O. Bolòs & J. Vigo, Butll. Inst. Catalana Hist. Nat. 38 (Sec. Bot. 1): 89 (1974).

Salsola lanata Pallas — Map 593.

Climacoptera lanata (Pallas) Botsch.

Generic delimitation and nomenclature. V.P. Botschantzev, Sbornik rabot akad. V.N. Sukaczevy 75-let.: 112 (1956); see also S.K. Czerepanov, Additam. Corr. ad Fl. URSS: 186—187 (Leningrad 1973).

Salsola turcomanica Litv. — Map 594.

Climacoptera turcomanica (Litv.) Botsch.

Generic delimitation and nomenclature. V.P. Botschantzev, Sbornik rabot akad. V.N. Sukaczevy 75-let.: 112 (1956); see under *Salsola lanata.*

Salsola crassa Bieb. — Map 594.

Climacoptera crassa (Bieb.) Botsch.

Generic delimitation and nomenclature. V.P. Botschantzev, Sbornik rabot akad. V.N. Sukaczevy 75-let.: 112 (1956); see under *Salsola lanata.*

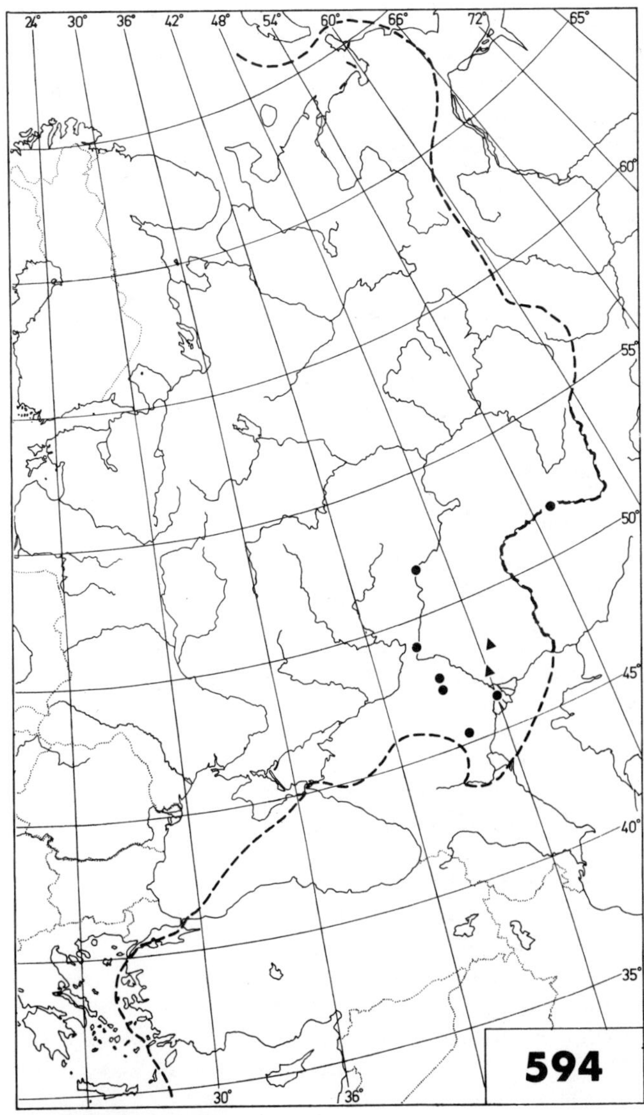

● = **Salsola crassa**
▲ = **S. turcomanica**

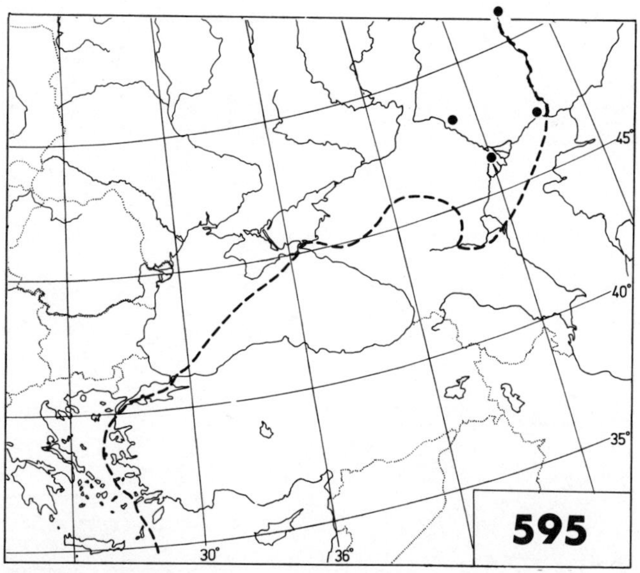

Salsola nitraria

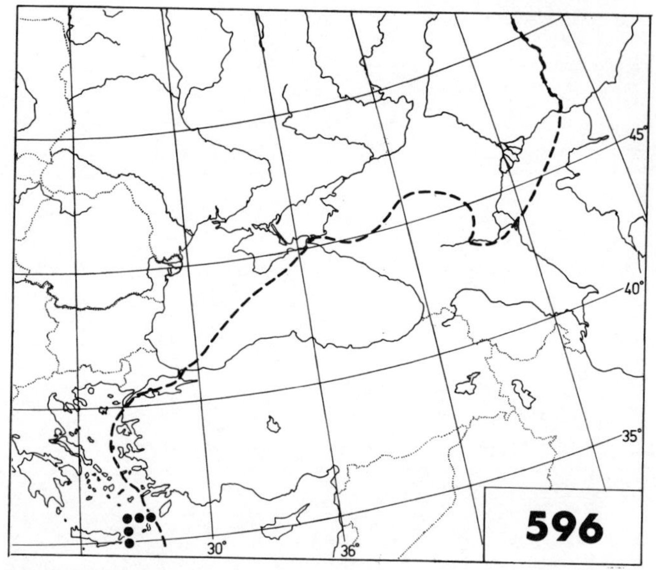

Salsola carpatha

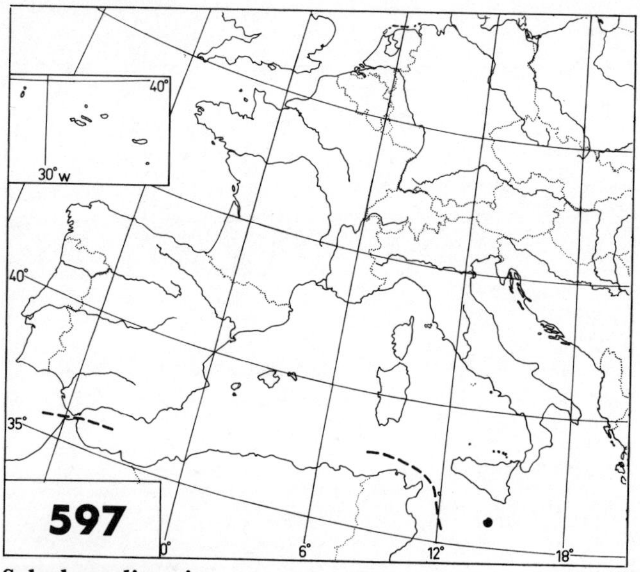

Salsola melitensis

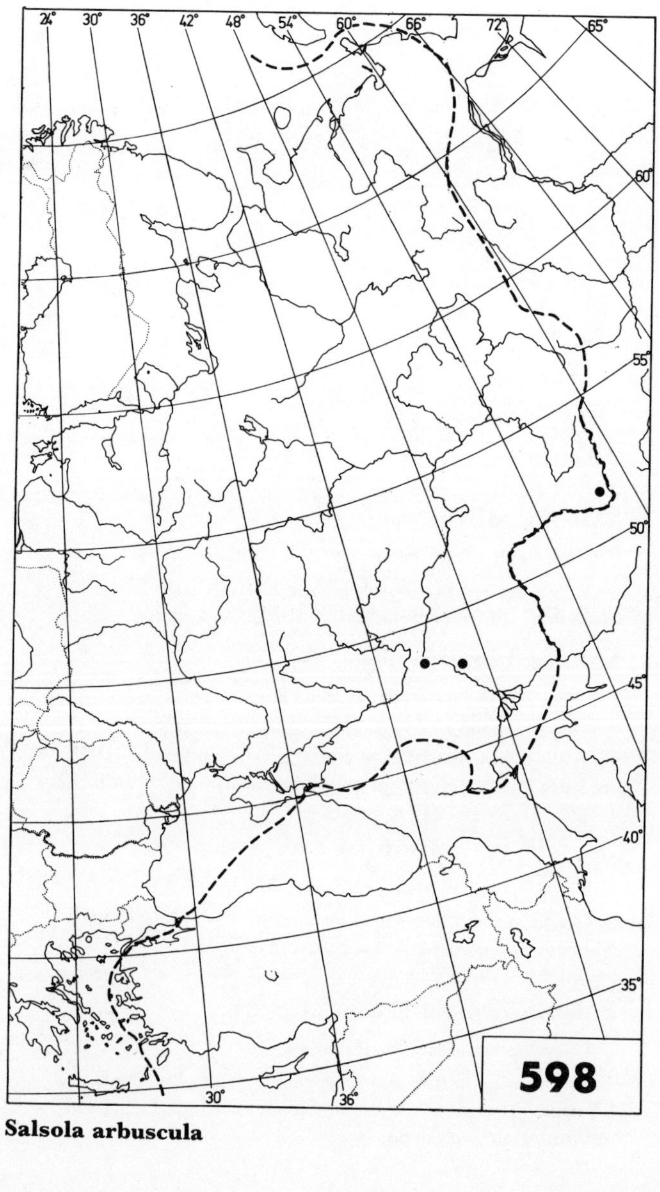

Salsola arbuscula

Salsola nitraria Pallas — Map 595.

Salsola affinis C.A. Meyer

Climacoptera affinis (C.A. Meyer) Botsch.

Generic delimitation and nomenclature. V.P. Botschantzev, Sbornik rabot akad. V.N. Sukaczevy 75-let.: 114 (1956); see under *Salsola lanata.*

Notes. Type localities: »Hab. in collibus aridis versus fluvium Tschu; crescit quoque in desertis ad mare Caspium». Not mapped.

Salsola nodulosa (Moq.) Iljin

Notes. Rs(E) not confirmed (?Rs(E) in Fl. Eur.), and thus the species not mapped.

Salsola carpatha P.H. Davis — Map 596.

Notes. Gr added (not given in Fl. Eur.).

Total range. Endemic to Europe.

Salsola melitensis Botsch. — Map 597.

Taxonomy. Related to the Algerian *Salsola zygophylla* Batt. & Trabut: V. Botschantzev, Nov. Syst. Pl. Vasc. (Leningrad) 13: 95—96 (1976).

Notes. Si (the species not mentioned in Fl. Eur.).

Total range. Endemic to the Isle of Gozo, Malta.

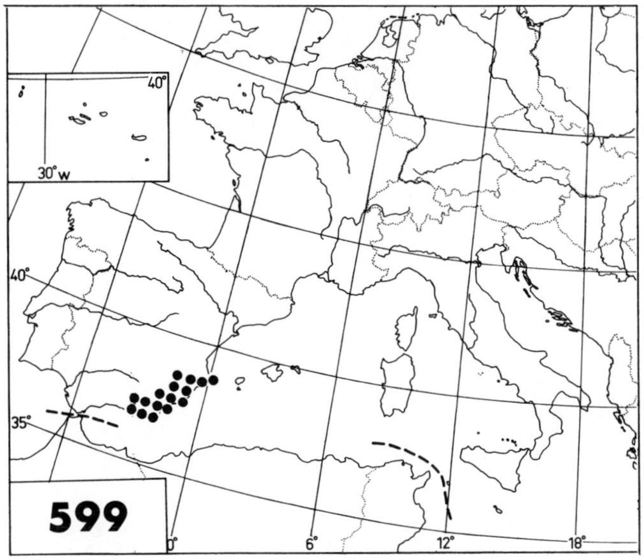

Salsola genistoides

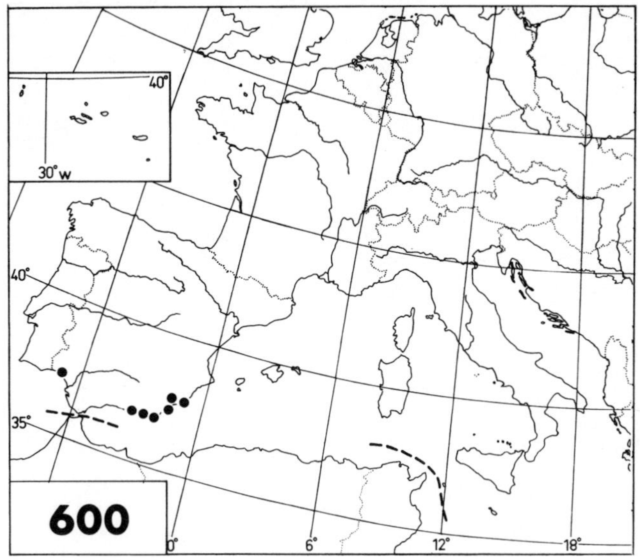

Salsola webbii

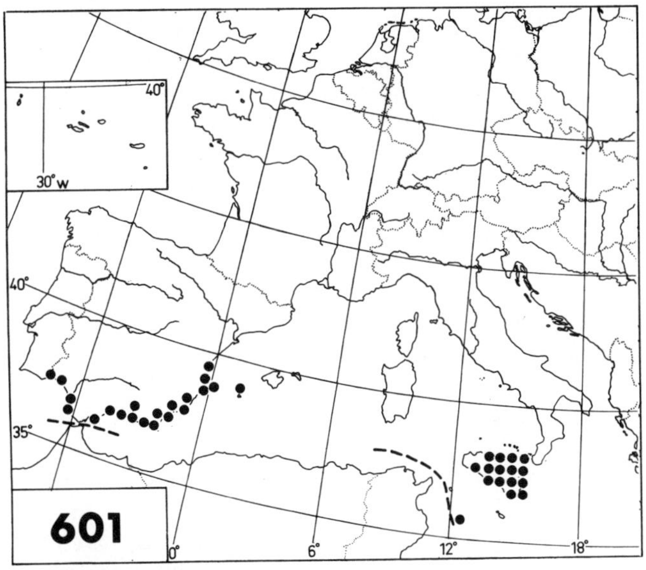

Salsola verticillata

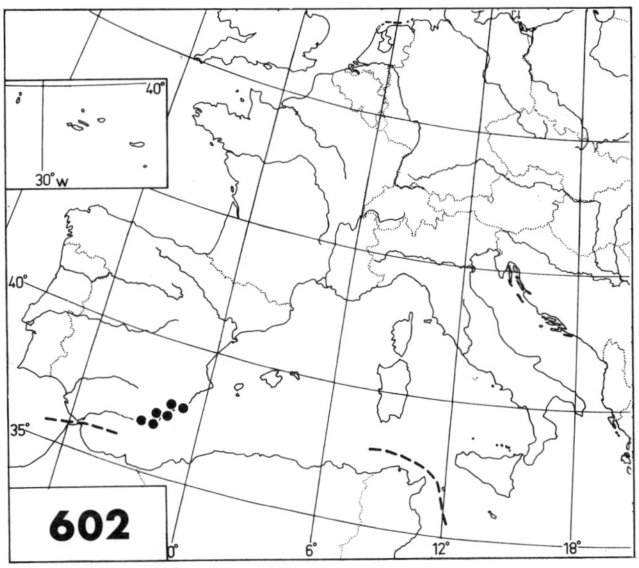

Salsola papillosa

Salsola arbuscula Pallas — Map 598.
Notes. Rs(C) added (not given in Fl. Eur.).

Salsola genistoides Juss. ex Poiret — Map 599.
Caroxylon tamariscifolium Moq., pro p.; Salsola tamariscifolia Lag., pro min. p.
Total range. Endemic to Europe.

Salsola webbii Moq. — Map 600.

Salsola verticillata Schousboe — Map 601.
Salsola longifolia sensu Willk.; S. oppositifolia Desf. non Pallas; S. sieberi auct.
Notes. Bl added (not given in Fl. Eur.): J. Duvigneaud, Soc. Échange Pl. Vasc. Europe Occ. Bassin Médit. 17(suppl.): 16 (1979).

Salsola papillosa (Cosson) Willk. — Map 602.
Total range. Endemic to Europe.

Salsola vermiculata L. — Map 603.
Salsola tamariscifolia (L.) Lag.
Taxonomy. P. Aellen in Fl. Turkey 1967: 334 (under *Salsola laricina*).
Total range. E.J. Jäger, Flora 160: 234 (1971) (incl. *Salsola aegaea*).

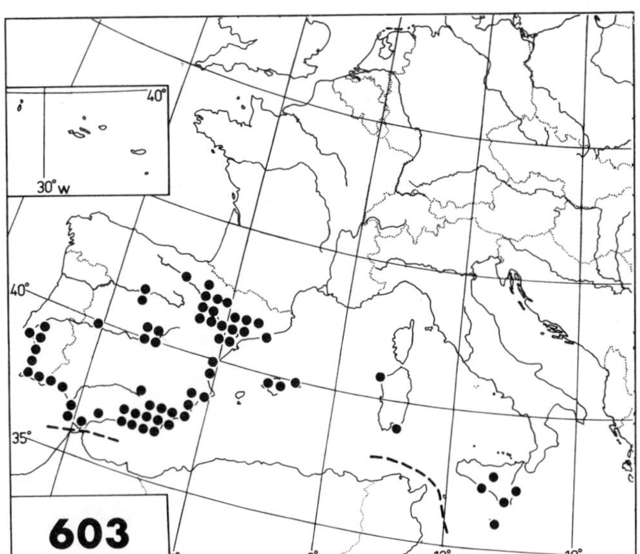

Salsola vermiculata

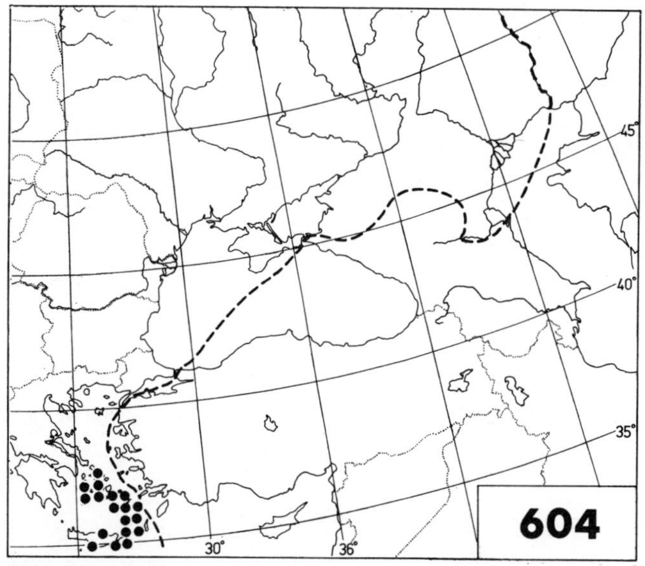

Salsola aegaea

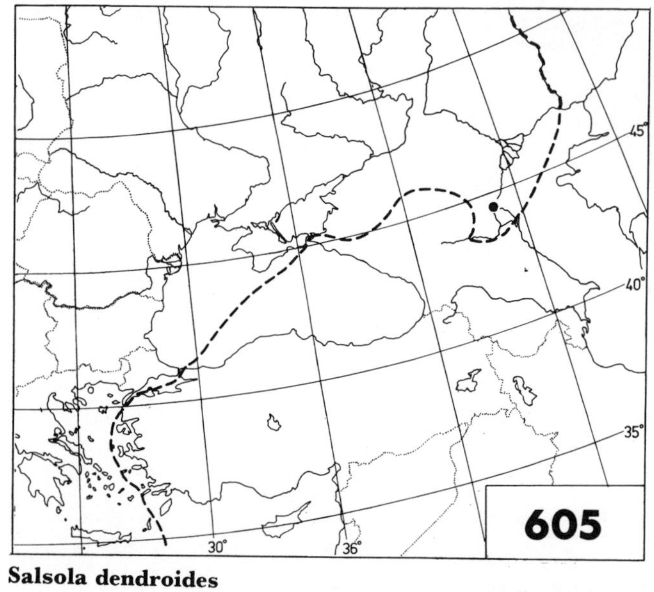

Salsola dendroides

Salsola aegaea Rech. fil. — Map 604.

Salsola vermiculata auct. balcan. non L.

Total range. H. Runemark, Bot. Not. 122: 123 (1969); cf. E.J. Jäger, Flora 160: 234 (1971) (under *Salsola vermiculata*), and W. Greuter, Ann. Mus. Goulandris 1: 30 (1973).

Salsola dendroides Pallas — Map 605.

Salsola brevifolia D'Urv.; S. verrucosa Bieb.

Total range. E.J. Jäger, Flora 160: 234 (1971).

Salsola laricina Pallas — Map 606.

Taxonomy. P. Aellen in Fl. Turkey 1967: 334.
Total range. E.J. Jäger, Flora 160: 234 (1971).

Noaea mucronata (Forskål) Ascherson & Schweinf. — Map 607.

Noaea spinosissima (L. fil.) Moq.; Salsola mucronata Forskål

Notes. Cr added (not given in Fl. Eur.), Tu confirmed (?Tu in Fl. Eur.): Fl. Turkey 1967: 336.

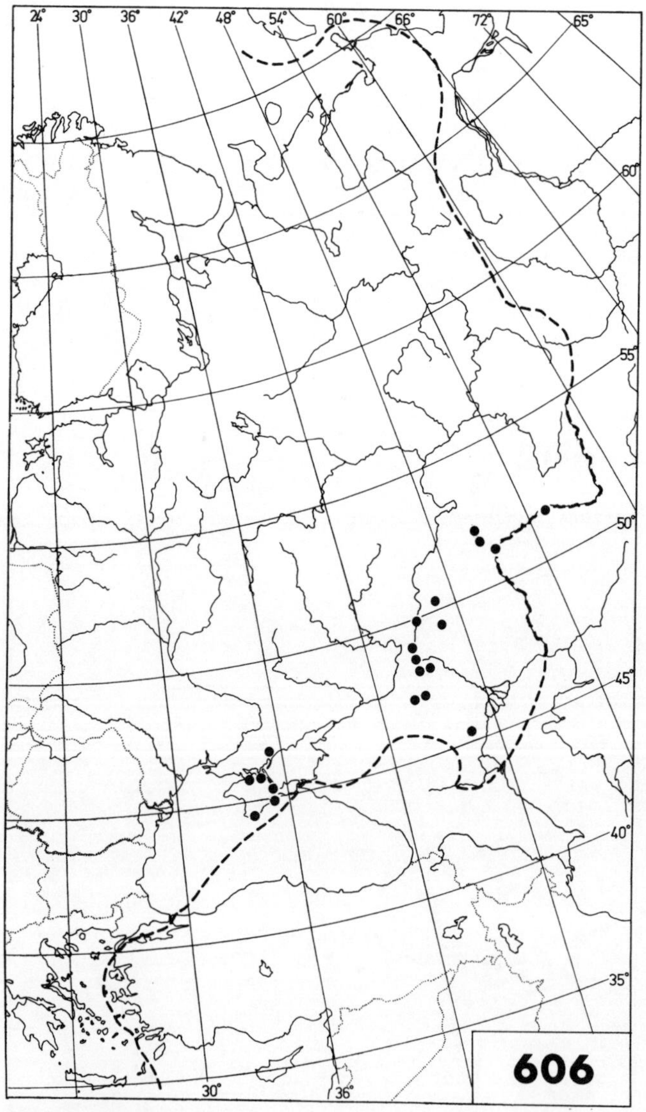

Salsola laricina

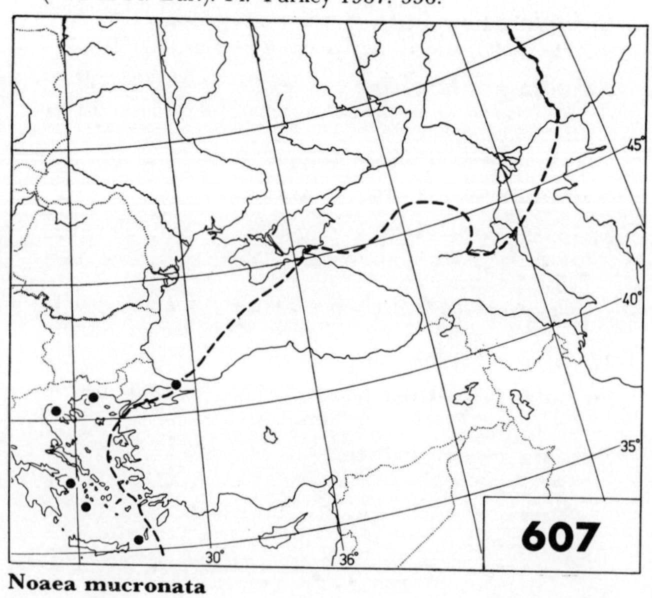

Noaea mucronata

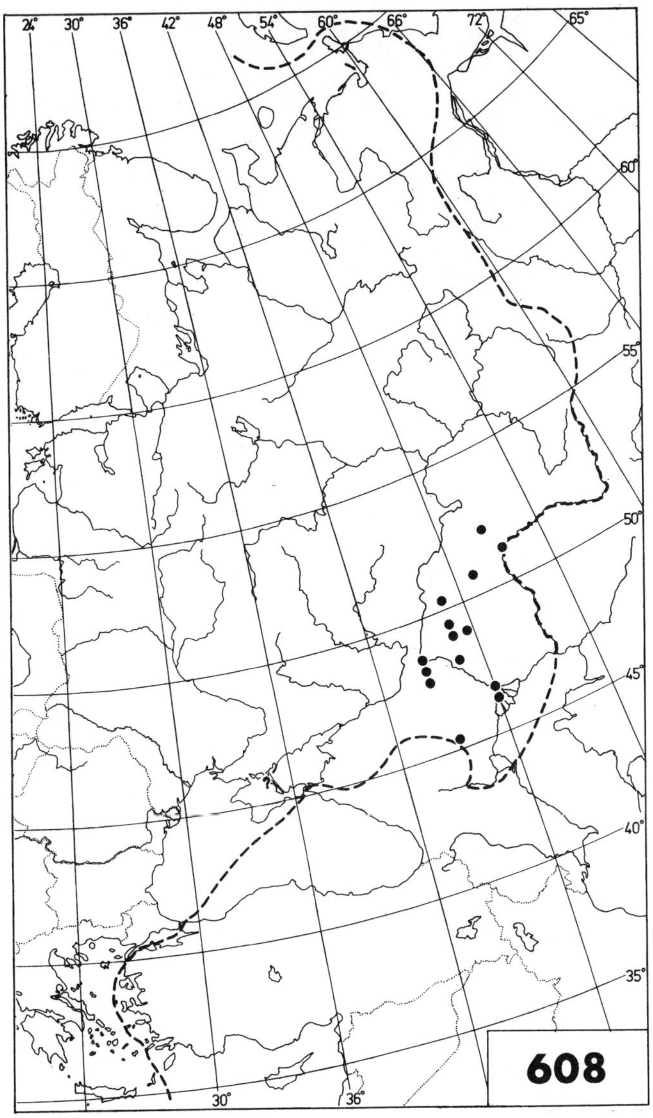

Ofaiston monandrum

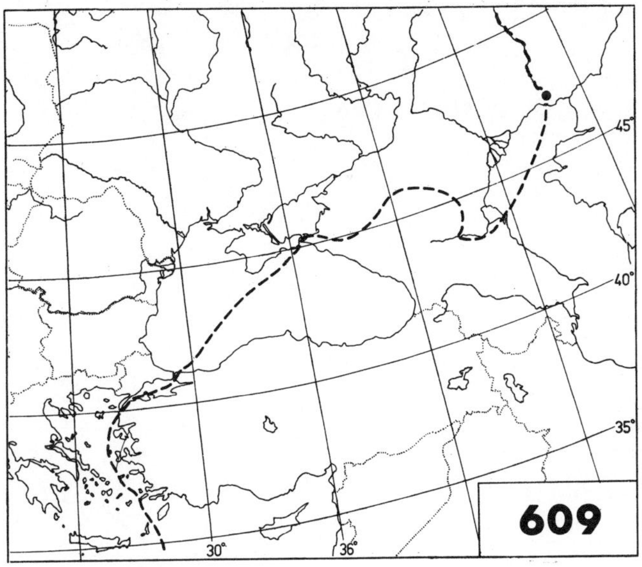

Girgensohnia oppositifolia

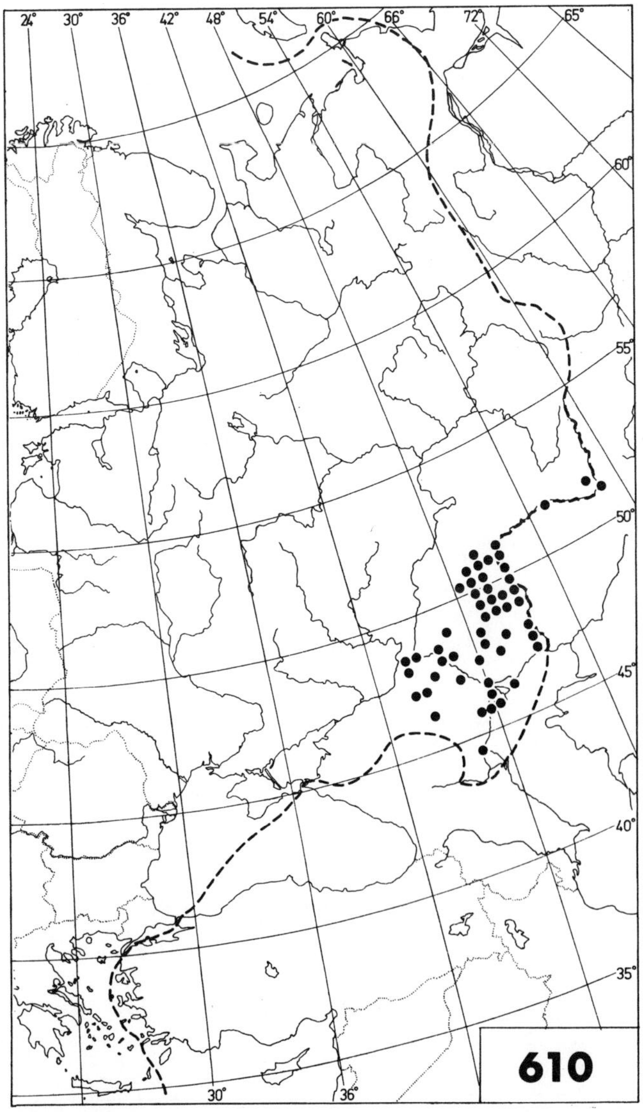

Anabasis salsa

Ofaiston monandrum (Pallas) Moq. — Map 608.

Notes. ?Rs(K) added (not given in Fl. Eur.); the locality reported by Rubtsev 1972: 137.

Girgensohnia oppositifolia (Pallas) Fenzl — Map 609.

Anabasis salsa (C.A. Meyer) Bentham ex Volkens — Map 610

Nomenclature. The corrected author's citation is according to S.K. Czerepanov, Additam. Corr. Fl. URSS, p. 183 (Leningrad 1973).

Notes. Rs(C) added (not given in Fl. Eur.): I.F. Musaev, Areali Rast. Fl. SSSR 3: 132 (Leningrad 1976).

Total range. E.M. Lavrenko & N.I. Nikolskaya, Bot. Žurn. 48: 1751 (1963); E.J. Jäger, Feddes Repert. 79: 252 (1968); I.F. Musaev, Areali Rast. Fl. SSSR 3: 132—133 (Leningrad 1976).

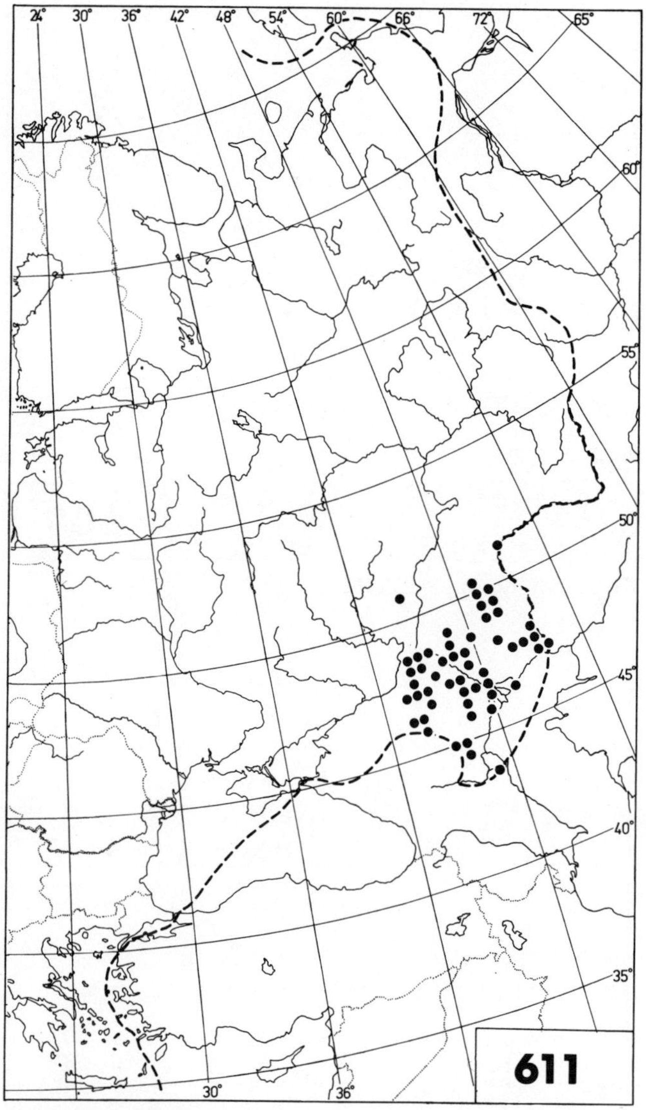

Anabasis aphylla

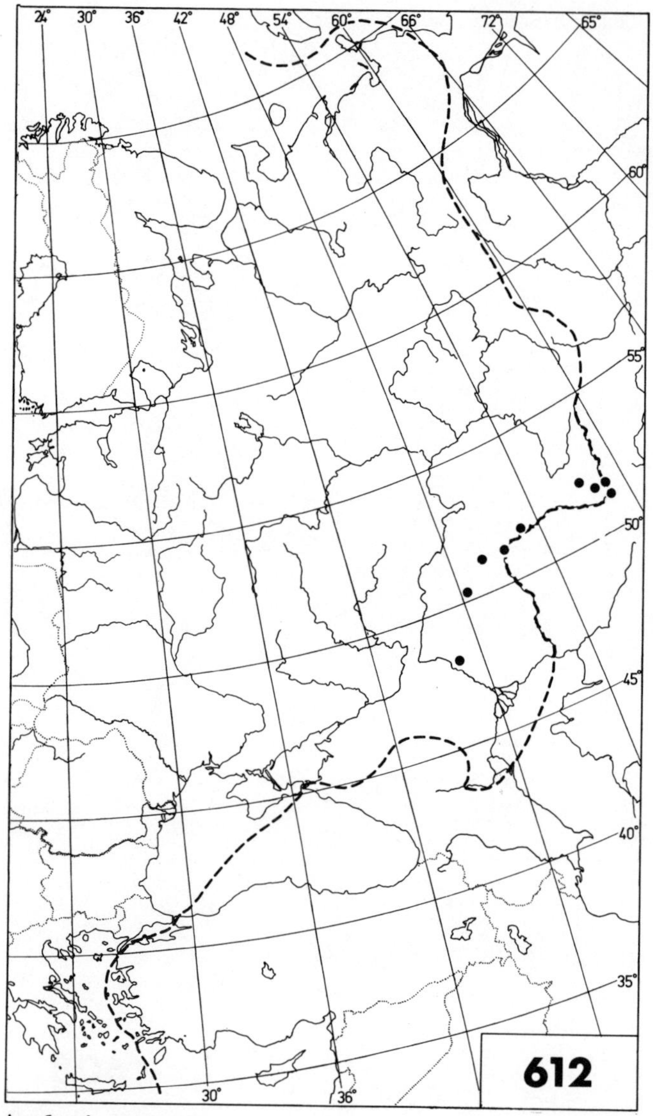

Anabasis cretacea

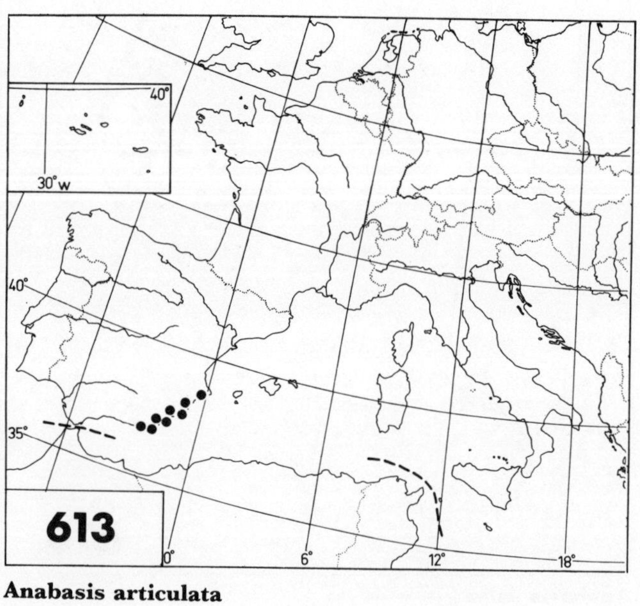

Anabasis articulata

Anabasis aphylla L. — Map 611.

Total range. H. Walter, Einführung in die Phytologie III(2): 187 (1954); I.F. Musaev, Vestn. Leningrad. Univ. (Biol.) 1966(3): 23 (1966) and Areali Rast. Fl. SSSR 3: 116—117 (Leningrad 1976); Atlas Lekarstv. Rast. SSSR: 64 (Moskva 1976).

Anabasis cretacea Pallas — Map 612.

Notes. Rs(C) added (not given in Fl. Eur.): I.F. Musaev, Areali Rast. Fl. SSSR 3: 124 (Leningrad 1976).
Total range. I.F. Musaev, Areali Rast. Fl. SSSR 3: 124 (Leningrad 1976).

Anabasis articulata (Forskål) Moq. — Map 613.

Anabasis hispanica Pau; A. mucronata (Lag.) C. Vicioso; Salsola articulata Forskål

Hammada scoparia (Pomel) Iljin — Map 614.

Caroxylon articulatum Moq.; Haloxylon articulatum (Moq.) Bunge; H. scoparium Pomel; H. tamariscifolium sensu Pau; Hammada articulata (Moq.) O. Bolòs & Vigo; Salsola articulata Cav. non Forskål

Nomenclature. M. Zohary, Fl. Palaest. 1: 164 (1966); O. Bolòs & J. Vigo, Butll. Inst. Catalana Hist. Nat. 38 (Sec. Bot. 1): 89 (1974); V. Botschantzev, Nov. Syst. Pl. Vasc. (Leningrad) 13: 102—103 (1976).

Nanophyton erinaceum (Pallas) Bunge — Map 615.

Petrosimonia brachiata (Pallas) Bunge — Map 616.

Notes. Rs(W) mapped according to Fl. URSR 1952: 400.

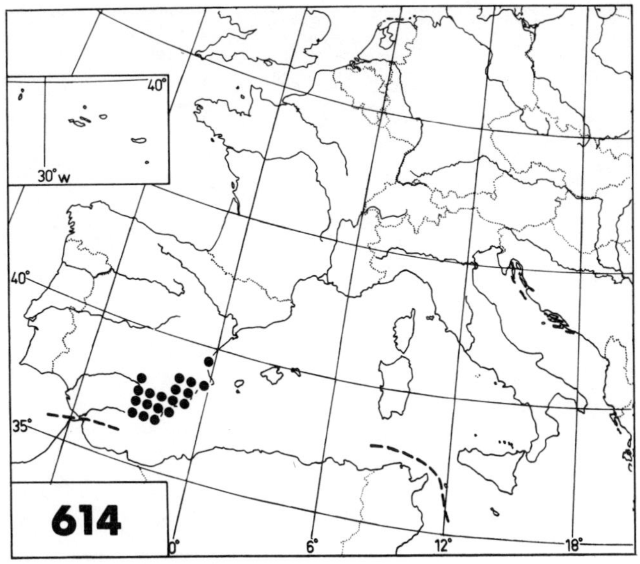

Hammada scoparia

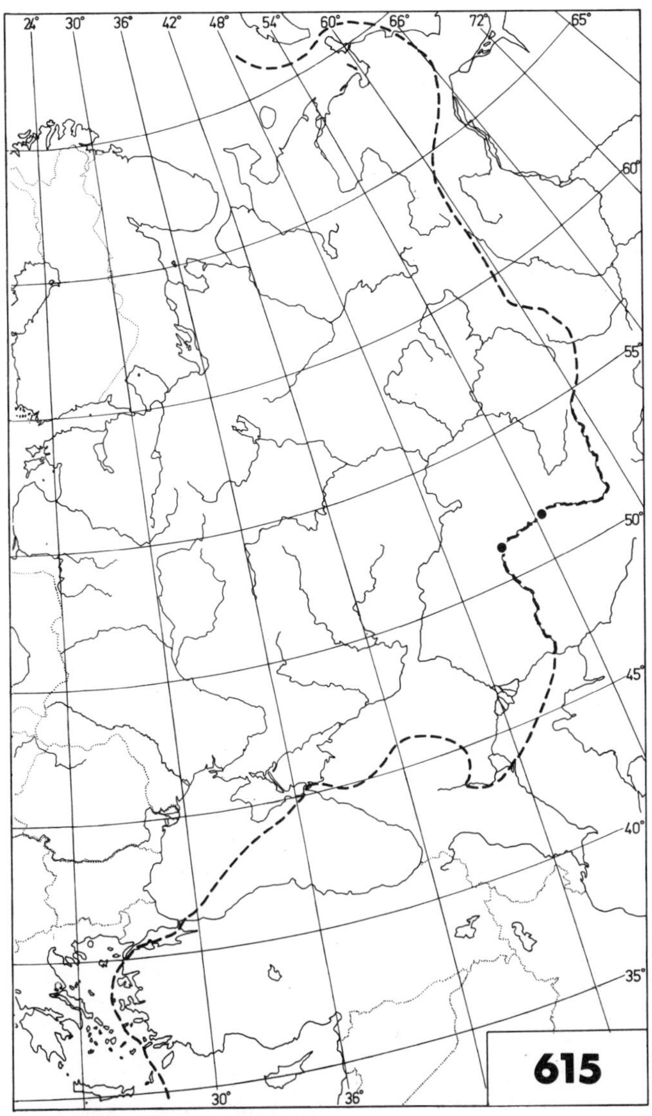

Nanophyton erinaceum

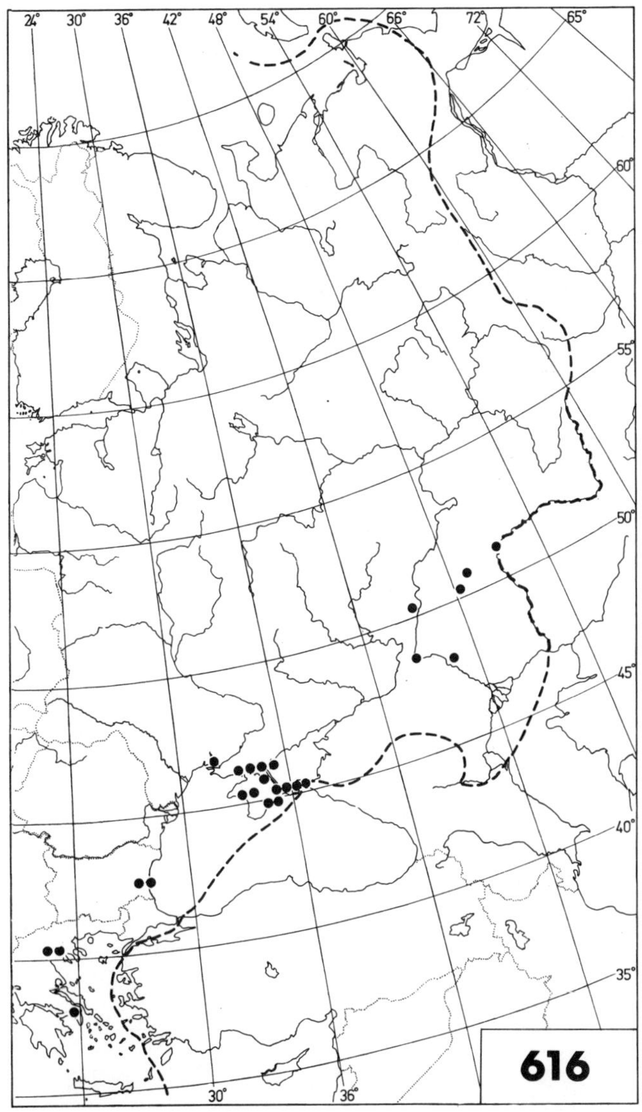

Petrosimonia brachiata

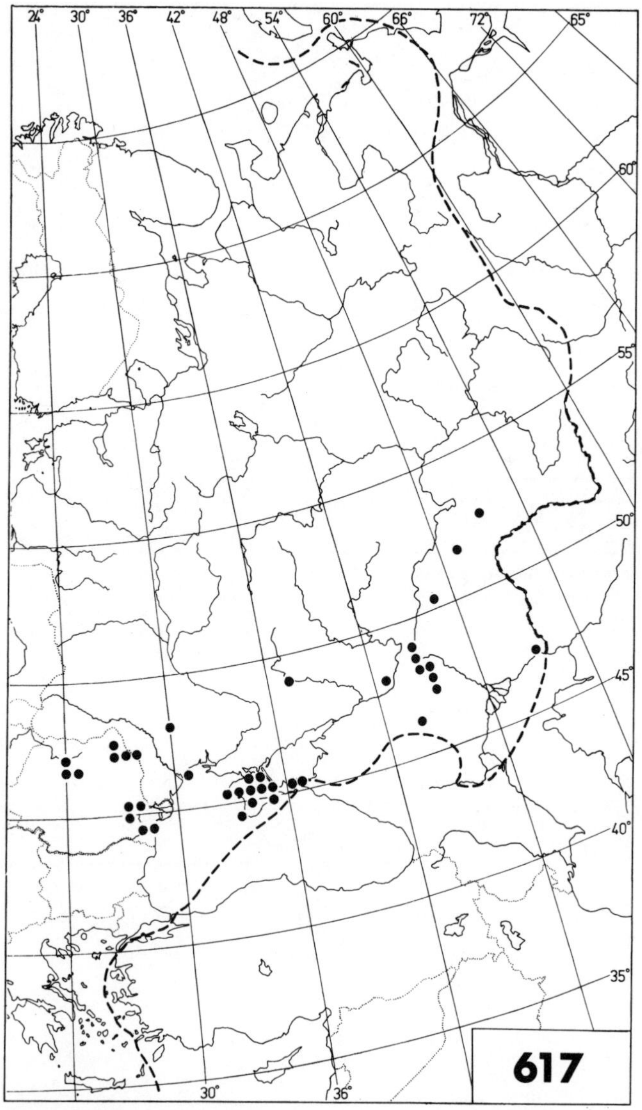

Petrosimonia triandra

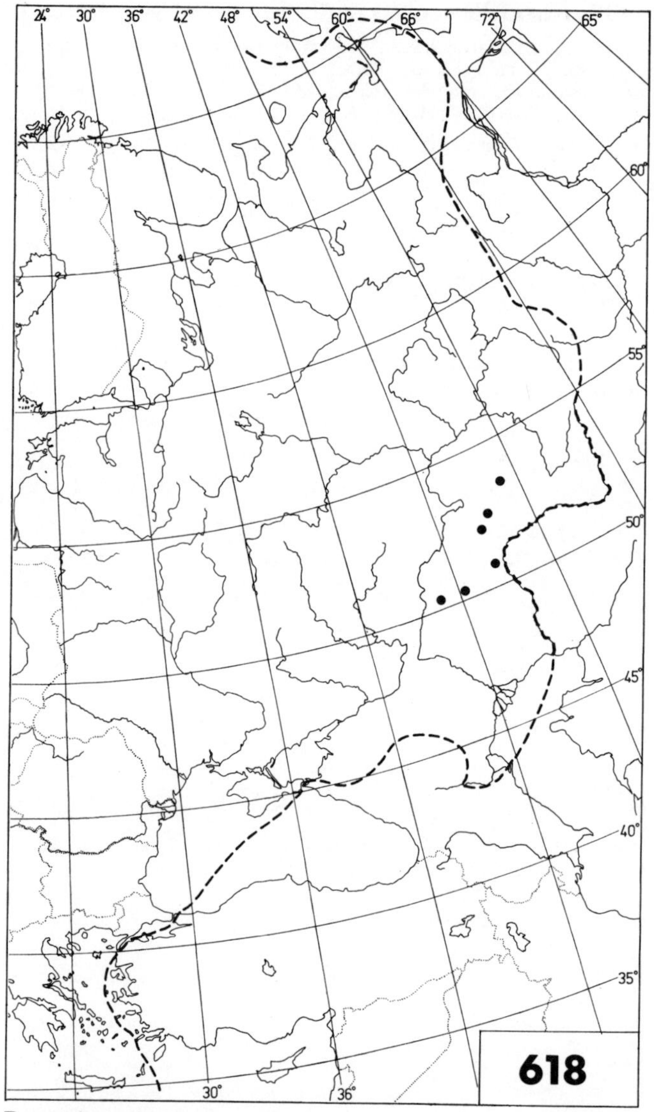

Petrosimonia litwinowii

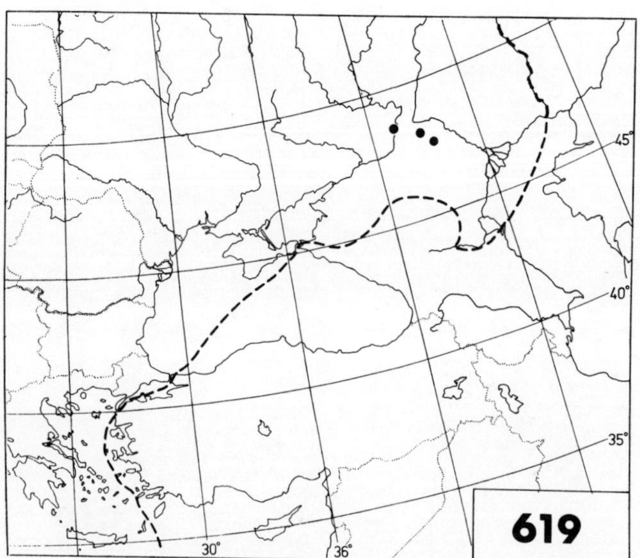

Petrosimonia monandra

Petrosimonia triandra (Pallas) Simonkai — Map 617.

Petrosimonia volvox (Pallas) Bunge

Notes. Rs(W) mapped according to Fl. URSR 1952: 397.

Petrosimonia litwinowii Korsh. — Map 618.

Petrosimonia monandra (Pallas) Bunge — Map 619.

Petrosimonia oppositifolia (Pallas) Litv. — Map 620.

Petrosimonia crassifolia auct.

Notes. Bu omitted (given in Fl. Eur.): B.A. Kuzmanov, Candollea 34: 17 (1979). Rs(W) mapped according to FL. URSR 1952: 398.

Petrosimonia brachyphylla (Bunge) Iljin — Map 620.

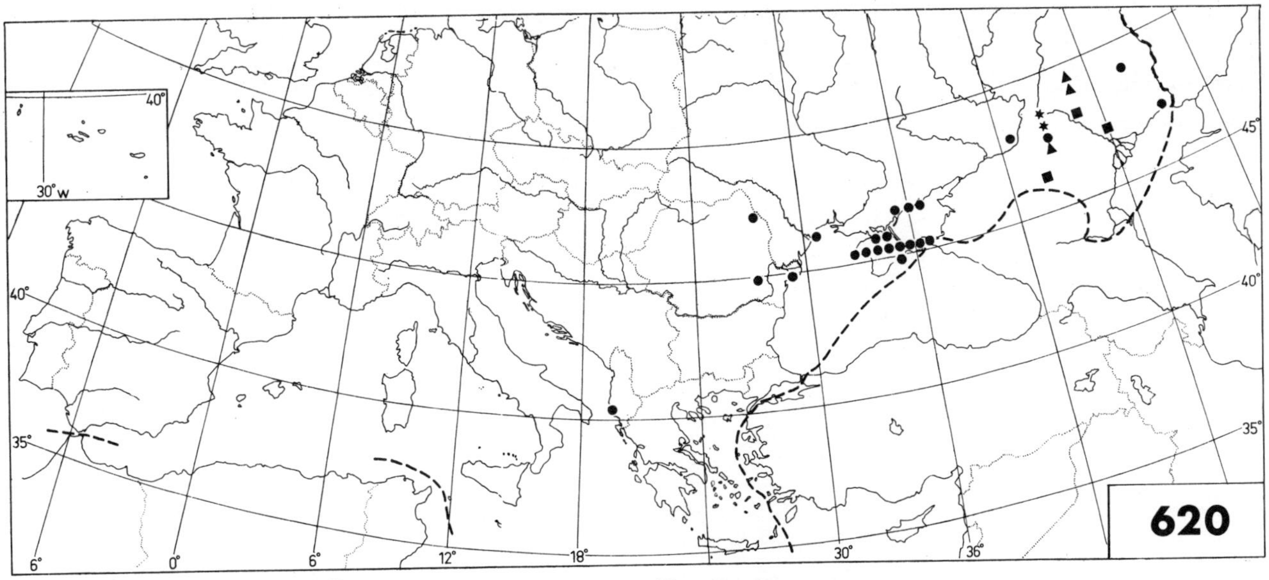

● = Petrosimonia oppositifolia
▲ = P. brachyphylla
■ = P. glaucescens
✶ = P. oppositifolia + P. glaucescens

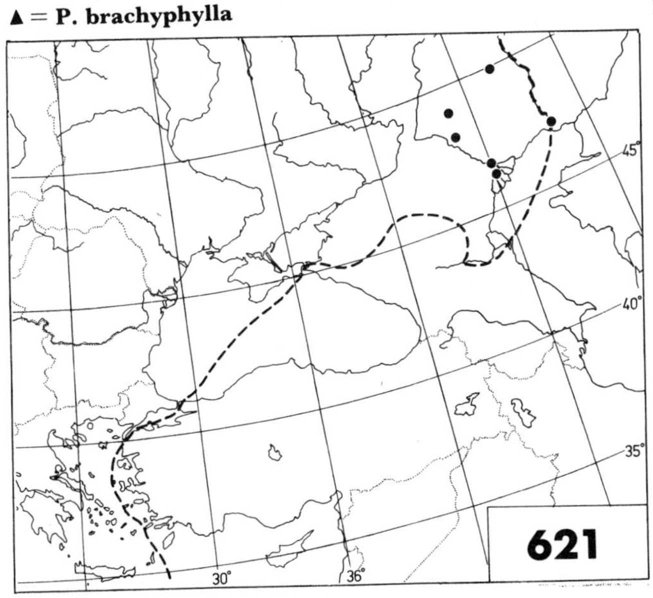

Halimocnemis sclerosperma

Petrosimonia glaucescens (Bunge) Iljin — Map 620.

Petrosimonia glauca sensu Boiss.

Halimocnemis sclerosperma (Pallas) C.A. Meyer — Map 621.

Halogeton sativus (L.) Moq. — Map 622.

Taxonomy and nomenclature. See next species.

Halogeton glomeratus (Bieb.) C.A. Meyer — Map 622.

Taxonomy and nomenclature. »Very doubtfully distinct from 1 [*Halogeton sativus*]» (Fl. Eur.). If the two are united, *H. glomeratus* (Bieb.) C.A. Meyer is the valid species name for the resultant taxon.

Notes. Rs(E). »S.E. Russia and W. Kazakhstan» (Fl. Eur.).

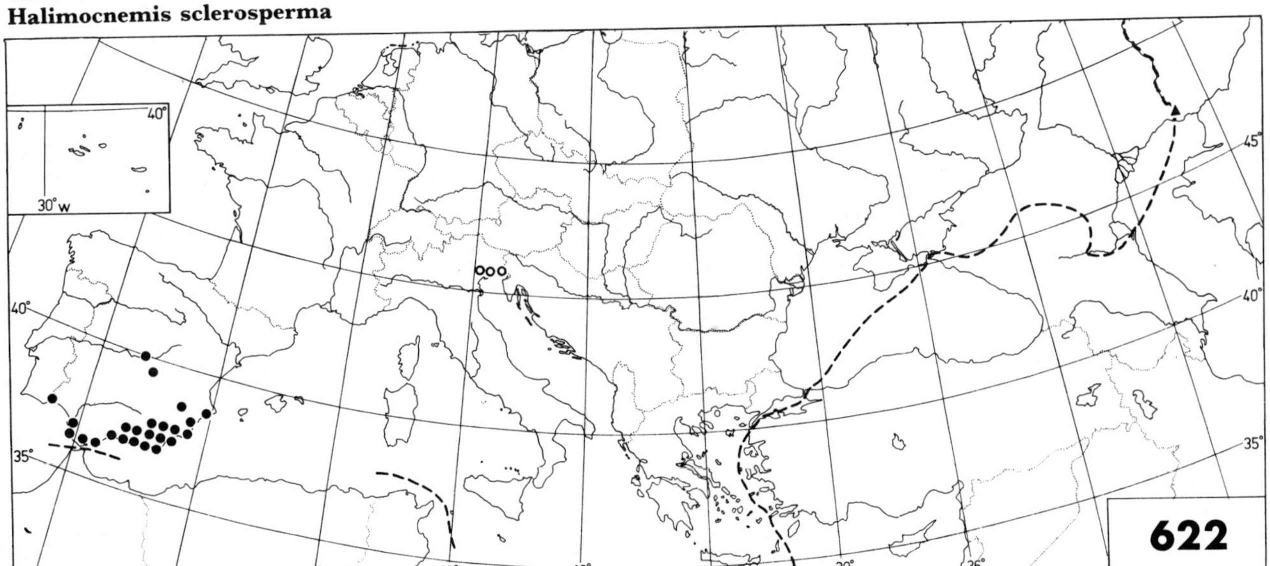

● = Halogeton sativus
▲ = H. glomeratus

AMARANTHACEAE

Celosia argentea L.

Incl. Celosia cristata L.; C. argentea var. cristata (L.) O. Kuntze

Notes. »Widely cultivated for ornament and frequently occurring as a casual in S. Europe» (Fl. Eur.). Not mapped.

Amaranthus hybridus L., s. lato — Map 623.

Amaranthus hypochondriacus + A. paniculatus + A. patulus + A. powellii

Taxonomy and nomenclature mainly follow J.D. Sauer, Ann. Missouri Bot. Garden 54: 103—137 (1967); see also F. Ehrendorfer, Liste Gefässpfl. Mitteleur. 2. Aufl., p. 15 (Stuttgart 1973).

Notes. The publication of a collective map was motivated by differences in the taxonomical treatment and serious nomenclatural discrepancies, especially the latter.

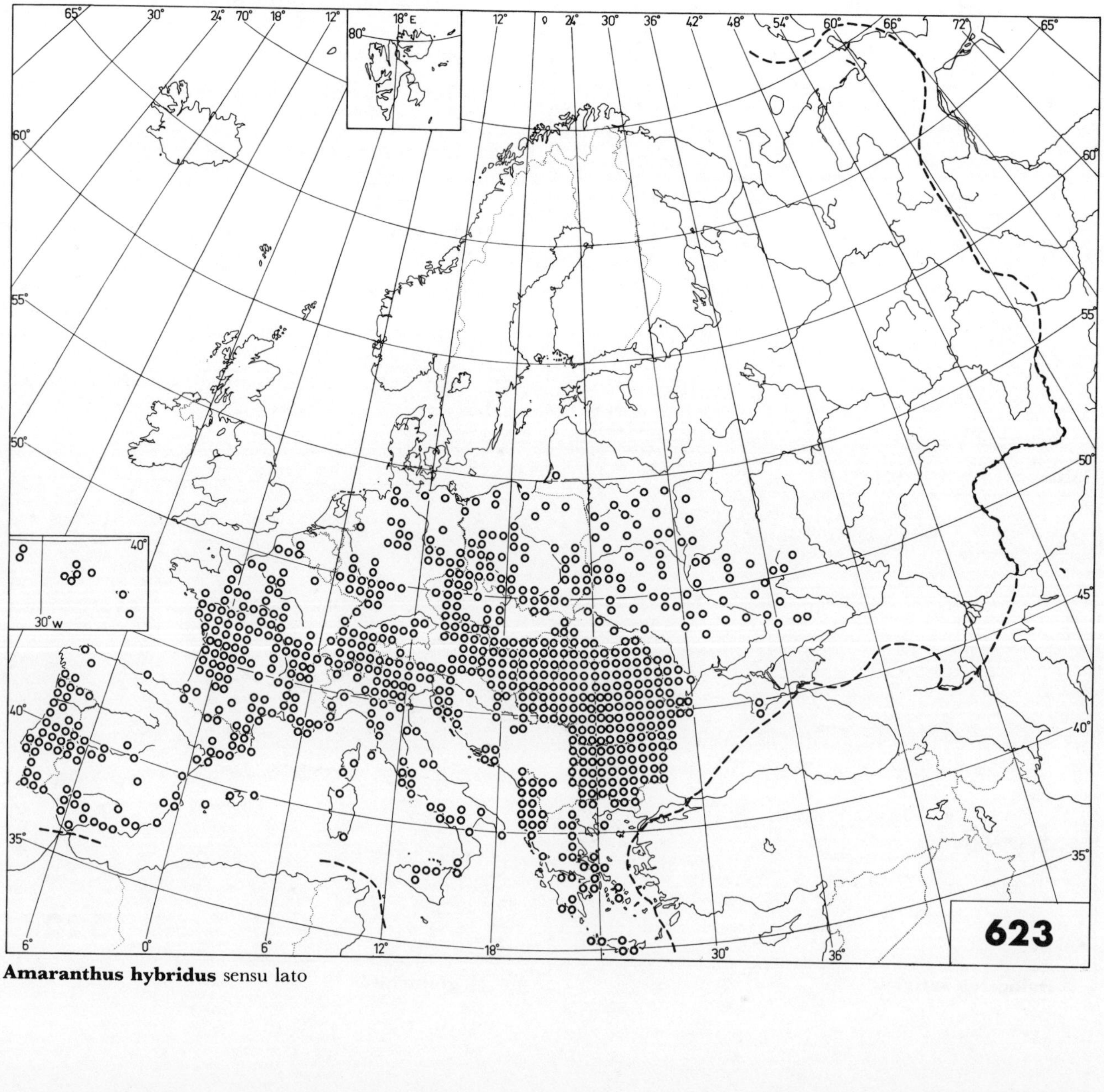

Amaranthus hybridus sensu lato

Amaranthus powellii S. Watson — Map 624.

Amaranthus hybridus sensu Fl. Eur., non L.; A. chlorostachys Willd. var. pseudoretroflexus Thell. Incl. A. bouchonii Thell., excl. A. hypochondriacus L.

Taxonomy. J.D. Sauer, Ann. Missouri Bot. Garden 54: 108 (1967); R. Soó, Feddes Repert. 83: 137 (1972).

Notes. [Bl, Bu, Cz, He and Rs(K)] added (not given in Fl. Eur.). Records from Cr and Gr are identified to the collective taxon only (given in Fl. Eur. as *Amaranthus hybridus*), as are also those from Rs(C) and most of Rs(W). Regarded as casual in Da and Ho ([Da, Ho] in Fl. Eur.). Many of the records from Ga are referable to *A. bouchonii.* The northern limit of the established occurrence seems to be poorly known, and thus provisional.

Native of tropical and subtropical America.

Amaranthus patulus Bertol. — Map 625.

Amaranthus cruentus sensu Fl. Eur., non L.; A. hybridus L., s. str., nomen ambig.

Taxonomy. J.D. Sauer, Ann. Missouri Bot. Garden 54: 108—109 (1967).

Notes. Regarded as only casual in Au ([Au] in Fl. Eur.). No records available from Az and Rs(E) ([Az, Rs(E)] in Fl. Eur.). [Bl, Co, Cz, He and Lu] added (not given in Fl. Eur.). Records from Cr, Rm and Rs(C) are identified to the collective taxon only (given in Fl. Eur. as *Amaranthus cruentus*) as are also those from Gr (not given in Fl. Eur.). Most of the Central European occurrences may be more or less casual.

Native of tropical and subtropical America.

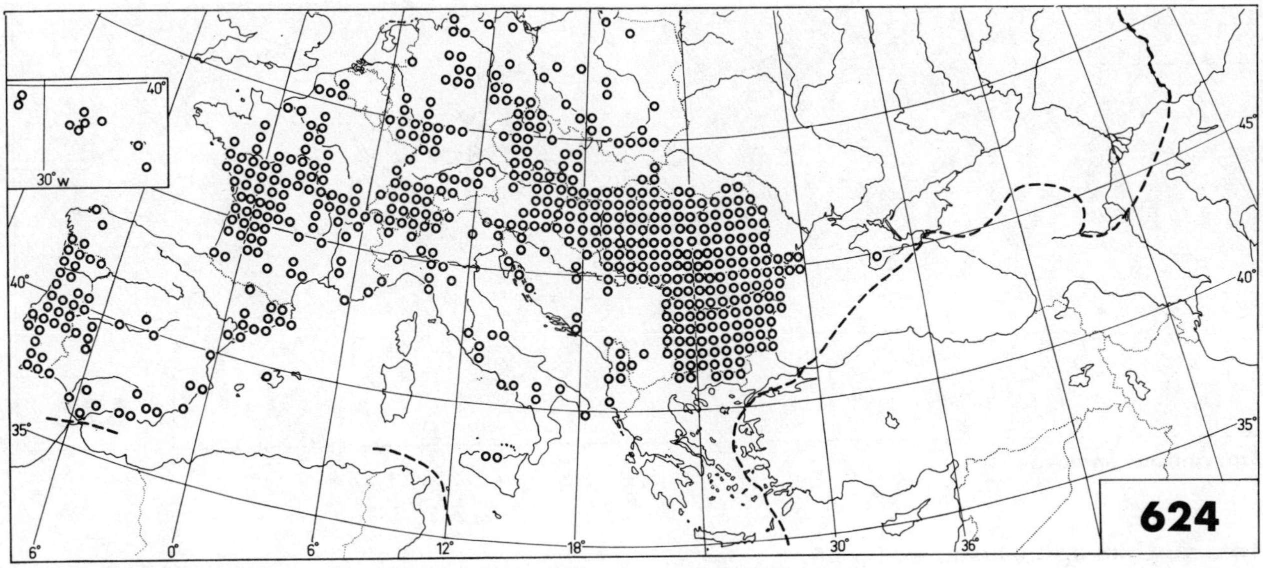

Amaranthus powellii

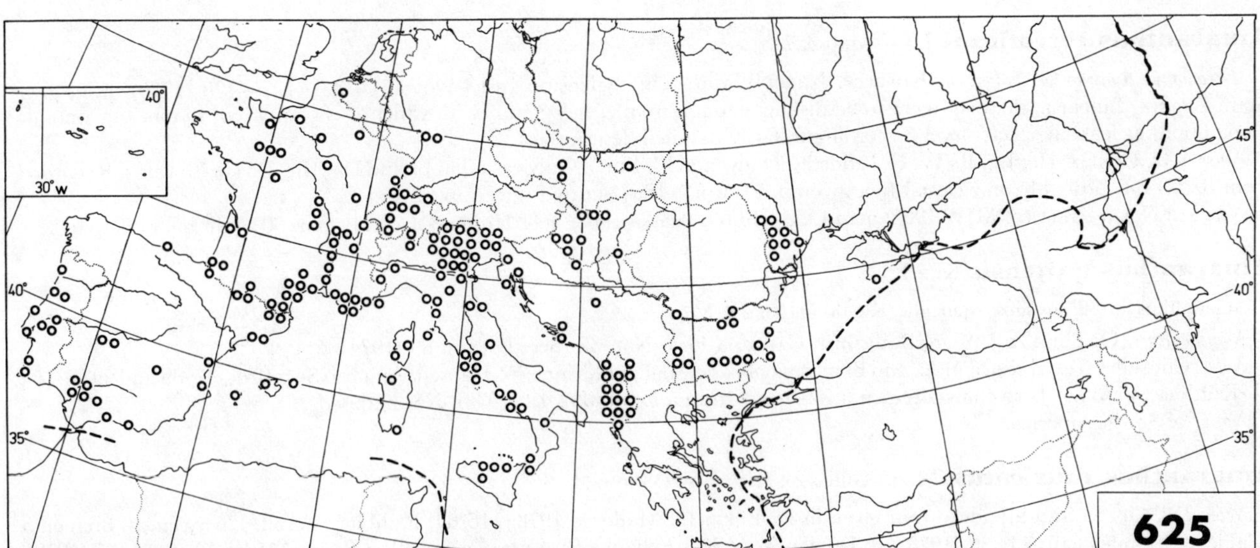

Amaranthus patulus

Amaranthus hypochondriacus L.

Amaranthus chlorostachys Willd. var. erythrostachys (Moq.) Aellen; A. hybridus sensu Fl. Eur., pro p., non L.

Taxonomy. J.D. Sauer, Ann. Missouri Bot. Garden 54: 110—112 (1967).

Notes. Reported from parts of C. and S.E. Europe but apparently only casual (not recognized in Fl. Eur.). Not mapped.

Originated in cultivation in America.

Amaranthus paniculatus L.

Amaranthus cruentus L., s. str., nomen ambig.

Taxonomy. J.D. Sauer, Ann. Missouri Bot. Garden 54: 122—123 (1967).

Notes. Possibly becoming locally established in S. and C. Europe. »Commonly cultivated and sometimes occurring as a casual» (Fl. Eur.); see also F. Ehrendorfer, Liste Gefässpfl. Mitteleur. 2. Aufl., p. 15 (Stuttgart 1973). Not mapped.

Originated in cultivation in America.

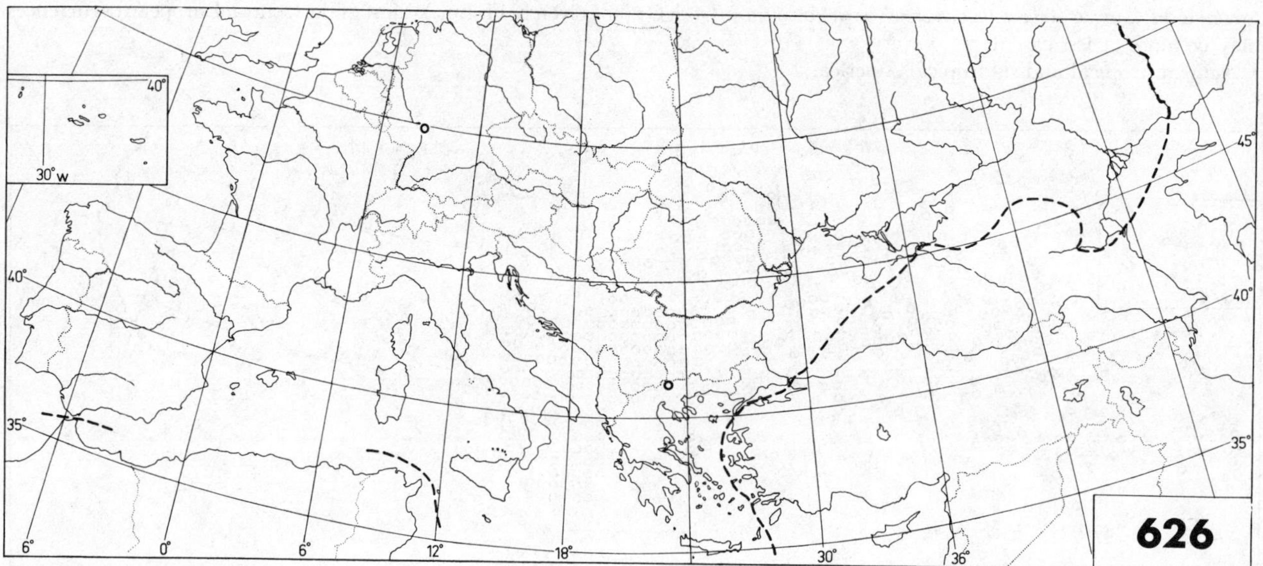

Amaranthus spinosus

Amaranthus spinosus L. — Map 626.

Notes. [Bu] and [Ge] added (the species not mentioned in Fl. Eur.). Possibly not truly naturalized.

Native of tropical America (?).

Amaranthus retroflexus L. — Map 627.

Taxonomy. Amaranthus bulgaricus Kovacev, Naucni Trud. Viss. Selkostop Inst. (Plovdiv) 19(3): 127—128 (1970), from near Stara Zagora, Bulgaria, does not seem to be distinct enough from *A. retroflexus* (with which it is compared) to merit taxonomic recognition, at least at species level (according to fairly detailed description).

Notes. [Az, Co, Cz, He, Lu, Rs(W, K)] added (not given in Fl. Eur.). Only casual in Br and Da ([Br, Da] in Fl. Eur.). Recorded from Rs(N) but probably only casual (not given in Fl. Eur.). [Sa] omitted (given in Fl. Eur.).

Native of North America. MJW 1965: map 135d; Hultén Alaska 1968: 463; Hultén CP 1971: map 301 (all: holarctic part).

Amaranthus quitensis Kunth

Amaranthus retroflexus subsp. quitensis (Kunth) O. Bolòs & Vigo

Nomenclature. O. Bolòs & J. Vigo, Butll. Inst. Catalana Hist. Nat. 38 (Sec. Bot. 1): 89 (1974).

Notes. Considered established in Az and Bl (given only as casual in Fl. Eur.): A. Hansen, Anuario Soc. Brot. 37: 89 and 99 (1971); H. Kuhbier, Veröff. Überseemus. Bremen 5 (Cahiers Optima Leafl. 83): 12 (1978). Not mapped.

Native of South America.

Amaranthus muricatus (Moq.) Gillies ex Hicken — Map 628.

Notes. [Bl, Gr, Sa and Si] added (not given in Fl. Eur.): Fl. Mallorca 1978: 116; H. Kuhbier, Veröff. Überseemus. Bremen 5 (Cahiers Optima Leafl. 83): 12 (1978); P. Duvigneaud, Soc. Échange Pl. Vasc. Europe Occ. Bassin Médit. 17(suppl.): 5 (1979); M. Sortino, Lav. Ist. Bot. Giard. Col. Palermo 24: 203—208 (1970).

Native of temperate South America.

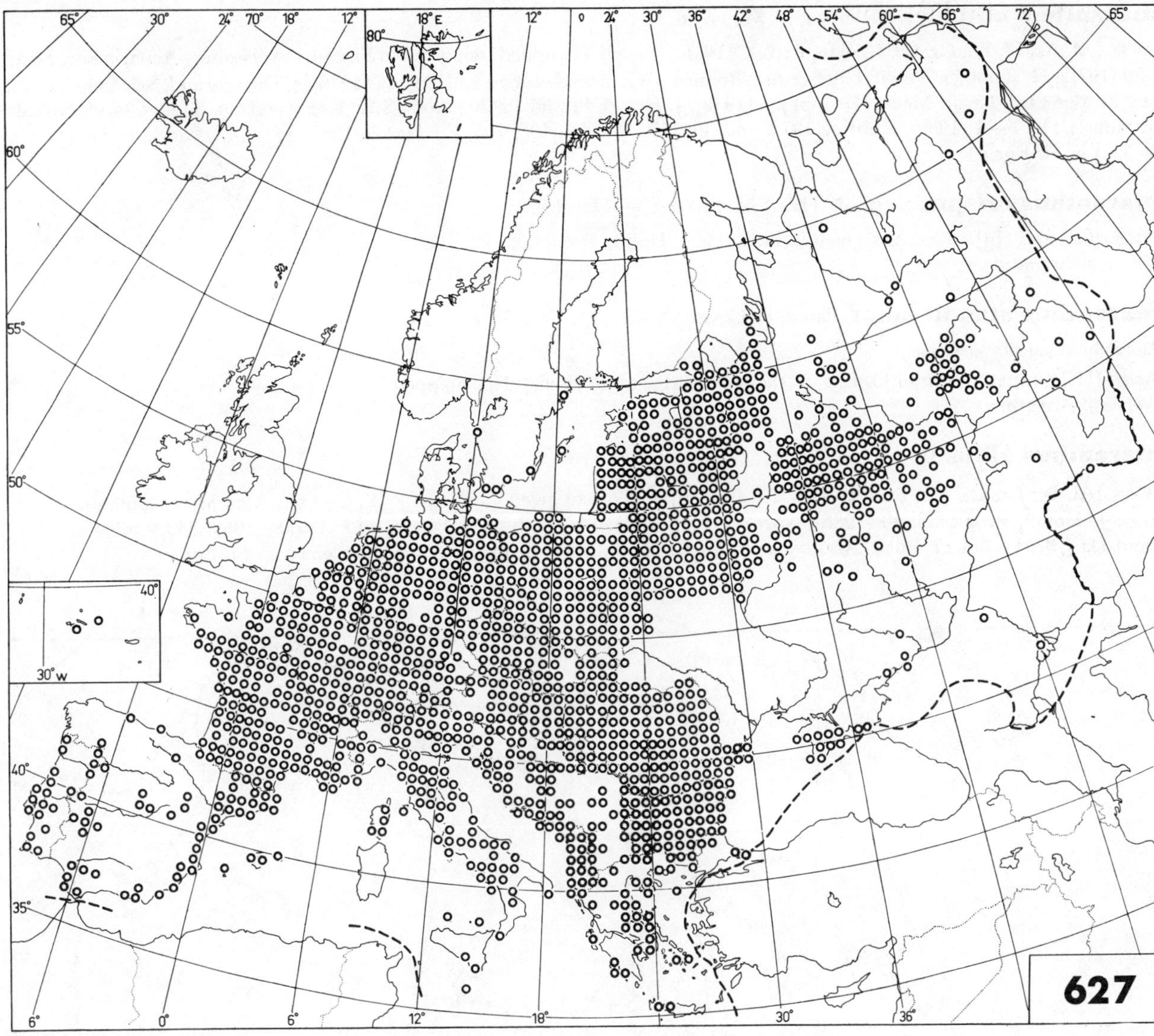

Amaranthus retroflexus

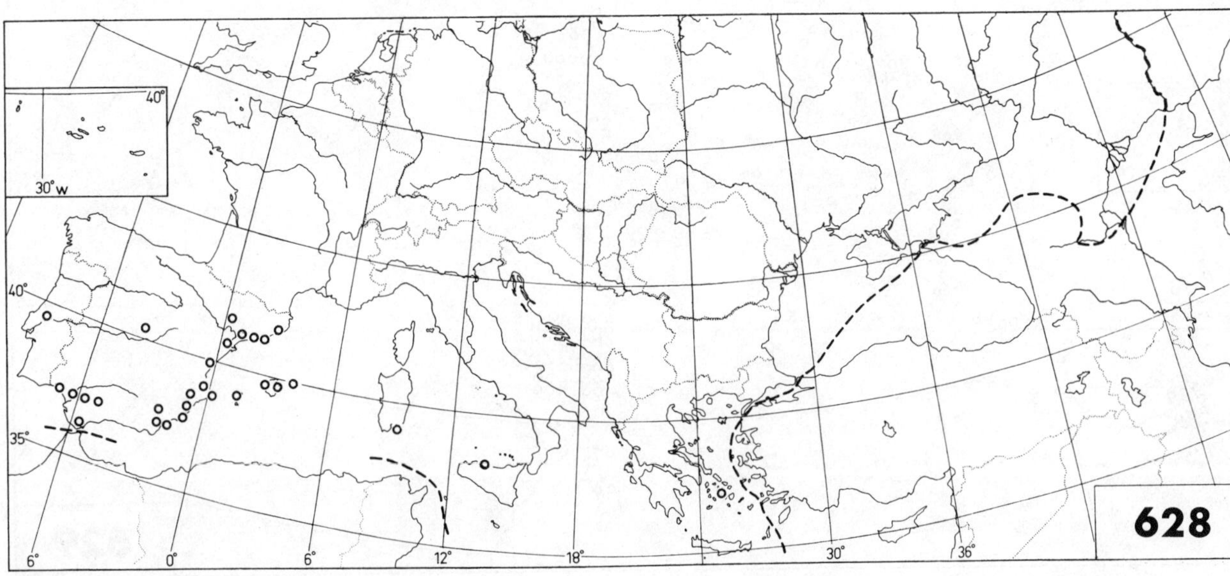

Amaranthus muricatus

Amaranthus blitoides S. Watson — Map 629.

Notes. [Al, Az, Bl, Bu, Cr, Cz, Gr, Ho, Rs(B, C, W, K), Si and Tu] added (not given in Fl. Eur.): A. Hansen, Anuario Soc. Brot. 37: 89 (1971); H. Kuhbier, Veröff. Überseemus. Bremen 5 (Cahiers Optima Leafl. 83): 12 (1978); J. Duvigneaud, Soc. Échange Pl. Vasc. Europe Occ. Bassin Médit.17(suppl.): 5 (1979); J. Holub, Preslia 38: 80 (1966); S.K. Kozhevnikova, Bjull. Gosud. Nikitsk Bot. Sada 4(11): 3—4 (1969); Rubtsev 1972: 140; Fl. URSR 1952: 406, 408. Fl. Turkey 1967: 343. Only casual in Br and Da ([Br, Da] in Fl. Eur.).

Amaranthus crispus (Lesp. & Thév.) N. Terracc. — Map 630.

Notes. [Cz] and [Ju] added (not given in Fl. Eur.): J. Holub, Preslia 38: 80 (1966).
Native of Argentina.

Amaranthus standleyanus Parodi ex Covas

Amaranthus vulgatissimus auct.

Notes. [Hs] omitted (given in Fl. Eur. as the only European territory). Not mapped.
Native of Argentina.

Amaranthus albus L. — Map 631.

Notes. [Al, Be, Cr, Cz, He, Ju, Lu, Rs(B, K) and Tu] added (not given in Fl. Eur.): W. Greuter, Ann. Mus. Goulandris 1: 30 (1973); J. Holub, Preslia 38: 80 (1966); Nova Fl. Port. 1971: 103; Rubtsev 1972: 140; Fl. Turkey 1967: 343. Only casual in Br and Da ([Br, Da] in Fl. Eur.). Also recorded for Rs(N) but most probably casual (not given in Fl. Eur.).

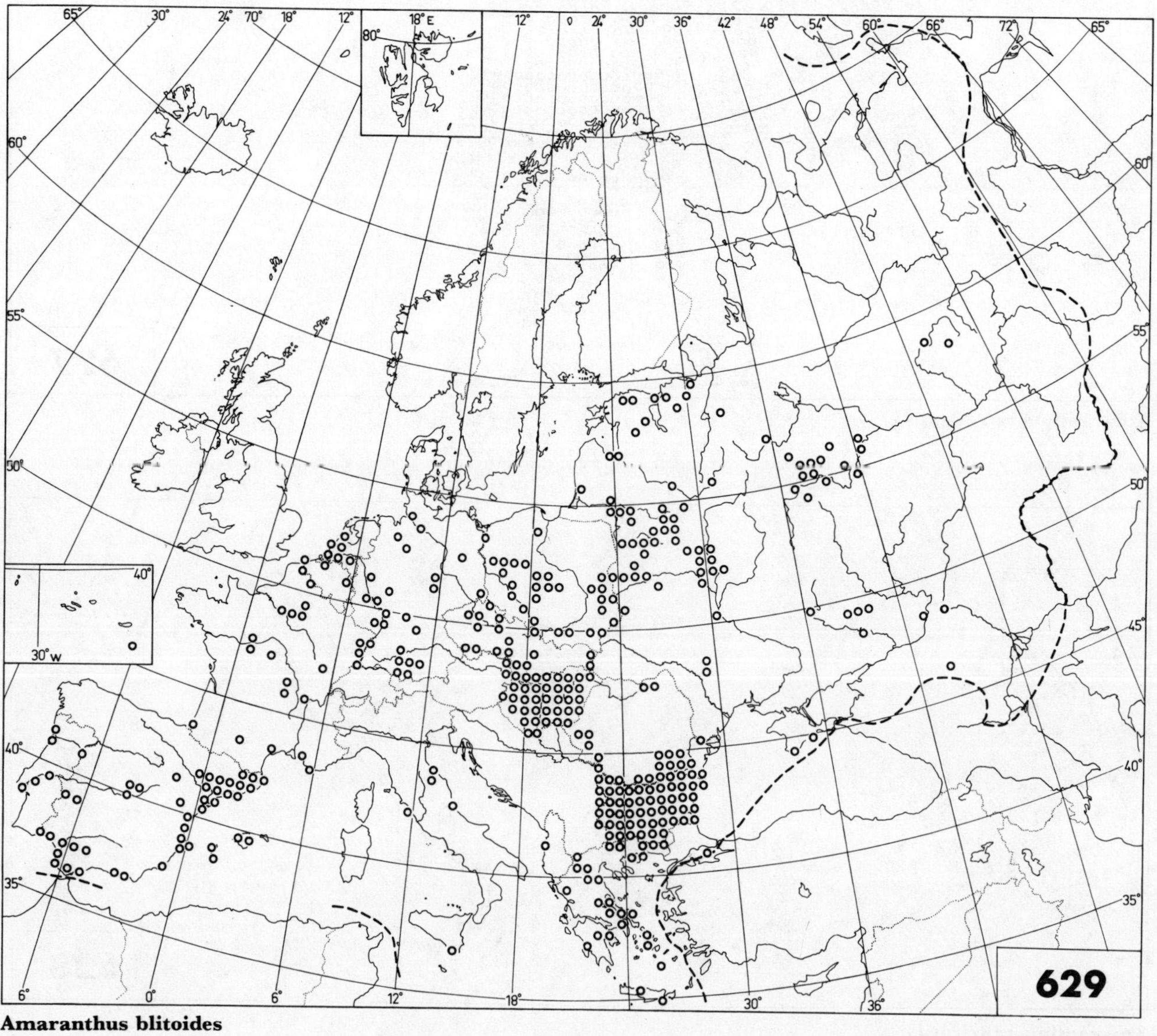

Amaranthus blitoides

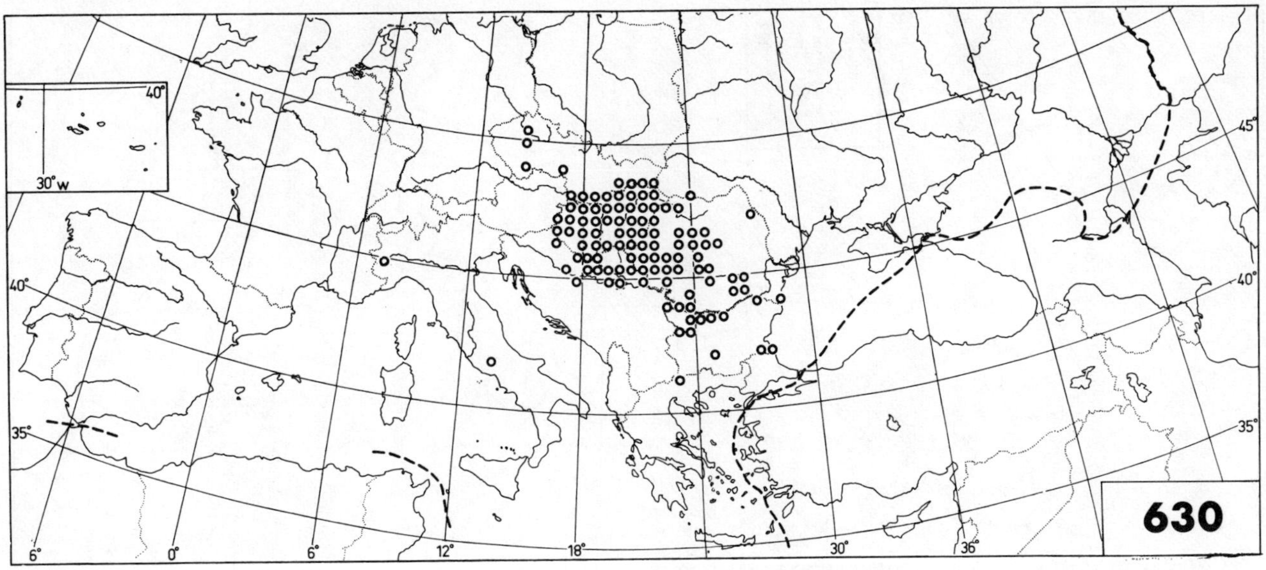

Amaranthus crispus

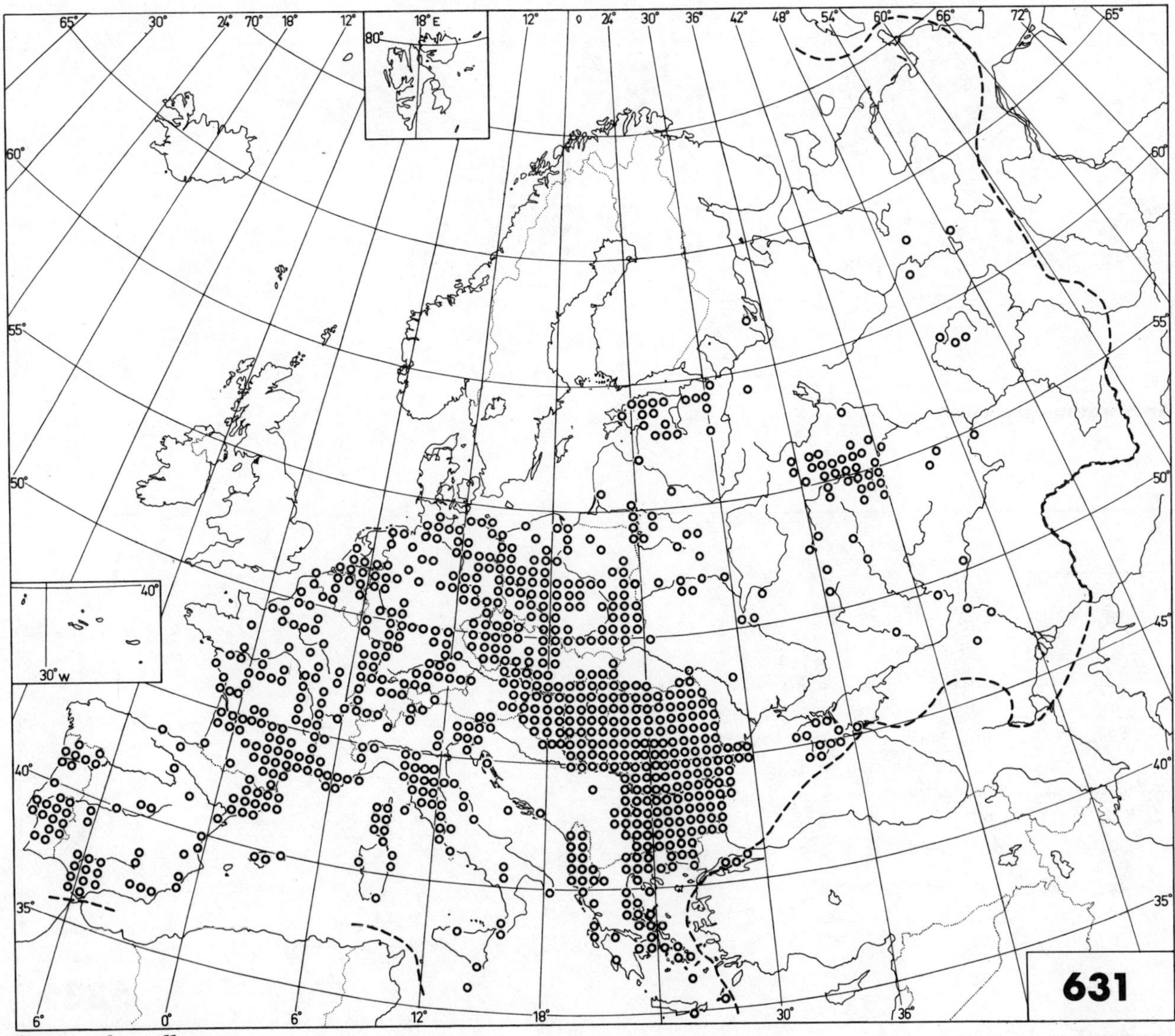

Amaranthus albus

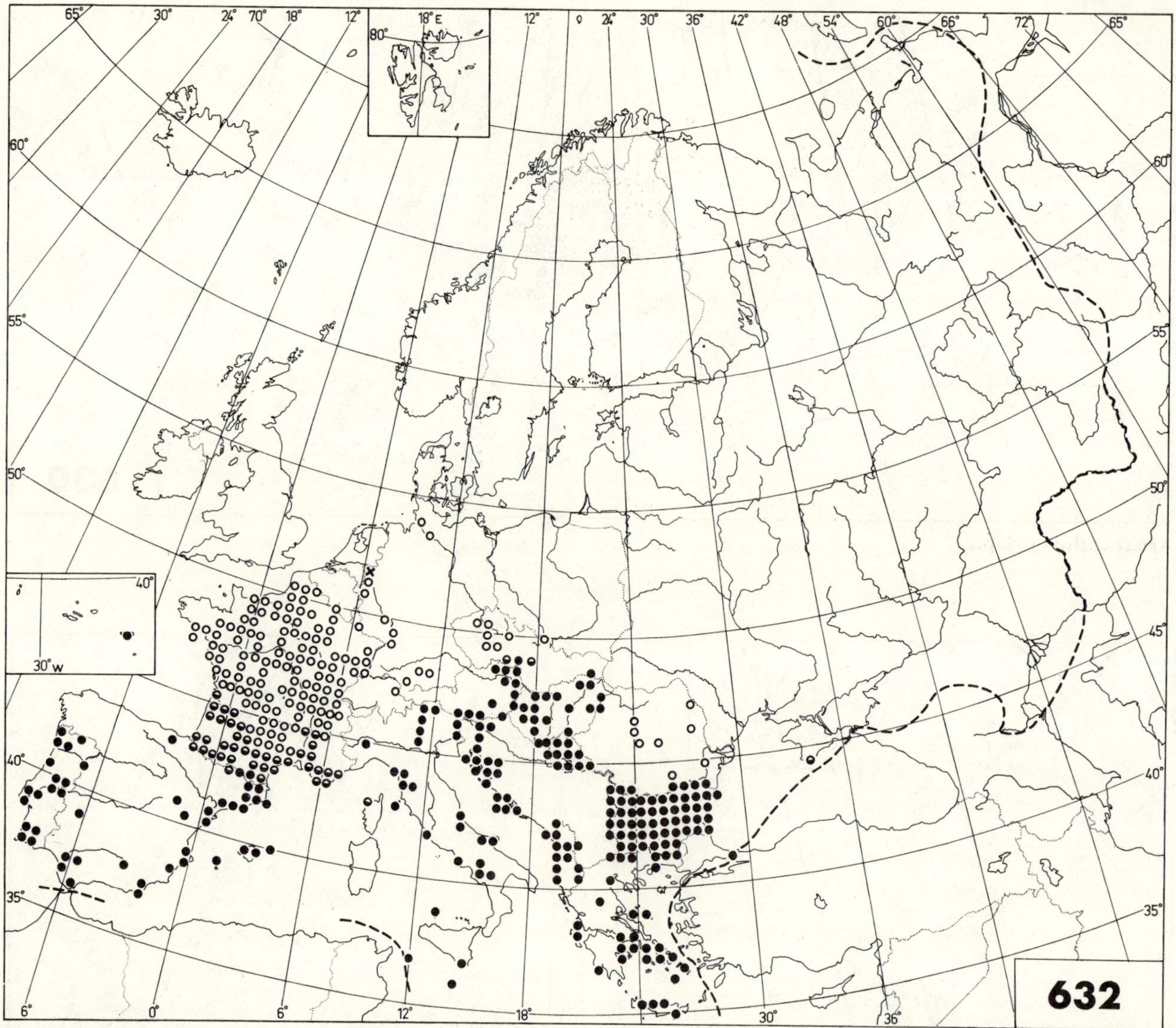

Amaranthus graecizans

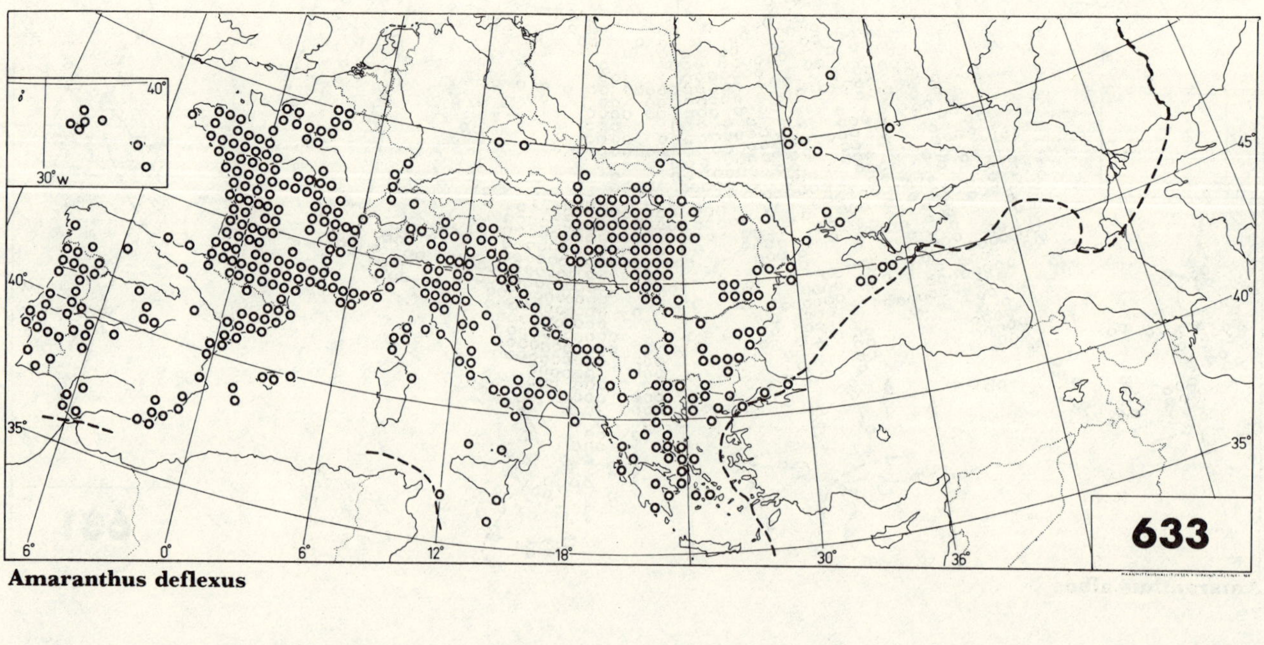

Amaranthus deflexus

Amaranthus graecizans L. — Map 632.

Amaranthus angustifolius Lam.; A. blitum auct. non L.; A. sylvestris Vill.

Notes. Considered native in Au and Hu ([Au, Hu] in Fl. Eur.). Az, [Be], *Co, *Cz, Gr, [He] and Rs(*K) added (not given in Fl. Eur.): J. Holub, Preslia 38: 80 (1966); Rubtsev 1972: 140. Doubtful native in Ga, and only casual in Po (Ga, [Po] in Fl. Eur.). [Rm] mapped according to Fl. Rep. Pop. Rom. 1: 599 (1952). Sa omitted (given in Fl. Eur.).

Total range. Hultén Alaska 1968: 403 (holarctic part).

Amaranthus deflexus L. — Map 633.

Notes. Considered casual in Au ([Au] in Fl. Eur.). [Co, Cz, He, Lu and Rs(C)] added (not given in Fl. Eur.): J. Holub, Preslia 38: 80 (1966); Nova Fl. Port. 1971. 103; [Rs(E)] omitted (given in Fl. Eur.). The northern limit of established occurrence must be considered poorly known and approximative; cf. P. Aellen in Hegi 1959: 504.

Native of South America: P. Aellen in Hegi 1959: 504.

Amaranthus lividus L. — Map 634.

Amaranthus ascendens Loisel.; A. blitum L.; A. polygonoides L.

Taxonomy and nomenclature. R. Soó, Feddes Repert. 83: 137 (1972); F. Fillias, A. Gaulliez & M. Guédès, Taxon 29: 149—150 (1980). There has been extensive taxonomic and nomenclatural confusion with *Amaranthus graecizans* and *A. viridis.*

Notes. Considered native in Au, Hu and Rm ([Au, Hu, Rm] in Fl. Eur.). Bl, *Cz and *He added (not given in Fl. Eur.): J. Duvigneaud, Soc. Échange Pl. Vasc. Europe Occ. Bassin Médit. 17(suppl.): 5 (1979); J. Holub, Preslia 38: 80 (1966). Cr omitted, the records (like most of those from Gr) being referable to *Amaranthus viridis* (given in Fl. Eur.). Only casual in Da and Fe ([Da, Fe] in Fl. Eur.). Doubtful native in Ge and Po ([Ge, Po] in Fl. Eur.). Revision of the material from S. Europe may show that much of it in fact belongs to *A. viridis.*

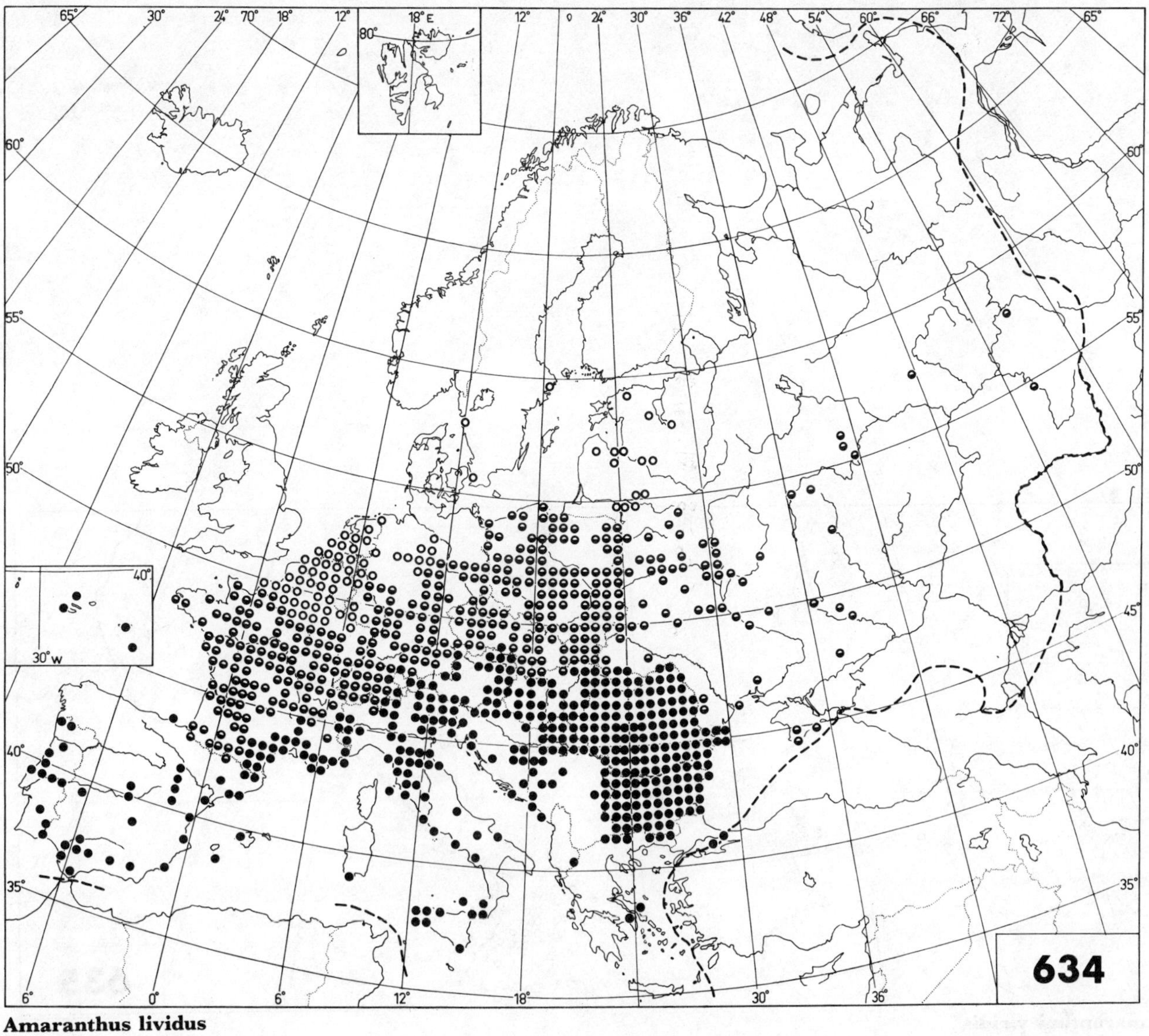

Amaranthus lividus

Amaranthus viridis L. — Map 635.

Amaranthus gracilis Desf.

Notes. [Az, Bl, Cr, Gr, Hs, It, Si] (the species not mentioned in Fl. Eur.): H. Kuhbier, Veröff. Überseemus. Bremen 5 (Cahiers Optima Leafl. 83): 12 (1978); J. Duvigneaud, Soc. Échange Pl. Vasc. Europe Occ. Bassin Médit. 17(suppl.): 5 (1979); W. Greuter, Mem. Soc. Brot. 24: 140 (1974); A. Cacciato, Ann. Bot. (Roma) 27: 221—229 (1962). Frequently confused with *Amaranthus lividus*, and certainly underrepresented on the map.

Native of South America (?): cf. P. Aellen in Hegi 1959: 503 (under *Amaranthus gracilis*).

Amaranthus tuberculatus (Moq.) Sauer

Acnida tuberculata Moq.

Taxonomy. J.D. Sauer, Madroño 13: 5—46 (1955).

Notes. »Since 1949, both sexes have been found repeatedly along railroads and on dumps near Graz»: J.D. Sauer, Madroño 21: 430 (1972), see also P. Aellen in Hegi 1959: 509. Considered not established, and not mapped (the species not mentioned in Fl. Eur.).

Native of the Midwestern United States.

Amaranthus rudis Sauer

Amaranthus tamariscinus auct. non Nutt.

Taxonomy. J.D. Sauer, Madroño 21: 426—428 (1972).

Notes. »A persistent colony has evidently been established near the harbor at Wageningen, where both males and seed-bearing females were collected in 1961 and subsequent years»: J.D. Sauer, Madroño 21: 428 (1972). Considered not truly established, and not mapped (the species not mentioned in Fl. Eur.).

Native of the southeastern Great Plains, U.S.A.

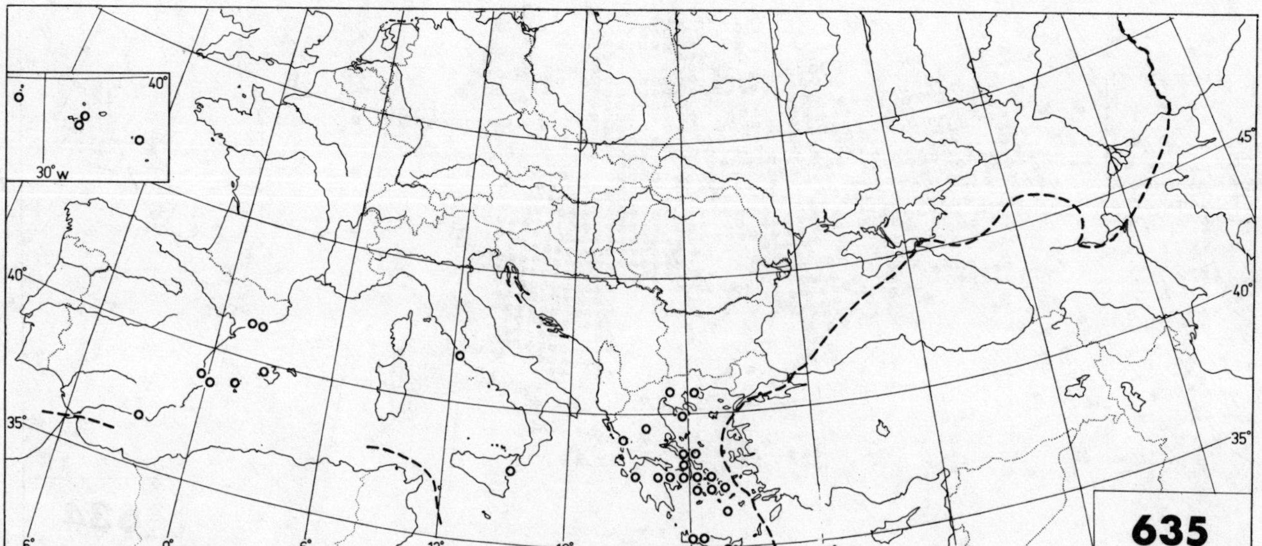

Amaranthus viridis

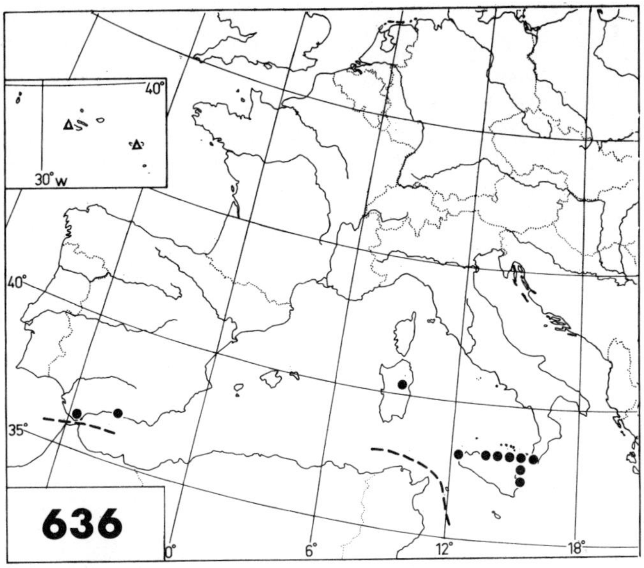

• = **Achyranthes sicula**
Δ = **A. aspera**

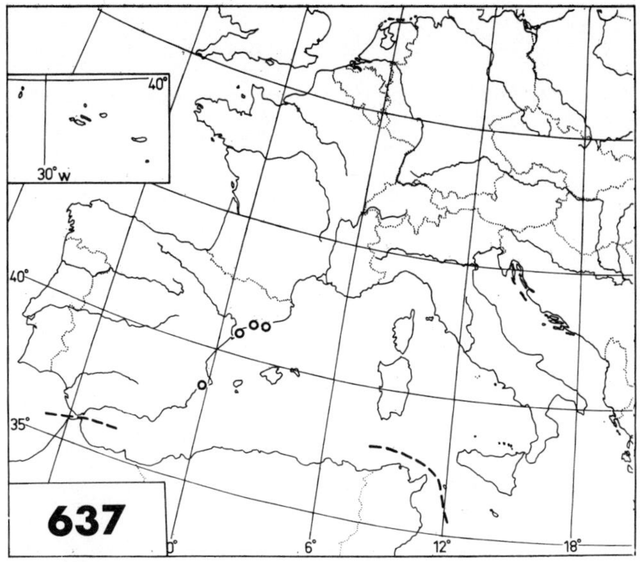

Alternanthera pungens

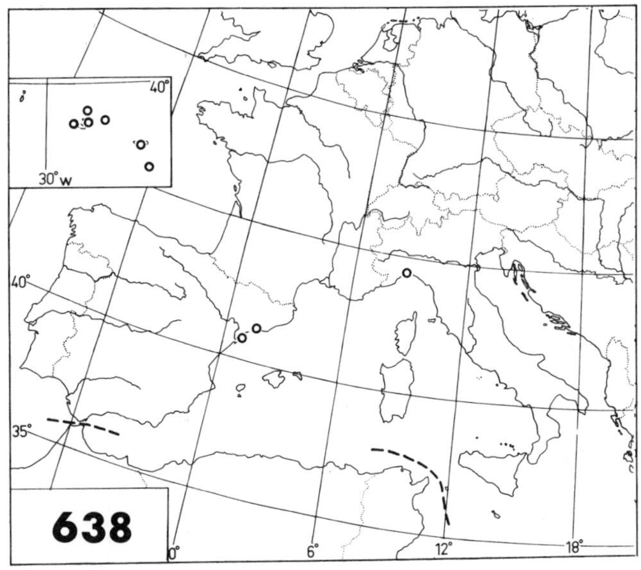

Alternanthera caracasana

Achyranthes aspera L. — Map 636.

Notes. [Az]: A. Hansen, Bol. Soc. Brot. 46: 226 (1972); possibly not fully naturalized. »More or less naturalized in S. Europe» (Fl. Eur.).
Native of the tropics.

Achyranthes sicula (L.) All. — Map 636.

Achyranthes argentea Lam.; A. aspera var. sicula L.

Alternanthera pungens Kunth — Map 637.

Alternanthera achyrantha (L.) Swartz, pro p.; A. repens auct.

Notes. [Hs]: R. Folch i Guillèn, El poblament vegetal... el coll d'Alforja i l'Ebre, pp. 141—142 (Barcelona 1976). »Perhaps naturalized in S.W. Europe» (Fl. Eur.).
Native of temperate South America.

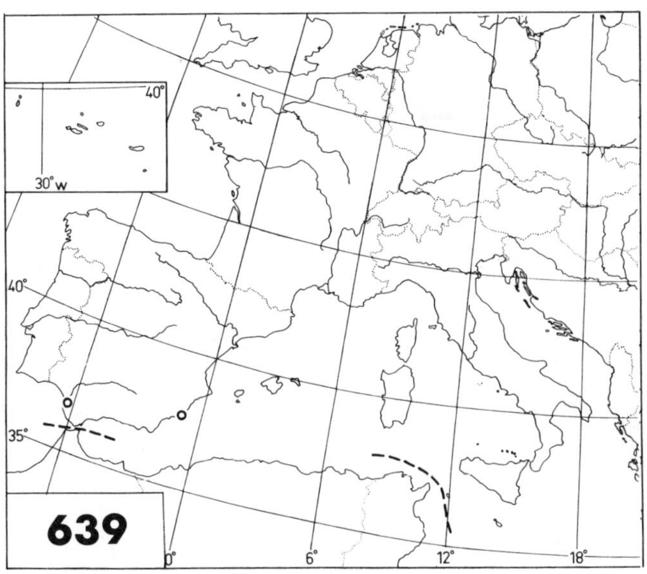

Alternanthera nodiflora

Alternanthera caracasana Humb., Bonpl. & Kunth — Map 638.

Achyranthes peploides (Humb. & Bonpl.) Britton; Alternanthera achyrantha (L.) Swartz, pro p.; A. peploides (Humb. & Bonpl.) Urban; A. pungens auct. non Kunth; A. repens auct.

Nomenclature. T.M. Pedersen, Darwiniana 14: 431 (1967); R. Folch i Guillèn, El poblament vegetal... el coll d'Alforja i l'Ebre, p. 141 (Barcelona 1976).
Notes. [Hs] confirmed ([?Hs] in Fl. Eur.): R. Folch i Guillèn, Collect. Bot. (Barcelona) 10: 182 (1976). [It] added (not given in Fl. Eur.).
Native of West Indies and Central America.

Alternanthera nodiflora R. Br. — Map 639.

Native of Australia, India and Africa.

NYCTAGINACEAE

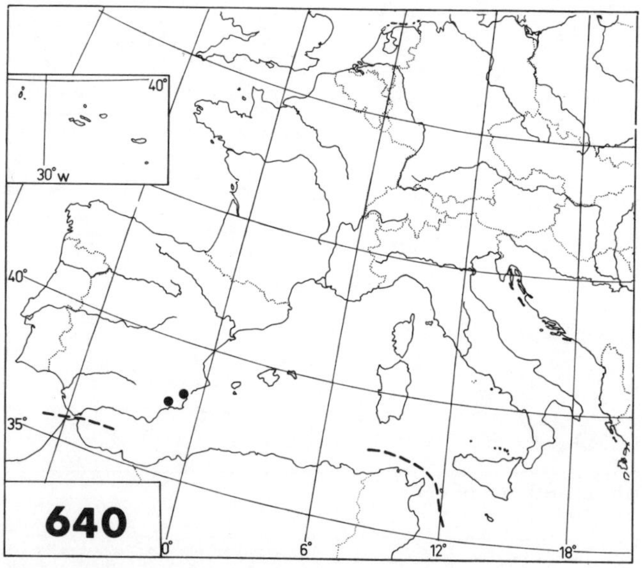

Commicarpus plumbagineus

Commicarpus plumbagineus (Cav.) Standley — Map 640.

Boerhavia plumbaginea Cav.

Mirabilis jalapa L.

Incl. Mirabilis dichotoma L.; M. odorata L.

Notes. [Bl] and [Bu] added (not given in Fl. Eur.): H. Kuhbier, Veröff. Überseemus. Bremen 5 (Cahiers Optima Leafl. 83): 12 (1978); »subespontània» according to Fl. Mallorca 1978: 119. Only casual in Ga ([Ga] in Fl. Eur.), and probably elsewhere in Europe. Not mapped.
Native of tropical America.

Mirabilis longiflora L.

Notes. Only casual in Cz and Ga (given as naturalized in Fl. Eur.). Not mapped.
Native of Mexico.

Oxybaphus nyctagineus (Michx) Sweet — Map 641.

Mirabilis nyctaginea (Michx) McMillan; M. oxybaphus Bordzil.

Notes. The status of all the records, whether truly naturalized or only casual, should perhaps be checked.
Native of North America.

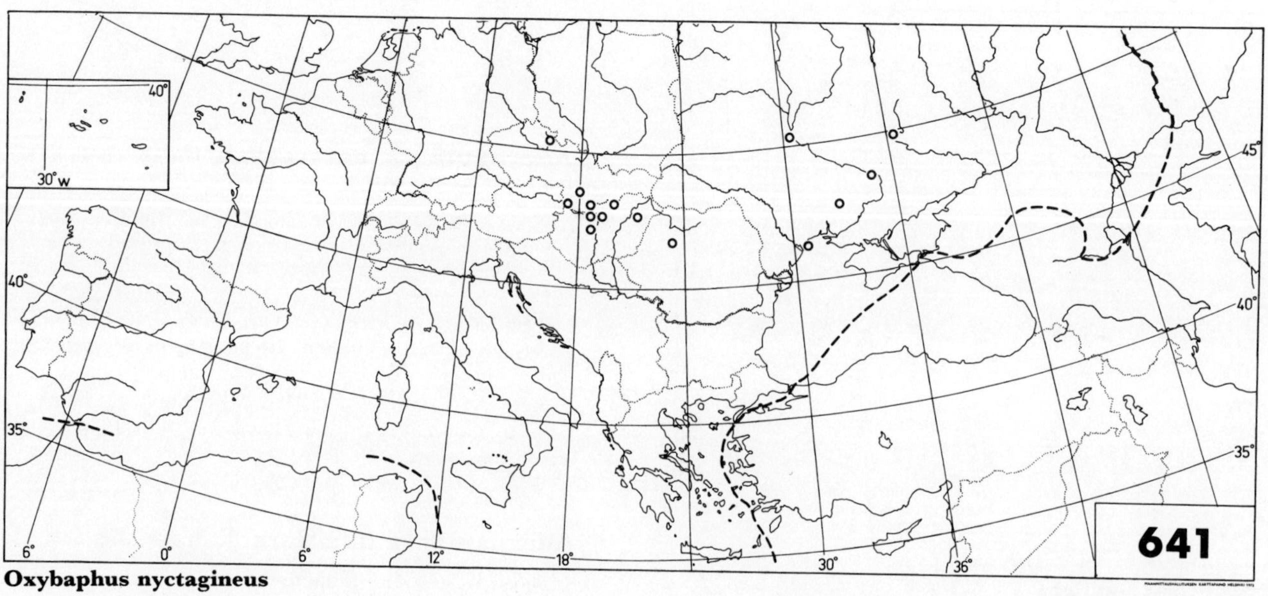

Oxybaphus nyctagineus

PHYTOLACCACEAE

Phytolacca americana L. — Map 642.

Phytolacca decandra L.

Notes. Sa omitted ([Sa] in Fl. Eur.), [Ge] and [Rs(W)] added (not given in Fl. Eur.).
Native of North America.

Phytolacca esculenta Van Houtte — Map 643.

Notes. [Cz] and [Ho] added (not given in Fl. Eur.): V. Skalický, Preslia 44: 364—369 (1972).
Native of E. Asia.

Phytolacca dioica L.

Notes. »Planted for ornament and shade in the Mediterranean region and locally naturalized» (Fl. Eur.). Not mapped.
Native of temperate and subtropical South America.

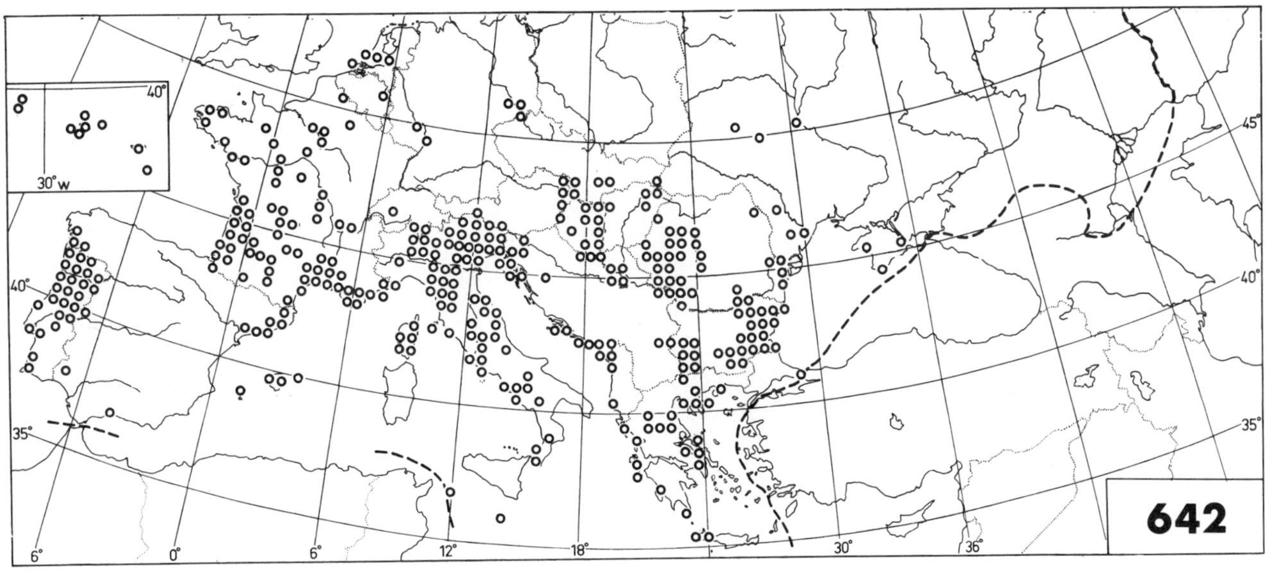

Phytolacca americana

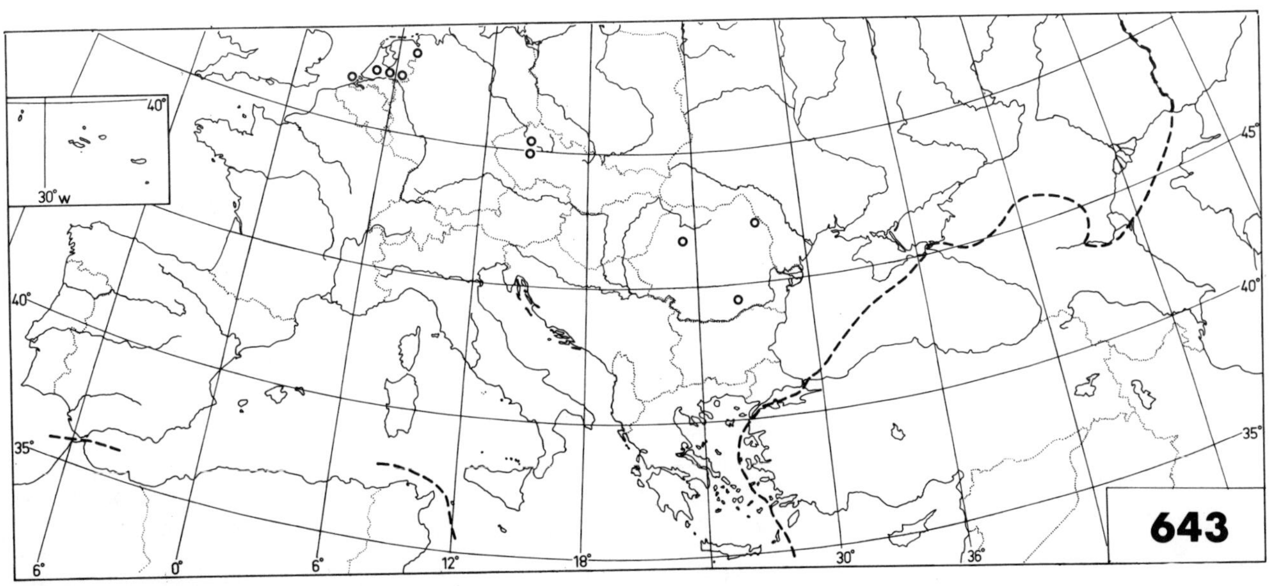

Phytolacca esculenta

AIZOACEAE

Galenia secunda Sonder — Map 644.

Notes. [Hs], first recorded in 1961 (the species not mentioned in Fl. Eur.): E.F. Galiano & S. Silvestre, Lagascalia 5: 94 (1975), L. Arvidsson & M. Lidén, Bot. Not. 130: 130 (1977).

Native of South America.

Aizoon hispanicum L. — Map 645.

Notes. Bl confirmed (?Bl in Fl. Eur.): Fl. Mallorca 1978: 125—126; J. Duvigneaud, Soc. Échange Pl. Vasc. Europe Occ. Bassin Médit. 17(suppl.): 5 (1979).

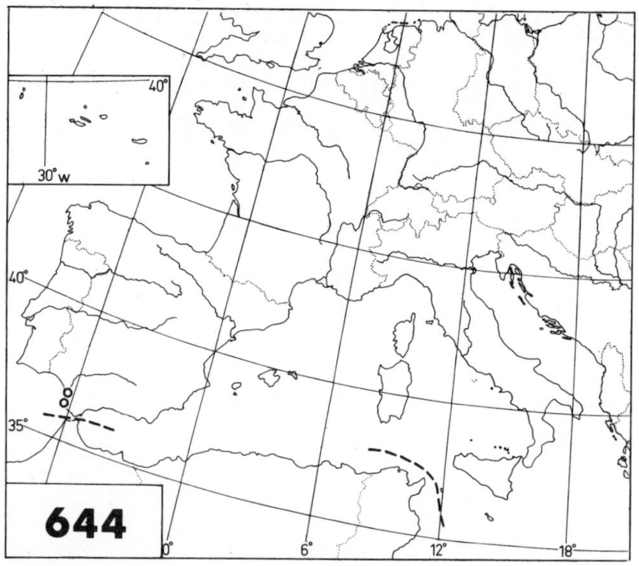

Galenia secunda

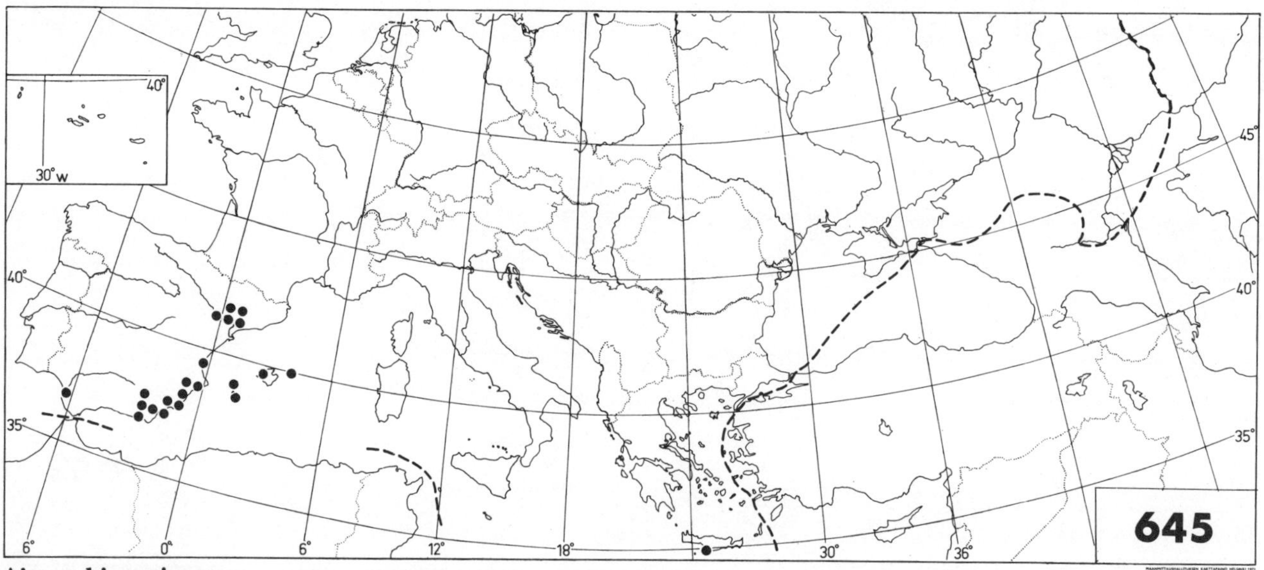

Aizoon hispanicum

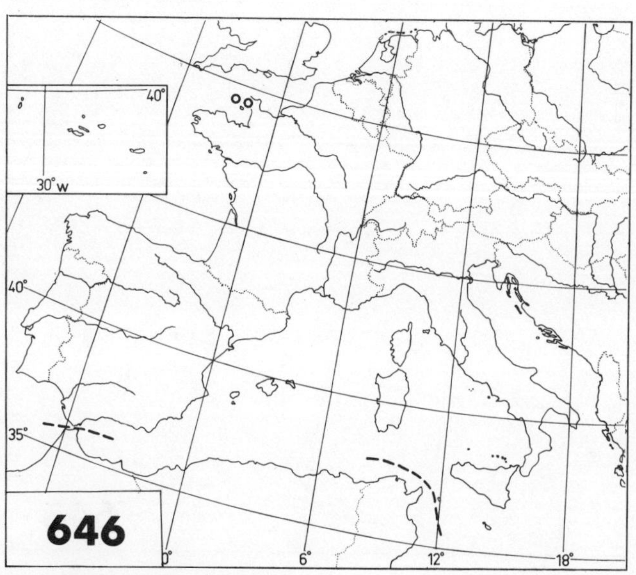

Carpobrotus glaucescens

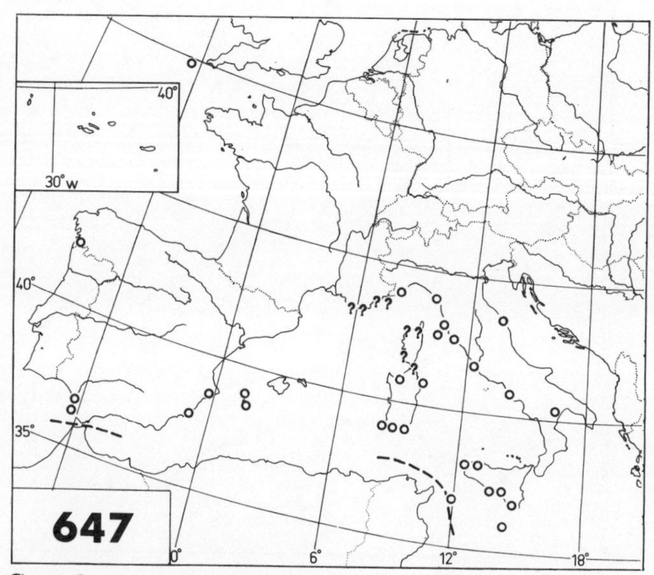

Carpobrotus acinaciformis

Sesuvium portulacastrum L.

Notes. Once found on the left bank of the mouth of the River Tagus, not recorded from there since 1888 ([Lu] in Fl. Eur.). Not mapped.

Carpobrotus glaucescens (Haw.) Schwantes — Map 646.

Notes. [Ga] (the Channel Islands) (the species not mentioned in Fl. Eur.).
Native of Australia.

Carpobrotus acinaciformis (L.) L. Bolus — Map 647.

Mesembryanthemum acinaciforme L.

Notes. [Bl, Br, ?Co and Si] added (not given in Fl. Eur.): H. Kuhbier, Veröff. Überseemus. Bremen 5 (Cahiers Optima Leafl. 83): 12 (1978); J. Duvigneaud, Soc. Échange Pl. Vasc. Europe Occ. Bassin Médit. 17(suppl.): 5 (1979); J.E. Lousley, Fl. Isles Scilly, p. 145 (Newton Abbot 1971); Fl. France 1973: 331. Presence in Ga doubtful ([Ga] in Fl. Eur.). No reliable records of naturalized occurrence available for Gr ([Gr] in Fl. Eur.). The records for Lu and possibly also those for Co and Ga may refer to *Carpobrotus edulis*; see Fl. France 1973: 331, and Nova Fl. Port. 1971: 106—107.
Native of South Africa.

Carpobrotus edulis (L.) N.E. Br. — Map 648.

Mesembryanthemum edule L.

Notes. [Al, Bl, Co and Si] added (not given in Fl. Eur.): Fl. France 1973: 331. [Be] omitted (given in Fl. Eur.).
Native of South Africa.

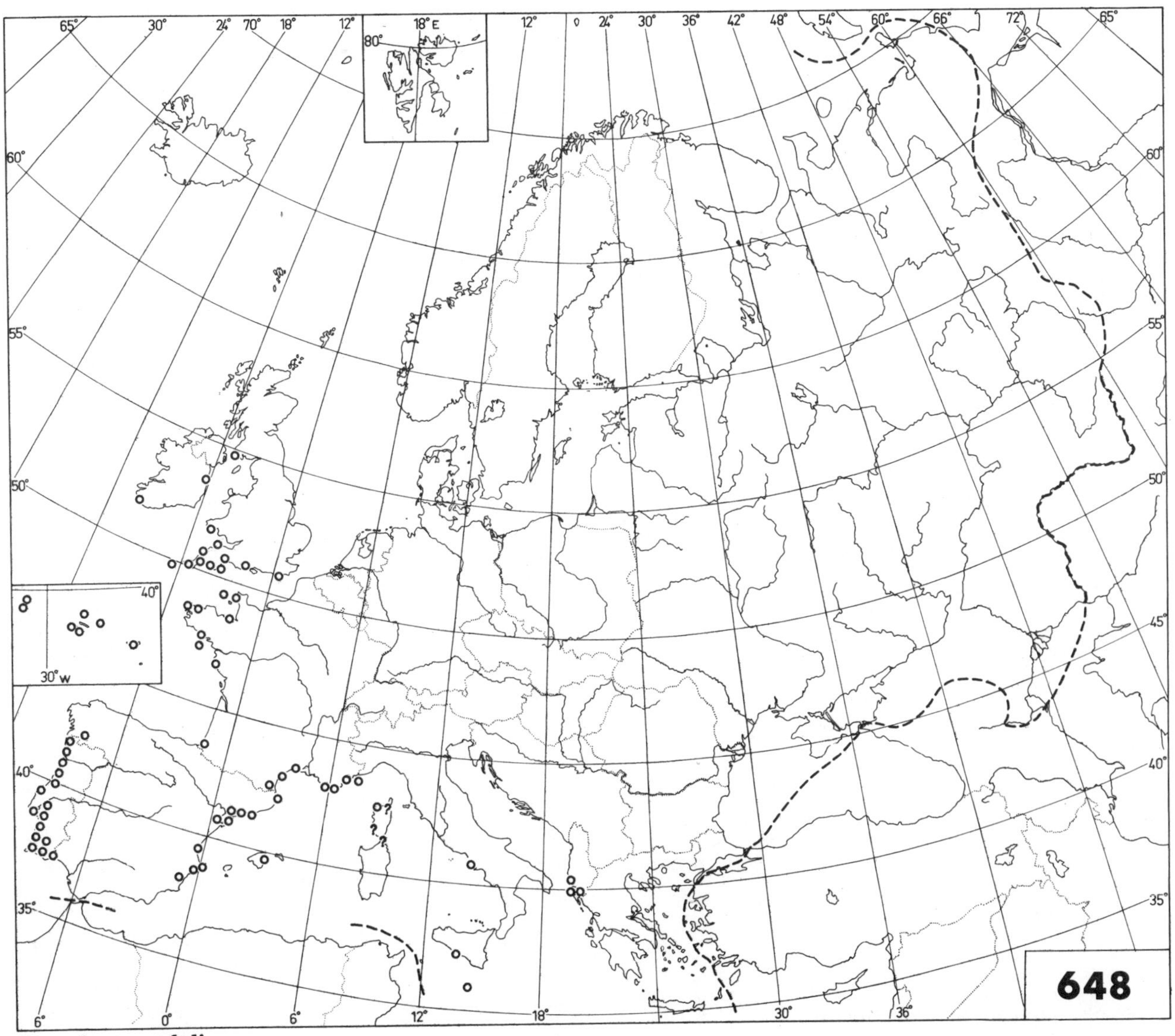

Carpobrotus edulis

Carpobrotus chilensis (Molina) N.E. Br.

Mesembryanthemum aequilaterale auct. non Haw.; M. chilense Molina

Notes. Known only from cultivation in Hs ([Hs] in Fl. Eur.). There is thus no evidence that the species is truly naturalized in Europe. Not mapped.

Native of W. America.

Mesembryanthemum nodiflorum L. — Map 649.

Cryophytum nodiflorum (L.) N.E. Br.; Gasoul nodiflorum (L.) Rothm.

Notes. Considered established alien in Lu (given as native in Fl. Eur.): Nova Fl. Port. 1971: 107.

Mesembryanthemum crystallinum L. — Map 650.

Cryophytum crystallinum (L.) N.E. Br.; Gasoul crystallinum (L.) Rothm.

Notes. Native status doubtful in Cr and Gr, and considered established alien in Lu (given as native in Fl. Eur.): Nova Fl. Port. 1971: 107.

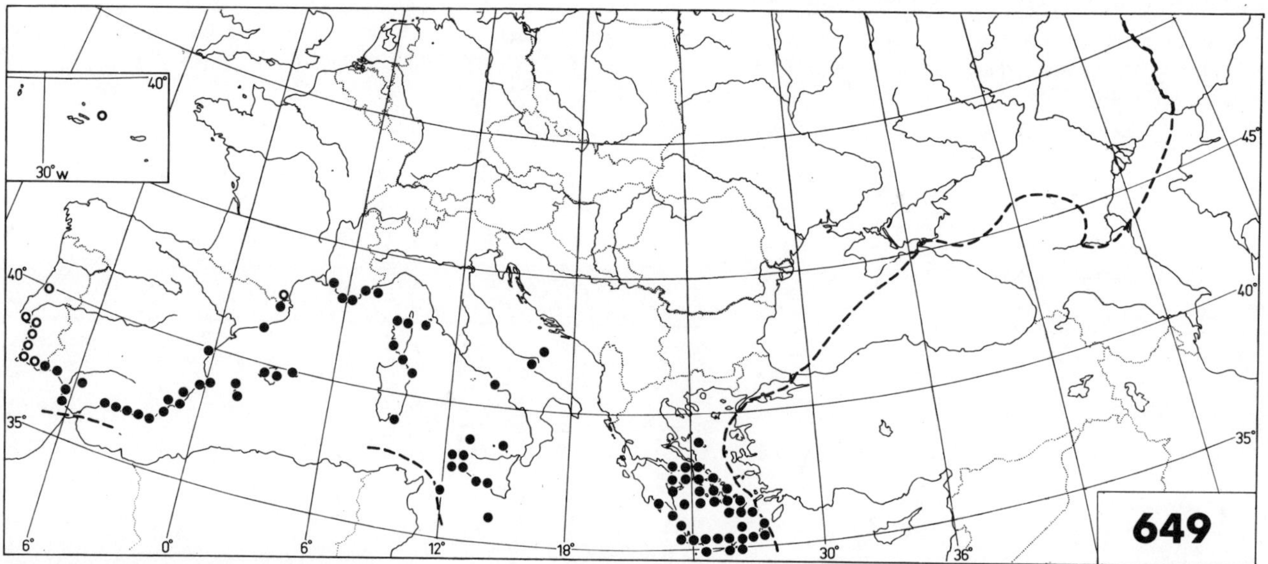

Mesembryanthemum nodiflorum

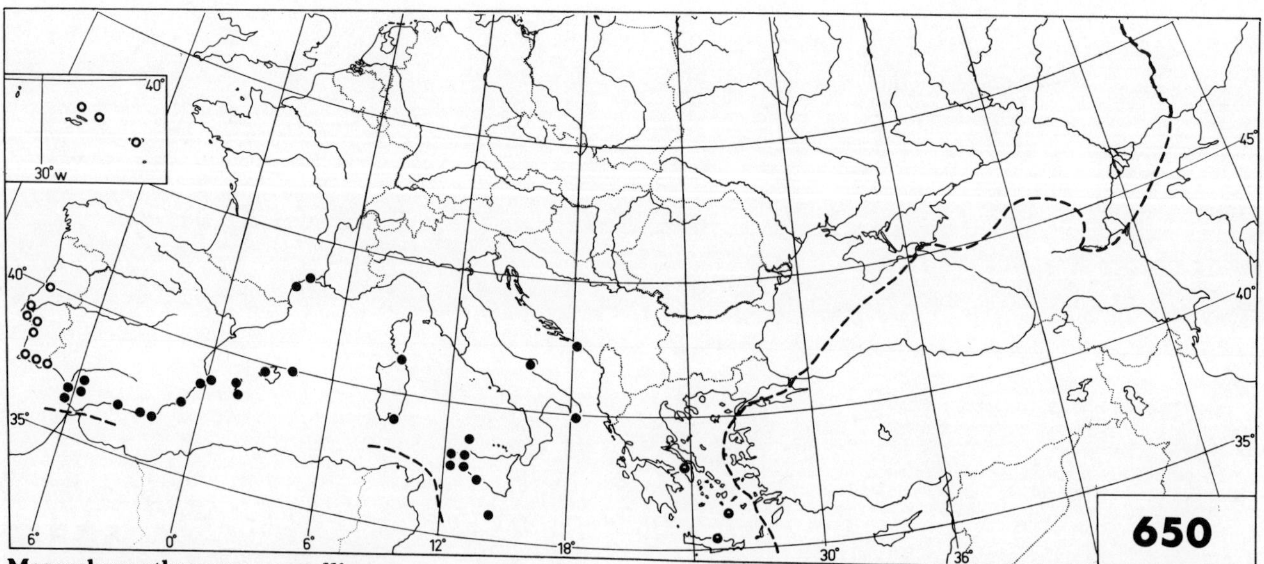

Mesembryanthemum crystallinum

Drosanthemum candens (Haw.) Schwantes — Map 651.

Notes. [Br] omitted (given in Fl. Eur.), the material in question being referable to *Drosanthemum floribundum.*
Native of South Africa.

Drosanthemum floribundum (Haw.) Schwantes — Map 651.

Notes. [Br, Ga] (the species not mentioned in Fl. Eur.): J.E. Lousley, Fl. Isles Scilly, p. 145—146 (Newton Abbot 1971), and in P.S. Green (ed.), Plants: wild and cultivated, p. 83 (Cambridge 1973); D. McClintock, Wild Flowers of Guernsey (London 1975). For Br, given as *Drosanthemum candens* in Fl. Eur.
Native of South Africa.

Oscularia deltoides (Miller) Schwantes

Notes. [Br] (the species not mentioned in Fl. Eur.): J.E. Lousley, Fl. Isles Scilly, p. 146 (Newton Abbot 1971), and in P.S. Green (ed.), Plants: wild and cultivated, p. 83 (Cambridge 1973). Not mapped.
Native of South Africa.

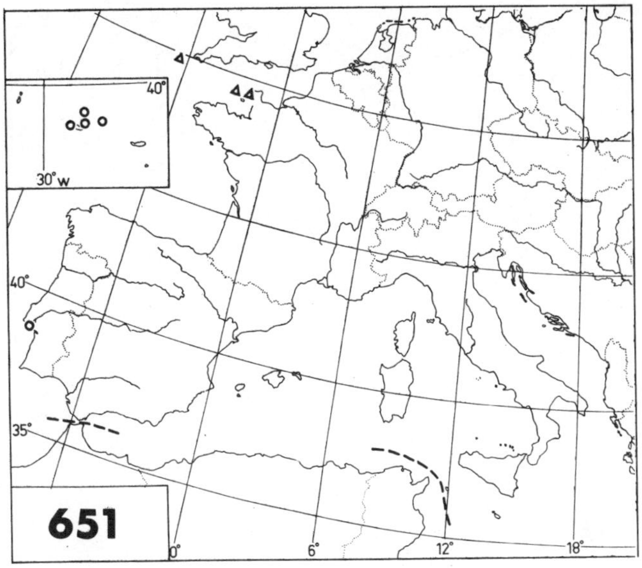

○ = **Drosanthemum candens**
Δ = **D. floribundum**

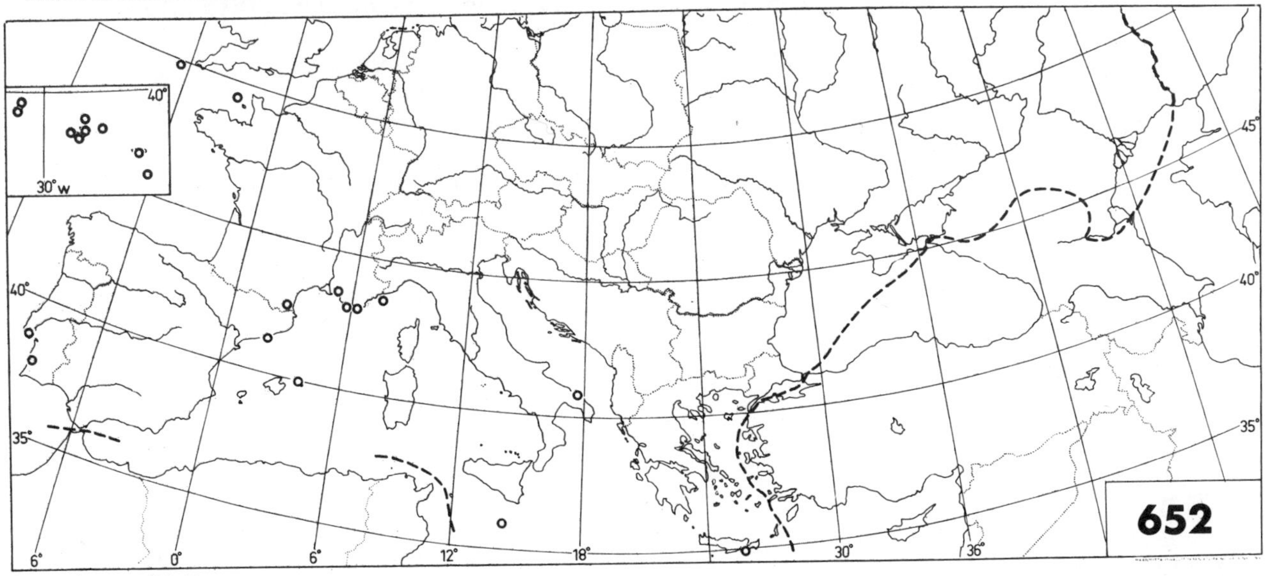

Aptenia cordifolia

Aptenia cordifolia (L. fil.) N.E. Br. — Map 652.

Mesembryanthemum cordifolium L. fil.

Notes. [Bl, Cr and Si] added (not given in Fl. Eur.): J. Duvigneaud, Soc. Échange Pl. Vasc. Europe Occ. Bassin Médit. 17(suppl.): 5 (1979). In Br, the species »shows no tendency to spread by natural means, but persists for a time where planted»: J.E. Lousley, Fl. Isles Scilly, p. 146—147 (Newton Abbot 1971), and in P.S. Green (ed.), Plants: wild and cultivated, p. 83—85 (Cambridge 1973).
Native of South Africa.

Disphyma crassifolium (L.) L. Bolus — Map 653.

Mesembryanthemum cordifolium L. fil.

Notes. [Az, Br and Ga] added (not given in Fl. Eur.): Nova Fl. Port. 1971: 106; J.E. Lousley, Fl. Isles Scilly, p. 148 (Newton Abbot 1971), and in P.S. Green (ed.), Plants: wild and cultivated, p. 83 (Cambridge 1973).
Native of South Africa.

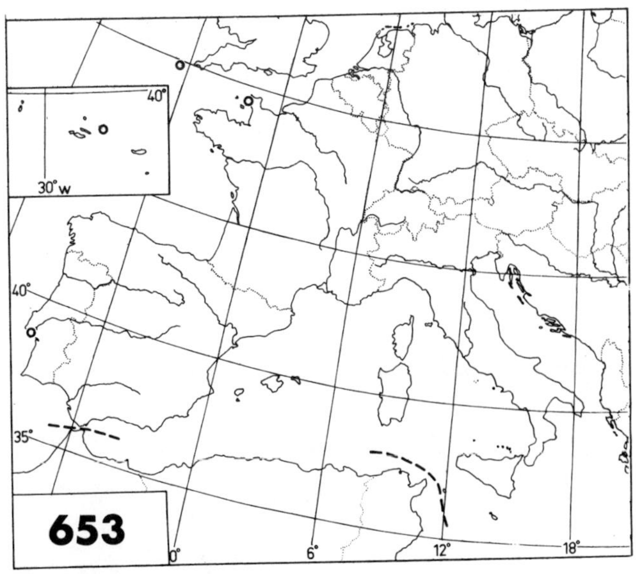

Disphyma crassifolium

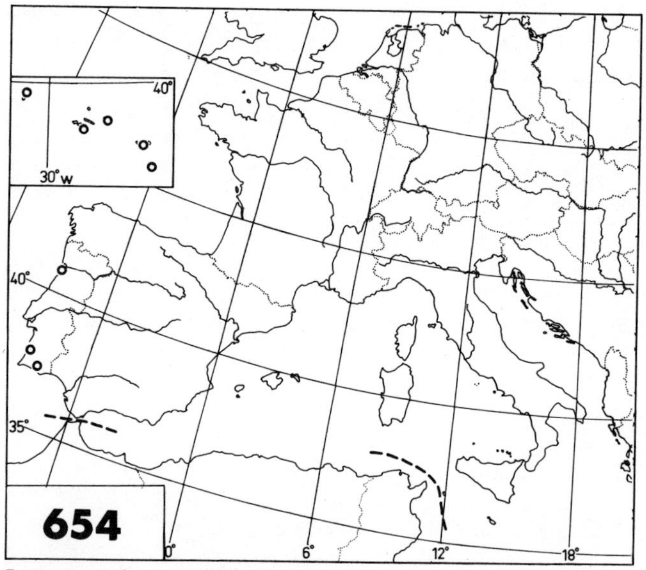

Lampranthus multiradiatus

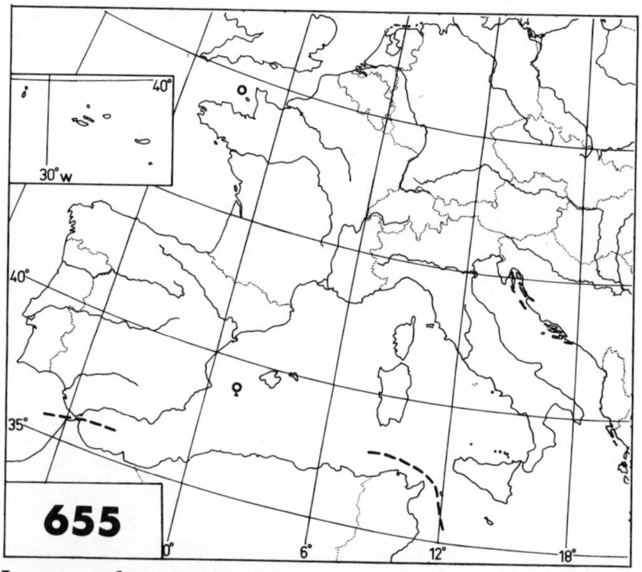

Lampranthus roseus

Erepsia heteropetala (Haw.) Schwantes — Map 656.

Mesembryanthemum heteropetalum Haw.

Notes. [Br] (the species not mentioned in Fl. Eur.). »Spreads freely from seed»: J.E. Lousley, Fl. Isles Scilly, p. 148 (Newton Abbot 1971), and in P.S. Green (ed.), Plants: wild and cultivated, p. 83 (Cambridge 1973).
Native of South Africa.

Ruschia caroli (L. Bolus) Schwantes

Notes. [Br] (the species not mentioned in Fl. Eur.): J.E. Lousley, Fl. Isles Scilly, p. 147 (Newton Abbot 1971), and in P.S. Green (ed.), Plants: wild and cultivated, p. 85 (Cambridge 1973). Not mapped.
Native of South Africa.

Lampranthus glaucus (L.) N.E. Br.

Notes. Not considered truly naturalized in Lu ([Lu] given in Fl. Eur. as the only European territory). Not mapped.
Native of South Africa.

Lampranthus falciformis (Haw.) N.E. Br.

Mesembryanthemum falciforme Haw.

Notes. [Br] (the species not mentioned in Fl. Eur.): J.E. Lousley, Fl. Isles Scilly, p. 147 (Newton Abbot 1971); T.A.W. Davis, Bot. Soc. Brit. Isles News 19: 17 (1978). Not mapped.
Native of South Africa.

Lampranthus multiradiatus (Jacq.) N.E. Br. — Map 654.

Notes. [Az, Lu] (the species not mentioned in Fl. Eur.): Nova Fl. Port. 1971: 106.
Native of South Africa.

Lampranthus roseus (Willd.) Schwantes — Map 655.

Notes. [Bl, Ga: the Channel Islands] (the species not mentioned in Fl. Eur.): H. Kuhbier, Veröff. Überseemus. Bremen 5 (Cahiers Optima Leafl. 83): 12 (1978).
Native of South Africa.

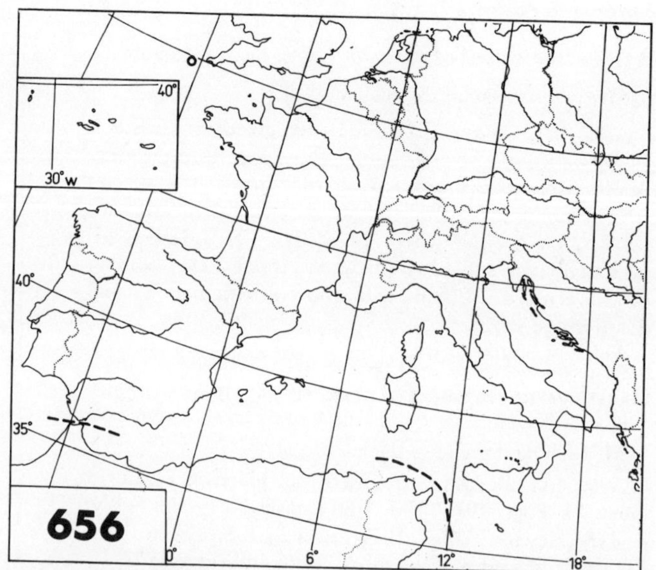

Erepsia heteropetala

MOLLUGINACEAE

Mollugo verticillata L. — Map 657.

Notes. [Az] and [Ga] added (not given in Fl. Eur.). *Native* of tropical America.

Mollugo cerviana (L.) Ser. — Map 658.

Notes. [Ga] added (not given in Fl. Eur.). Probably extinct in It (given as present in Fl. Eur.). Rs(C, K) omitted (given in Fl. Eur.); reported from Rs(K) by Rubtsev 1972: 141. Rs(W) underrepresented.

Glinus lotoides L. — Map 659.

Mollugo lotoides (L.) Arcangeli

Notes. Bu added (not given in Fl. Eur.): T. Georgiev in Fl. Reipubl. Pop. Bulg. 1966: 266; Si omitted (given in Fl. Eur.).

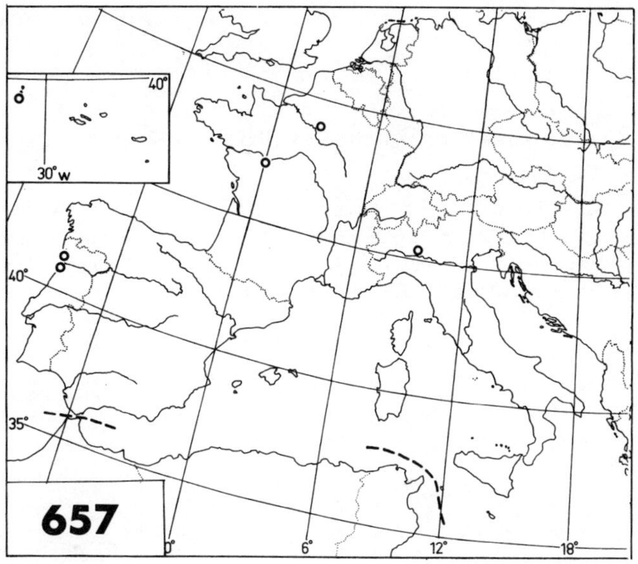

Mollugo verticillata

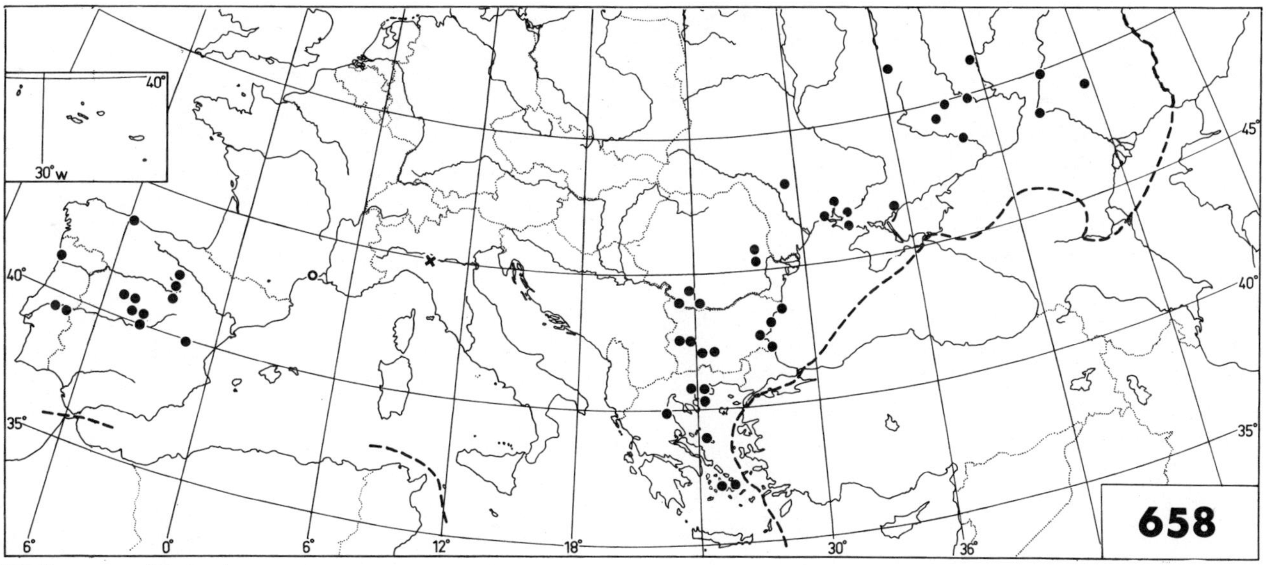

Mollugo cerviana

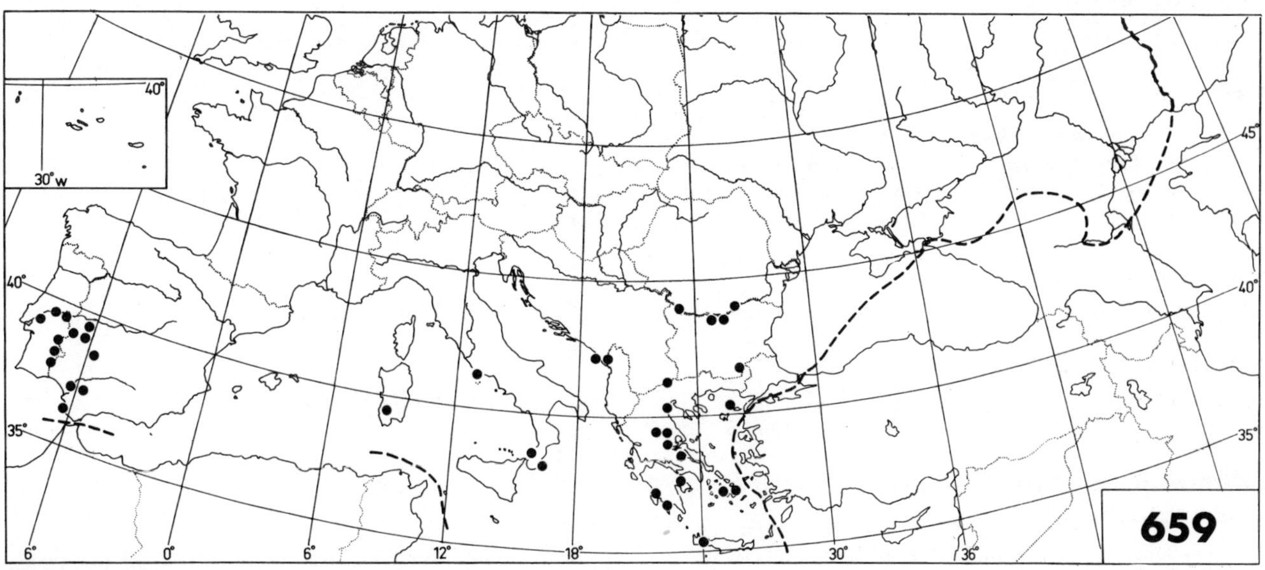

Glinus lotoides

TETRAGONIACEAE

Tetragonia tetragonoides (Pallas) O. Kuntze

Tetragonia expansa Murray

Notes. Only temporary escape from cultivation in Az and Lu ([Az, Lu] in Fl. Eur.). Considered »naturalizada» in Bl (not given in Fl. Eur.): Fl. Mallorca 1978: 126. It is questionable whether the species is truly naturalized anywhere in Europe. Commonly cultivated as a vegetable, naturalized in S.W. Europe» (Fl. Eur.). Not mapped.
Native of Australia and New Zealand.

PORTULACACEAE

Portulaca oleracea L.

Notes. *Cr added (not given in Fl. Eur.). Native status doubtful in whole of Europe (given as native in Fl. Eur.).

P. oleracea subsp. **oleracea** — Map 660.

Portulaca oleracea subsp. silvestris (Montandon) Thell.

Notes. *Al, *Au, *Az, [Be], *Bl, [Br], *Bu, *Co, *Cr, *Cz, *Ga, *Ge, *Gr, *He, [Ho], *Hs, *Hu, *It, *Ju, *Lu, [Po], *Rm, Rs(*C, *W, *K, *E), *Sa, *Si, *Tu. »Widespread weed of cultivation in S. & C. Europe, rare and casual in N. Europe» (Fl. Eur.).
Native of tropical — subtropical Asia or N. Africa, according to Hegi 1979: 1199.

P. oleracea subsp. **sativa** (Haw.) Čelak.

Notes. »Cultivated as a vegetable in S. & S.E. Europe and occasionally occurring as an escape» (Fl. Eur.). Not mapped.

Portulaca grandiflora Hooker — Map 661.

Notes. [Bu, Cz, Ge, It, Rm]. »Occasionally established... in S. & S.C. Europe» (Fl. Eur.; no territories listed).
Native of Argentina.

Talinum paniculatum (Jacq.) Gaertn.

Talinum patens (L.) Willd.

Notes. »Subspontaneous» in Az (the species not mentioned in Fl. Eur.): A. Hansen, An. Soc. Brot. 40: 15 (1974). Not mapped.
Native of tropical and subtropical America.

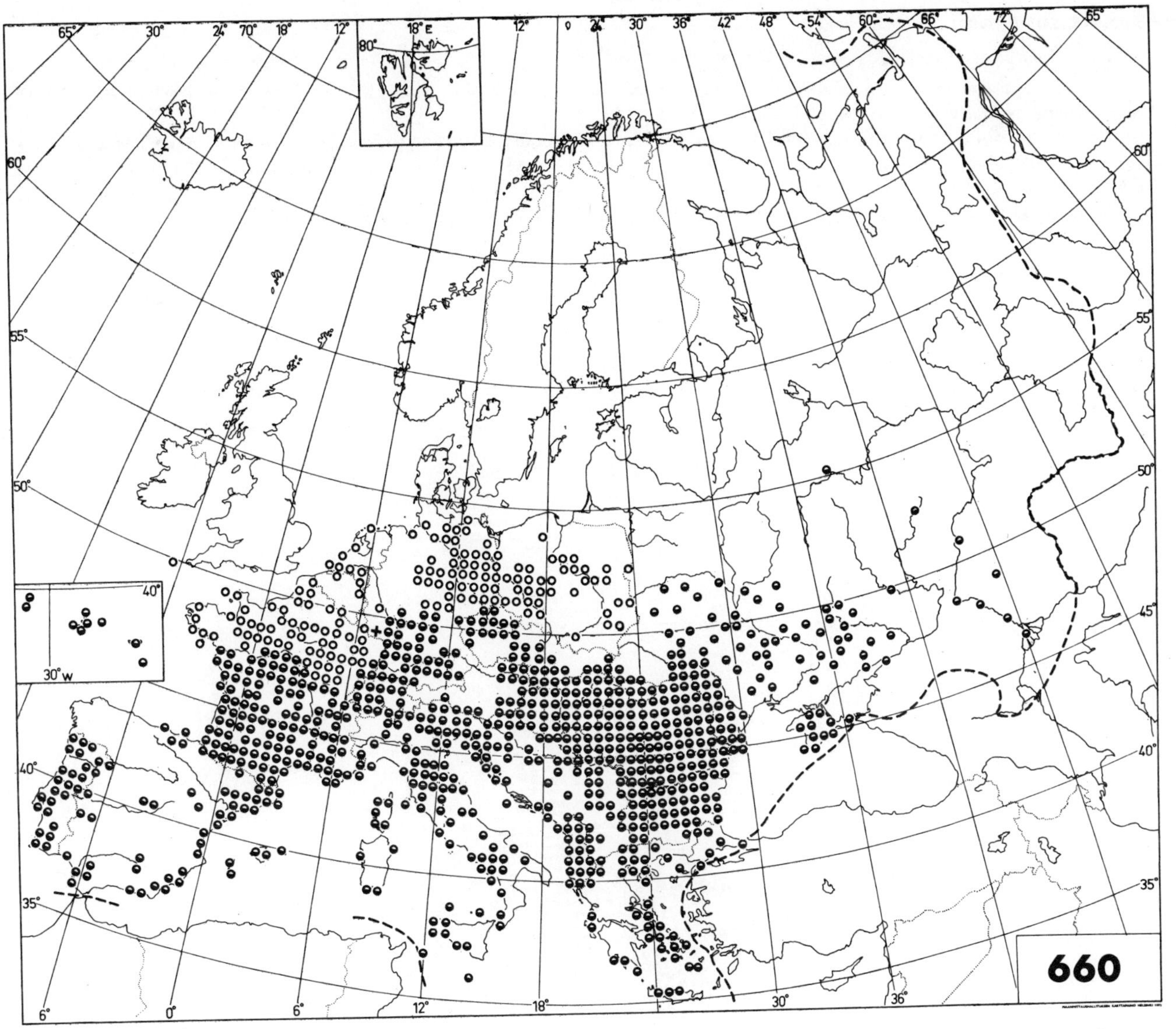

Portulaca oleracea subsp. **oleracea**

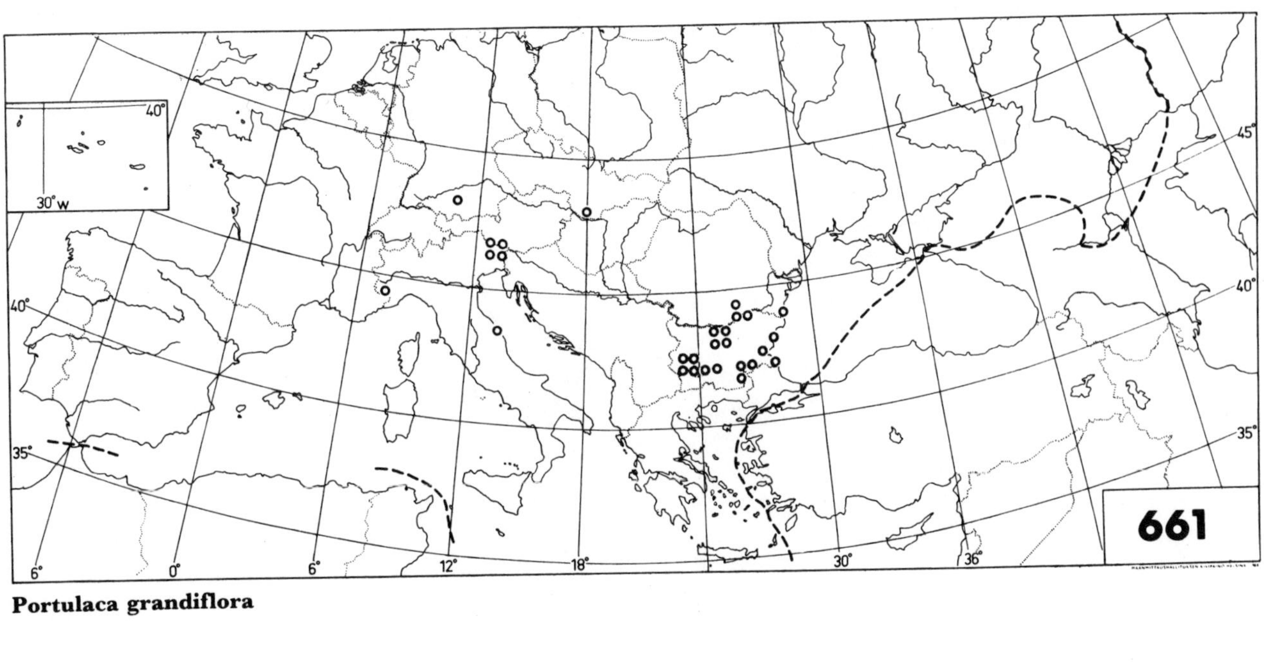

Portulaca grandiflora

Montia fontana L. — Map 662.

Montia rivularis C.C. Gmelin; M. verna Necker

Taxonomy and nomenclature. D.M. Moore, Bot. Not. 116: 16—30 (1963); J. Holub, Acta Horti Bot. Pragensis 1963: 56—57 (1964); Yu. P. Kozhevnikov, Bot. Žurn. 64: 693—699 (1979).

Notes. The distribution of the subspecies in C. (and S.) Europe is mainly according to data provided and/or checked by H. Jage (Kemberg/Wittenberg). Cr and Tu added (not given in Fl. Eur.): W. Greuter, Ann. Mus. Goulandris 1: 30 (1973); Fl. Turkey 1967: 14. Sb omitted (given in Fl. Eur.).

Total range. Hultén AA 1958: map 239; Hultén Alaska 1968: 410 (holarctic part); Hegi 1979: 1215.

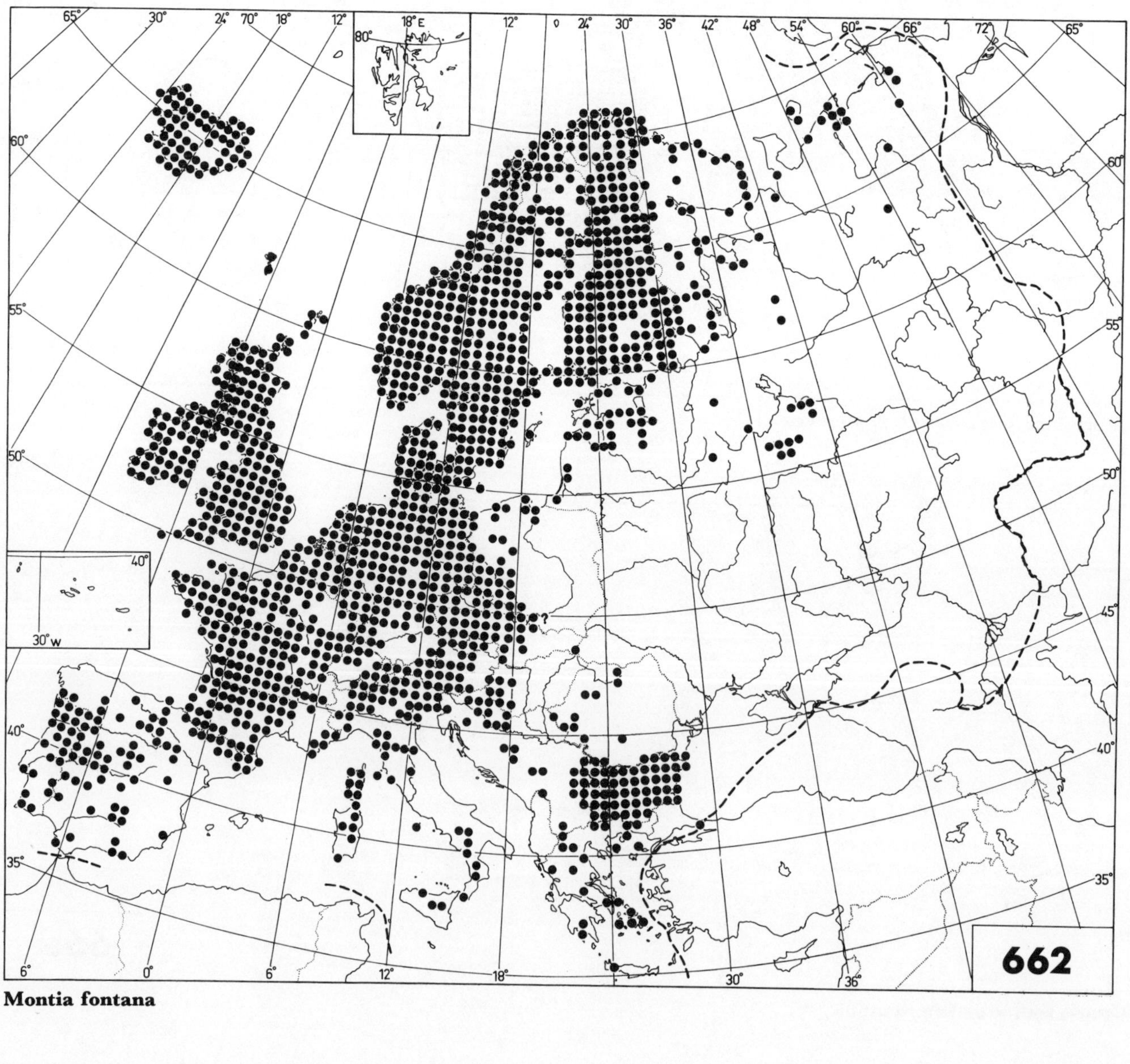

Montia fontana

M. fontana subsp. **fontana** — Map 663.

Montia lamprosperma Cham.; M. rivularis auct.

Notes. Au, Be, Br, Cz, Da, Fa, Fe, Ga, Ge, Hb, He, Ho, Hs, Is, It, Ju, No, Po, Rs(N, B, C), ?Sa, Su. »N. & C. Europe; mountains of S. Europe» (Fl. Eur.).

Total range. MJW 1965: map 136a (incl. subsp. *variabilis*).

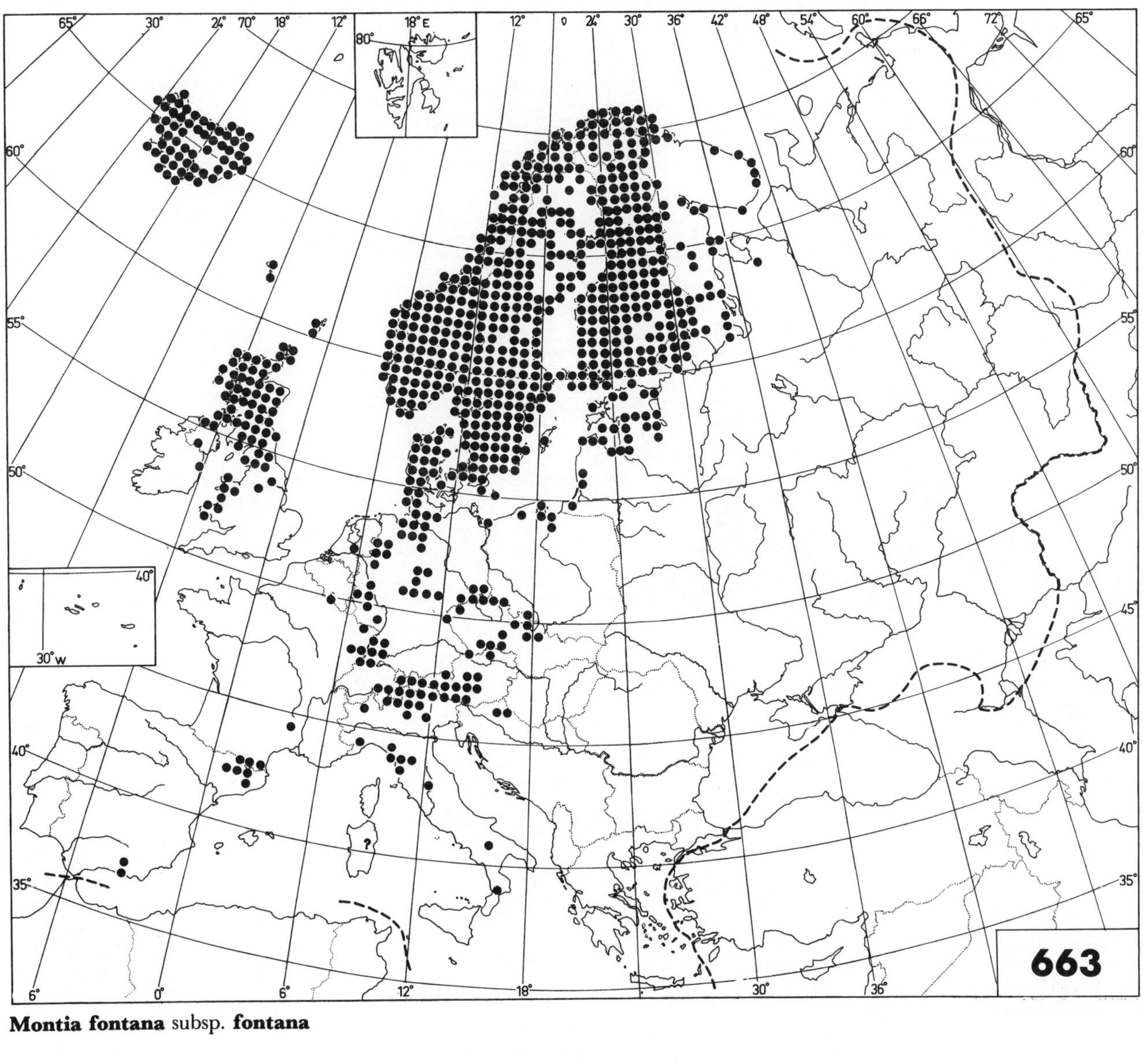

Montia fontana subsp. **fontana**

M. fontana subsp. **variabilis** Walters — Map 664.

Montia funstonii Rydb. ampl. Munz; M. rivularis auct.

Notes. Au, Be, Br, Cz, Da, Fa, Ga, Ge, Hb, He, Ho, Hs, Is, It, Po, Rs(B), Sa, Si, Su. »W. & C. Europe; detailed distribution not yet known» (Fl. Eur.).

M. fontana subsp. **amporitana** Sennen — Map 664.

Montia fontana subsp. intermedia (Beeby) Walters; M. hallii (A. Gray) Greene; M. limosa Decker; M. lusitanica Samp.; M. rivularis auct.

Notes. Au, Be, Br, Co, Cz, Ga, Ge, Gr, Hb, He, Ho, Hs, Is, It, Ju, Lu, Po, Rm, Sa, Si, Su. »W. & S.W. Europe; apparently rare in C. Europe, but detailed distribution not yet known» (Fl. Eur.).

Total range. MJW 1965: map 136a.

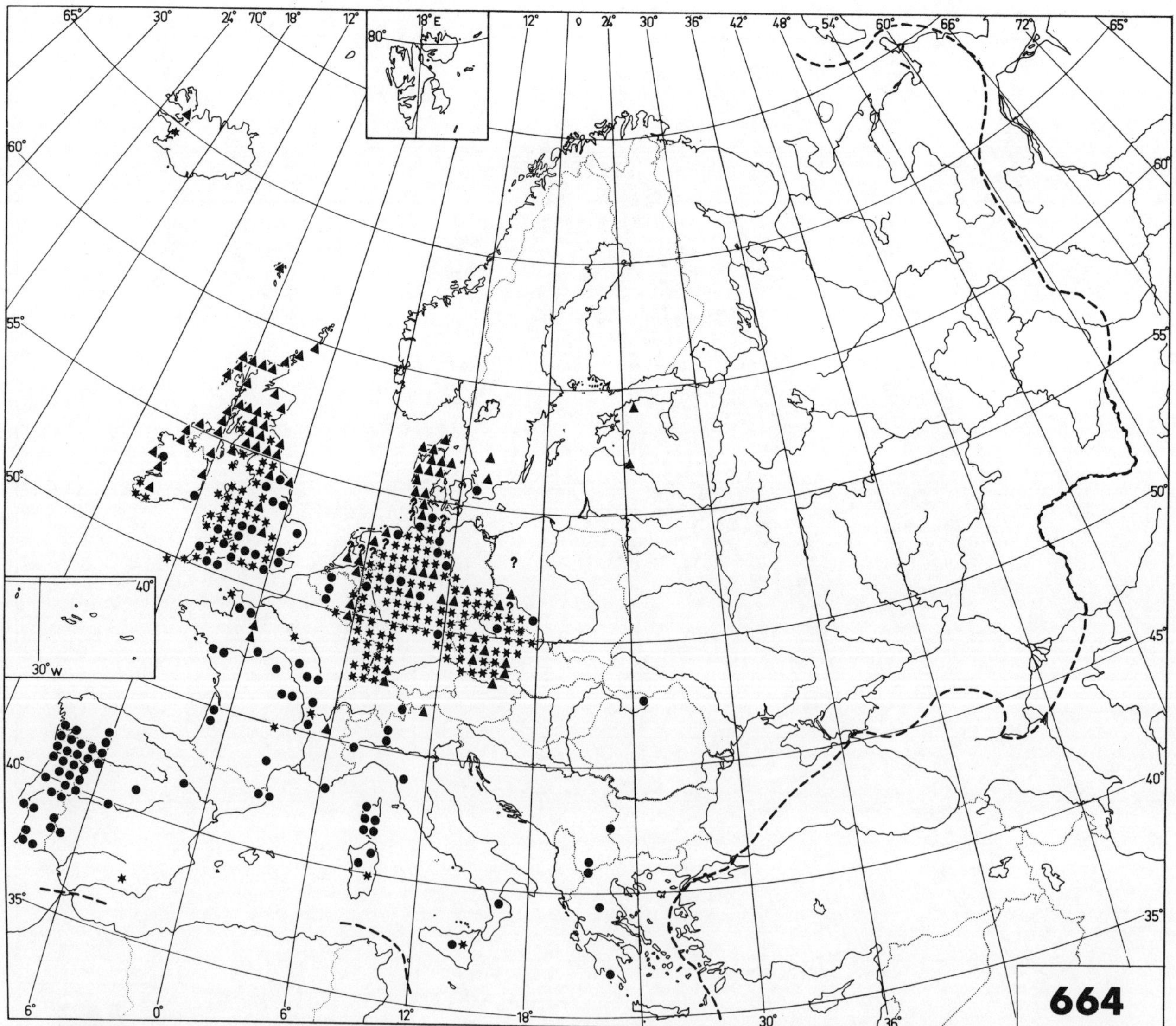

● = **Montia fontana** subsp. **amporitana**

▲ = **M. fontana** subsp. **variabilis**

✶ = subsp. **amporitana** + subsp. **variabilis**

M. fontana subsp. **chondrosperma** (Fenzl) Walters — Map 665.

Montia arvensis Wallr.; ? M. fontana subsp. minor Schübler & Martens; M. minor auct. non C.C. Gmelin; M. verna auct.

Nomenclature. J. Holub, Acta Horti Bot. Pragensis 1963: 57 (1964) and Preslia 38: 80 (1966); Hegi 1979: 1219; Yu. P. Kozhevnikov, Bot. Žurn. 64: 693—699 (1979).

Notes. Be, Br, Bu, Cr, Cz, Da, Ga, Ge, Gr, Hb, He, Ho, Hs, Hu, It, Ju, Po, Rm, ?Rs(B), Sa, Si, Su, Tu. Presence in Co doubtful (given as present in Fl. France 1973: 251, but all the material seen is subsp. *amporitana*). »Europe, northwards to Scotland and S. Sweden» (Fl. Eur.).

Total range. MJW 1965; map 136a.

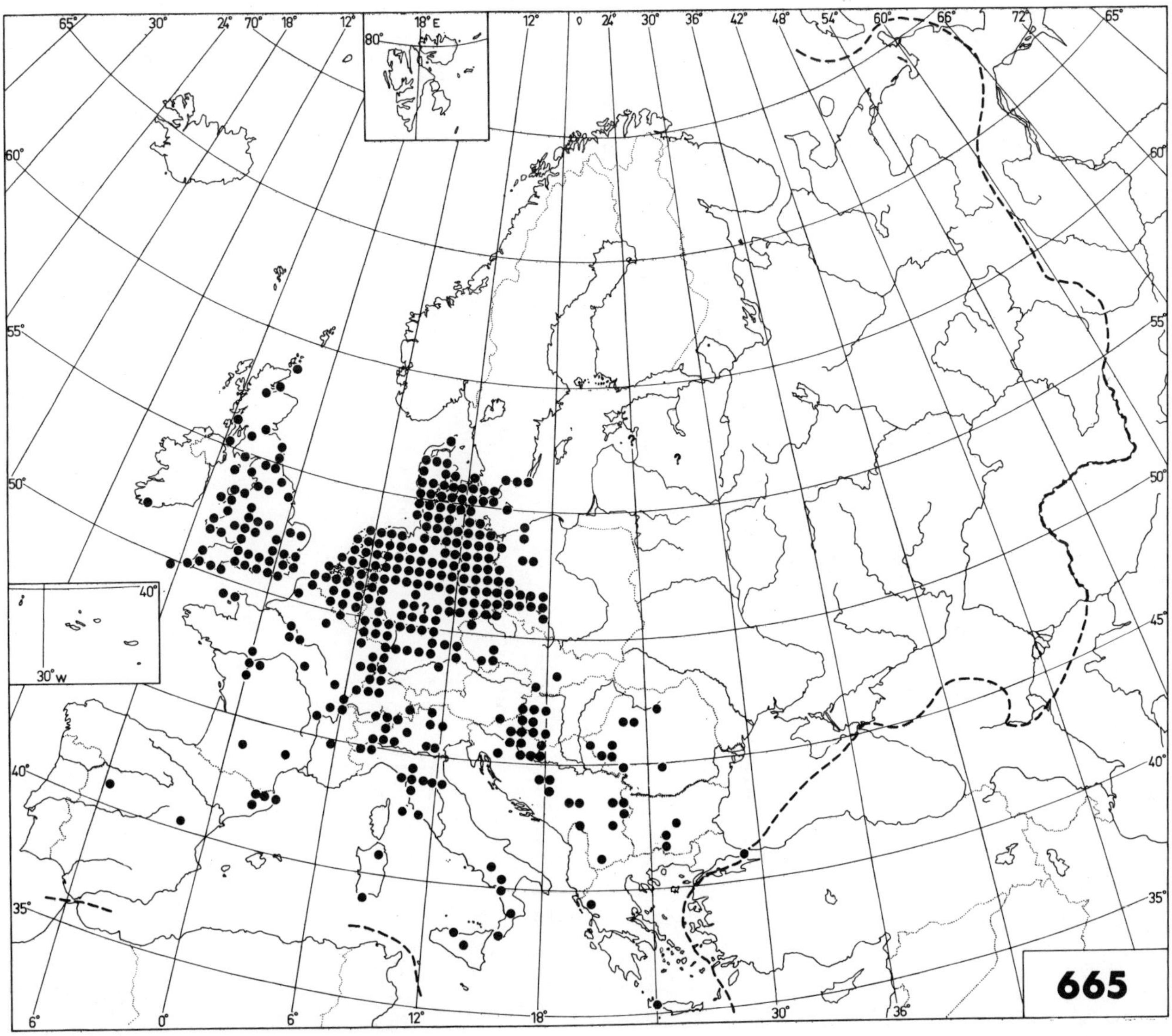

Montia fontana subsp. **chondrosperma**

Claytonia perfoliata Donn ex Willd. — Map 666.

Montia perfoliata (Donn ex Willd.) Howell

Generic delimitation of *Claytonia.* R.J. Davis, Brittonia 18: 285—303 (1966); Ö. Nilsson, Grana Palynol. 7: 279—363 (1967); J McNeill, Canad. Jour. Bot. 53: 789—809 (1975).

Taxonomy. J.M. Miller, Syst. Bot. 1: 20—34 (1976).

Notes. [Cz], [Hb] and [Su] added (not given in Fl. Eur.), casual in He and Hu ([He] and [Hu] in Fl. Eur.).

Native of W. North America: Hultén Alaska 1968: 409; Hegi 1979: 1209.

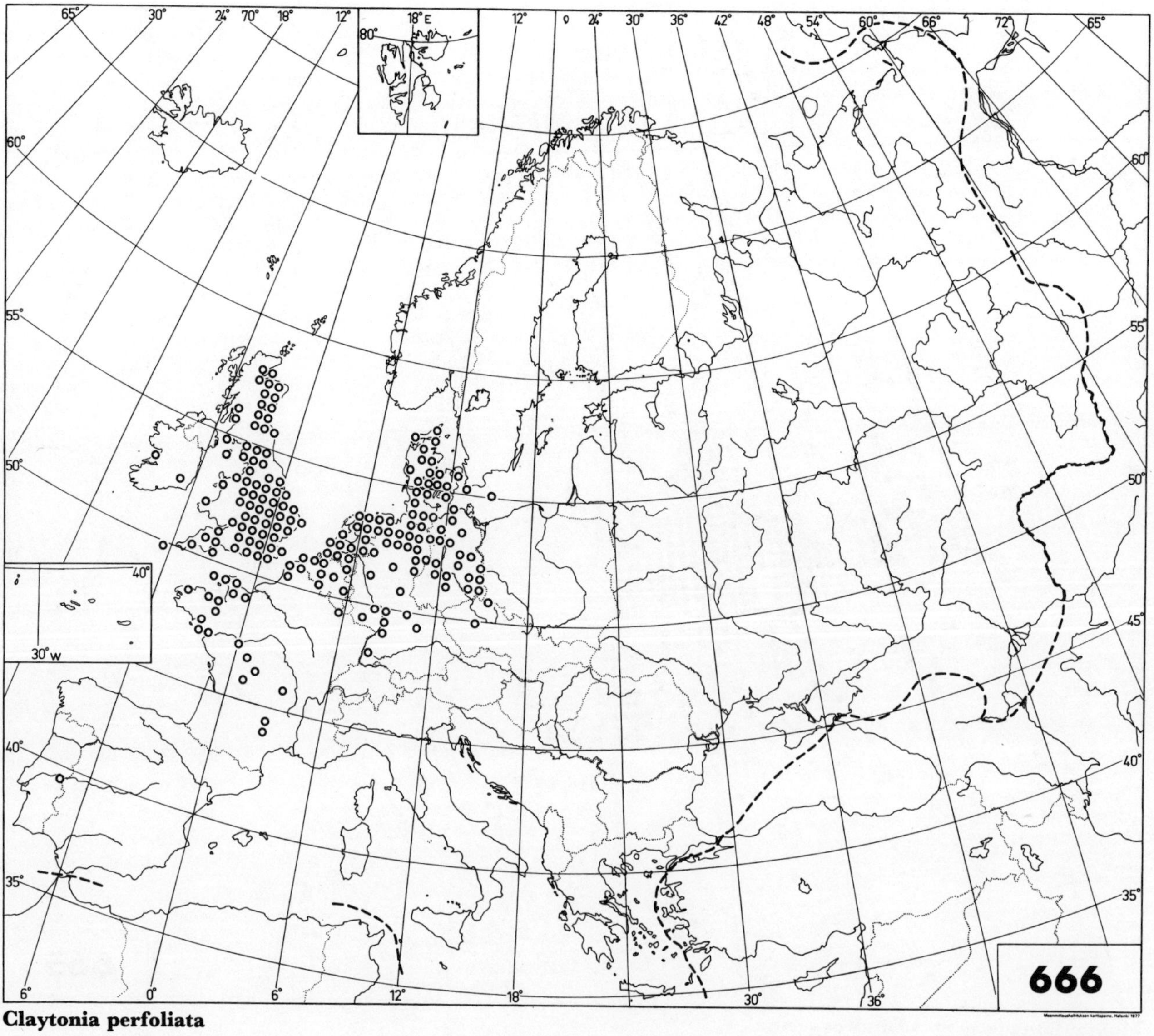

Claytonia perfoliata

Claytonia sibirica L. — Map 667.

Claytonia alsinoides Sims; Montia sibirica (L.) Howell

Taxonomy. E.V. Volkova, Nov. Syst. (Leningrad) 1965: 114—117; J. Holub, Preslia 47: 317—330 (1975).

Notes. [Bu, Cz, Fa, Fe, Ga, Hb, Ho and No] added (not given in Fl. Eur.): Fl. Reip. Pop. Bulg. 1966: 271; B.A. Kuzmanov & S.I. Kozuharov, Boissiera 19: 322 (1971); J. Holub, Preslia 47: 317 and 329 (1975); P.M. Jørgensen, Blyttia 33: 63 (1975).

Native of W. North America.

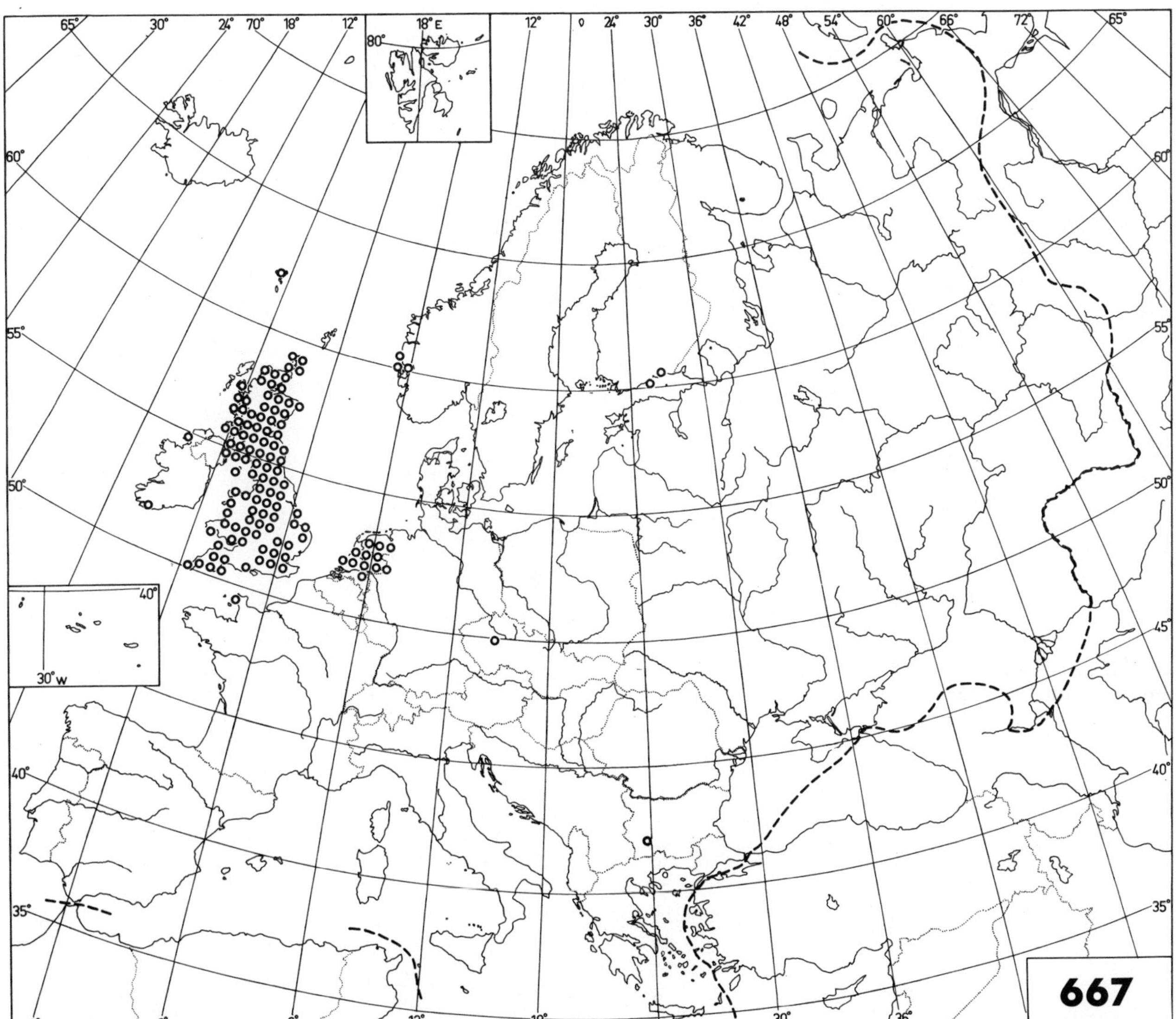

Claytonia sibirica

BASELLACEAE

Basella alba L.

Notes. »Widely cultivated... in S. Europe and occasionally escapes, but does not seem to be naturalized» (Fl. Eur.). Not mapped.
Native of tropical Asia (?).

Boussingaultia cordifolia Ten. — Map 668.

Boussingaultia baselloides auct. non Humb., Bonpl. & Kunth

Notes. [Ju] added (not given in Fl. Eur.), [Rs(W, E)] omitted (given in Fl. Eur.).
Native of subtropical South America.

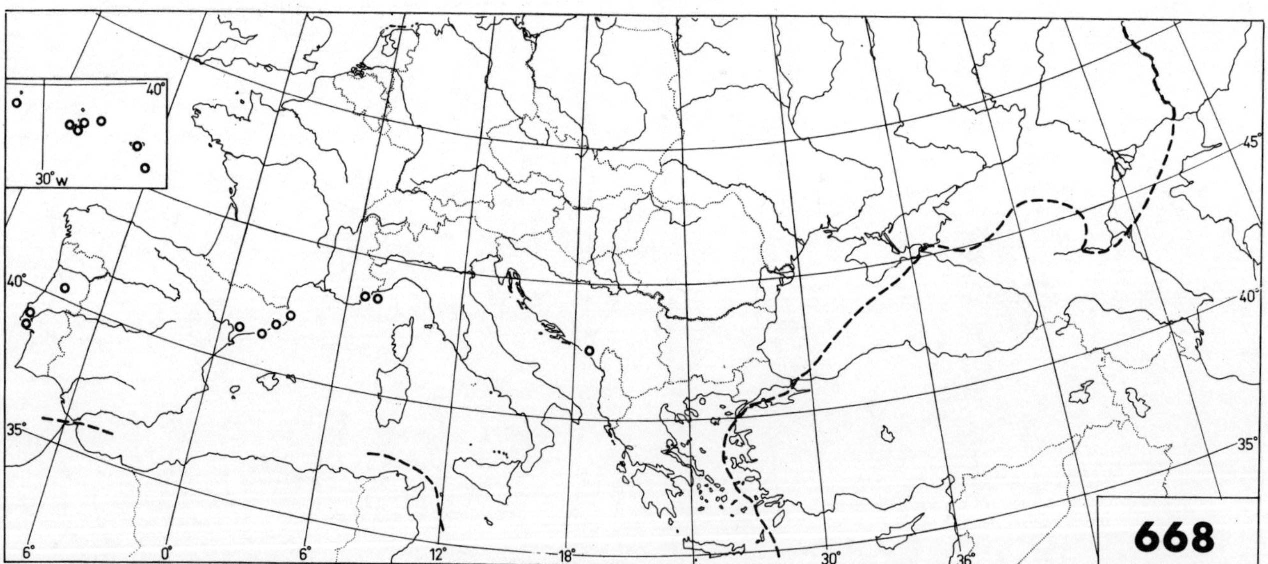

Boussingaultia cordifolia

INDEX TO VOL. 3

Names appearing in the text or as synonyms, and their respective page numbers, are given *in italics.*

INDEX TO VOL.4

Names appearing in the text or as synonyms, and their respective page numbers, are given *in italics*.

INDEX TO VOL.5

Names appearing in the text or as synonyms, and their respective page numbers, are given *in italics.*

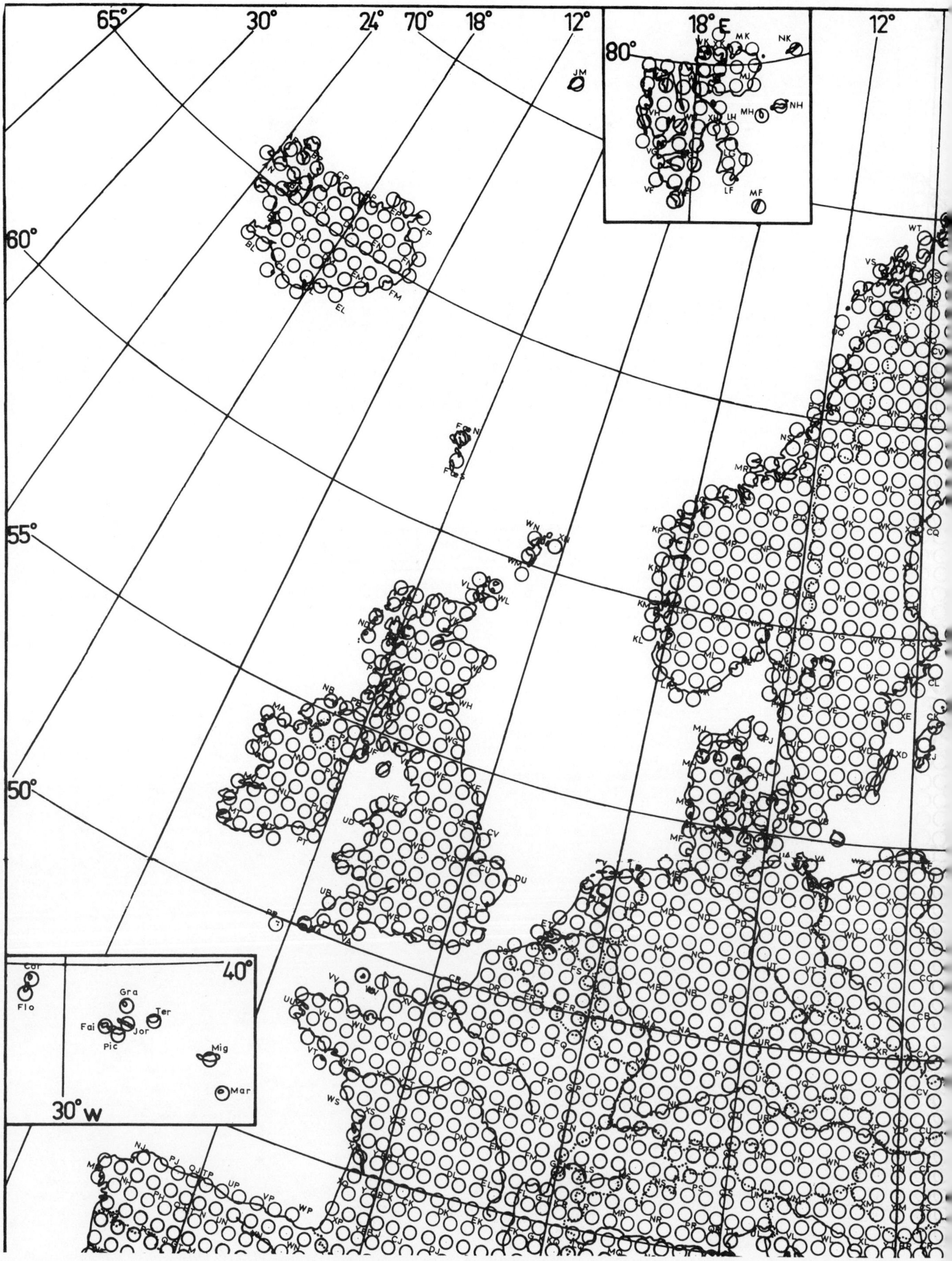
65°
30°
24°
70°
18°
12°
18°E
80°
12°
60°
55°
50°
40°
30°W
Cor
Flo
Gra
Ter
Fai
Jor
Pic
Mig
Mar
JM
MK
NK
NH
MH
MF

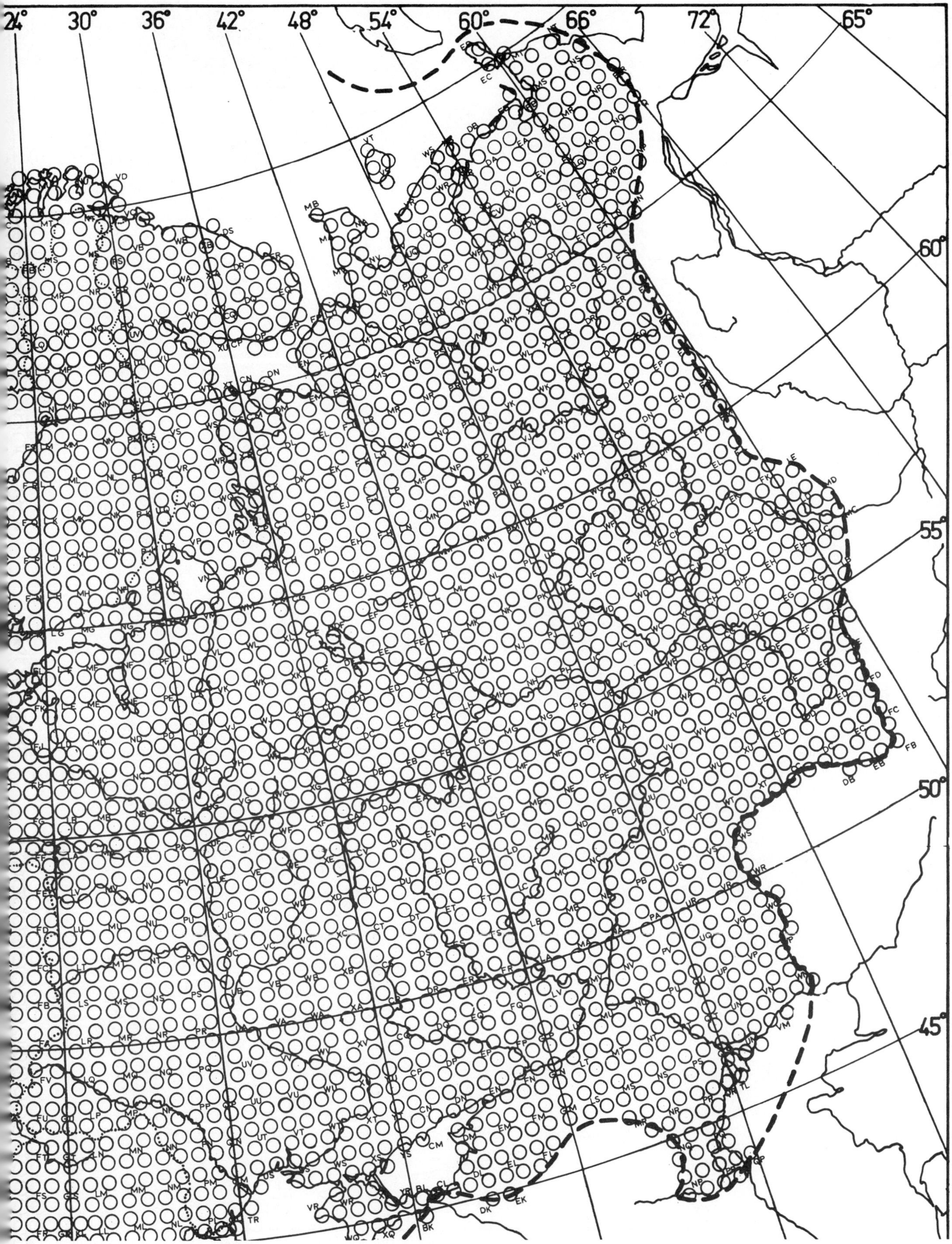

24°
30°
36°
42°
48°
54°
60°
66°
72°
65°
60°
55°
50°
45°

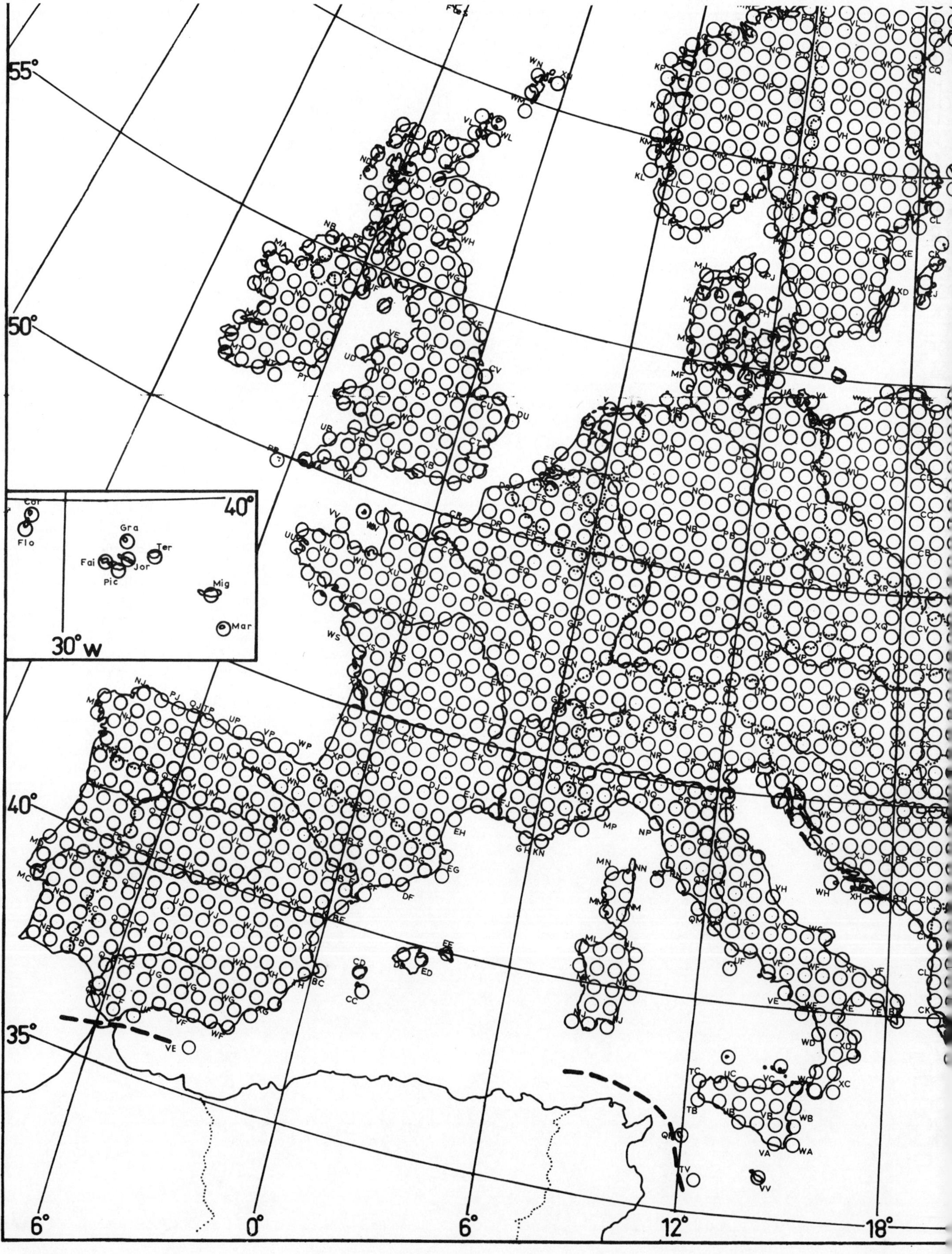
55°
50°
40°
35°
40°
Cor
Flo
Gra
Ter
Fai
Jor
Pic
Mig
Mar
30°W
6°
0°
6°
12°
18°

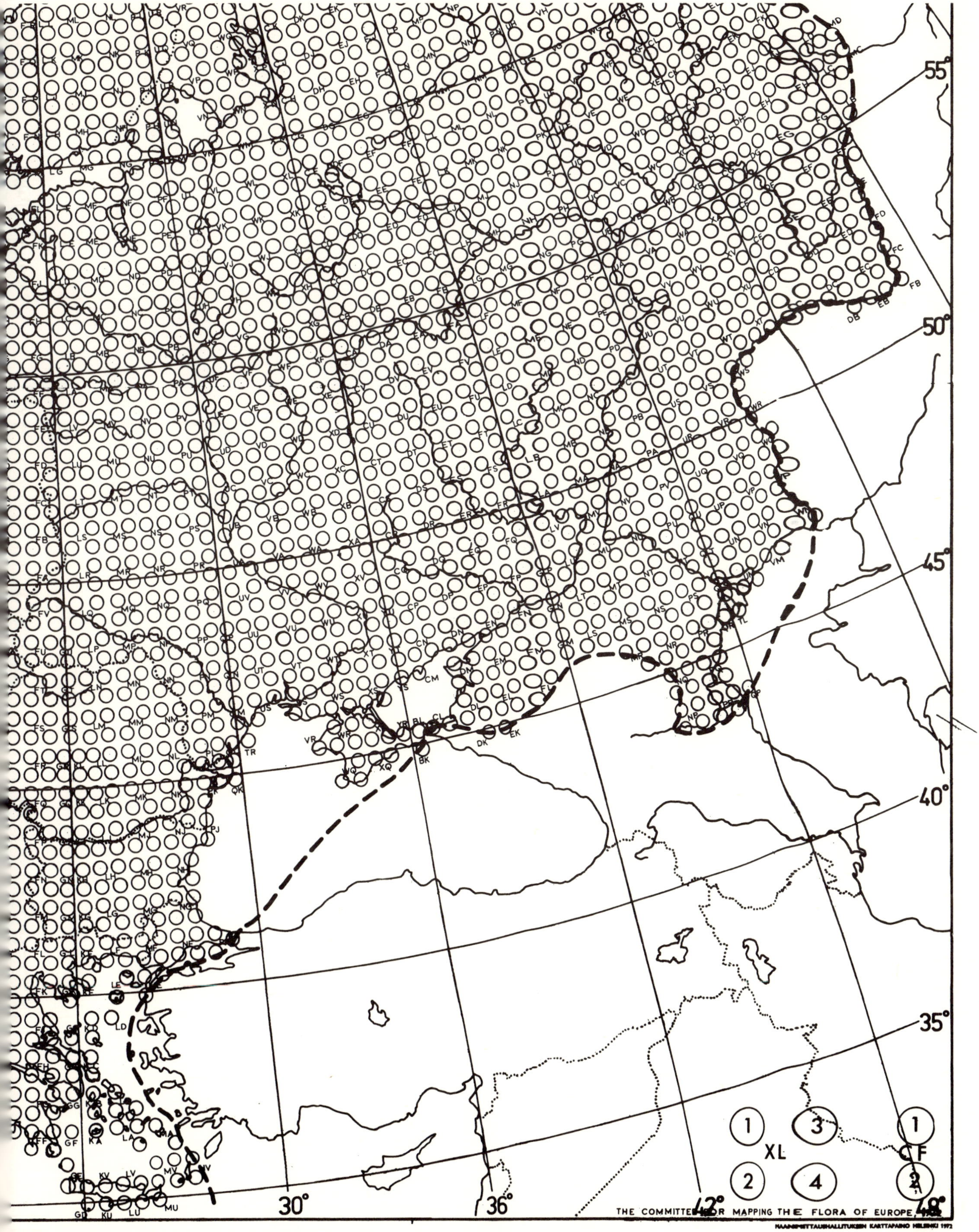
55°
50°
45°
40°
35°
30°
36°
42°
48°
1 3 1
XL CF
2 4 2
THE COMMITTEE FOR MAPPING THE FLORA OF EUROPE, 1972
MAANMITTAUSHALLITUKSEN KARTTAPAINO HELSINKI 1972